Molecular BIOTECHNOLOGY

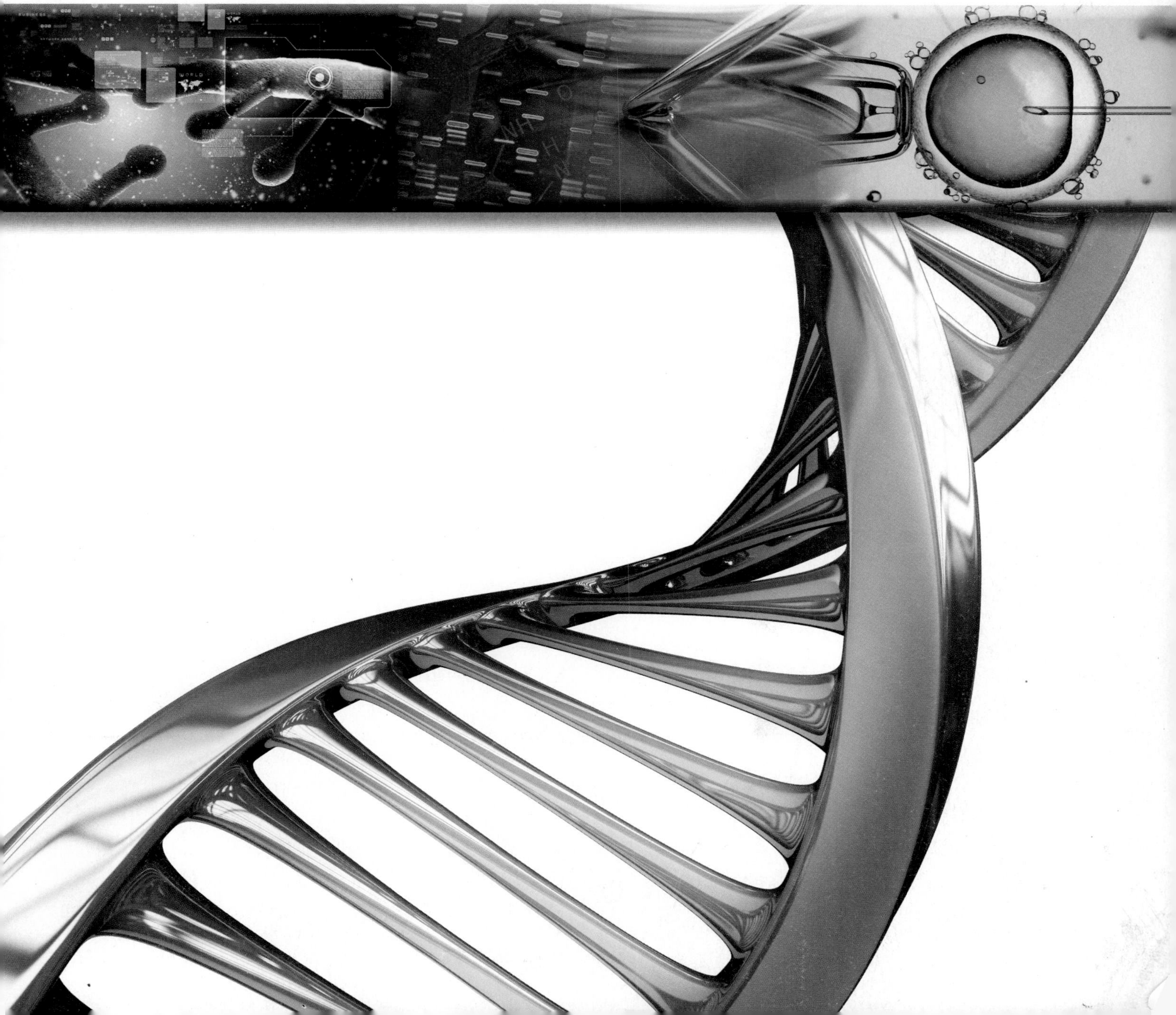

Molecular BIOTECHNOLOGY

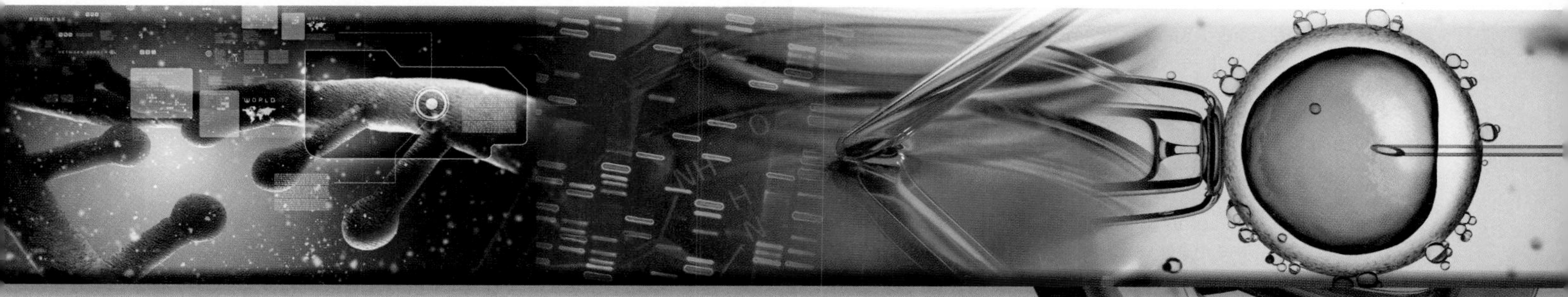

Carolyn A. Dehlinger, MAS, MS
Professor of Biological Sciences
Keiser University
Jacksonville, Florida

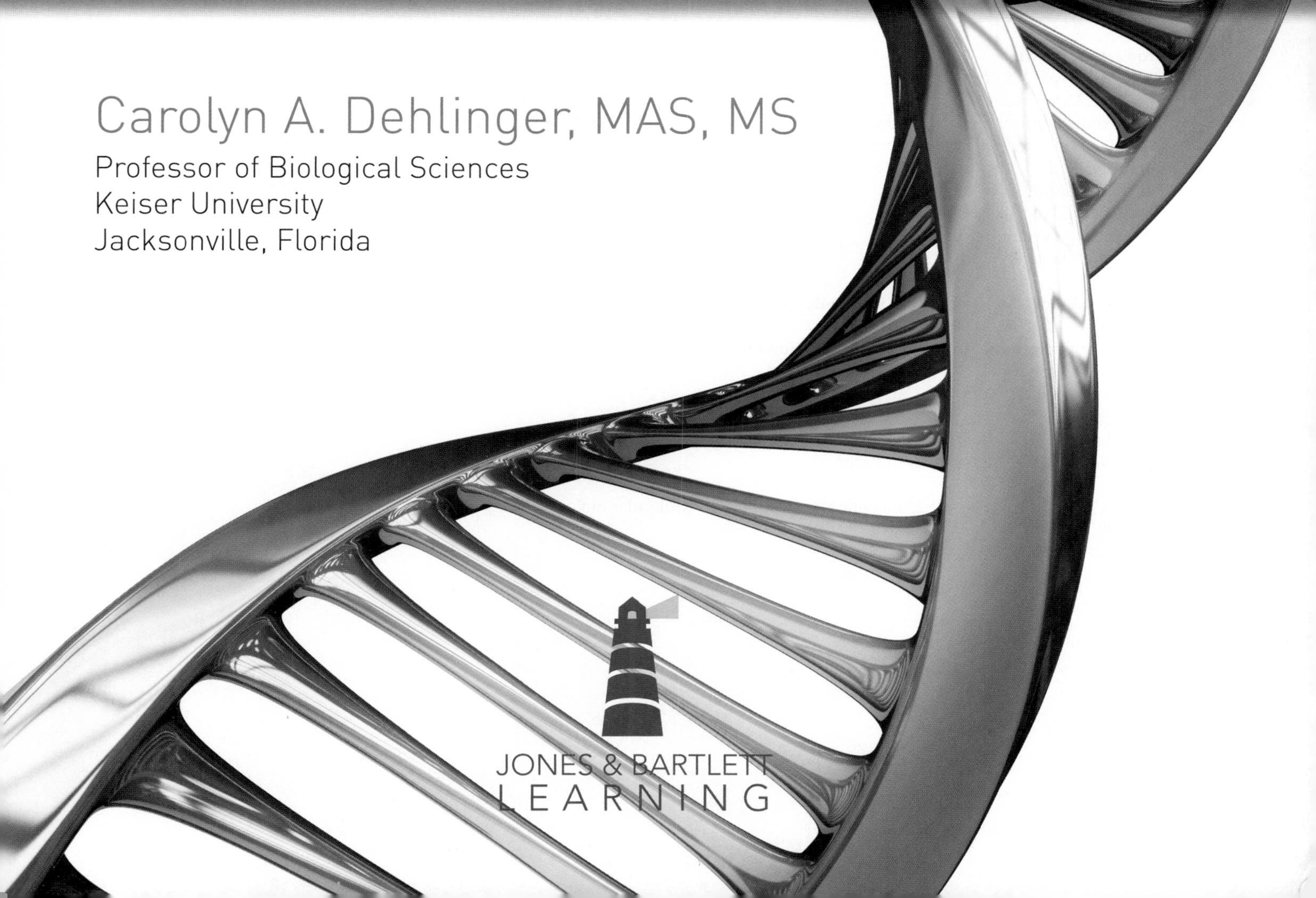

JONES & BARTLETT
LEARNING

World Headquarters
Jones & Bartlett Learning
5 Wall Street
Burlington, MA 01803
978-443-5000
info@jblearning.com
www.jblearning.com

Jones & Bartlett Learning books and products are available through most bookstores and online booksellers. To contact Jones & Bartlett Learning directly, call 800-832-0034, fax 978-443-8000, or visit our website, www.jblearning.com.

Substantial discounts on bulk quantities of Jones & Bartlett Learning publications are available to corporations, professional associations, and other qualified organizations. For details and specific discount information, contact the special sales department at Jones & Bartlett Learning via the above contact information or send an email to specialsales@jblearning.com.

Production Credits
Chief Executive Officer: Ty Field
President: James Homer
Chief Product Officer: Eduardo Moura
Executive Publisher: William Brottmiller
Publisher: Cathy L. Esperti
Editorial Assistant: Raven Heroux
Production Editor: Jill Morton
Production Assistant: Talia Adry
Marketing Manager: Lindsay White
Manufacturing and Inventory Control Supervisor: Amy Bacus
Composition: Cenveo Publisher Services
Cover Design: Kristin E. Parker
Manager of Photo Research, Rights and Permissions: Lauren Miller
Cover Image: Background, © Iaroslav Neliubov/ShutterStock, Inc.; Left to right, © Sergey Nivens/ShutterStock, Inc., © isak55/ShutterStock, Inc., © koya979/ShutterStock, Inc.
Printing and Binding: Courier Companies
Cover Printing: Courier Companies

To order this product, use ISBN: 9781284031409

Library of Congress Cataloging-in-Publication Data
Dehlinger, Carolyn, author.
Molecular biotechnology / by Carolyn A. Dehlinger.
p. ; cm.
Includes bibliographical references and index.
ISBN 978-1-284-05783-6 (pbk. : alk. paper)
I. Title.
[DNLM: 1. Biotechnology. 2. Genetic Engineering. 3. Molecular Biology. W 82]
QH390
572.8'38—dc23
2014008000

6048

Printed in the United States of America
18 17 16 15 14 10 9 8 7 6 5 4 3 2 1

Brief Contents

Chapter 1 The Emergence of Molecular Biotechnology 1

Chapter 2 The Molecular Biotechnology Industry Today 40

Chapter 3 Governmental Regulation of Molecular Biotechnology 56

Chapter 4 Bioinformatics: Genomics, Proteomics, and Phenomics 74

Chapter 5 Industrial Biotechnology 102

Chapter 6 Life Sciences and Health Care 126

Chapter 7 Environmental Biotechnology and Conservation 160

Chapter 8 Agriculture and Food Production 178

Chapter 9 Forensics and Biodefense 194

Chapter 10 Evo Devo: The Biotechnology of Evolution and Development 214

Chapter 11 The Biotechnology of Anthropology 244

Chapter 12 The Future of Biotechnology 262

Appendix 1 Genome Structure: DNA, Genes, and Chromosomes 272

Appendix 2 Basics of Gene Expression and DNA Replication 277

Appendix 3 A Primer in Classical Genetics 283

Contents

Preface ix
About the Author xv
Acknowledgments xvi
Reviewers xvii

Chapter 1 The Emergence of Molecular Biotechnology 1

What Is Molecular Biotechnology? 3
How Did We Get Here? The Path to Molecular Biotechnology 5
Notable Projects 33

Chapter 2 The Molecular Biotechnology Industry Today 40

Applying Molecular Biotechnology to Modern Lifestyles 41
Molecular Biotechnology Industry Practices 49
Careers in Molecular Biotechnology 51
Focus on Careers **51**

Chapter 3 Governmental Regulation of Molecular Biotechnology 56

Regulatory Oversight: The Federal Agencies 57
Beyond Regulation: National Institutes of Health Guidelines 60
Regulation of Genetically Modified Organisms 62
Focus on Careers **APHIS Inspectors 66**

Chapter 4 Bioinformatics: Genomics, Proteomics, and Phenomics 74

Bioinformatics 75
Genomics 76
Proteomics 85
Focus on Careers **Biostatisticians 76**

Chapter 5 Industrial Biotechnology 102

Commercial Products of Industrial Biotechnology 103
Commercial Processes of Industrial Biotechnology 113

Chapter 6 Life Sciences and Health Care 126

Genetic Counseling and Gene Therapy 127
Pharmaceuticals and Therapeutics 143
Regenerative Medicine 151
Focus on Careers **Genetic Counselors 127**

Chapter 7 Environmental Biotechnology and Conservation 160

Environmental Biotechnology 161
Conservation Biotechnology 172
Focus on Careers **Bioremediation Project Scientists 164**

Chapter 8 Agriculture and Food Production 178

Agriculture Biotechnology 179
Food Biotechnology 190

Chapter 9 Forensics and Biodefense 194

Forensics 195
Biodefense 204
Focus on Careers **Forensic Scientists 201**

Chapter 10 Evo Devo: The Biotechnology of Evolution and Development 214

Evolution 215
Development 231
Focus on Careers **Evo Devo Biologists 240**

Chapter 11 The Biotechnology of Anthropology 244

Becoming Human 246
Divergence from Other Primates 246
The Hominin Lineage 250
Focus on Careers **Paleoanthropologists 245**

Chapter 12 The Future of Biotechnology 262

Regulatory Status and Economic Impact 263
Industry Forecasts 264
Bioethics and Risk 268

Appendix 1 Genome Structure: DNA, Genes, and Chromosomes 272

Chromosome Structure 272
Deoxyribonucleic Acid Structure 272

Ribonucleic Acid Structure 273
Genes 273
Key Terms 275

Appendix 2 Basics of Gene Expression and DNA Replication 277

The Genetic Code 277
Protein Synthesis 278
DNA Replication 281
Key Terms 281

Appendix 3 A Primer in Classical Genetics 283

Genotypes and Phenotypes 283
Monohybrid Crosses 283
Dihybrid Crosses 283
X-Linked Crosses 283
Key Terms 286

Glossary 289

Index 295

Preface

Connecting the Dots

In this first edition of *Molecular Biotechnology*, history and science are interwoven into a narrative yet technical account of humanity taking control of its destiny. This may seem like an overzealous or romantic description for a science textbook. Nonetheless my primary goal in writing these pages was to bring color into an otherwise black-and-white subject. Biotechnology exists at the intersection of scientific research and the business of profit. It can be taught strictly as an arsenal of laboratory techniques and smattering of government regulations. Many existing publications follow this precept. Yet in the educational environment of today, relevance and meaning hold the attention of learners to a much higher degree than facts alone. To that end, *Molecular Biotechnology* places a strong emphasis on cultural and personal aspects, such as bioethical issues and the diverse career paths in this expanding field. It seeks to connect the dots in exploring the past, describing the present, and considering the future of biotechnology.

As early as the introduction of agriculture, humanity has sought to control its fate by controlling nature. Biotechnology is an extension of this. After millennia of plant breeding and animal husbandry, this past century has experienced an unprecedented acceleration of scientific knowledge, edging out even the Age of Enlightenment. Discovery of the genetic material, DNA, its structure and transmission have led to the Modern Synthesis of scientific understanding. With the Modern Synthesis, the molecular details of evolution and genetics were elucidated. While Charles Darwin conceived the steps of natural selection leading to adaptation, he could not begin to explain exactly how organisms changed over time or how traits were passed through generations. Biotechnology has answered these questions. The Chromosomal Theory of Inheritance and the Genetic Code laid the foundation in this regard. From there, we have gone from describing molecular events to orchestrating them. Instead of selecting traits to develop plant varieties and breed animals, we can select genes. Furthermore, the advances taking place concurrently in computer science and information technology have allowed biotechnologists to store and process enormous amounts of genetic data. Without that ability, we would not be where we are today. What a serendipitous series of events!

We live in a time when the average person is likely to be more influenced and affected by science than ever before. In particular, most of us have daily contact with biotechnology in one form or another. At minimum, the food we eat and the medicine we take are unmistakenly the results of biotechnology. Along with this, the concept of scientific literacy is creeping into our collective mindset. Ironically, it is also a time when well-established and almost universally recognized theories are subject to harsh criticism by the scientifically illiterate. As an educator, combatting this paradox is of utmost importance. Among the maligned theories, evolution and climate change are at the forefront. Reading headlines and hearing voice clips in mainstream media would suggest these heavily supported paradigms of science are merely unfounded suggestions cherry-picked from a whole realm of possibilities. Perhaps this is not new at all, considering the words of Darwin in *The Descent of Man*:

> Ignorance more frequently begets confidence than does knowledge: it is those who know little, not those who know much, who so positively assert that this or that problem will never be solved by science.

To become scientifically literate, one must join the ranks of those who know much. By knowing much, one becomes an informed citizen, a discerning consumer, and an enfranchised member of society. That is the inspiration for *Molecular Biotechnology*.

It would be an act of hubris to suggest this or that problem will always be solved by science. But the possibilities before us are truly mind-boggling. We are at the precipice of personalized medicine based on genetic make-up. We are seeking to protect and expand the world's food supply through genetic engineering and cloning. We are improving products and processes by making them faster, cheaper, and of better quality. We are cleaning up pollution and preserving endangered species. We are exonerating wrongly accused felons through DNA fingerprinting. We are determining how genes and other molecules interact to create the diversity of life on this planet. We are even unraveling the mysteries of our own past by delving deeper into the evolution of *Homo sapiens*.

In using this textbook, be mindful of its spirit and intent: connecting the dots. Historical narrative, bioethical discussion, and career preparation are the main focus. Technical information is offered as contextual detail. The laboratory methods utilized in biotechnology must be learned by any learner entering this profession, but this is not a lab manual. Foundational knowledge in the areas of chemistry, genetics, and molecular biology is undeniably beneficial. For those lacking that knowledge or those in need of a refresher, the appendices provide summaries of the most basic and most necessary information. The technical aspects of biotechnology are also heavily presented in graphics and images with the intention of simplifying concepts. This subject can be complex, challenging, and even overwhelming. This textbook was purposefully designed to be none of these things. Instead, the aim was to create a textbook covering a relatively advanced subject in a familial tone and with an approachable presentation. I hope you will agree.

The Student Experience

Learning Objectives: The learning objectives identify the knowledge and competencies that a student needs to master through reading and comprehension of the chapter material. To use the objectives effectively, they should be reviewed before and after reading the chapters. This review will help students focus on the major knowledge points in each chapter and facilitate answering the questions and completing the activities at the end of each chapter.

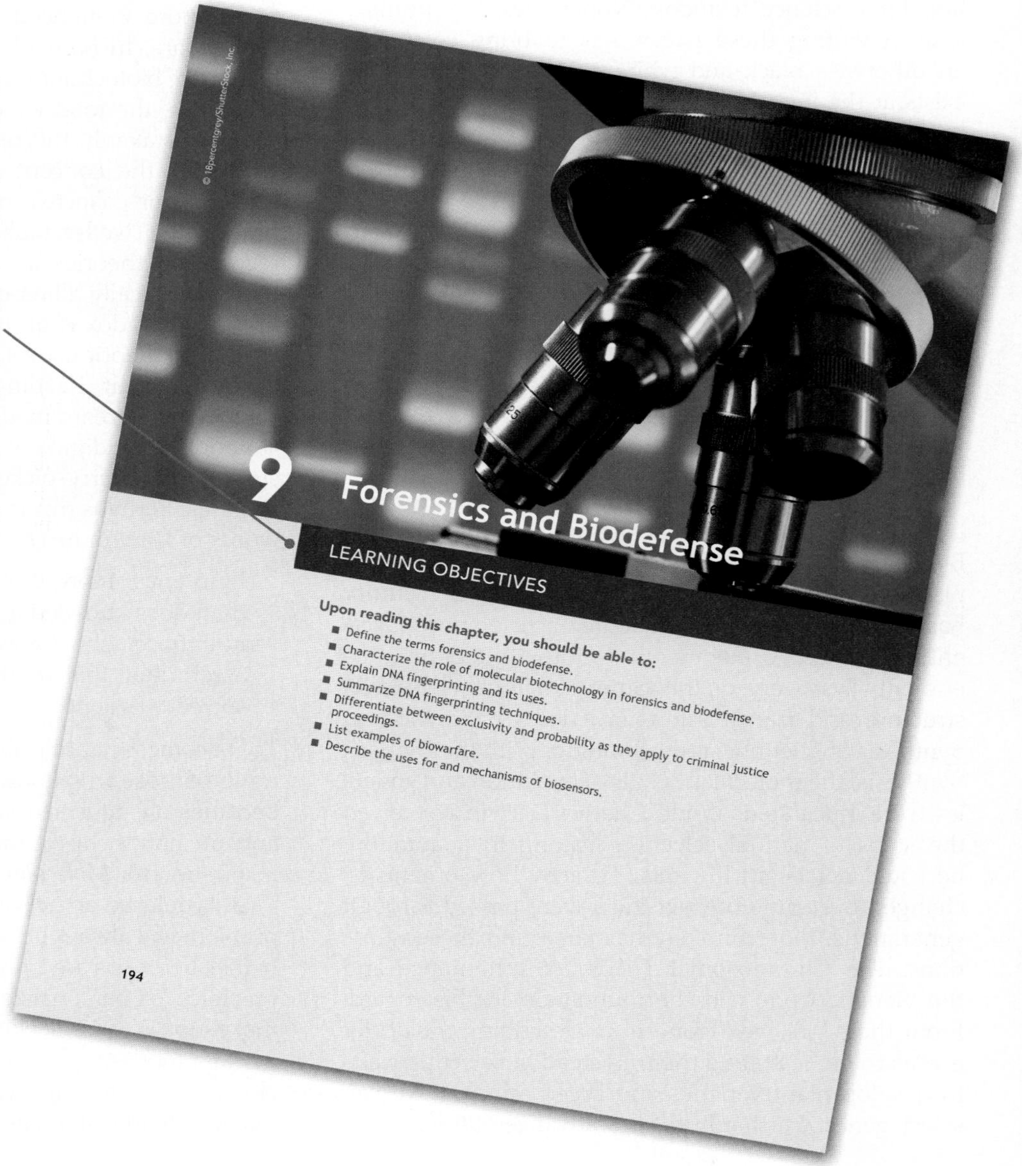

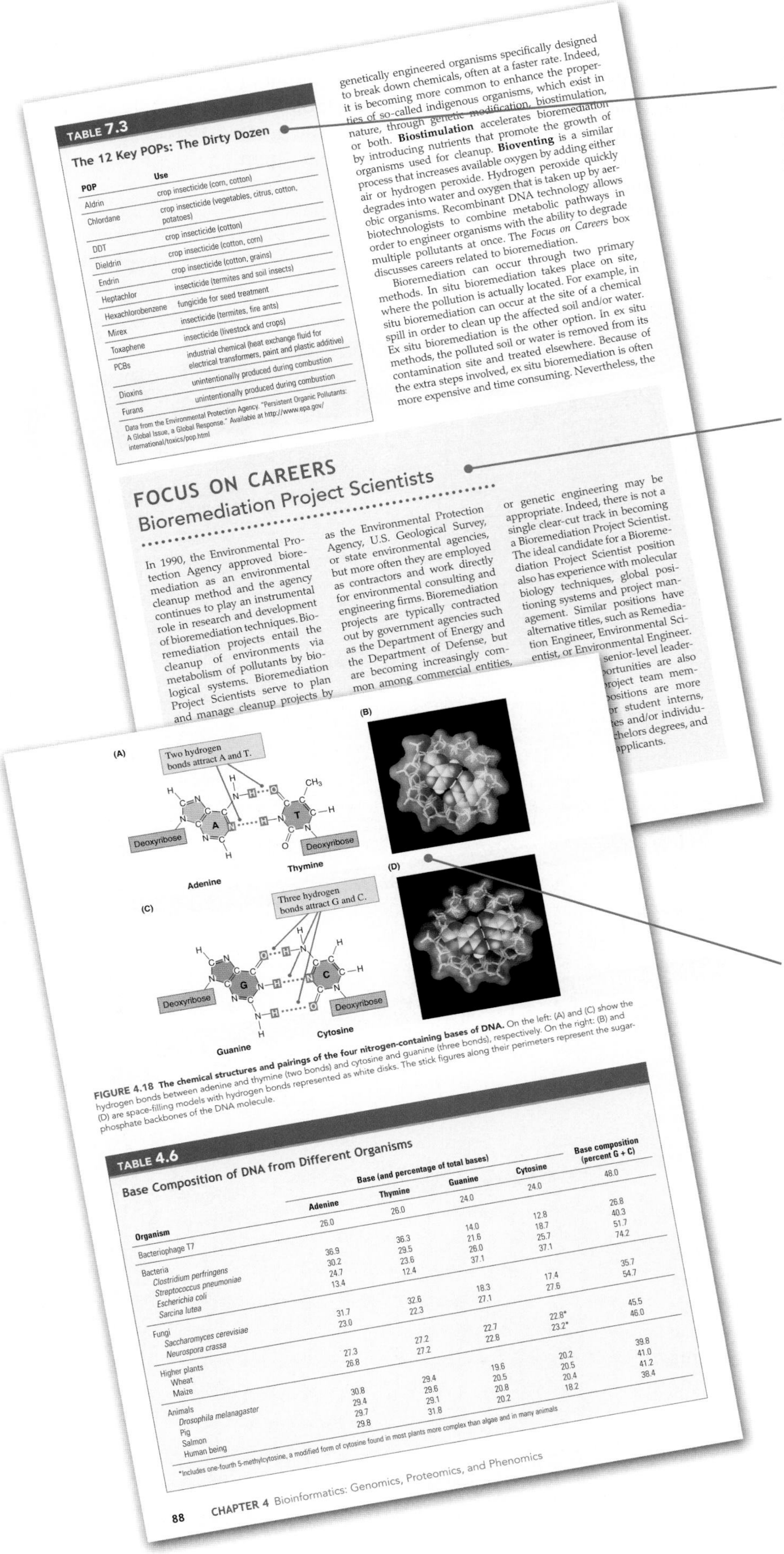

Instructional Tables: Useful tables organize important information from trusted sources into easy-to-read lists. These tables provide current data to help summarize key material.

***Focus on Careers* Boxes:** These informative boxes introduce students to exciting and cutting-edge career opportunities in the area of molecular biotechnology. The information contained within these boxes can help students formulate their educational needs to achieve their career aspirations.

Full-color Illustrations: Figures illustrate key molecular processes and gene sequences, making difficult concepts easier to understand.

End-of-Chapter Summary: Chapter Summaries appear at the end of each chapter and contain a comprehensive summary of the main points in the chapter. These provide students with a great reference tool and study guide to help them determine if they comprehended the key competencies and core knowledge in each chapter.

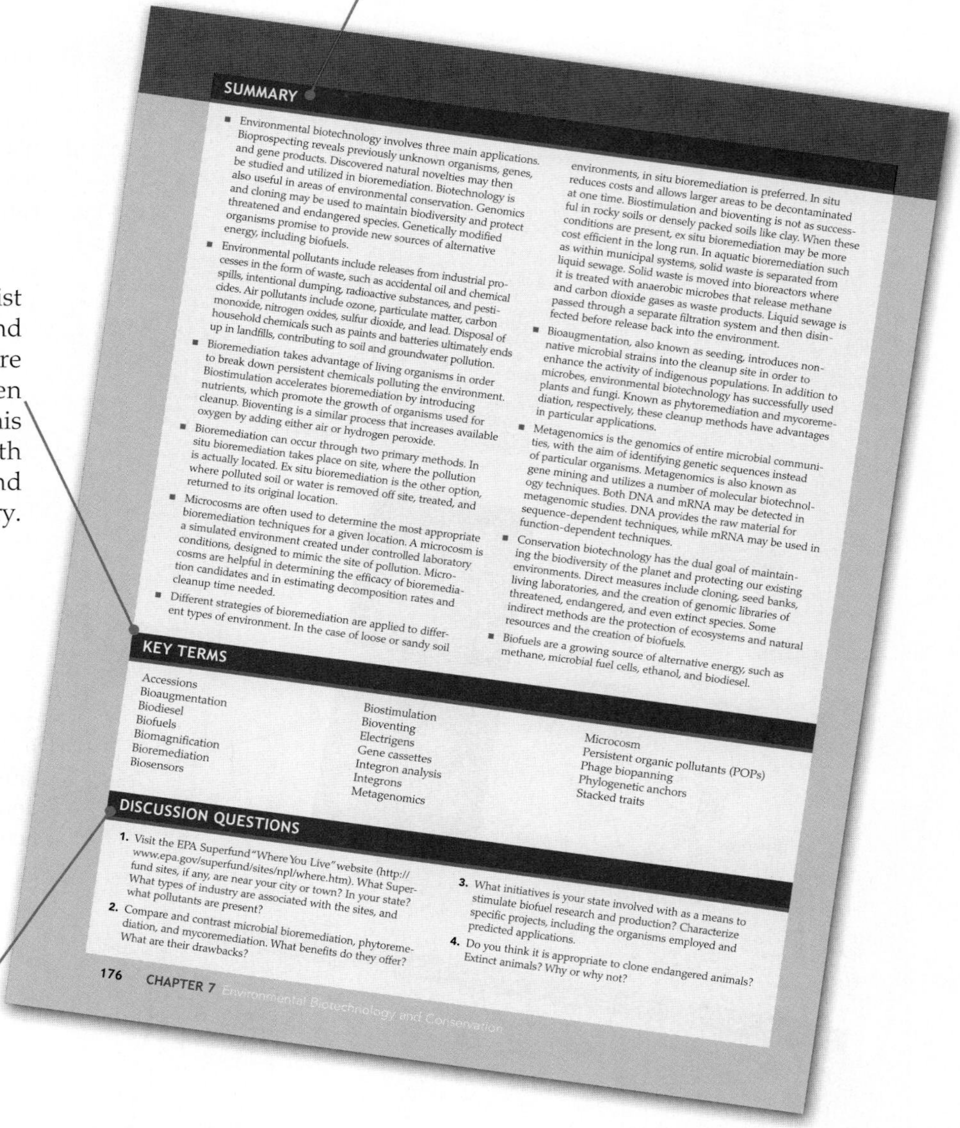

SUMMARY

- Environmental biotechnology involves three main applications. Bioprospecting reveals previously unknown organisms, genes, and gene products. Discovered natural novelties may then be studied and utilized in bioremediation. Biotechnology is also useful in areas of environmental conservation. Genomics and cloning may be used to maintain biodiversity and protect threatened and endangered species. Genetically modified organisms promise to provide new sources of alternative energy, including biofuels.
- Environmental pollutants include releases from industrial processes in the form of waste, such as accidental oil and chemical spills, intentional dumping, radioactive substances, and pesticides. Air pollutants include ozone, particulate matter, carbon monoxide, nitrogen oxides, sulfur dioxide, and lead. Disposal of household chemicals such as paints and batteries ultimately ends up in landfills, contributing to soil and groundwater pollution.
- Bioremediation takes advantage of living organisms in order to break down persistent chemicals polluting the environment. Biostimulation accelerates bioremediation by introducing nutrients, which promote the growth of organisms used for cleanup. Bioventing is a similar process that increases available oxygen by adding either air or hydrogen peroxide.
- Bioremediation can occur through two primary methods. In situ bioremediation takes place on site, where the pollution is actually located. Ex situ bioremediation is the other option, where polluted soil or water is removed off site, treated, and returned to its original location.
- Microcosms are often used to determine the most appropriate bioremediation techniques for a given location. A microcosm is a simulated environment created under controlled laboratory conditions, designed to mimic the site of pollution. Microcosms are helpful in determining the efficacy of bioremediation candidates and in estimating decomposition rates and cleanup time needed.
- Different strategies of bioremediation are applied to different types of environment. In the case of loose or sandy soil environments, in situ bioremediation is preferred. In situ reduces costs and allows larger areas to be decontaminated at one time. Biostimulation and bioventing is not as successful in rocky soils or densely packed soils like clay. When these conditions are present, ex situ bioremediation may be more cost efficient in the long run. In aquatic bioremediation such as within municipal systems, solid waste is separated from liquid sewage. Solid waste is moved into bioreactors where it is treated with anaerobic microbes that release methane and carbon dioxide gases as waste products. Liquid sewage is passed through a separate filtration system and then disinfected before release back into the environment.
- Bioaugmentation, also known as seeding, introduces nonnative microbial strains into the cleanup site in order to enhance the activity of indigenous populations. In addition to microbes, environmental biotechnology has successfully used plants and fungi. Known as phytoremediation and mycoremediation, respectively, these cleanup methods have advantages in particular applications.
- Metagenomics is the genomics of entire microbial communities, with the aim of identifying genetic sequences instead of particular organisms. Metagenomics is also known as gene mining and utilizes a number of molecular biotechnology techniques. Both DNA and mRNA may be detected in metagenomic studies. DNA provides the raw material for sequence-dependent techniques, while mRNA may be used in function-dependent techniques.
- Conservation biotechnology has the dual goal of maintaining the biodiversity of the planet and protecting our existing environments. Direct measures include cloning, seed banks, living laboratories, and the creation of genomic libraries of threatened, endangered, and even extinct species. Some indirect methods are the protection of ecosystems and natural resources and the creation of biofuels.
- Biofuels are a growing source of alternative energy, such as methane, microbial fuel cells, ethanol, and biodiesel.

KEY TERMS

Accessions
Bioaugmentation
Biodiesel
Biofuels
Biomagnification
Bioremediation
Biosensors
Biostimulation
Bioventing
Electrigens
Gene cassettes
Integron analysis
Integrons
Metagenomics
Microcosm
Persistent organic pollutants (POPs)
Phage biopanning
Phylogenetic anchors
Stacked traits

DISCUSSION QUESTIONS

1. Visit the EPA Superfund "Where You Live" website (http://www.epa.gov/superfund/sites/npl/where.htm). What Superfund sites, if any, are near your city or town? In your state? What types of industry are associated with the sites, and what pollutants are present?
2. Compare and contrast microbial bioremediation, phytoremediation, and mycoremediation. What benefits do they offer? What are their drawbacks?
3. What initiatives is your state involved with as a means to stimulate biofuel research and production? Characterize specific projects, including the organisms employed and predicted applications.
4. Do you think it is appropriate to clone endangered animals? Extinct animals? Why or why not?

176 CHAPTER 7 Environmental Biotechnology and Conservation

Key Terms: A comprehensive list of key terms is listed at the end of each chapter. These terms are also bolded in the chapter when the term is first introduced. This list can be used to help with comprehension and to expand students' professional vocabulary.

Discussion Questions: The purpose of the discussion questions at the end of each chapter is to provide the students with feedback regarding their mastery of the chapter's content. These critical thinking style questions help students apply their knowledge in an individual assignment or as part of a group project.

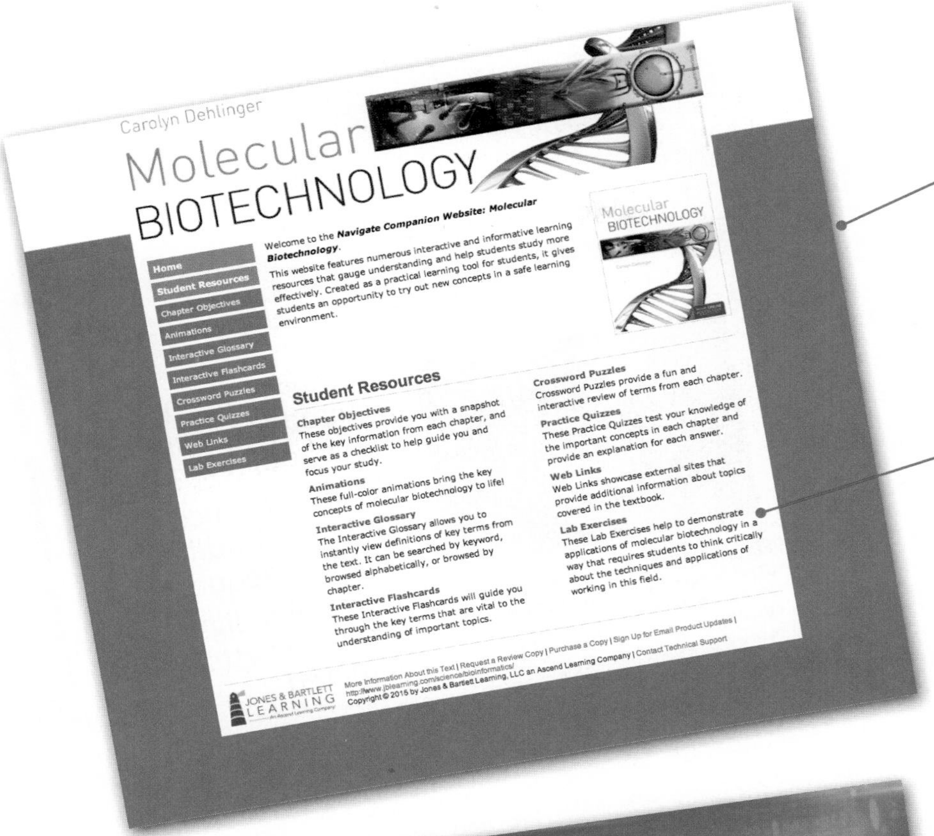

Companion Website: Online resources, including practice quizzes, weblinks, and an interactive glossary with flashcards, are available to help students study.

Lab Exercises: The Student Lab Exercises contain a lab activity for each chapter. These labs can be used as an applied study tool, assigned as a classroom activity, or as an individual homework assignment. The Student Lab Exercises are an excellent way for students to apply knowledge and gain practical lab skills.

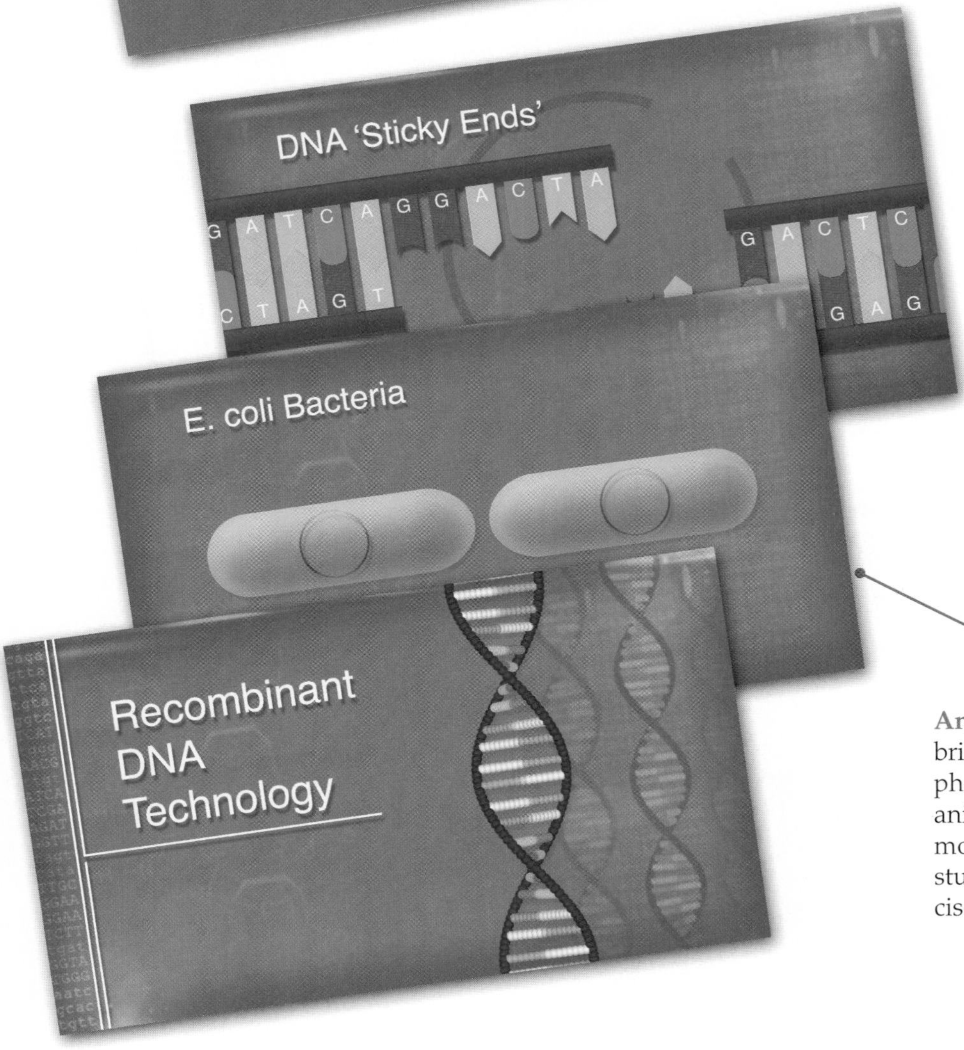

Animations: Engaging animations bring fascinating molecular biology phenomena to life! Each interactive animation guides students through molecular processes and gauges student's understanding with exercises and assessment questions.

Teaching Tools

Jones & Bartlett Learning offers an array of resources to save instructors valuable time in preparation and instruction of this course. Additional information and review copies of any of the following items are available through your Jones & Bartlett Learning Account Specialist or by going to go.jblearning.com/MolecularBiotechnology.

Lecture Outlines in PowerPoint format: This presentation package provides lecture notes and images from the text for each chapter of *Molecular Biotechnology*. Instructors with the Microsoft PowerPoint software can customize the outlines, art, and order of presentation to suit their course needs.

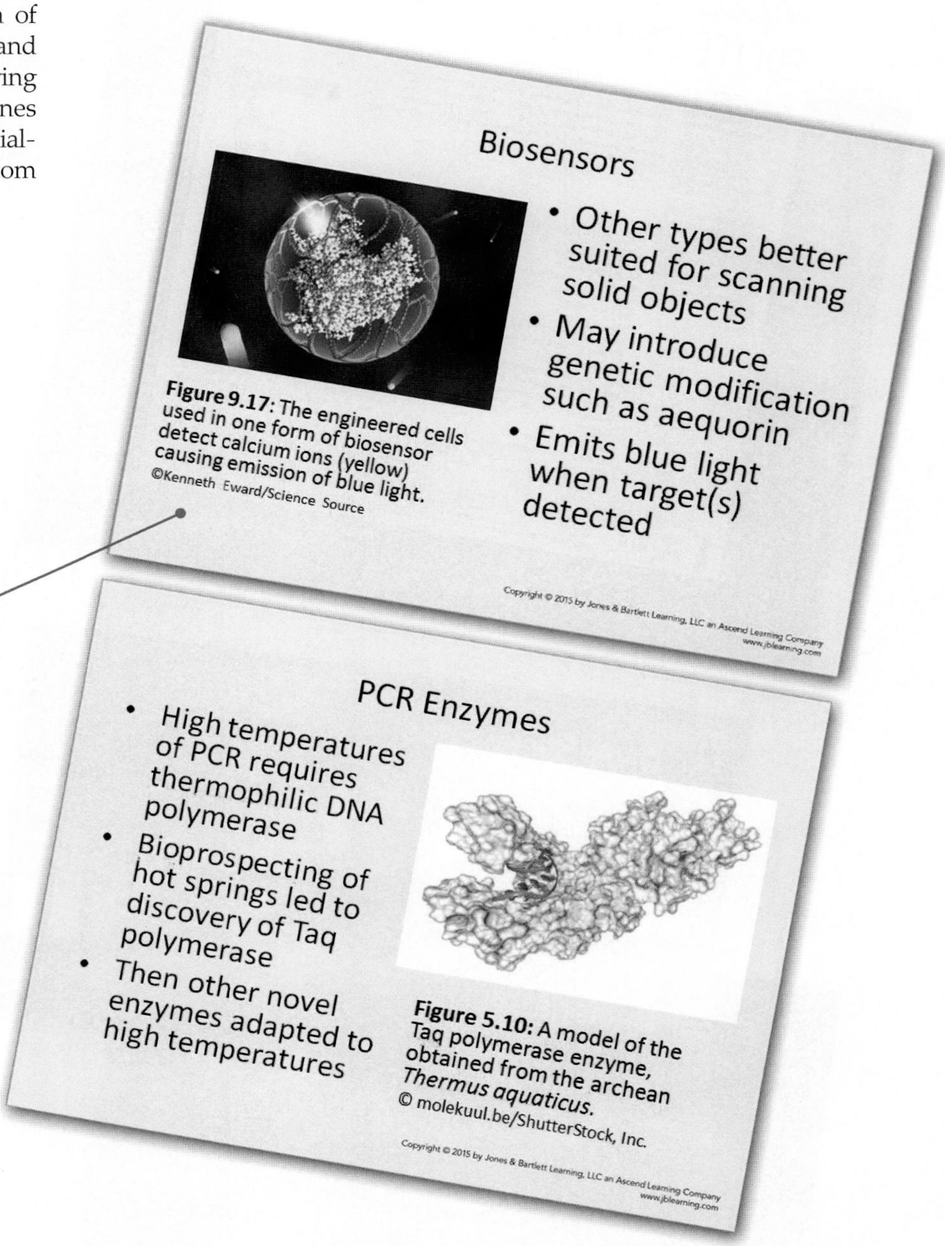

Sample Syllabus: This text file is provided to offer instructors a sample course outline. Instructors can customize the syllabus to tailor their course needs and schedule.

Test Bank: This text file provides instructors with hundreds of exam questions and their answers for each chapter of *Molecular Biotechnology.*

About the Author

Carolyn A. Dehlinger is a professor of Biological Sciences at Keiser University in Jacksonville, Florida. She received her Master of Applied Science in Technology Management from the University of Denver and her Master of Science in Biology from Mississippi State University. She completed her undergraduate degree at the University of Florida. She teaches students in the Biomedical Science and Biotechnology degree programs in addition to students enrolled in non-majors biology courses. At Keiser, Carolyn played an integral role in the development of the Biomedical Science degree curriculum and has served on multiple National Science Foundation grant projects aimed at delivering biotechnology education to both students and educators. Carolyn serves as a subject matter expert with the American Council on Education in Washington, DC. In this role, she evaluates curriculum and determines their eligibility for accreditation. She is a proud member of the Honorable Order of Kentucky Colonels, an honor bestowed by the governor of Kentucky for community service. Carolyn is a practitioner and registered teacher of yoga and enjoys spending her free time with her husband Charles, their two golden retrievers, Alice and Ivan, and their calico cat, Chloe.

Acknowledgments

I am happy to acknowledge the assistance of all the faculty reviewers who participated in this project. The knowledgeable contributions provided by these scholars helped to remove erroneous remarks and include useful elaboration on material. Their suggestions collectively improved the content and flow of this book and helped develop a student-centered presentation.

I would like to acknowledge the talented individuals at Jones & Bartlett Learning for providing support and guidance throughout the creation of this book. Publisher, Cathy Esperti provided leadership, vision, and style. Casa Price, the sales representative for Keiser University, was instrumental in facilitating the initial steps of turning an idea into a reality and encouraging me toward authoring. Erin O'Connor, the Acquisitions Editor in charge of this project, has offered irreplaceable advice and assistance and served as a leading figure throughout the entire process. Editorial Assistant Raven Heroux remained patient and helpful while organizing the many details of the project. I am grateful to the art and production team members, including Jill Morton, Mary Colleen Liburdi, Talia Adry, Lauren Miller, and Joanna Lundeen for creating outstanding illustrations and obtaining the necessary permissions for the images within this book. I appreciate the permissions granted on behalf of the authors, editors, and publishers of all copyrighted material.

My colleague at Keiser University, Professor of Biological Sciences George Ealy, MD, was a tremendous source of advice and motivation. You are a truly inspiring mentor, Colonel!

Lastly, I must thank my friends and family, especially my husband Charles, for generously donating the borrowed time I took to complete this project. Your endless support and understanding cannot be underestimated and will never be forgotten. Nothing you would take, everything you gave.

—Carolyn A. Dehlinger

Reviewers

The author and Publisher would like to thank the following individuals for their service as reviewers of this book during its various stages:

Giles Bolduc, Massasoit Community College

Christopher Bush, Florida State College at Jacksonville

James J. Cheetham, Carleton University

Marilyn Cruz-Alvarez, Florida Gulf Coast University

George Ealy, Keiser University Jacksonville

Matthew J. Farber, University of the Sciences

Rocco Fiandaca, Grayson Scientific Consultants

Philip Gibson, Gwinnett Technical College

Misty Gish, Owensboro Community and Technical College

Daniel Kainer, Lone Star College-Montgomery

Kiran J. Kaur, El Centro College

Parul Khurana, Indiana University East

Bridgette Kirkpatrick, Collin College

Deanna Langfitt, Southwest Tennessee Community College

Ryan Lee, Indiana University

Edwin Li, Saint Joseph's University

Nancy Magill, Indiana University

William McClain, Keiser University Jacksonville

David Morris, The George Washington University

Jennifer L. Myers, University of Houston

Kathryn M. Rafferty, University of Cincinnati

Steven Roof, Fairmont State University

Laura Grayson Roselli, Burlington County College

Larissa Parsley Walker, University of Mobile

Caroline Zitzke, Florida State College at Jacksonville

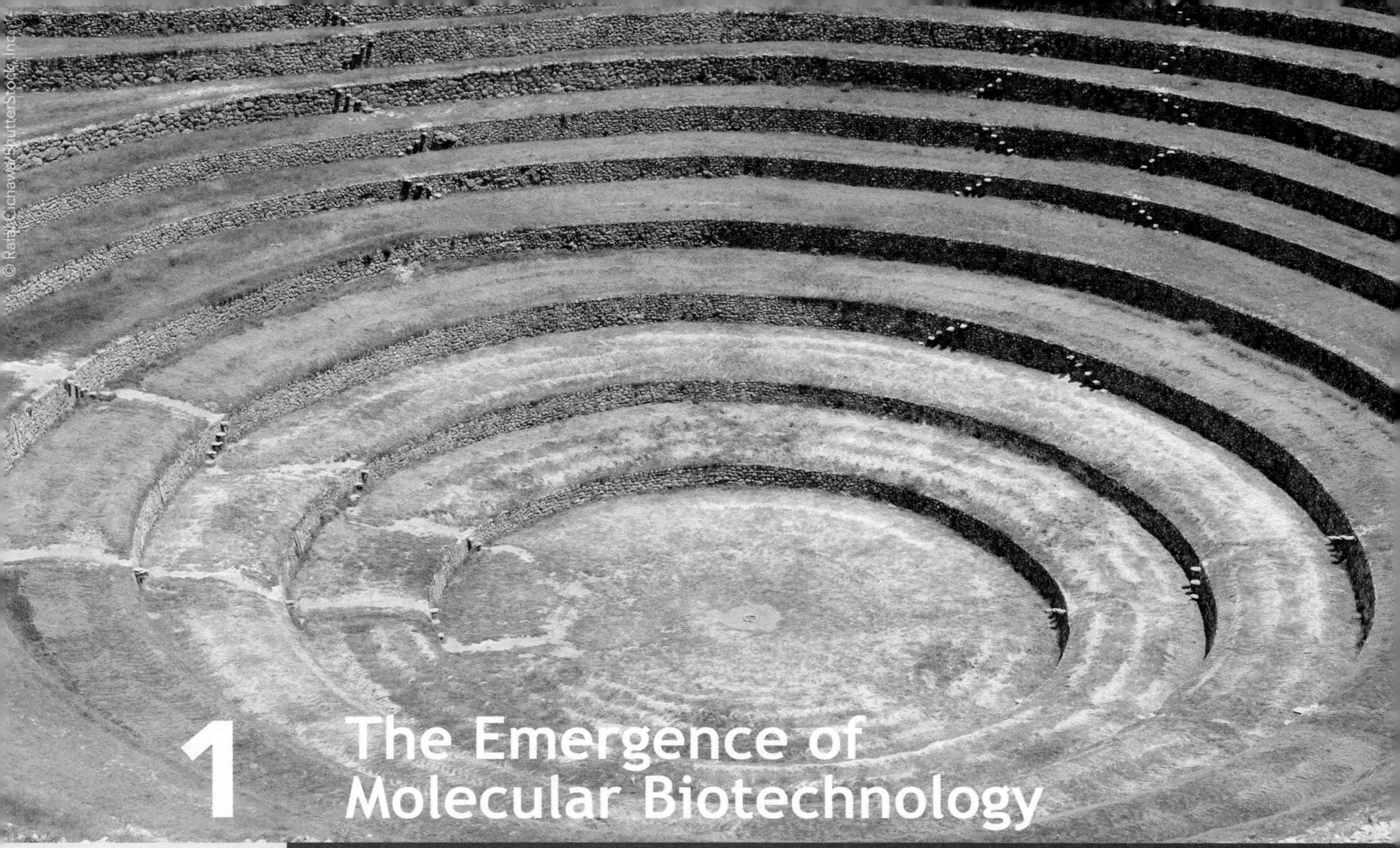

1 The Emergence of Molecular Biotechnology

LEARNING OBJECTIVES

Upon reading this chapter, you should be able to:

- Define biotechnology and relate it to the specialized field of molecular biotechnology.
- Appraise the interplay of science and technology in the field of molecular biotechnology.
- Describe the scientific contributions of Gregor Johann Mendel.
- Explain the process of natural selection and identify those responsible for its conception.
- Illuminate the significance of the modern synthesis.
- Differentiate genes, DNA, and chromosomes in the description of the chromosomal theory of inheritance.
- Summarize the experiments that identified DNA as the molecular basis of encoding and transmitting genetic information.
- Recall complementary base pairing and relate Chargaff's Rules to the structure of DNA.
- Associate the genetic code with the central dogma of molecular biology.
- Relate restriction endonucleases and plasmids to the production of recombinant DNA technology.
- Explain the role of the Asilomar Conference in the biotechnology industry.
- Describe the processes of DNA sequencing and the polymerase chain reaction.
- Discuss the roles of genomics and proteomics in notable biotechnology projects.

Ruins of an agricultural terrace from the Incan civilization of ancient Peru.

Ten thousand years ago, *Homo sapiens* embarked on an experiment that forever changed humanity. By establishing the practice we now call agriculture, our species took control of its food supply, and ultimately its future. No longer would we be subject to a hunting and gathering lifestyle, where nomadic behavior and merely subsisting were the norm. We began quite literally sowing the seeds of civilization that have led us to today's advanced and complex societies. We possess an enormous capacity to manipulate the essence of life—genetic material. Our first attempts to domesticate wild plant and animal species have allowed us to further refine living beings better suited for our needs through selective breeding. The artificial selection of crops and livestock was our first foray into bioengineering (**FIGURE 1.1**).

All these millennia later, we are not limited to the confines of breeding for particular traits within a single species. We now live in a world where traits can jump species, thanks to our knowledge of molecular biotechnology. Once again we engage in an experiment that has already permanently changed the direction of our species. Within the United States, as of June 2012, 88% of our corn, 93% of our cotton, and 94% of our soybeans are genetically modified organisms, or GMOs (**FIGURE 1.2**). These are rather astounding statistics if you consider the first GMOs were only introduced in 1996.

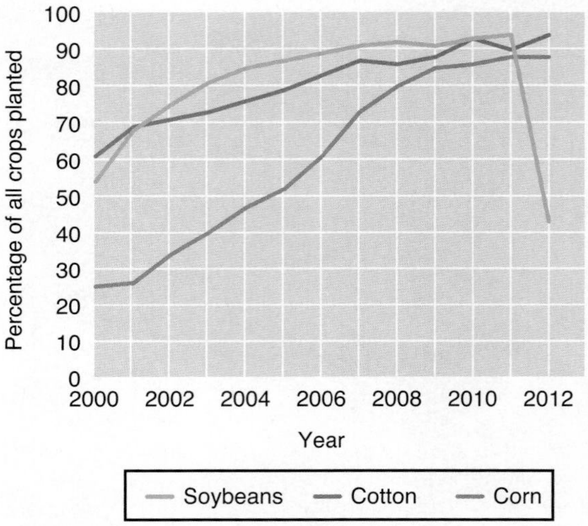

FIGURE 1.2 The percentage of total crops planted in the United States that are any genetically modified variety has increased each year since their first introduction. The most planted genetically modified crops in the United States are corn, cotton, and soybeans. Data from ERS/USDA

Worldwide, agricultural fields composed of GMOs have increased 100-fold in less than 20 years, from 1.7 million hectares to 170 million hectares (**FIGURE 1.3**). While the United States has by far the greatest amount

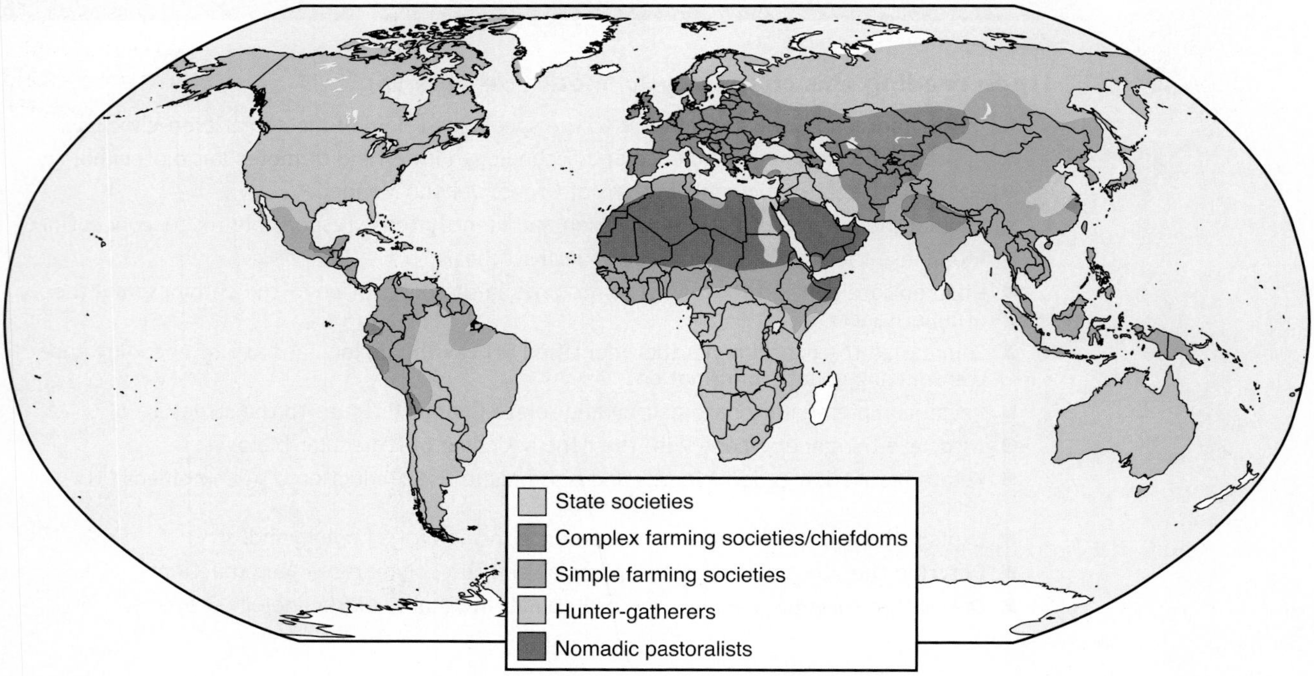

FIGURE 1.1 Distribution of the world's peoples based upon food source, as of 1000 BCE. Note the regions in yellow are dominated by hunter-gatherers, who have yet to transition into bioengineering. The remaining regions include involvement of varying degrees in agriculture and animal husbandry.

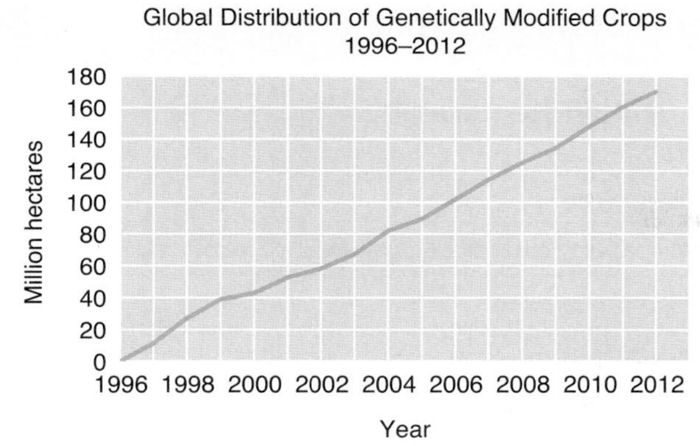

FIGURE 1.3 Worldwide agricultural fields planted with genetically modified crops have increased 100-fold since 1996, from 1.7 million hectares to over 170 million hectares in 2012.

FIGURE 1.4 Karl Ereky (1878–1952) coined the term "biotechnology" in 1917.

of GMO agricultural acreage, there are 28 countries planting biotech crops (**TABLE 1.1**).

This is just one example of the myriad ways molecular biotechnology has changed the living world. Molecular biotechnology allows us to explore a branch of knowledge that has altered the nature of life as we know it; a technological pursuit full of both inspirational promise and unsettling risk.

What Is Molecular Biotechnology?

In 1917, a Hungarian engineer named Karl Ereky coined the term "biotechnology." Ereky defined biotechnology as "all lines of work by which products are produced from raw materials with the aid of living things." Ereky was a futurist; he envisioned a historical age of humans built upon biotechnology, just as there had previously been a stone age, a bronze age, and an iron age (**FIGURE 1.4**). Well before 1917, humanity had already begun these "lines of work." We knew how to ferment fruits and grains into alcohol with the aid of a little microbe called yeast. We could preserve milk in the form of yogurt with the help of lactic acid. Pathogenic microbes were used as inoculants to prevent disease. Once these types of applications reached the commercial realm on a mass scale, the biotechnology industry was born.

Starting in the 1960s, the term "**biotechnology**" has been applied to the industrial production of goods and services by processes utilizing biological organisms and systems. Molecular biotechnology focuses on the areas of this enterprise that exist at the molecular level. The molecules of concern include nucleic acids and proteins; we are analyzing and manipulating **genes**, the functional units of heredity that code for proteins and other functional products. Molecular biotechnology has its roots in industrial applications such as food and medicine production, but we will see that it has branched out well beyond commerce. As our intimacy with biological molecules grows, areas of criminal justice and defense, environmental conservation, and anthropology benefit from molecular biotechnology as well. The subject is illuminating our understanding of the origins of life and the developmental processes which contribute to the tremendous biodiversity found on Earth. We are pursuing no less of a goal than to decipher the mystery of life. Just as agriculture allowed humans to take control of our food source (and therefore our future), so too has molecular biotechnology redefined our place in the natural world and the vast number of future possibilities yet to come (see **FIGURE 1.5**).

"Okay—is there anybody ELSE whose homework ate their dog?"

FIGURE 1.5 While biotechnology has advanced greatly in a relatively short period of time, this scenario is still the stuff of satire. © www.CartoonStock.com

TABLE 1.1

Global Area of Biotech Crops in 2012: By Country (Million Hectares)**

Rank	Country	Area (million hectares)	Biotech Crops
1	USA*	69.5	Maize, soybean, cotton, canola, sugarbeet, alfalfa, papaya, squash
2	Brazil*	36.6	Soybean, maize, cotton
3	Argentina*	23.9	Soybean, maize, cotton
4	Canada*	11.6	Canola, maize, soybean, sugarbeet
5	India*	10.8	Cotton
6	China*	4.0	Cotton, papaya, poplar, tomato, sweet pepper
7	Paraguay*	3.4	Soybean, maize, cotton
8	South Africa*	2.9	Maize, soybean, cotton
9	Pakistan*	2.8	Cotton
10	Uruguay*	1.4	Soybean, maize
11	Bolivia*	1.0	Soybean
12	Philippines*	0.8	Maize
13	Australia*	0.7	Cotton, canola
14	Burkina Faso*	0.3	Cotton
15	Myanmar*	0.3	Cotton
16	Maxico*	0.2	Cotton, soybean
17	Spain*	0.1	Maize
18	Chile*	<0.1	Maize, soybean, canola
19	Colombia	<0.1	Cotton
20	Honduras	<0.1	Maize
21	Sudan	<0.1	Cotton
22	Portugal	<0.1	Maize
23	Czech Republic	<0.1	Maize
24	Cuba	<0.1	Maize
25	Egypt	<0.1	Maize
26	Costa Rica	<0.1	Cotton, soybean
27	Romania	<0.1	Maize
28	Slovakia	<0.1	Maize
	Total	**170.3**	

* 18 biotech mega-countries growing 50,000 hectares, or more, of biotech crops
** Rounded off to the nearest hundred thousand
Data from International Service for the Acquisition of Agri-biotech Applications

How Did We Get Here? The Path to Molecular Biotechnology

Understanding the landscape of molecular biotechnology requires that we first explore the steps we've taken to get here. The first bioengineers knew traits were inherited, but the leaps forward that have taken place in the last century are the direct result of our having figured out the genetic basis of that heritability. Let's look at the individuals and organizations that have contributed to molecular biotechnology, both as a discipline and as an industry, as well as their major discoveries and projects.

The Science: Major Discoveries and Influential Players

As a bioengineer, it bodes well to be familiar with the history of biotechnology. Each discovery is a chapter in a larger story. The researchers outlined below have each played a role in the culmination of the biotechnology industry as it exists today, with each discovery paving the way to the next. It is a fine example of the piggybacking, collaborative nature of science.

Gregor Johann Mendel Fathers the Study of Genetics

As the discipline of science established itself in the Age of Enlightenment, the subject of biology and its cousin natural history were dominated by descriptive pursuits. "Naturalists" focused on the physical form of organisms, a field we know as **morphology**. The *what* was preferred over the *why*. There was little concern with why an organism is the way it is. This focus shifted in the mid to late 19th century. In Brno, Czech Republic, a monk named Gregor Johann Mendel altered the course of biology as he sought to decipher the 10,000 years of civilization fostered by agriculture (**FIGURE 1.6**).

FIGURE 1.6 A sculpture of Gregor Johann Mendel (1822–1884) at the Augustinian Abbey of St. Thomas in Brno, Czech Republic. © Luis Dafos/Alamy

FIGURE 1.7 Mendel planted over 33,500 pea plants in this garden during his eight years of genetic research. Courtesy of Daniel Hartl.

As a botanist, Mendel used garden peas (*Pisum sativum*) as his model organism to perform a variety of genetic studies from 1856–1863. He crossed true-breeding plants (those that are homozygous for a particular trait) in order to determine the patterns of heredity (**FIGURE 1.7**). With a strong background in mathematics, Mendel was able to apply statistical analysis to his data and establish predictable ratios exhibited in offspring, based upon the parental types. He recognized that traits exhibited in an organism are passed on via the gametes of each parent, and so segregate independently of each other, establishing his Laws of Segregation and Independent Assortment (see **BOX 1.1**).

We now know that traits are passed on via genes, although Mendel labeled them "factors." Genes are molecular units of heredity; most code for proteins but some code for RNA molecules. While he published his findings, "Experiments with Plant Hybrids," in 1866, it was not until a few decades later that the value of his work was heralded. Three European scientists, Hugo de Vries, Carl Correns, and Erich von Tschermak, rediscovered Mendel's paper in 1900 after having reached similar conclusions to those of Mendel through their own research. Their collective rediscovery of Mendel's research sparked an influx of interest in heredity at the molecular level. Mendel, now considered the father of genetics, sparked the scientific community to ask why instead of what. Today, the study of how genes are passed from parents to offspring is known as **Mendelian genetics**. It is also referred to as transmission genetics because it relates to how genes are transmitted from one generation to the next.

Natural Selection and Evolution: Alfred Russel Wallace and Charles Darwin

Even the progenitors of evolutionary theory initiated their scientific careers in a descriptive rather than explanatory manner. Charles Darwin practiced as a

Box 1.1 Mendel's Laws of Heredity

- Law of Segregation: Mendel's law of segregation states the following conditions of heredity through sexual reproduction.
  1. All individuals possess two factors for each trait.
  2. The two factors segregate during the formation of gametes, resulting in gametes with only one factor for each trait.
  3. The process of fertilization unites factors from each parent, resulting in offspring with two factors for each trait.

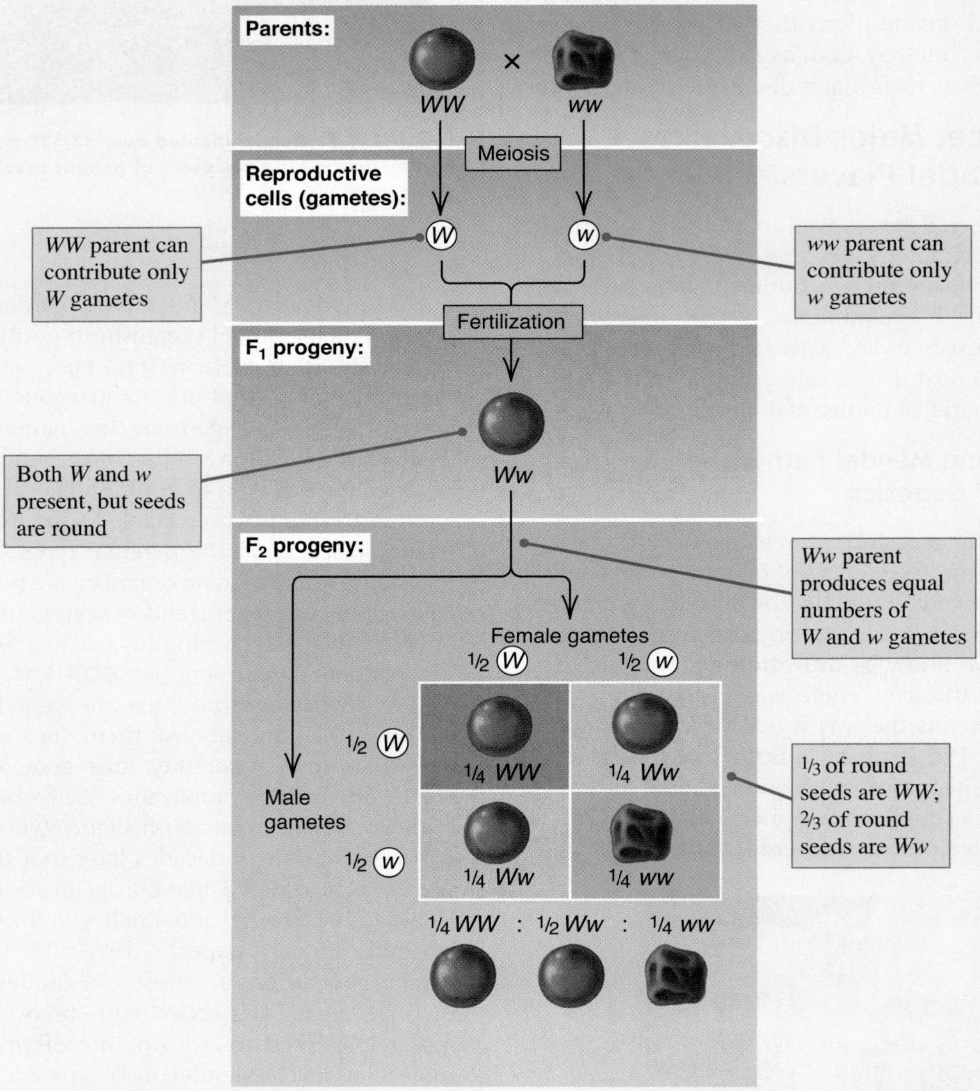

FIGURE 1 A monohybrid cross demonstrates the principles of Mendel's Law of Segregation. Each parent has two factors for the wrinkled trait. Gametes contain one factor for the wrinkled trait. Upon fertilization, progeny of both the F_1 and F_2 generations contain two factors for the wrinkled trait.

- Law of Independent Assortment: The two factors for a given trait segregate independently from other sets of two factors for separate traits. This allows all possible combinations of factors in the resulting gametes.

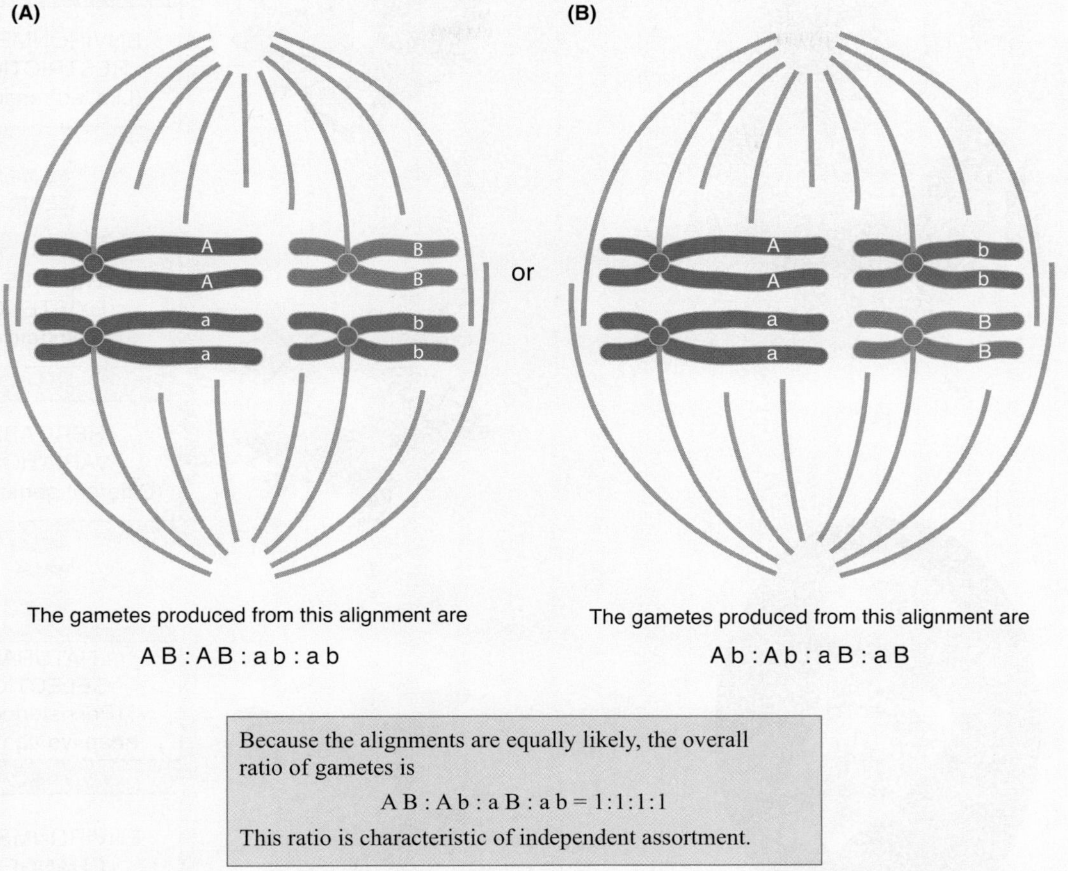
(A)
(B)
A
A
a
a
B
B
b
b
or
A
A
a
a
b
b
B
B
The gametes produced from this alignment are
AB : AB : ab : ab
The gametes produced from this alignment are
Ab : Ab : aB : aB
Because the alignments are equally likely, the overall ratio of gametes is
AB : Ab : aB : ab = 1:1:1:1
This ratio is characteristic of independent assortment.

FIGURE 2 During gamete formation, chromosome pairs sort independently, resulting in all possible combinations of factors from the parents. In this example, the orientation of chromosomes may result in the combinations seen on the left *OR* the combinations seen on the right.

naturalist; that was his assigned task on the voyage of the HMS Beagle (**FIGURE 1.8A**). During his circumnavigation of the planet his myriad observations opened a new world to him, essentially forcing him to think beyond description into the field of explanation. As a contemporary of Mendel, Darwin became equally concerned with descent, although his primary interest laid in understanding how organisms change over time. In 1858 he proposed a primary mechanism of evolution with Alfred Russel Wallace: natural selection. **Natural selection** is the gradual change of organisms over time that leads to adaptations to the environment. Wallace was a fellow naturalist who similarly traveled to exotic locales, sparking his own questions pertaining to the origins of the biodiversity he witnessed (**FIGURE 1.8B**).

However, the concept of natural selection serves as an explanation for change at the organismal level only. Variety within a species and the competition individual members engaged in for survival were the sole means of evolutionary change according to Darwin and Wallace (**FIGURE 1.9**). Without the molecular details available to them, Darwin and Wallace were unable to find any other plausible explanation for change in organisms over time.

The modern synthesis reconciled the link between survival of the fittest and descent with modification, which are based on phenotypes, to classical genetics, which includes genotypes as well. The **modern synthesis** is the currently accepted paradigm of evolutionary biology that bridges gaps between microevolution, occurring at the genetic level, and macroevolution, occurring at the species level. Natural selection has held up as an evolutionary mechanism, but it is now evident molecular events are also at work, such as mutation,

(A)

(B)

FIGURE 1.8 (A) Charles Darwin (1809–1892). (B) Alfred Russel Wallace (1823–1913). © pictore/iStockphoto.com; Reproduced from Alfred Russell Wallace (1889) Darwinism, London and New York: Macmillan and Co.

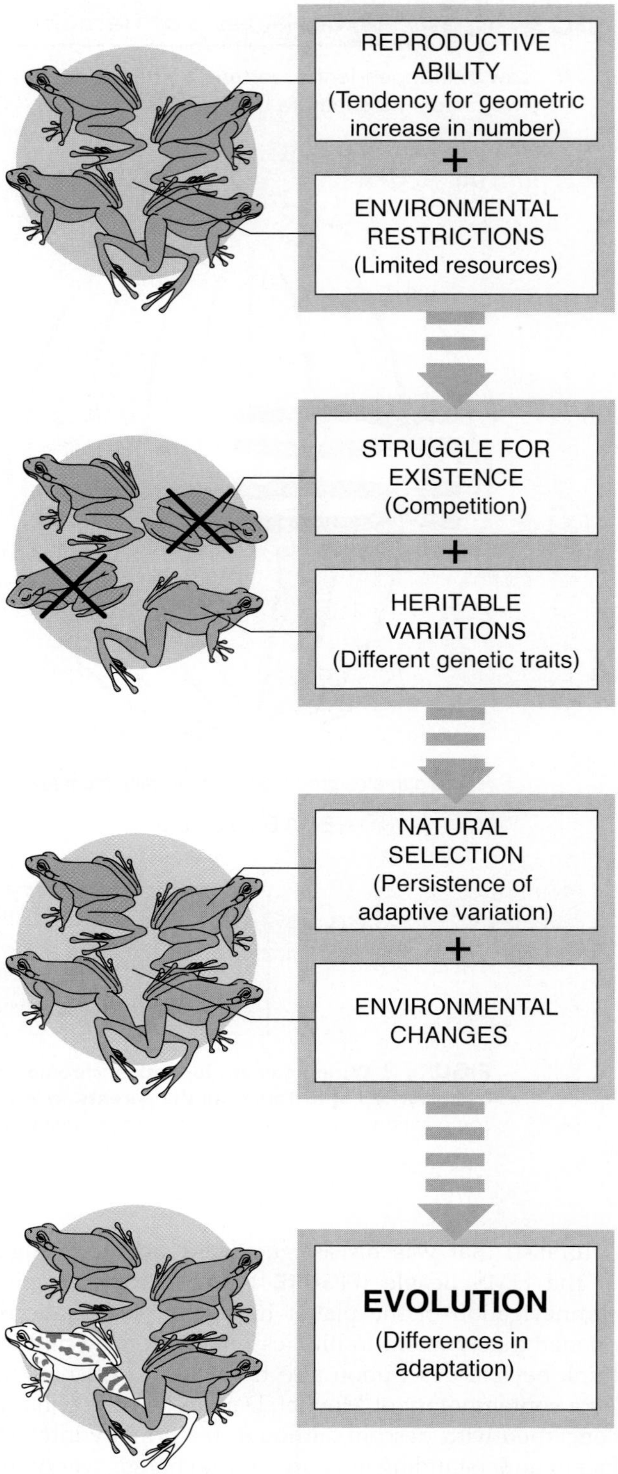

FIGURE 1.9 The line of reasoning introduced by Darwin and Wallace as the evolutionary mechanism of natural selection. Adapted from a table in Wallace, A.R., 1889. *Darwinism: An Exposition of the Theory of Natural Selection with Some of Its Applications.* Macmillan, London.

gene migration, and genetic drift. Molecular events that contribute to evolution can operate at both the population level and the level of an individual organism. The modern synthesis has created a hierarchical understanding of evolution with one mode of action having a cascade effect upon the next level (**TABLE 1.2**).

Johann Friedrich Miescher Discovers DNA

Not all of the seminal discoveries leading to molecular biotechnology were intended for that purpose; the discovery of deoxyribonucleic acid (DNA) is one such

TABLE 1.2

The Modern Synthesis Created a Hierarchical Interpretation of Evolutionary Mechanisms

Level of Action	Genetic	Organismal	Population
Mode of Action	-substitution of alleles	-individual variation	-gene migration -genetic drift
	-changes in gene regulation	-natural selection	-adaptive radiation -speciation
	-mutation	-adaptation	

case. Today, the extraction of DNA from cells is simple enough to perform with materials from around the house. The process was first introduced into the scientific community in the late 19th century. In 1869, Swiss medical doctor Johann Friedrich Miescher isolated an acidic, nitrogen- and phosphate-rich substance from the nuclei of white blood cells while undergoing research at the University of Tübingen in Germany (**FIGURE 1.10**). Miescher's primary research interest was to isolate and characterize molecular components of cells. Miescher performed experiments that ruled out the possibility that the substance was a protein or a lipid and also found it lacked any sulfur. He was sufficiently confident he had discovered a novel molecule and called the substance "nuclein," as it was located in the nucleus of each cell. Nuclein became known more commonly as nucleic acid. The **nucleic acids** include DNA as well as ribonucleic acid (RNA).

FIGURE 1.10 Johann Friedrich Miescher (1844–1895).
Courtesy of Friedrich Miescher Institute

Miescher and members of his laboratory studied a wide variety of cells and determined nuclein was a common feature to all nuclei, leading him to hypothesize that it must be a critical component in the function of the nucleus, such as cell division. He also briefly considered the idea that nuclein might play a role in the transmission of hereditary material, which he later (incorrectly) rejected. Miescher would go to his grave believing proteins were the molecules of heredity, thus never fully comprehending the importance of his discovery.

The Chromosome Theory of Inheritance: Walter Sutton, Theodor Boveri, and Thomas Hunt Morgan

Aware that Gregor Mendel had previously investigated heredity, American zoologist Walter Sutton was familiar with Mendel's proposition that traits are passed down through heritable factors. Sutton proposed that these factors were passed down through generations on chromosomes, but was unable to prove his hypothesis, at least not directly. However, he did prove the physical mechanism behind Mendel's laws of heredity. Sutton successfully carried out cytological experiments with grasshoppers supporting what would become known as the **Chromosome Theory of Inheritance**. This unifying theory of genetics ascertains that chromosomes are the carriers of genetic information, that is, genes (**FIGURE 1.11**). His work showed that chromosomes exist in distinct pairs and segregate during the formation of gametes:

> "I may finally call attention to the probability that the association of paternal and maternal chromosomes in pairs and their subsequent separation during the reducing division...may constitute the physical basis of the Mendelian law of heredity."
>
> *Walter Sutton (1877–1916)*

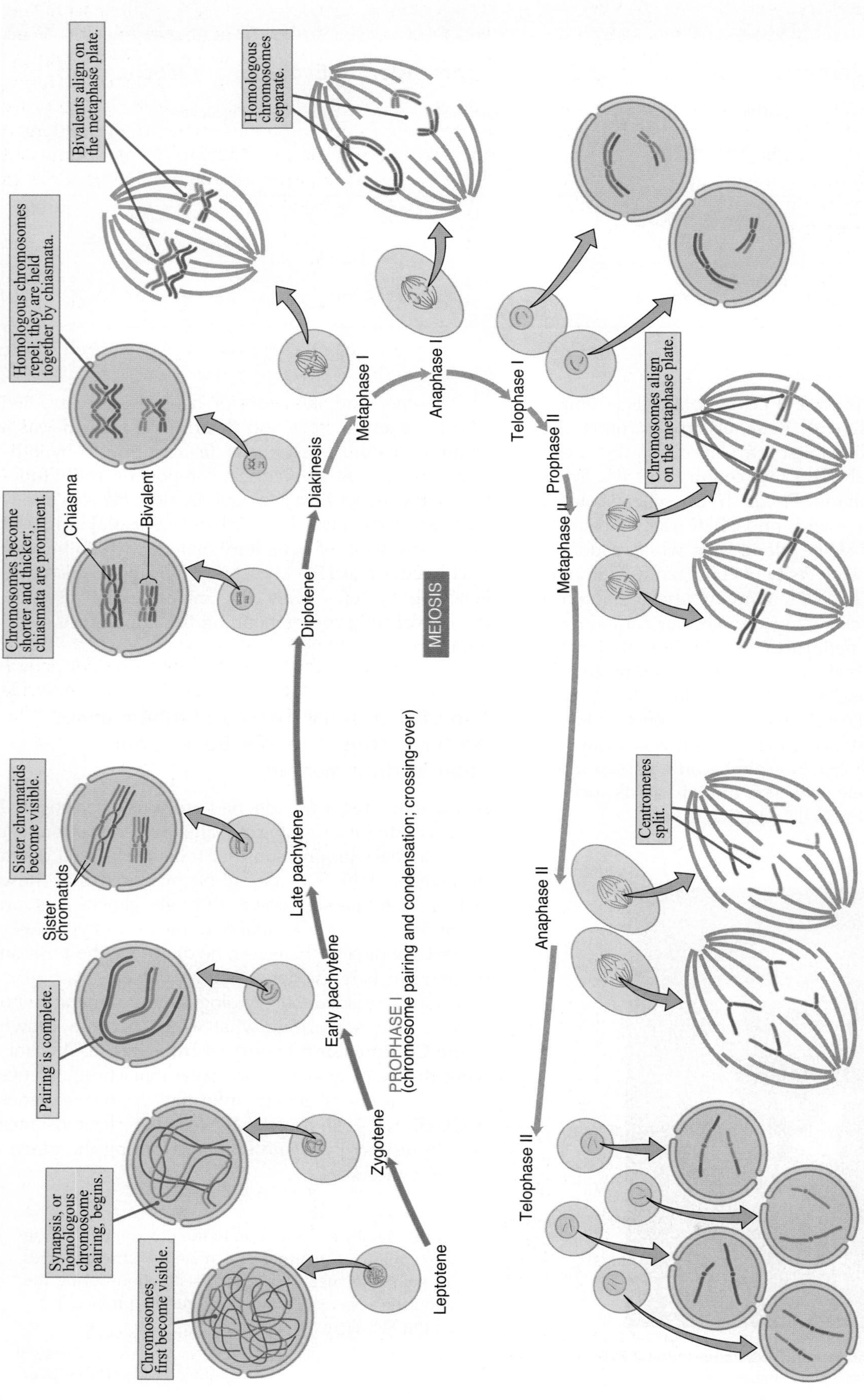

FIGURE 1.11 Sutton's words describe the events taking place during meiosis, the formation of gametes. Here the behavior of chromosomes during meiosis is outlined.

Sutton's first round of experiments in 1902 therefore provided evidence for Mendel's Law of Segregation. Subsequent investigations by Sutton in 1903 led to his conclusion that chromosome pairs orient at random during meiosis, supporting Mendel's earlier proposition on independent assortment. German cytologist Theodor Boveri reached similar conclusions as Sutton at approximately the same time, and to this day the Chromosome Theory of Inheritance is often referred to as the Sutton-Boveri hypothesis (see **FIGURE 1.12A** and **B**).

Thomas H. Morgan and Walter Sutton were both researchers at Columbia University in the laboratory of Edmund B. Wilson. Morgan came to Columbia in 1904 as an experimental zoologist investigating embryology. Soon after his interests shifted to species variation and he thus began experimentation with *Drosophila melanogaster*, the fruit fly. His lab at Columbia, established in 1911, became known as the "Fly Room." Once again a major discovery was to provide evidence for another purpose: Morgan sought to explain the genetic basis of evolution. He was not directly concerned with how genes are transmitted, although this information would be critical for his larger goal of determining how organisms change over time. Surprisingly, Morgan dismissed the idea that chromosome behavior determined patterns of heredity. Nevertheless, through his work he greatly expanded Mendel's assertions and, in 1910, successfully proved the Sutton-Boveri hypothesis. Genes are in fact carried on chromosomes (**FIGURE 1.13**).

Morgan and his students at Columbia would go on to pioneer the process of gene mapping and determination of genetic linkage, as well as describing the processes of crossing over and nondisjunction during meiosis. Morgan's years in the Fly Room would prove to be his greatest accomplishment. In 1933, Morgan and two of his former students, Alfred Sturtevant and Calvin Bridges, were awarded the Nobel Prize in Physiology or Medicine (**FIGURE 1.14**).

Protein or Nucleic Acid?

For decades a debate continued as to which molecule is responsible for transmitting genetic information. Miescher himself, the man who first isolated DNA, rejected his initial hypothesis that the molecule he discovered was the basis of heredity. Rather, he held the belief that proteins acted as the genetic material. He was not alone. Proteins were known to exist in many forms, while Miescher's work indicated DNA was the same no matter what the source organism may be. Proponents for protein as the genetic material argued the uniform nature of DNA across cell types and different species made it too simplistic to be the source of the seemingly endless diversity evident in nature. As proteins are a widely diverse class of molecules, many scientists thought proteins were a more reasonable candidate as the carriers of genetic information. DNA was not universally recognized as the molecule containing genetic material until nearly a century after

(A) (B)

FIGURE 1.12 (A) Walter Sutton (1877–1916) and (B) Theodor Boveri (1862–1915) shared credit in the Sutton-Boveri Hypothesis, which later became the Chromosomal Theory of Inheritance. Sources: Courtesy of the University of Kansas; Courtesy of Hans Stubbe, *History of Genetics*

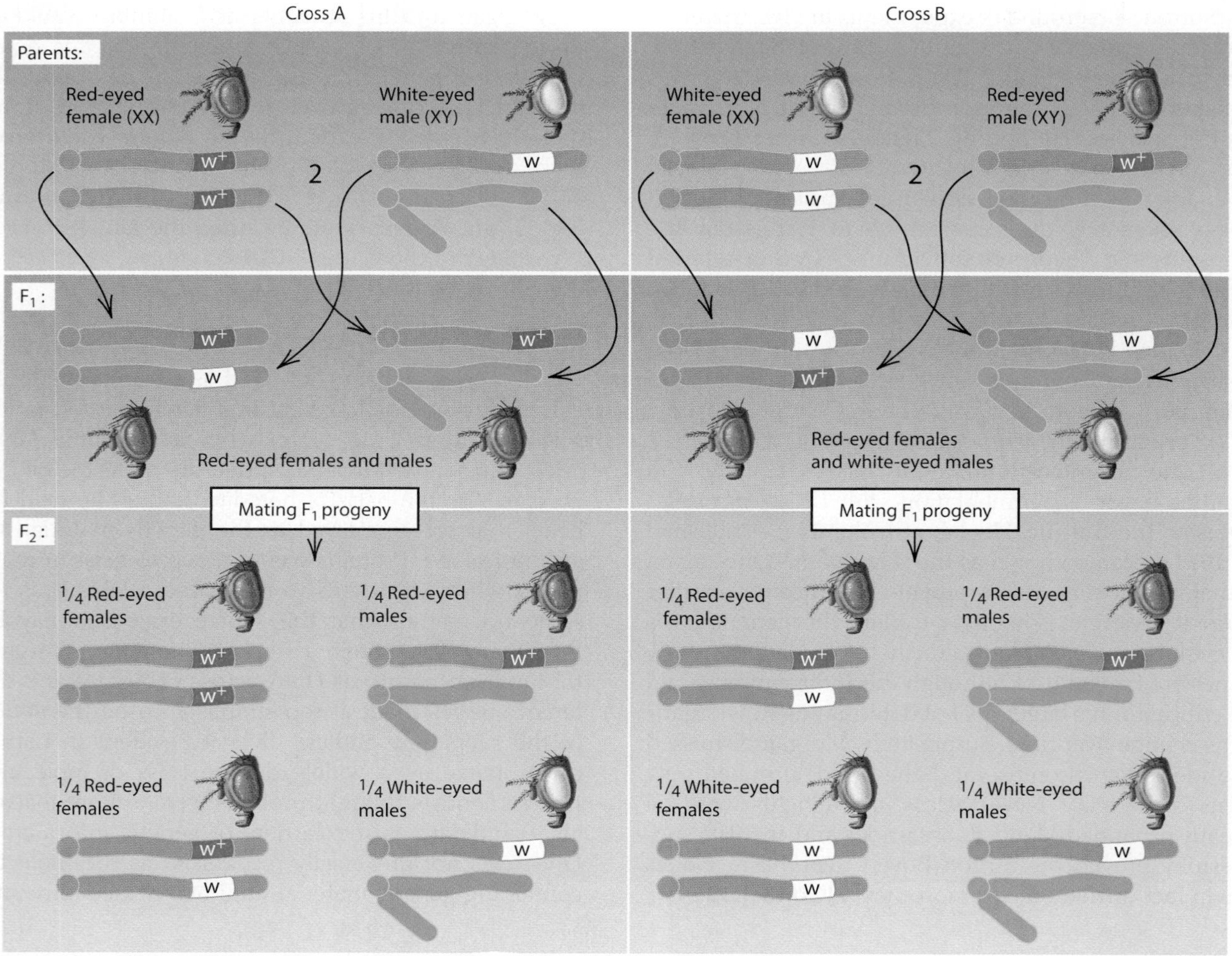

FIGURE 1.13 Crosses A and B demonstrate the Chromosomal Theory of Inheritance as observed by Morgan in his fruit fly test subjects. This particular cross supported Morgan's assertion that the white eye mutation observed in members of his population is due to the inheritance of an eye color gene carried on the X chromosome.

FIGURE 1.14 Morgan and the Fly Room research team celebrate the return of Alfred Sturtevant from World War I at Columbia University in 1919. Morgan is on the far right in the back row. Sturtevant is in the right foreground, leaning back in his chair. Courtesy of the Archives, California Institute of Technology.

Miescher's discovery. Two sets of experiments solidified the function of DNA and quashed the protein versus DNA debate once and for all.

Avery, MacLeod, and McCarty Experiments

In the early 1940s, Oswald T. Avery, Colin MacLeod, and Maclyn McCarty were co-researchers at the Rockefeller Institute interested in settling the protein versus DNA debate. Previous studies in the 1920s showed that pathogenic bacteria ("smooth" colonies of *Streptococcus pneumoniae*) had the ability to transform harmless strains of bacteria ("rough" colonies of *S. pneumoniae*) into pathogenic variants, even when the original pathogens had been killed. The experiments of Frederick Griffith confirmed that **bacterial transformation**, a change in the bacteria's properties, is possible (**FIGURE 1.15**). Bacteria are able to uptake exogenous genetic material from their surrounding environment.

Twenty years later, the Rockefeller team assumed there must be some chemical component driving the transformation; perhaps the chemical was actually a gene. In their experiments, Avery, MacLeod, and McCarty isolated the "transforming principle" from *Streptococcus pneumoniae* and put it through a series of treatments. First, the transforming principle was treated with protein-digesting enzymes. The enzymes had no effect. But when the transforming principle was subjected to treatment with a DNA-digesting enzyme, it was destroyed. The scientists concluded that the transforming principle must be DNA, which would indicate genes must likewise be comprised of DNA (**FIGURE 1.16**). Regardless, these experiments did not solidify DNA as the genetic molecule; many scientists still suspected proteins were the carriers of genetic information.

Hershey and Chase Experiments

In 1952, Alfred Hershey and his research assistant Martha Chase sought to settle the oft-debated question once and for all. Their experiments at Cold Spring Harbor Laboratory served as an extension of the work being performed at the Rockefeller Institute. Acting on the assertion that bacteria can exchange genetic material, Hershey and Chase chose *Escherichia coli* as their subject organism. Previous work in electron microscopy suggested bacterial viruses, or bacteriophages, have similar transformative properties and could likewise alter the genetic make-up of its host. Hershey and Chase decided to concurrently study an *E. coli* bacteriophage, T2 (**FIGURE 1.17**). A **bacteriophage**, or simply "phage," is a virus that infects bacterial cells.

In their experiments, Hershey and Chase utilized radioactive isotope tracers as a tool to track transmission

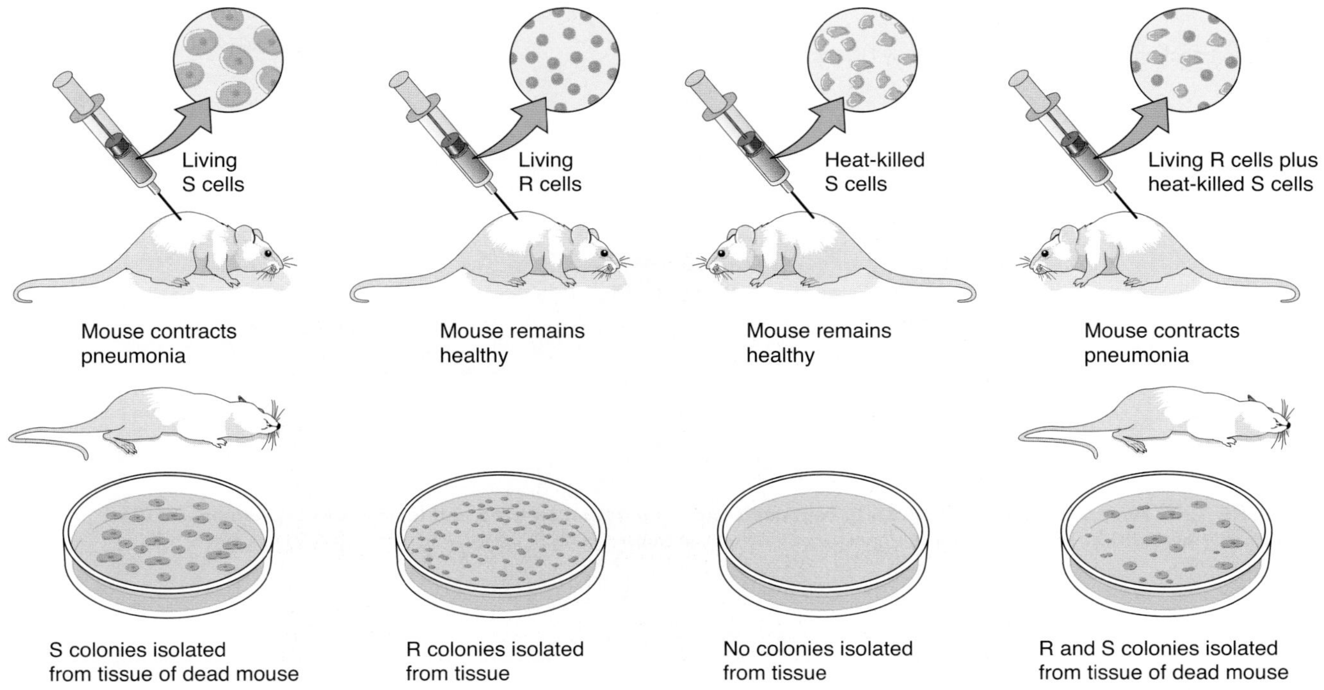

FIGURE 1.15 In 1928, Frederick Griffith's experiments with S and R strains of *Streptococcus pneumoniae* proved the possibility of bacterial transformation.

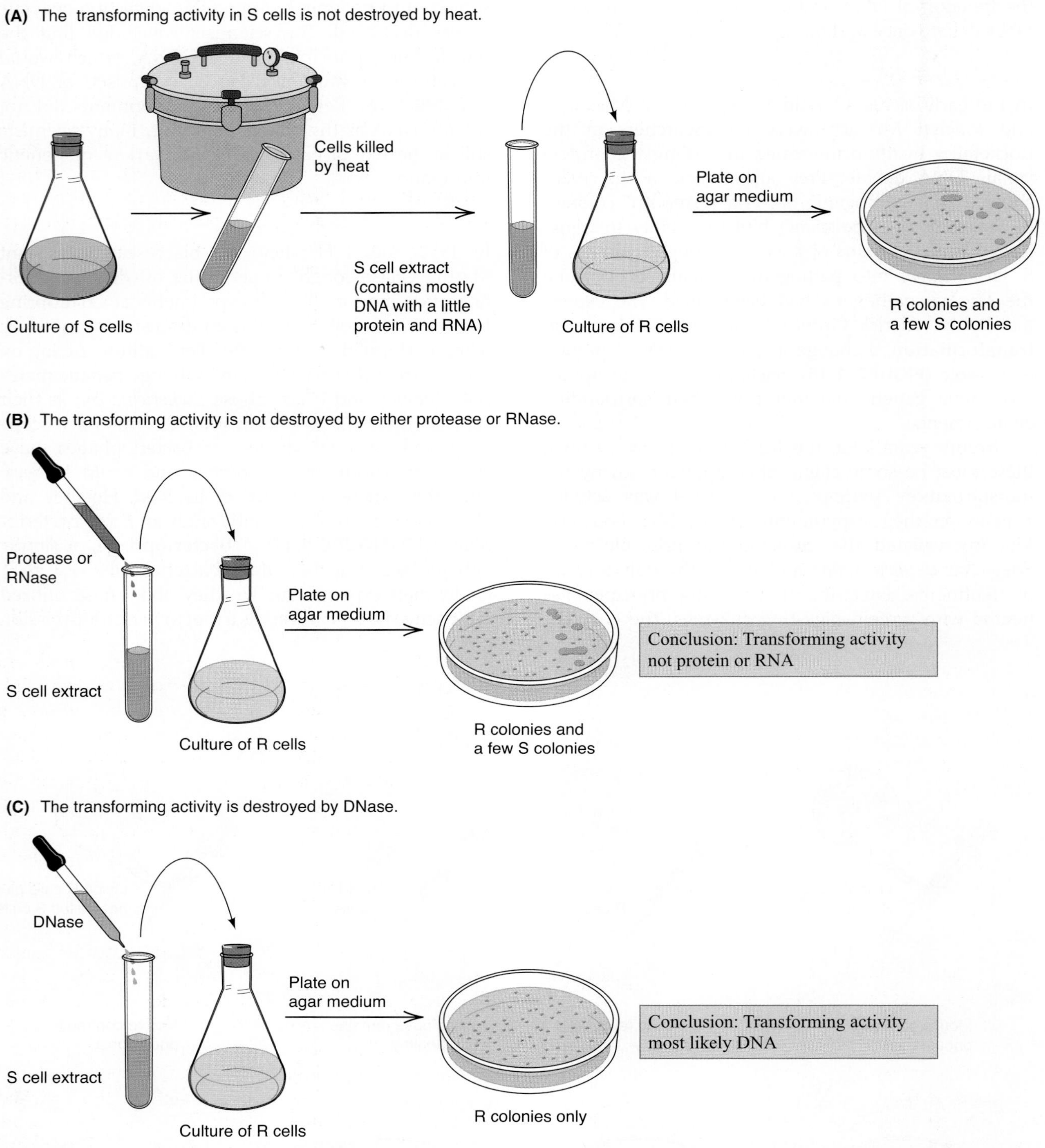

FIGURE 1.16 Avery, MacLeod, and McCarty continued experimentation on the S and R strains of *S. pneumoniae* used previously by Griffith. This experiment demonstrated DNA is the transforming principle which changed R cells into S cells.

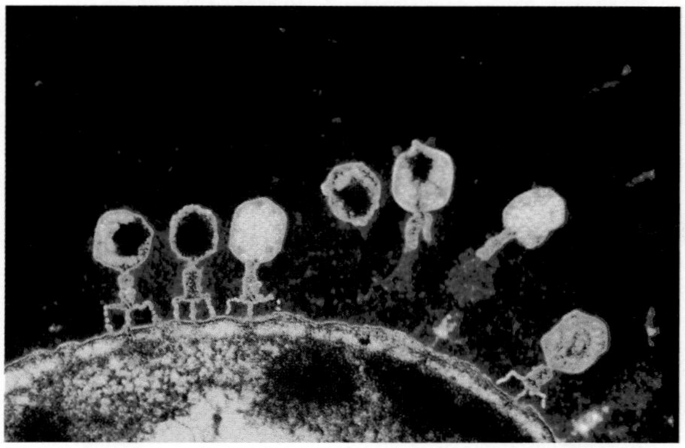

FIGURE 1.17 An electron micrograph of T2 bacteriophage infecting an *E. coli* bacterial cell. © Oliver Meckes/E.O.S./MPI Tubingen/Science Source

of genetic material between *E. coli* and T2. First a tracer was inserted in the protein coat of T2 and observed as T2 infected an *E.coli* host cell. The tracer remained outside of *E.coli* cells and did not "infect" the host. Such a result suggested protein was not a transforming agent and therefore could not be genetic material. To confirm this initial conclusion, Hershey and Chase then inserted a tracer into the DNA of the T2 virus. The bacteriophage was then allowed to infect an *E. coli* organism. During this process, the two researchers observed the tracer enter and integrate itself into the host cell. This second experiment made it clear to Hershey and Chase that DNA—not protein—must be the genetic molecule (**FIGURE 1.18**). With additional evidence, the scientific community could not deny that DNA transmits genetic information from parent to offspring. Hershey would go on to receive the 1969 Nobel Prize in Physiology or Medicine for collaborations with Max Delbrück and Salvador Luria involving the genetic structures and replication mechanisms of viruses.

Chargaff's Rules

In the late 1940s, Erwin Chargaff and his research team at Columbia University performed experiments on the physical nature of DNA. Chargaff built upon the knowledge that DNA is a polymer consisting of nucleotide monomers. Each nucleotide consists of a deoxyribose sugar, a phosphate group, and one of four nitrogenous bases: adenine (A) and guanine (G), collectively known as the **purines**, and thymine (T) and cytosine (C), the **pyrimidines** (**FIGURE 1.19A** and **B**). In 1950, Chargaff was able to determine that these four bases exist in varying proportions within different organisms, yet adenine and thymine are always present in a 1:1 ratio. His team found guanine and cytosine exhibited the same 1:1 ratio. These ratios became known as **Chargaff's Rules**: A always pairs with T and C always pairs with G (**FIGURE 1.19C**).

DNA Is a Double Helix: Watson, Crick, Franklin, and Wilkins

It is quite a list of names associated with the "discovery" of DNA, but from our discussion thus far it should be evident that DNA was already isolated and analyzed by the time the molecular structure was determined. The nucleotide building blocks and Chargaff's Rules both acted as significant clues in the search for the chemical structure of DNA. James Watson and Francis Crick receive the most credit for determining the structure of the DNA molecule; they are household names. The historical footnotes paint a more blurred picture. Further complicating the identity of the true discoverers of DNA are a number of contradicting accounts of the following events.

Two separate research teams contributed to the discovery of DNA structure in 1953. At King's College in London, Rosalind Franklin and Maurice Wilkins, along with their assistants Alec Stokes, Ray Gosling, and Herbert Wilson, produced X-ray crystallography images of DNA (**FIGURE 1.20**). From these images, they observed the major and minor grooves characteristic of the DNA molecule (**FIGURE 1.21**). Franklin and Wilkins independently concluded that DNA must be helical and double-stranded.

Nearby at the Cavendish Laboratory in Cambridge, American geneticist James Watson and English physicist Francis Crick had obtained the images produced at King's College in addition to some of Franklin's data. How these were obtained is uncertain. According to renowned chemist Linus Pauling, who was also working to determine DNA structure, "it was Wilkins' experimental work that put Watson and Crick on the right track." Additionally, the two had conversed with Erwin Chargaff, receiving first-hand information about his base pairing rules. According to Chargaff, "I believe that the double-stranded model of DNA came about as a consequence of our conversation."

Nevertheless, their knowledge of the polymer nature of DNA, Chargaff's Rules, and now the major and minor grooves seen in the King's College images allowed Watson and Crick to decipher the exact structure. They engaged in building three-dimensional models that took into account complementary base pairing as well as the periodicities created by the major and minor grooves. At last they built a model that satisfied all existing knowledge: a right-handed helical molecule composed of two sugar-phosphate strands running in opposite directions, linked together by the hydrogen bonding of complementary bases (**FIGURE 1.22**).

April 1953 saw the publication of three separate papers in the journal *Nature*: one authored by Watson and Crick, another by Wilkins, Stokes, and Wilson, and a third by Franklin and Gosling. Just a little more than a month later, Watson and Crick published another paper proposing quite accurately a mechanism for DNA

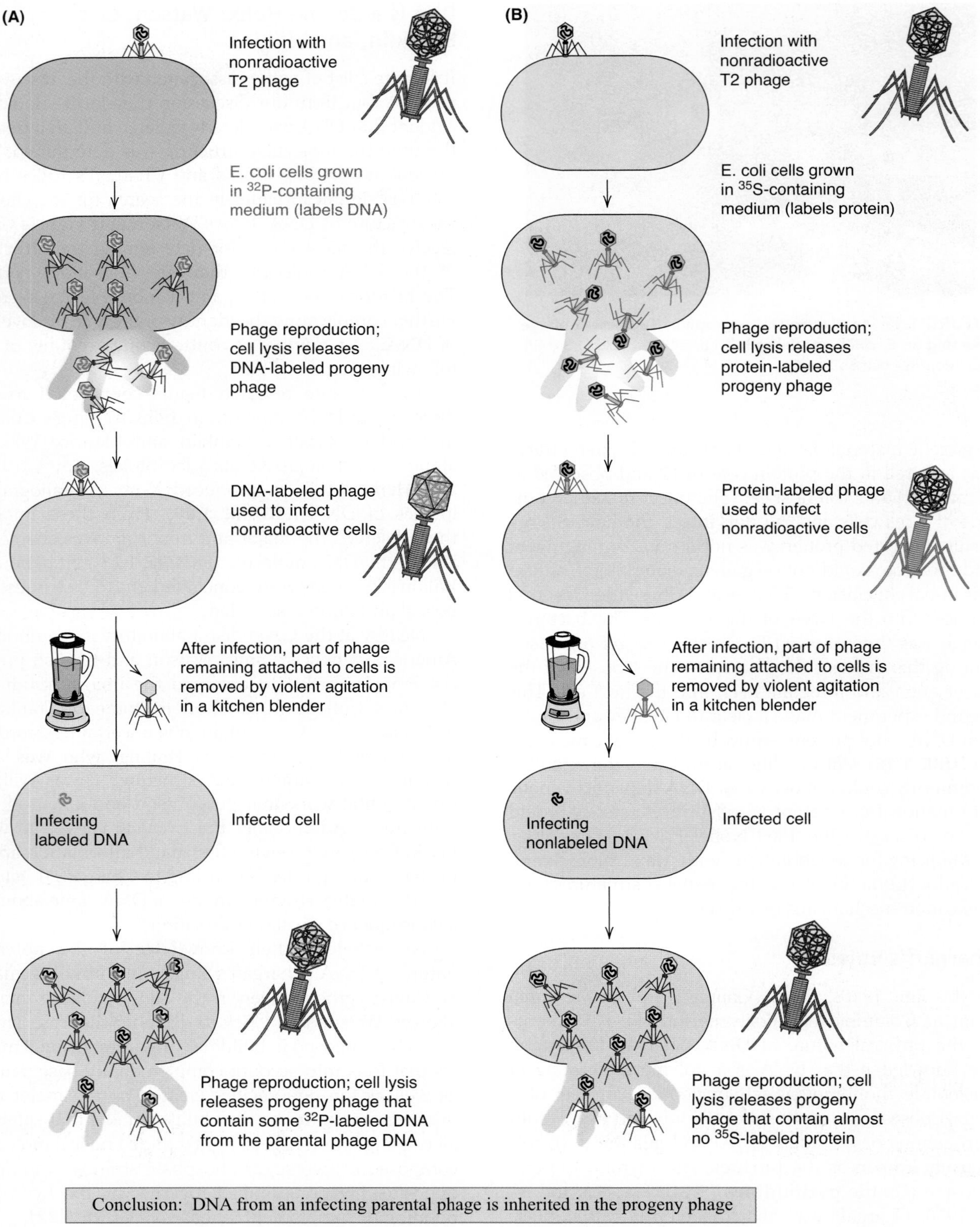

FIGURE 1.18 The Hershey-Chase experiments demonstrate DNA is the genetic material, not protein. (A) Radioactive DNA is tracked during T2 bacteriophage infection of *E. coli*. (B) Radioactive protein is tracked.

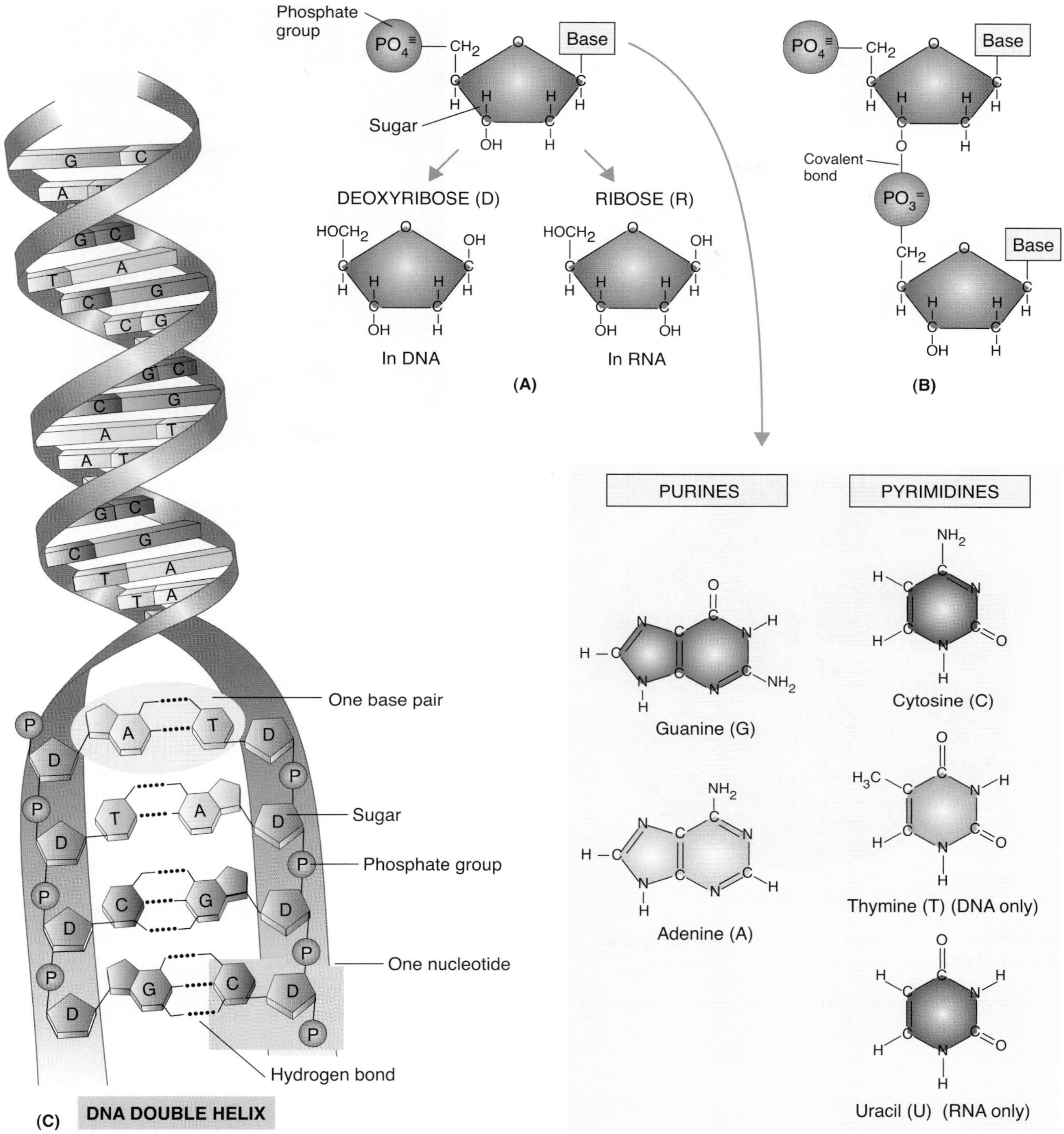

FIGURE 1.19 **(A) Each nucleotide monomer of the DNA molecule consists of a deoxyribose sugar, a phosphate group, and one of four nitrogenous bases. (B) The nitrogenous bases include the purines and the pyrimidines. (C) Chargaff's Rules: The DNA double helix is held together by the complementary base pairing of A with T and C with G.**

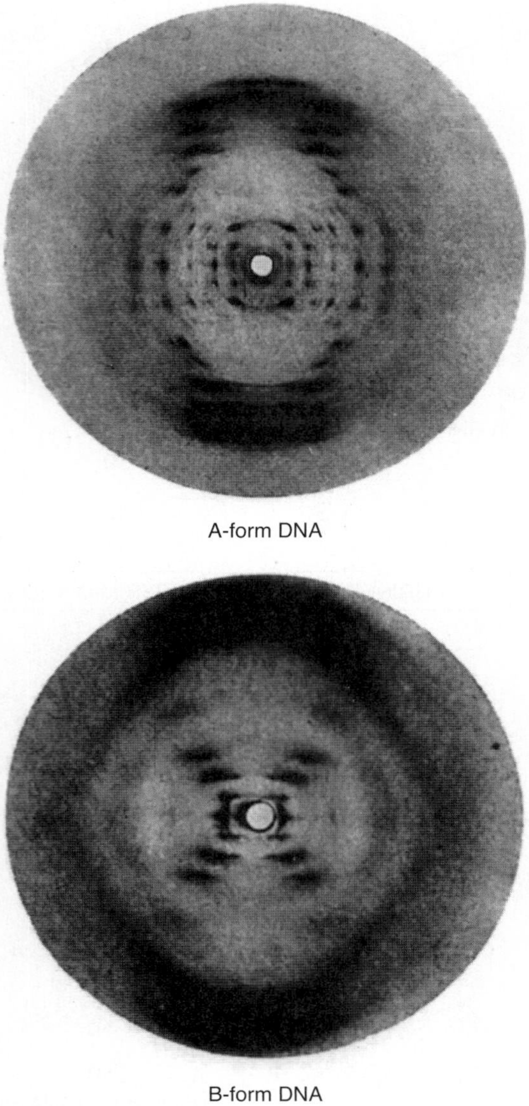

FIGURE 1.20 The King's College X-ray crystallography images of DNA. [Reproduced from R. E. Franklin and R. G. Gosling, *Acta Crystallographica 6* (1953): 673-677. Photos courtesy of the International Union of Crystallography.]

replication involving the predictability of complementary base sequences. In subsequent years, Maurice Wilkins continued research on the double helix in order to verify the work of Watson and Crick. This additional work led to his inclusion in receiving the Nobel Prize in Physiology or Medicine along with Watson and Crick in 1962. Franklin passed away in the interim and was excluded from the prize. Nobel rules forbid posthumous awards. Watson and Crick were immortalized in Cambridge with the words of James Watson etched in stone (**FIGURE 1.23**).

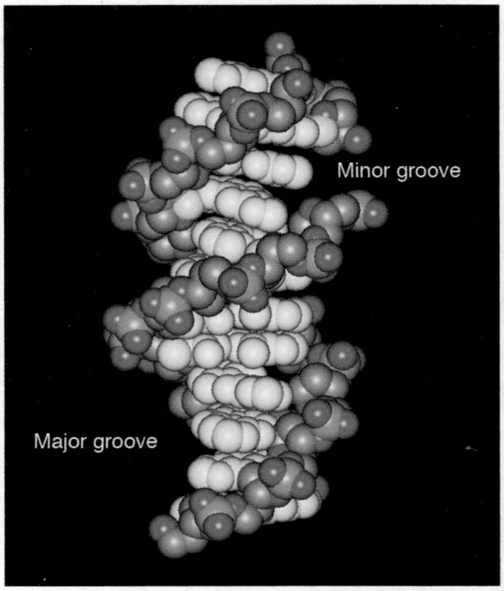

FIGURE 1.21 A model of DNA showing the major and minor groove patterns characteristic of the molecule. Source: Structure from Protein Data Bank 1BNA. H. R. Drew, et al., *Proc. Natl. Acad. Sci. USA* 78 (1981): 2179–2183. Prepared by B. E. Tropp.

> "The double helix is indeed a remarkable molecule. Modern man is perhaps 50,000 years old, civilization has existed for scarcely 10,000 years and the United States for only just over 200 years; but DNA and RNA have been around for at least several billion years. All that time the double helix has been there, and active, and yet we are the first creatures on Earth to become aware of its existence."
>
> *Francis Crick (1916–2004)*

Grandiose verbage aside, Crick was hardly underestimating the truth. It bears repeating that the field of biotechnology owes its gratitude to our understanding of nucleic acids, including their structure. As the first creatures to be aware of its existence, we have certainly put the information to a multitude of uses, both for our own benefit and that of the planet we inhabit.

The Genetic Code

The structure of DNA was a seminal discovery in its own right, but it still does not tell us how genetic information is conveyed by a cell and expressed in an organism. In light of the complementary base pairing between A's and T's and C's and G's, Watson and Crick predicated the idea of a genetic code in their 1953 *Nature* paper: "It has not escaped our notice that the specific pairing we have postulated immediately suggests a possible copying mechanism for the genetic material." This copying mechanism is observed in the

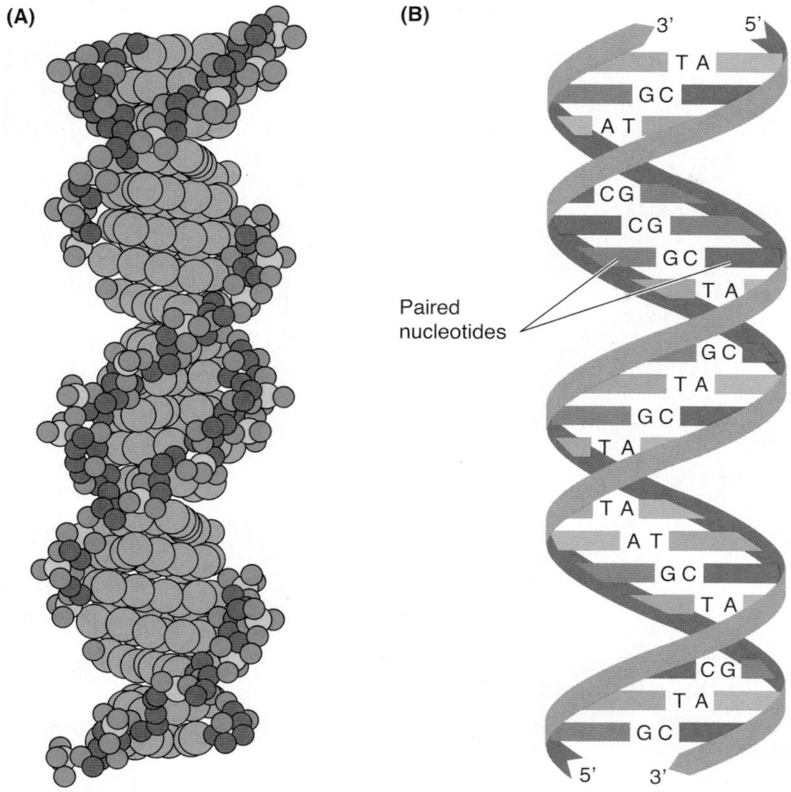

FIGURE 1.22 The molecular structure of DNA as described by Watson and Crick in 1953. (A) A space-filling model. (B) A diagram featuring complementary base pairing.

FIGURE 1.23 A plaque commemorating James Watson and Francis Crick on the Eagle Pub, Cambridge, England. According to Watson, "At lunch Francis winged into the Eagle to tell everyone within hearing distance that we had found the secret of life." © Awe Inspiring Images/ ShutterStock, Inc.

processes of DNA replication and protein synthesis (**FIGURE 1.24**).

It was Crick who formulated the **central dogma of molecular biology**: genetic information flows from DNA to RNA to a final protein product (**FIGURE 1.25**). Ribonucleic acid (RNA), a separate nucleic acid previously identified in the cytoplasm of cells, would therefore act as an intermediary to translate the information coded in a DNA nucleotide sequence into the amino acid sequence found in the resulting protein (**FIGURE 1.26A** and **B**). Therefore, much like the Rosetta Stone served as a key to translate ancient languages, a **genetic code** would provide the information to translate nucleotide sequences into amino acid sequences.

Part of the problem in solving the genetic code boiled down to basic mathematics and the concept of **parsimony** (the best possible explanation is the simplest one). Twenty amino acids were known to exist in proteins and four different nucleotides are present in DNA. If just four nucleotides act as the basis for a genetic code, clearly they could not code for all

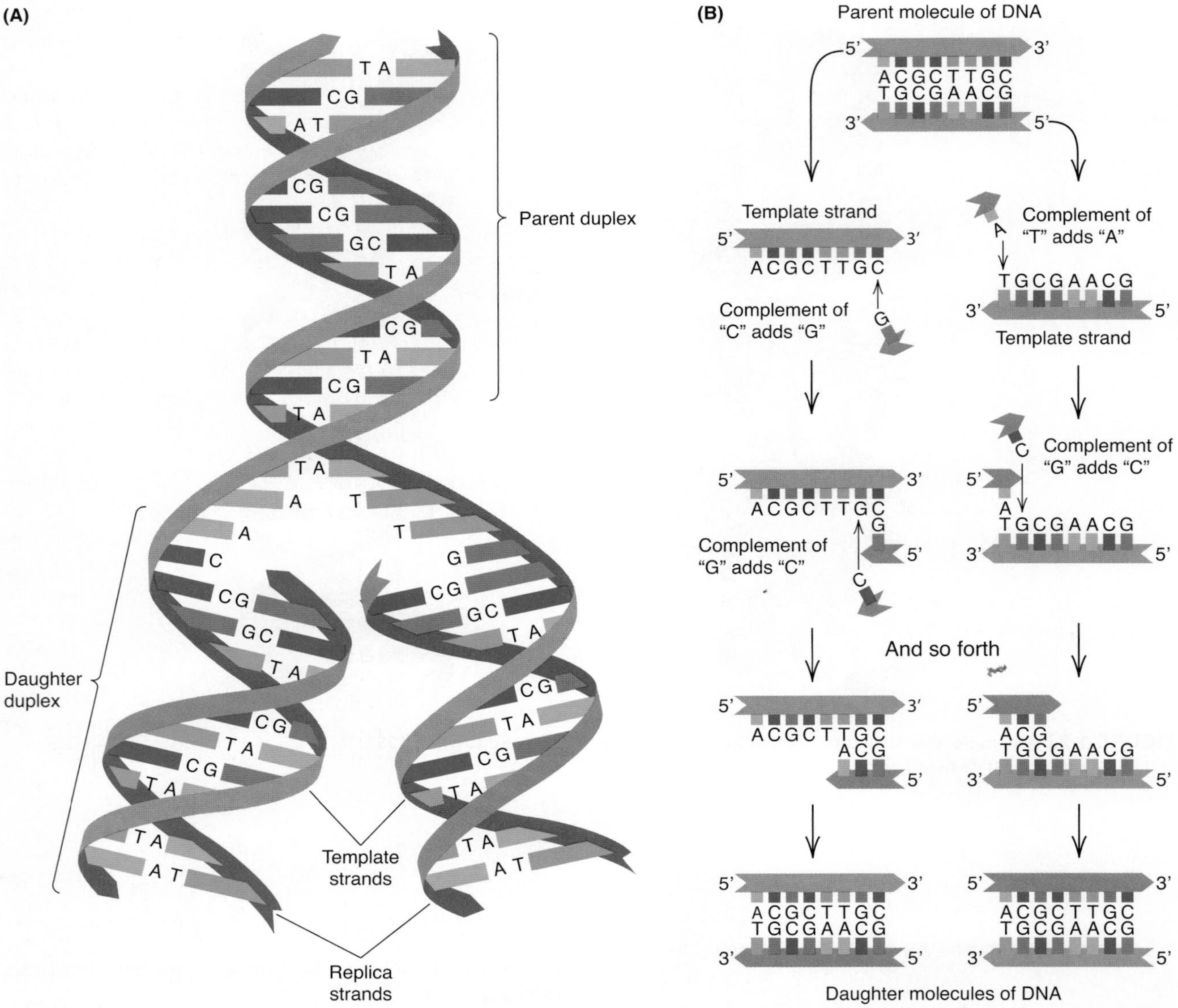

FIGURE 1.24 The "copying mechanism" predicted by Watson and Crick is the process of DNA replication. (A) The double helix separates and each parental strand acts as a template for two new daughter strands via complementary base pairing. (B) Detail of complementary base pairing in the formation of daughter molecules of DNA.

20 amino acids in a 1:1 ratio; this would only account for four amino acids. A two-nucleotide combination would only result in 16 amino acids, again not enough. A three-nucleotide sequence seemed plausible, because it would allow for 64 combinations, more than enough. A four-nucleotide sequence would be overly complex and not fit the parameters of parsimony. A three-nucleotide or "triplet" code appeared most likely. The triplets are referred to as **codons**. Aided by previous work by George Gamow, Crick published "Codes Without Commas" with J.S. Griffith and L.E. Orgel in 1957. This paper presents the assumption that if there are 64 codons, each amino acid must be encoded by more than one codon. Yet the paper was entirely theoretical and not based upon experimentation, so the potential of a three-nucleotide codon was not definitively proven.

Nine years later, the genetic code was broken. Researchers at the University of British Columbia, led by Har Gobind Khorana, and at the National Institutes of Health, led by Marshall Nirenberg, played a role. As RNA was the likely suspect involved in translating DNA into a protein, the research teams narrowed their focus to experiments with RNA (**FIGURE 1.27**). Every possible codon, or sequence of three nucleotides, was used to synthesize long repetitious strands of

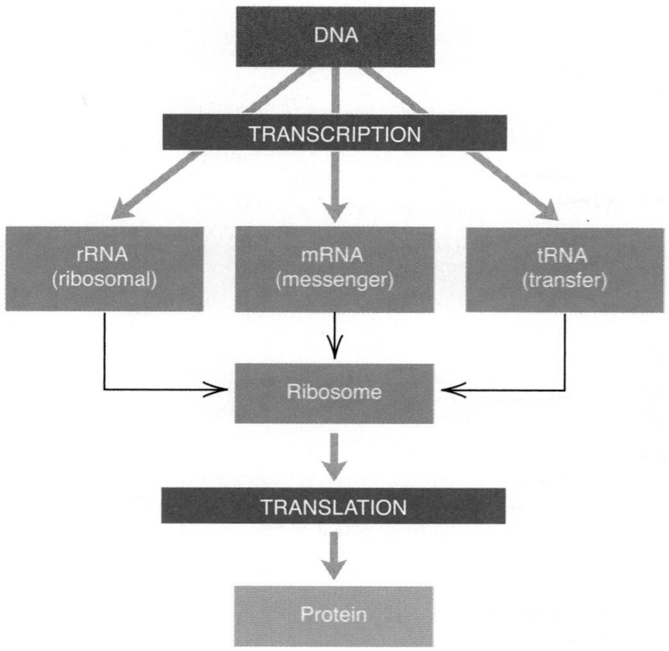

FIGURE 1.25 The "central dogma" of molecular biology asserts genetic information flows from DNA to RNA to protein.

RNA. For example, an RNA strand with a repeating sequence of GCA. As there are 64 possible codons, 64 RNA strands were synthesized. They created a synthetic cell environment composed of cytoplasmic molecules: amino acids, transfer RNAs, and ribosomes. Each of the 64 RNA strands were individually introduced one at a time into the cytoplasmic environment. If the "Codons Without Commas" prediction was correct, the researchers expected to see each type of repetitive RNA strand to result in protein composed of one amino acid repeating itself again and again. In the end, Marshall Nirenberg, Har Gobind Khorana, and their research teams determined each of the 20 amino acid monomers found in proteins is coded for by at least one codon. Because more than one codon could code for a particular amino acid, the genetic code is considered redundant, or **degenerate**. One "start" codon and three "stop" codons were also identified (**TABLE 1.3**). Start codons initiate the process of translation from RNA to protein, while the stop codons cease the process. All 64 possible codon combinations were accounted for and the "Codes Without Commas" hypothesis was supported. Nirenberg and Khorana shared the 1968 Nobel Prize for Physiology or Medicine with Robert Holley, who sequenced the first transfer RNA molecule and determined the structure of tRNA.

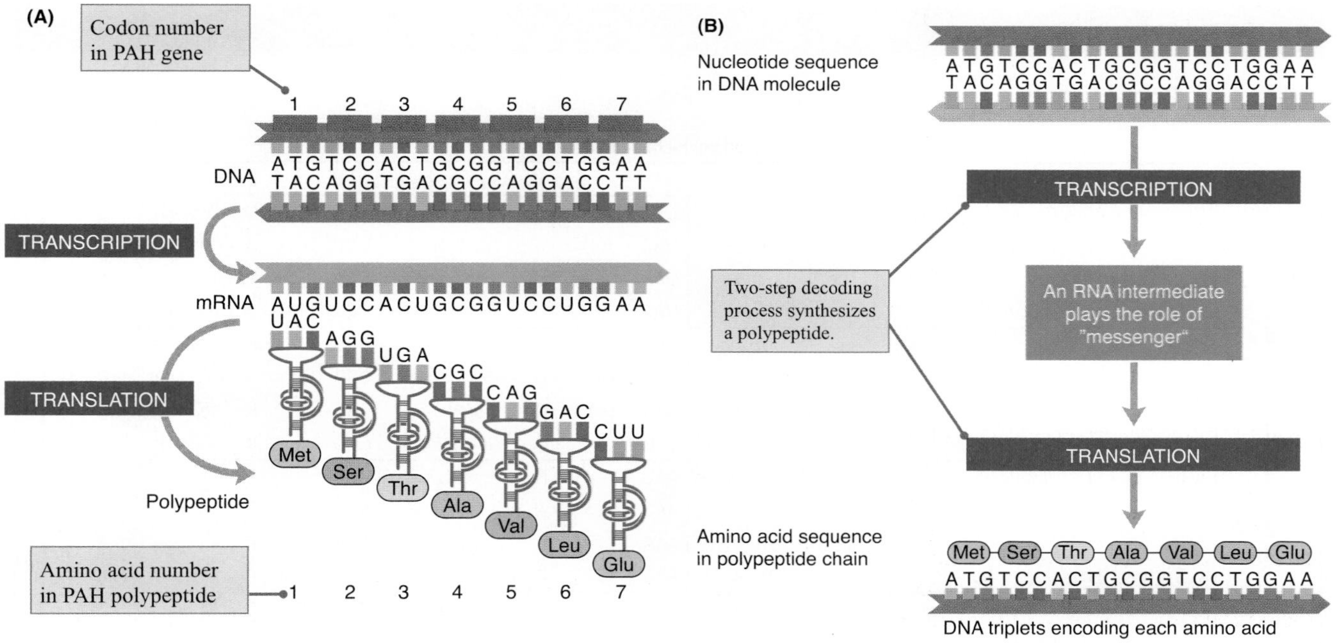

FIGURE 1.26 (A) A visual interpretation of transcription (DNA to RNA) and translation (RNA to protein) carried out during gene expression. (B) A closer look at gene expression, the process of protein synthesis. Nucleotide sequences and resulting amino acid sequences are featured. Adapted from D. Secko, *The Science Creative Quarterly*, 2007 (http://www.scq.ubc.ca/a-monks-flourishing-garden-the-basics-of-molecular-biology-explained/). Accessed February 5, 2014.

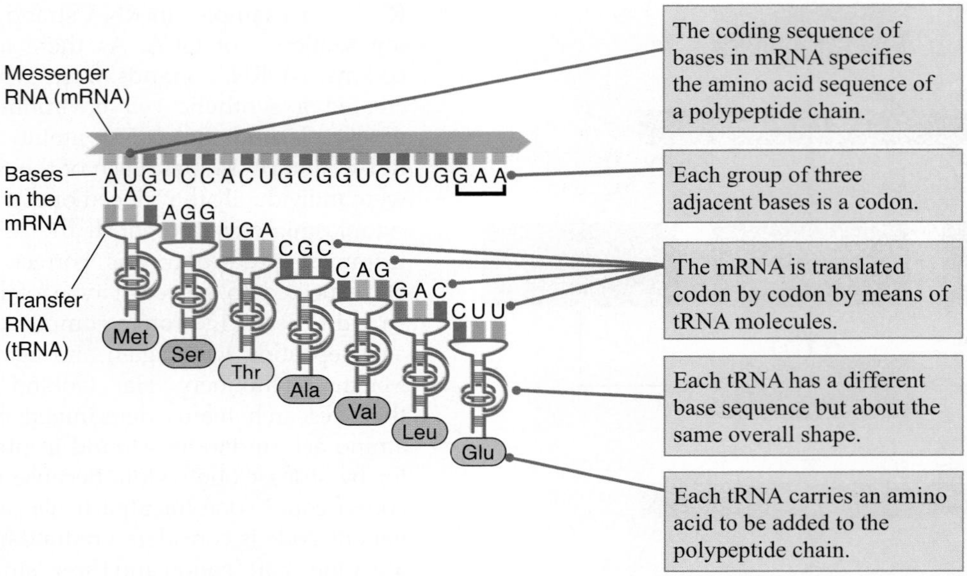

FIGURE 1.27 The role of messenger RNA (mRNA) in translation.

The Technology: An Industry Is Born

The events described thus far include those prior to the inception of biotechnology as an industry. Scientific knowledge was the primary pursuit. We now turn to those events that occurred once the industry took hold. The foundation of science is applied toward utilitarian purposes, establishing what we might collectively refer to as technology.

TABLE 1.3

Genetic Code

Second nucleotide in codon

First nucleotide in codon (5' end)	U				C				A				G				Third nucleotide in codon (3' end)
U	UUU	Phe	F	*Phenylalanine*	UCU	Ser	S	*Serine*	UAU	Tyr	Y	*Tyrosine*	UGU	Cys	C	*Cysteine*	U
	UUC	Phe	F	*Phenylalanine*	UCC	Ser	S	*Serine*	UAC	Tyr	Y	*Tyrosine*	UGC	Cys	C	*Cysteine*	C
	UUA	Leu	L	*Leucine*	UCA	Ser	S	*Serine*	UAA			Termination	UGA			Termination	A
	UUG	Leu	L	*Leucine*	UCG	Ser	S	*Serine*	UAG			Termination	UGG	Trp	W	*Tryptophan*	G
C	CUU	Leu	L	*Leucine*	CCU	Pro	P	*Proline*	CAU	His	H	*Histidine*	CGU	Arg	R	*Arginine*	U
	CUC	Leu	L	*Leucine*	CCC	Pro	P	*Proline*	CAC	His	H	*Histidine*	CGC	Arg	R	*Arginine*	C
	CUA	Leu	L	*Leucine*	CCA	Pro	P	*Proline*	CAA	Gln	Q	*Glutamine*	CGA	Arg	R	*Arginine*	A
	CUG	Leu	L	*Leucine*	CCG	Pro	P	*Proline*	CAG	Gln	Q	*Glutamine*	CGG	Arg	R	*Arginine*	G
A	AUU	Ile	I	*Isoleucine*	ACU	Thr	T	*Threonine*	AAU	Asn	N	*Asparagine*	AGU	Ser	S	*Serine*	U
	AUC	Ile	I	*Isoleucine*	ACC	Thr	T	*Threonine*	AAC	Asn	N	*Asparagine*	AGC	Ser	S	*Serine*	C
	AUA	Ile	I	*Isoleucine*	ACA	Thr	T	*Threonine*	AAA	Lys	K	*Lysine*	AGA	Arg	R	*Arginine*	A
	AUG	Met	M	*Methionine*	ACG	Thr	T	*Threonine*	AAG	Lys	K	*Lysine*	AGG	Arg	R	*Arginine*	G
G	GUU	Val	V	*Valine*	GCU	Ala	A	*Alanine*	GAU	Asp	D	*Aspartic acid*	GGU	Gly	G	*Glycine*	U
	GUC	Val	V	*Valine*	GCC	Ala	A	*Alanine*	GAC	Asp	D	*Aspartic acid*	GGC	Gly	G	*Glycine*	C
	GUA	Val	V	*Valine*	GCA	Ala	A	*Alanine*	GAA	Glu	E	*Glutamic acid*	GGA	Gly	G	*Glycine*	A
	GUG	Val	V	*Valine*	GCG	Ala	A	*Alanine*	GAG	Glu	E	*Glutamic acid*	GGG	Gly	G	*Glycine*	G

Codon

Three-letter and single-letter abbreviations

Restriction Endonucleases: Werner Arber, Daniel Nathans, and Hamilton Smith

The 1978 Nobel Prize in Physiology or Medicine was awarded to Werner Arber, Daniel Nathans, and Hamilton Smith for their discovery of **restriction endonucleases**. Also known as restriction enzymes, these molecules are a vital tool in recombining DNA for specific traits. In the process of their research, Arber, Nathans, and Smith found that restriction enzymes have the ability to cut up DNA into smaller, nonfunctional pieces. Swiss scientist Arber observed that certain bacteriophages were limited in their growth. He hypothesized their bacterial hosts possessed enzymes that could cut viral DNA and prevent replication of the bacteriophage. Smith isolated and characterized the first restriction enzyme to be well characterized, *Hind*III, named for the *Haemophilus influenza* bacterium from which it is derived (**FIGURE 1.28**).

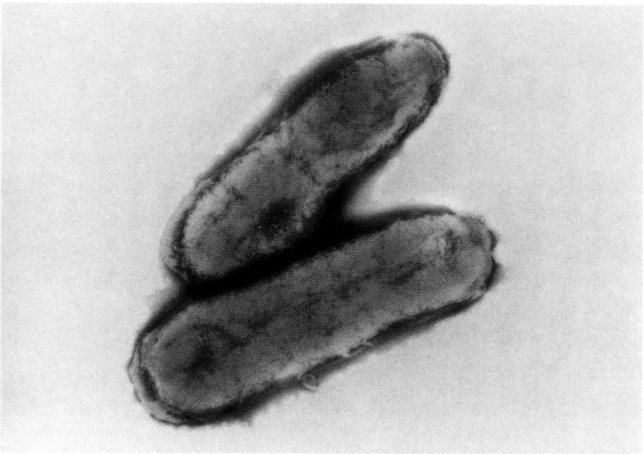

FIGURE 1.28 Scanning electron micrograph of the *Haemophilus influenza* bacterium. © CAMR/A. Barry Dowsett /Science Source

Since the 1970s hundreds of restriction enzymes have been isolated and characterized (**TABLE 1.4**). Each one targets a specific DNA sequence and cuts, or "cleaves" the DNA at what is known as the restriction site. Restriction enzymes are a vital tool in genetic engineering and recombinant DNA technology as they allow bioengineers to cut DNA, allowing them to subsequently "paste" in novel DNA sequences. This technology led to the development of recombinant DNA, discussed next.

TABLE 1.4

Restriction Endonucleases

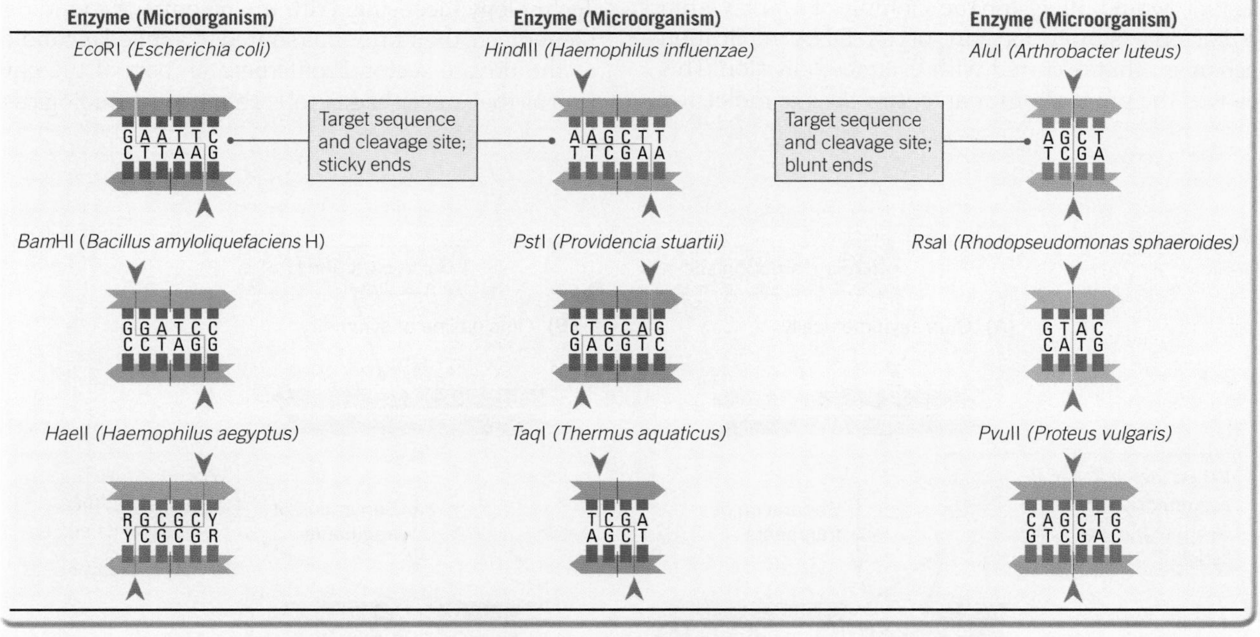

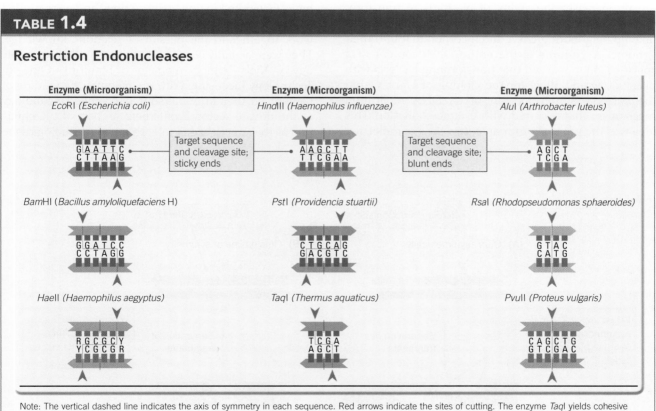

Note: The vertical dashed line indicates the axis of symmetry in each sequence. Red arrows indicate the sites of cutting. The enzyme *Taq*I yields cohesive ends consisting of two nucleotides, whereas the cohesive ends produced by the other enzymes contain four nucleotides. R and Y refer respectively to any complementary purines and pyrimidines.

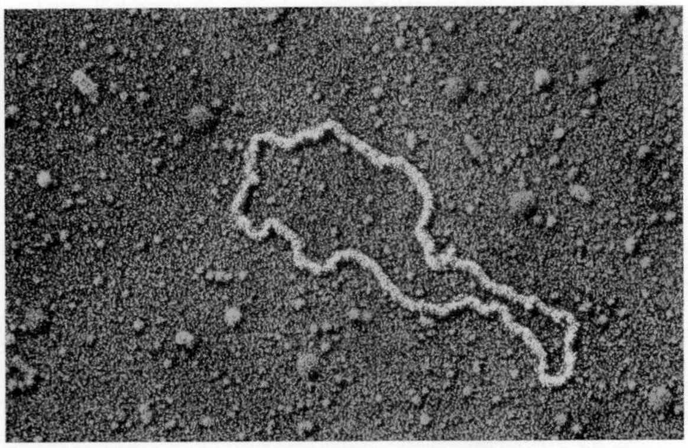

FIGURE 1.29 The pSC101 plasmid used by Stanley Cohen. Magnification ×30,000 at 35 mm. © Professor Stanley N. Cohen /Science Source

Recombinant DNA Technology: Stanley Cohen and Herb Boyer

Stanley Cohen was studying bacterial plasmids at Stanford University in the late 1960s. **Plasmids** are circular rings of DNA that exist separately from chromosomes (**FIGURE 1.29**). At the time, plasmids were being investigated for their clinical importance as many of them carry drug resistance genes. Protocols in his research led to the discovery that calcium chloride greatly increases the ability of plasmids to transform bacteria. In the process of transformation, bacterial cells take up neighboring DNA and incorporate it into their genome. A **genome** is the entire DNA content of a species. Cohen realized that transforming bacteria with plasmids would allow for the cloning of DNA within the bacteria, simply as part of its DNA replication mechanism that occurred with every cell division. This idea was the progenitor for an entire class of molecular biotechnology application, recombinant DNA technology, a technology that is not without its controversy (see Box 1.1).

Herbert Boyer, a professor at the University of California, San Francisco, was speaking at a conference attended by Cohen in 1972. The restriction enzyme *Eco*RI was the subject, in particular how the endonuclease generates sticky ends (**FIGURE 1.30**). The pair discussed combining plasmid transformation with restriction enzymes to produce a unique technology involving the recombination of DNA sequences. The concept of **recombinant DNA** was literally sketched out on a deli napkin. Cohen and Boyer hypothesized the technology would even allow for **transgenic** organisms, an organism containing genetic material from a species other than its own (**FIGURE 1.31**).

Recombinant DNA technology received one of the first ever patents in the field of biotechnology. Boyer and Cohen founded Genentech, Inc. with venture capitalist Robert Swanson in 1976. Genentech was the first biotechnology company listed on the New York Stock Exchange. The organization went on to develop the first pharmaceutical product produced by biotechnology in 1978: human insulin produced via recombinant DNA technology. The insulin was licensed to Eli Lilly in 1983 and commercialized as "humulin."

These events ushered in the era of biotechnology as a commercial industry. Prior to 1976, researchers typically worked in university settings and were funded by government grants. Boyer and Cohen began their work together in the pursuit of scientific knowledge. They were actually coaxed into patenting their technology by Niels Reimers, manager of Stanford's Office of Technology Licensing. With the patent and founding of Genentech, their priorities shifted to applied technology in the private sector. Profits became part of the equation as they never had before. Patented technologies are

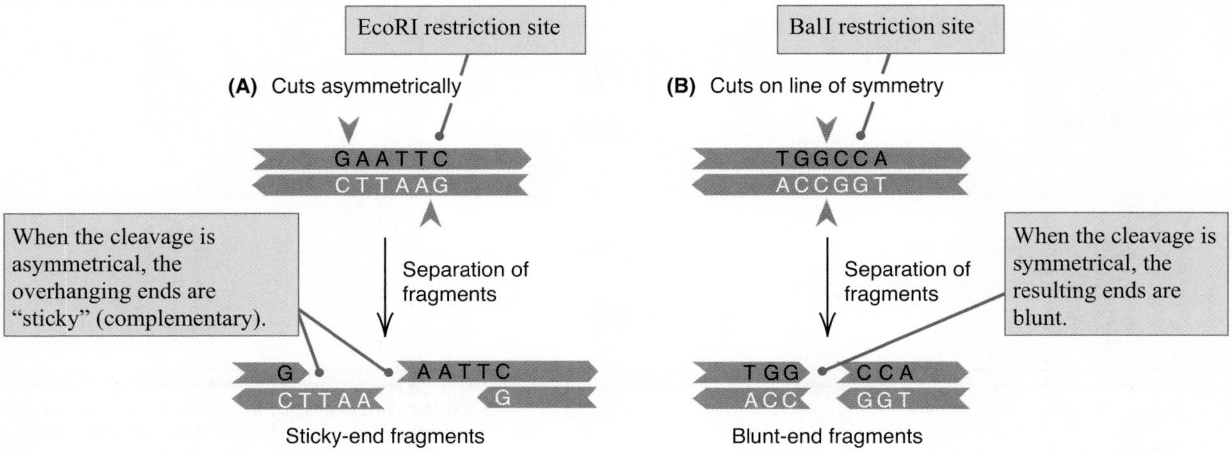

FIGURE 1.30 EcoRI cleaves DNA asymetrically, forming "sticky ends."

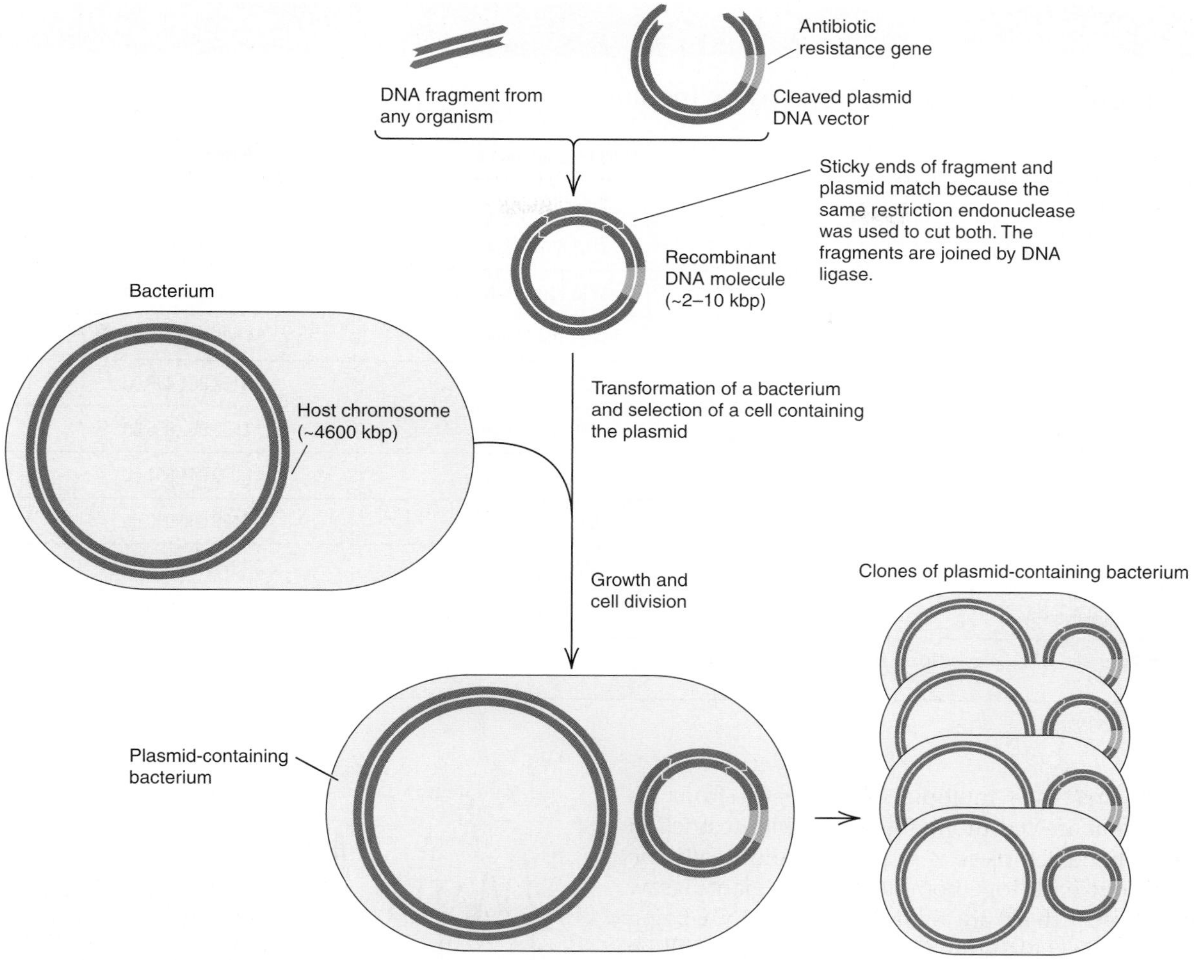

FIGURE 1.31 "Molecular Cloning Procedure" detailed by Stanley Cohen in 1975. Production of recombinant DNA plasmids using EcoRI. This technology led to the introduction of the first pharmaceutical product produced through recombinant DNA.

now a major revenue stream for academic institutions as it is standard to operate a technology licensing office (**TABLE 1.5**).

This transition created its own set of problems. Scientists began to straddle the two worlds and continue to do so; it is not unusual today for a scientist employed as a university professor to have ties to industry. In many cases it is preferred or expected. The potential for conflicts of interest are vast but considered part of the norm today. In the early years of the biotechnology industry, there was more emphasis on such conflicts (**FIGURE 1.32**).

FIGURE 1.32 The cover of *Chemical and Engineering News* on October 12, 1981. The scientist is trapped in a DNA molecule while being pulled by an industry executive on one side and an academic colleague on the other. Courtesy of Chemical and Engineering News

DNA Sequencing

DNA Sequencing is the process of determining the exact order of nucleotide bases within a DNA molecule. A DNA sequence therefore appears as a long string of A's, T's, C's, and G's (**FIGURE 1.33**). The technology behind this process originated in the 1970s. Several methods have been developed. The generalized

TABLE 1.5

Top 10 Universities by 2009 Licensing Income

		2009 License Income	2009 Research Exp.
1.	Columbia University	$154,257,579.00	$604,660,000.00
2.	Northwestern University	$161,591,544.00	$400,012,497.00
3.	New York University	$113,110,437.00	$308,834,000.00
4.	University of California System	$103,104,667.00	$4,686,598,210.00
5.	Wake Forest University	$95,636,362.00	$162,084,439.00
6.	University of Minnesota	$95,168,525.00	$590,880,956.00
7.	University of Washington/Wash. Res. Fdn	$87,339,905.00	$1,076,044,801.00
8.	University of Massachusetts	$70,553,428.00	$489,060,000.00
9.	Massachusetts Institute of Technology	$66,450,000.00	$1,375,073,000.00
10.	Stanford University	$65,054,187.00	$733,266,108.00

Data from: Association of University Technology Managers

process begins with multiple DNA copies from an individual that are cut into many pieces with restriction enzymes. Then each piece is sequenced and finally the pieces are put back together, much like a giant jigsaw puzzle. Because there are many copies of DNA being sequenced, it is statistically probable that there will be some repetition of the genome in the final sequence as well as some areas of the genome not being sequenced at all. To get a complete view of an individual genome, it is necessary to repeat the sequencing process a standard number of 28 times (called 28X). The 28X method

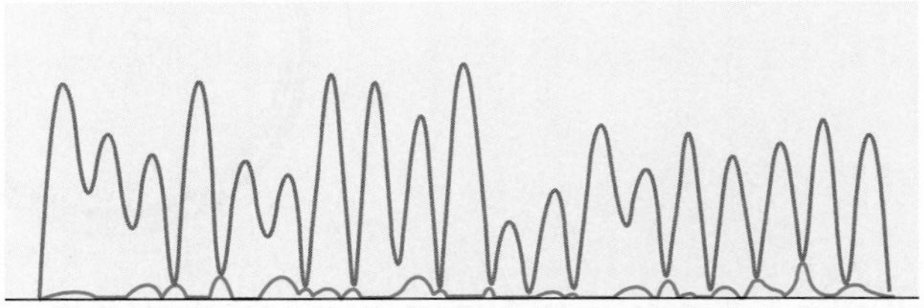

FIGURE 1.33 A computer-generated output of DNA sequence. The peak heights represent different nucleotide bases.

Box 1.2 The Asilomar Conference

Herbert Boyer and Stanley Cohen realized early on in their work that recombinant DNA technology had the potential to be misused. The eugenics movement in Germany and human experimentation performed by the Nazi party were very recent history. A brief moratorium on recombinant DNA technology was upheld from 1974 to 1975 based upon the suggestion of Stanford biochemist Paul Berg. Ten additional signatories on Berg's open letter included Boyer, Cohen, and James Watson.

Berg was also a primary organizer for the Asilomar Conference, held in a conference facility of the same name in February 1975. The conference was sponsored jointly by the National Academy of Sciences, the National Institutes of Health, and the National Science Foundation in response to the moratorium. The goal of the conference was to discuss the risks involved with recombinant DNA technology and to set forth guidelines on how the technology should and should not be applied. A final draft of the Asilomar guidelines was issued in June 1976 and was much more stringent than today. As technology has changed since the late 1970s, the guidelines have been repeatedly amended and relaxed.

If the Asilomar Conference succeeded in alleviating fears within the scientific community, it was not able to assuage mainstream media and by extension the public. *Time Magazine* featured the article entitled, "Tinkering with Life," in its April 1977 issue with a fearsome image of a DNA molecule with a serpent's head on its cover. Concerns revolving around the misuse of biotechnology, founded or not, persist all these decades later.

also ensures that both chromosome sets of a diploid person are accounted for.

DNA sequencing gives researchers the essential genetic information coded for in a given genome. Without this technology, the numerous sequencing projects described below would not be possible. Once a sequence is determined, researchers can analyze the data for a variety of purposes.

- Genomes can be sequenced and compared to track evolutionary relationships as well as genetic changes over time. Comparisons can be made within a single species or among different species (**TABLE 1.6**).
- Sequencing makes the identification of mutations possible (**FIGURE 1.34**). A **mutation** is a permanent alteration of genetic material. Once a mutation is located its effects can be determined, thus playing a role in our understanding of gene function and expression.
- Specific genes can be identified and the gene's products can be determined with use of the genetic code. Cataloging and characterizing proteins expressed in various organisms is known as **proteomics** (**TABLE 1.7**) Our understanding of health and disease will be forever changed thanks to DNA sequencing. This information is also valuable for genetic engineering, with applications in food production, biodefense, industrial processes, and green technologies.
- Forensic analysis of DNA includes sequencing in order to identify its origin (**FIGURE 1.35**).

Sequence 1: TAAAGACCATAGGAAATAAAGATAA
Sequence 2: TAACGACCAT–GGAAACAAAGATAA

FIGURE 1.34 Comparison of two DNA sequences identifies three single-nucleotide differences, indicating point mutations.

The cost to sequence DNA (per megabase, or 1,000,000 base pairs) has decreased dramatically since the advent of this technology. In September 2001 the cost was over $5,200, while the last available data from October 2013 has the cost at $0.057, representing more than 99% decrease (**FIGURE 1.36A**). Within the same time period, the cost to sequence an entire genome has gone from over $95 million to just over $5,000

TABLE 1.6

Genome Sizes

Genome	Approximate genome size in thousands of nucleotides	Form
Viruses		
MS2	4	Single-stranded RNA
Human immunodeficiency virus (HIV)	9	
Colorado tick fever virus	29	Linear double-stranded RNA
SV40	5	Circular double-stranded DNA
ϕX174	5	Circular single-stranded DNA; double-stranded replicative form
λ	50	Linear double-stranded DNA
Herpes simplex	152	
T2,T4,T6	165	
Smallpox	267	
Prokaryotes		
Methanococcus jannaschii	1,600	Circular double-stranded DNA
Escherichia coli K12	4,600	
Borrelia burgdorferi	910	Linear double-stranded DNA
Eukaryotes		Haploid chromosome number
Saccharomyces cerevisiae (yeast)	13,000	16
Caenorhabditis elegans (nematode)	97,000	6
Arabidopsis thaliana (wall cress)	100,000	5
Drosophila melanogaster (fruit fly)	180,000	4
Takifugu rubripes (fish)	400,000	22
Homo sapiens (human being)	3,000,000	23
Zea mays (maize)	4,500,000	10
Amphiuma means (salamander)	90,000,000	14

TABLE 1.7

Genome and Proteome Comparisons

Organism	Genome size, Mb[a] (approximate)	Number of genes (approximate)	Number of distinct proteins in proteome[b] (approximate)	Shared protein families
Hemophilus influenzae (causes bacterial meningitis)	1.9	1700	1400	
Saccharomyces cerevisiae (budding yeast)	13	6000	4400	3000
Caenorhabditis elegans (soil nematode)	100	20,000	9500	5000
Drosophila melanogaster (fruit fly)	120[c]	16,000	8000	7000[d]
Mus musculus (laboratory mouse)	2500	25,000	10,000	9900[f]
Homo sapiens (human being)	2900[e]	25,000	10,000	

[a] Millions of base pairs.
[b] Excludes "families" of proteins with similar sequences (and hence related functions).
[c] Excludes 60 Mb of specialized DNA ("heterochromatin") that has a very low content of genes.
[d] Based on similarity with sequences in messenger RNA (mRNA).
[e] For convenience, this estimate is rounded to 3000 Mb elsewhere in this book.
[f] Based on the observation that only about 1% of mouse genes lack a similar gene in the human genome, and *vice versa*.

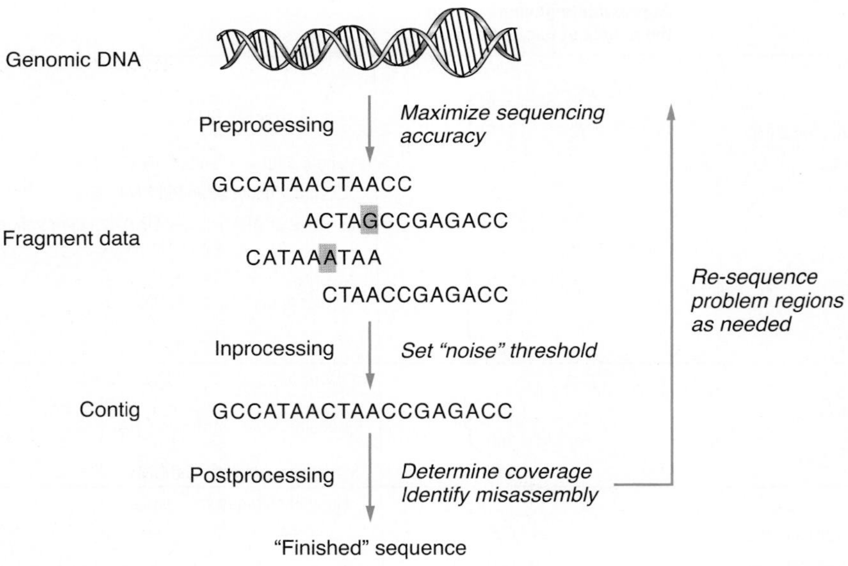

FIGURE 1.35 A generalized schematic of the DNA sequencing process. DNA samples are fragmented and sequenced, and then the sequences are rejoined in the contig sequence in order to produce a finished sequence. Several specific sequencing methods are currently employed.

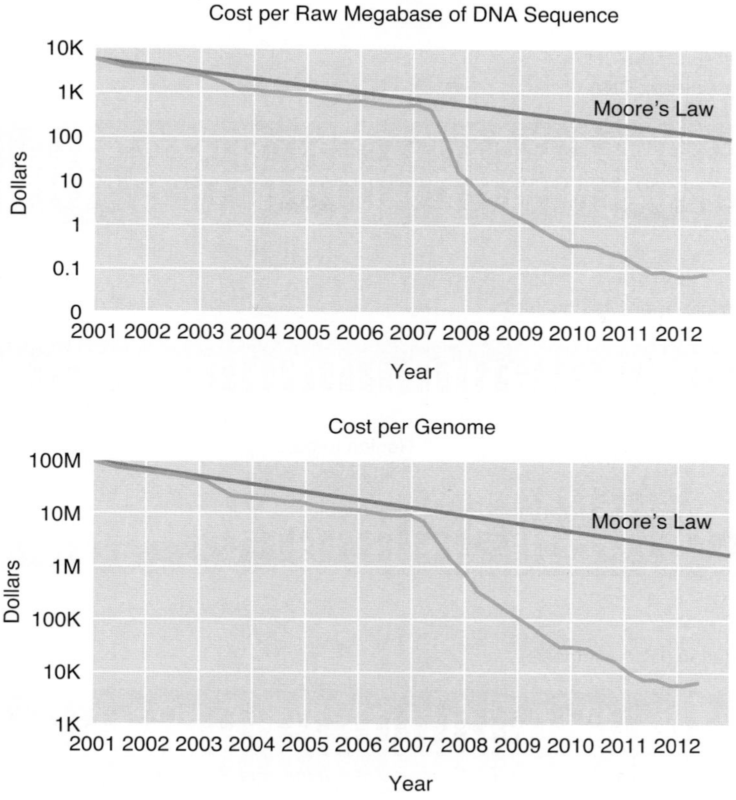

FIGURE 1.36 The cost of DNA sequencing has plummeted since 2001. (A) Sequencing costs per megabase. (B) Sequencing costs per genome. Reproduced from the NHGRI Genome Sequencing Program (GSP).

(**FIGURE 1.36B**). These drastic drops in costs make DNA sequencing more accessible than ever, essentially leveling the playing field for organizations of all sizes. It has gone from a highly cost-prohibitive technology afforded to government agencies and multinational corporations to a process that even smaller biotech companies and educational institutions can afford.

The Polymerase Chain Reaction

Kary Mullis invented the **polymerase chain reaction** in 1985 as an employee of the large biotech firm Cetus Corporation. Commonly known as PCR, the process amplifies a small fragment of DNA into millions of identical copies (**FIGURE 1.37A** and **B**). It follows the same concept as exponential growth, as it starts with one DNA segment and copies it into two, which both copy into four segments, and so on. Prior to the advent of this technology, the only way to make multiple copies of DNA was to insert the DNA sequence into bacteria, colonize the transformed cells, and extract the DNA of interest, a very laborious and time-consuming process. PCR is a relatively quick and easy process with numerous applications. It has become a staple technique in the biotechnology arsenal. For its tremendous value to the biotechnology industry, Mullis was awarded with the 1993 Nobel Prize for Chemistry.

Genetic Mapping

Genetic mapping characterizes the make-up of chromosomes. The process has two primary goals: physical mapping and genetic mapping (**TABLE 1.8**).

Much like DNA sequencing, genetic mapping will contribute to our understanding of health and disease in addition to the various applications described above. **Genetic mapping** takes the process a step further by categorizing the known DNA sequences into chromosomal fragments: genes, or coding regions, and the remaining DNA, or noncoding regions (**FIGURE 1.38** and **BOX 1.3**).

TABLE 1.8

Goals of Genetic Mapping

1. Physical mapping: Pinpoint the physical location of genes on a particular chromosome and the distance between them.
2. Genetic mapping: Determine the order or sequence in which the genes are situated.

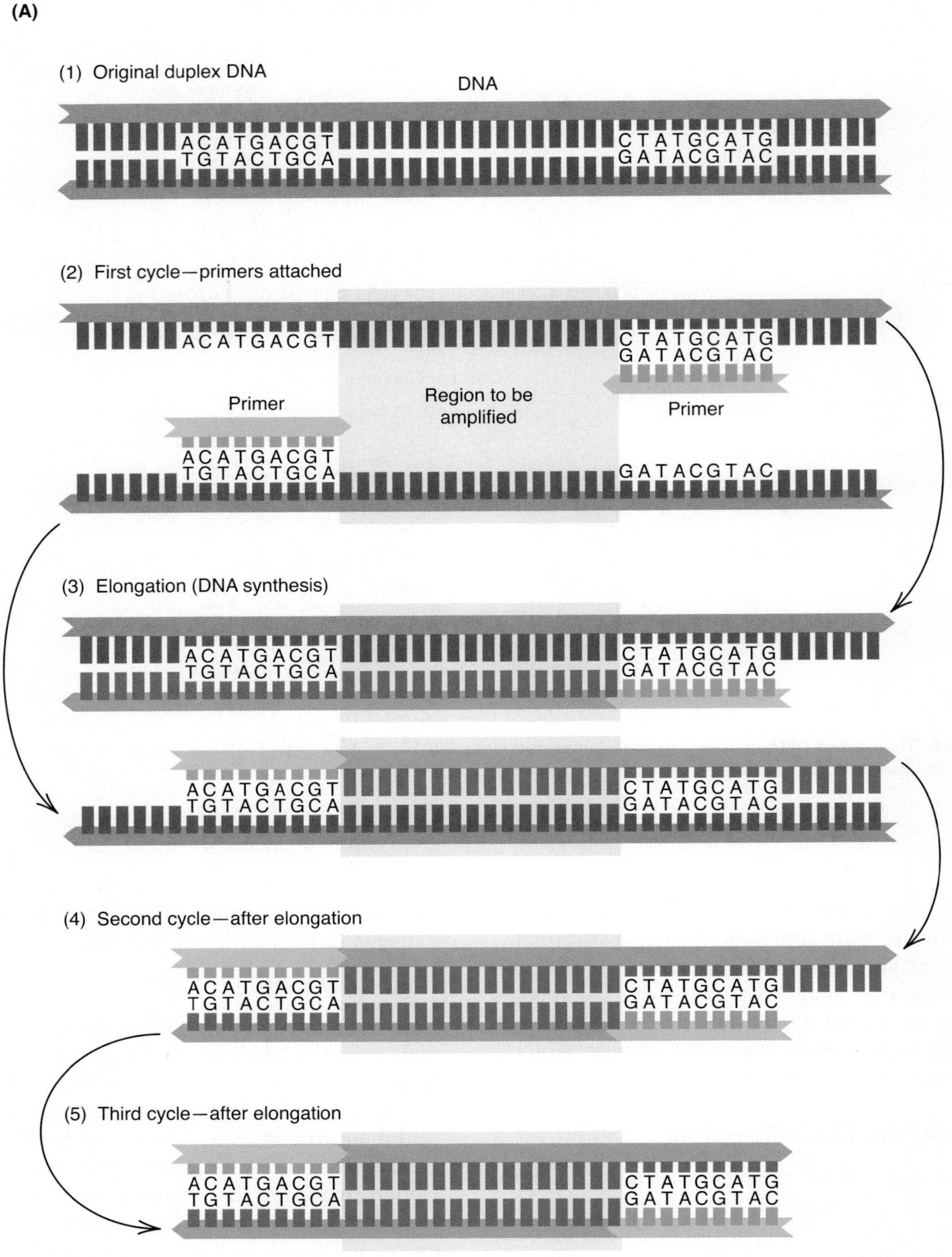

FIGURE 1.37 (A) The steps of the polymerase chain reaction. DNA is denatured, primers are annealed, and elongation occurs via complementary base pairing. (B) The manner in which PCR amplifies DNA samples follows an exponential pattern.

(B)

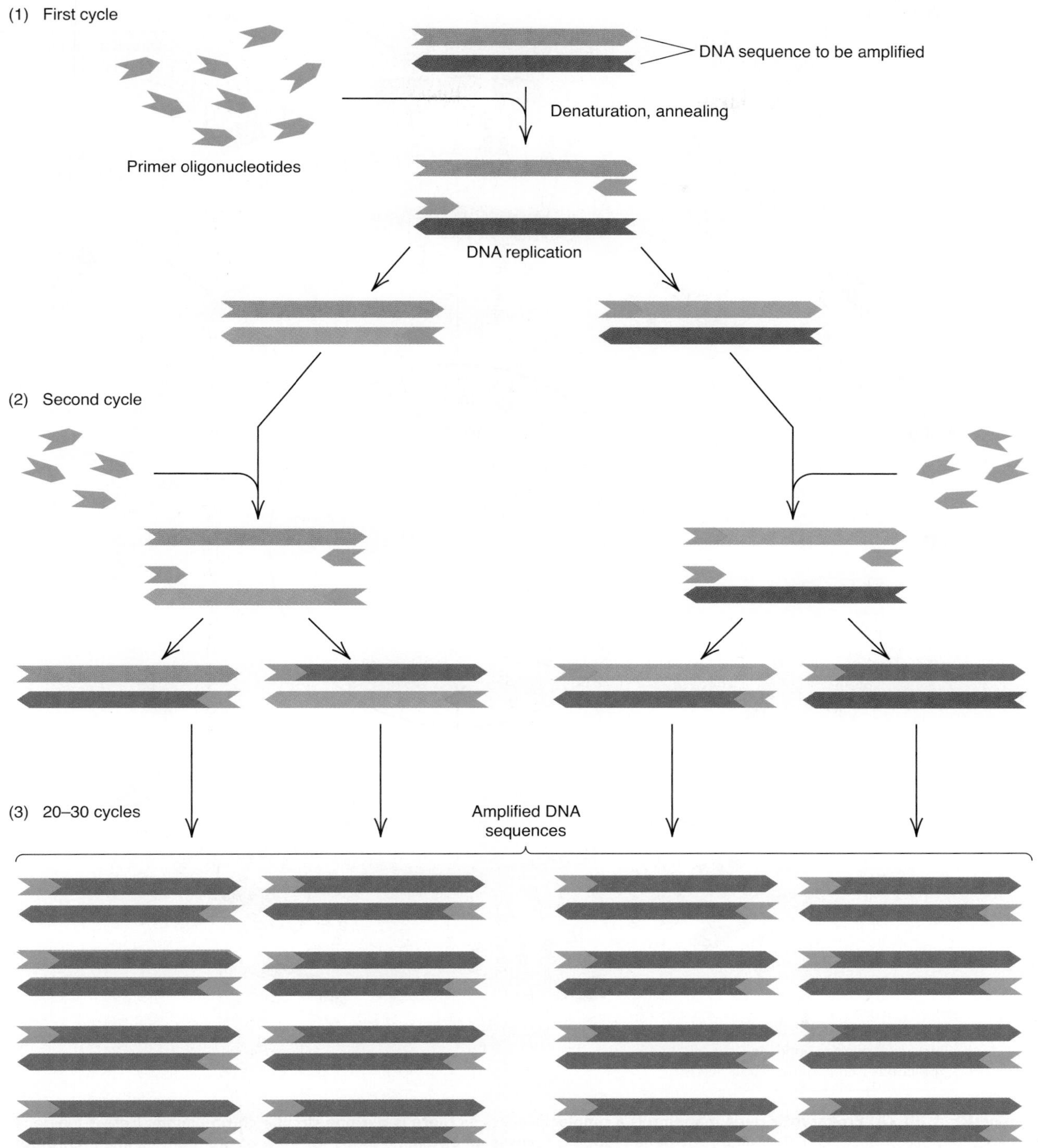

FIGURE 1.37 (A) The steps of the polymerase chain reaction. DNA is denatured, primers are annealed, and elongation occurs via complementary base pairing. (B) The manner in which PCR amplifies DNA samples follows an exponential pattern. (*continued*)

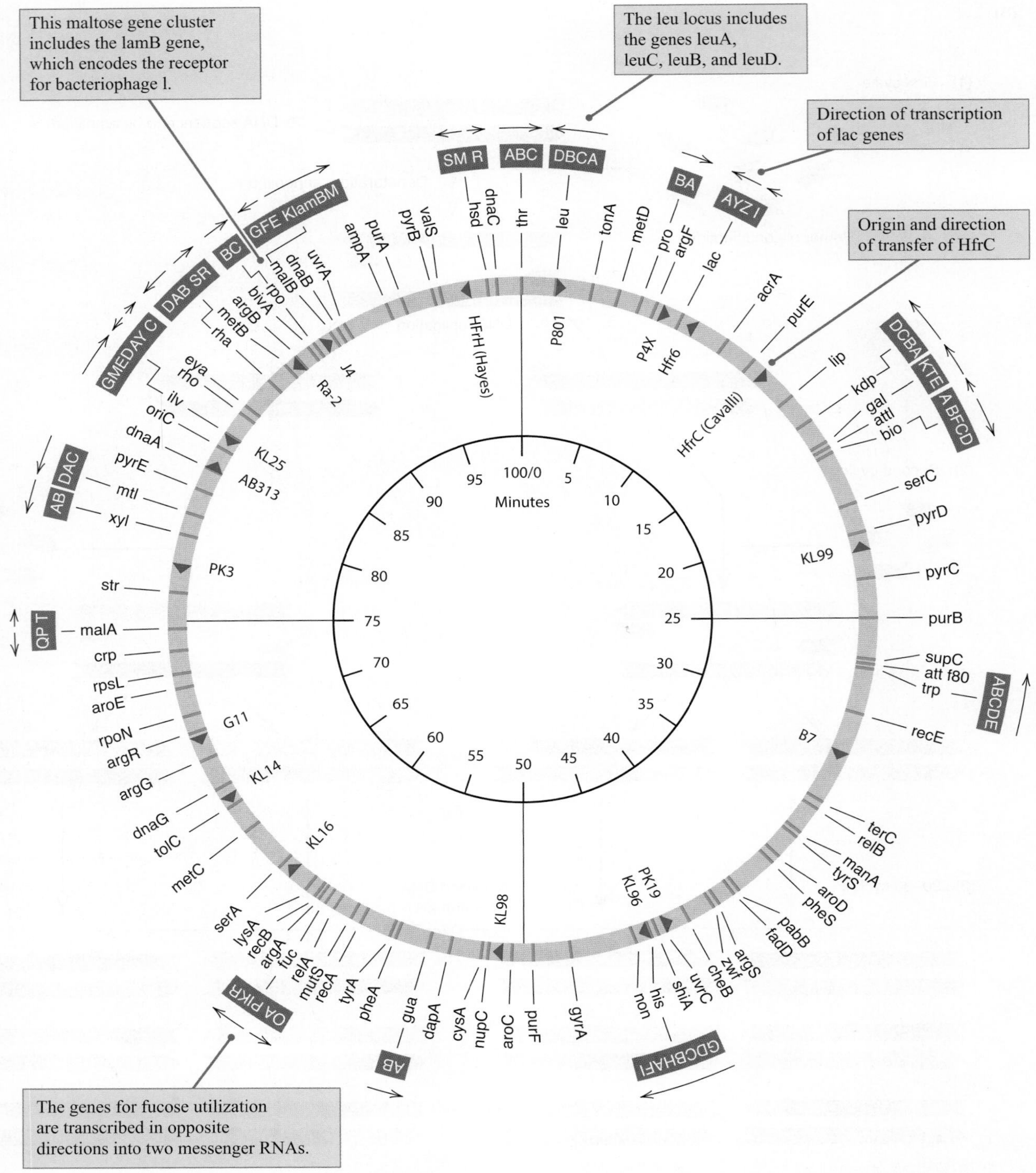

(A)

FIGURE 1.38 **(A) Prokaryotic gene mapping is a simpler process as there is but one circular chromosome.** Here is the genetic map of the bacterium *E. coli*. **(B) Eukaryotic gene mapping focuses on obtaining maps for each of the multiple linear chromosomes.** Here is the genetic map of the human X chromosome.

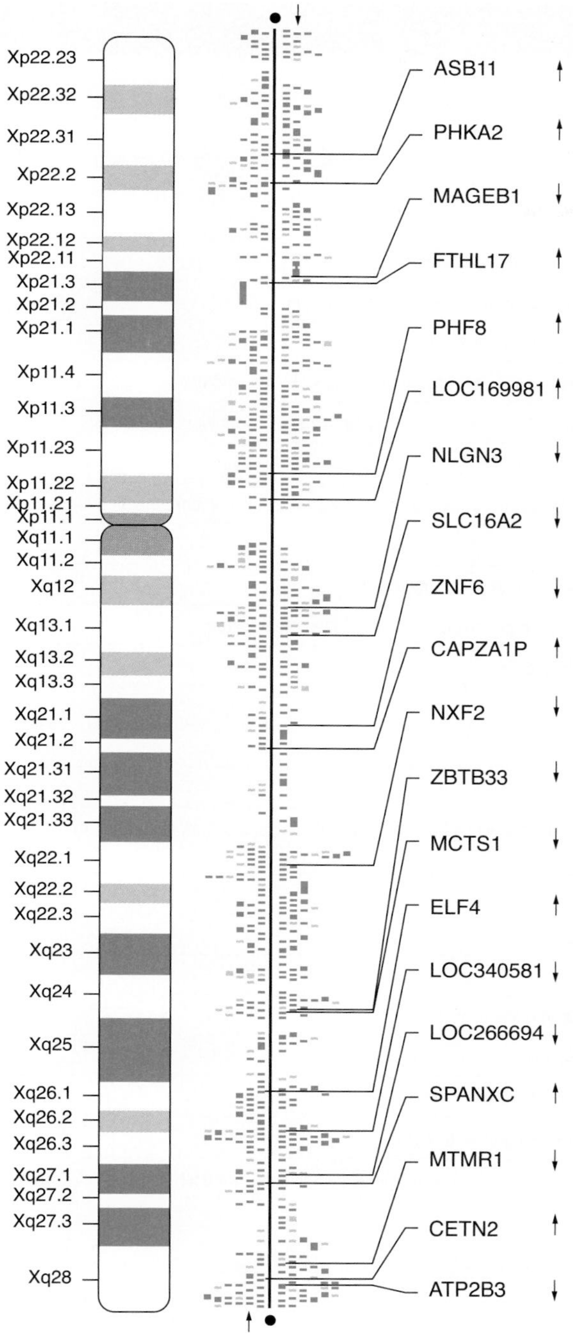

(B)

FIGURE 1.38 (A) Prokaryotic gene mapping is a simpler process as there is but one circular chromosome. Here is the genetic map of the bacterium *E. coli*. **(B) Eukaryotic gene mapping focuses on obtaining maps for each of the multiple linear chromosomes.** Here is the genetic map of the human X chromosome. (*continued*)

Box 1.3 Junk DNA Is Anything But

With over 3 billion base pairs in the human genome, scientists were surprised to find a comparatively small number of genes. Humans possess approximately 25,000 to 35,000 genes, representing about 2% of the genome. The other 98% was labeled "junk DNA" initially, as its role was unclear at first. It is now known that much of the so-called junk acts in the regulation of gene expression. The noncoding regions have the ability to turn genes on and off as well as modify the intensity of gene expression when a gene is turned on. This affects how much protein is produced and how often. This would clearly be especially useful in the development of an organism when timing and intensity orchestrate to establish a body plan for the growing organism. It continues to play a role throughout an organism's life as it interacts with its environment. In fact, a branch of biotechnology known as epigenomics (literally meaning "above the genome") seeks to clarify the relationship between noncoding DNA and environmental conditions. The ENCODE project recently released data analyses suggesting more than 80% of our noncoding DNA "serves some purpose, biochemically speaking." This DNA is clearly not junk.

Notable Projects

Many of the techniques described above have provided biotechnology with commercial successes. Not all areas of molecular biotechnology have profits as the bottom line. The projects highlighted below continue the tradition of scientific advancement in order to improve our knowledge. The projects all willingly publish their multitude of data for use by other scientists. The following projects are just a handful of those ongoing today. The National Institutes of Health alone currently funds over 30 projects related to molecular biotechnology. However, the projects below are among the most referenced and influential and provide a sample of the type of work being conducted. They represent collaborative efforts with one common goal: the betterment of humanity through science and technology. For a detailed timeline of biotechnology milestones, see **TABLE 1.9**.

Human Genome Project

A genome is all of the hereditary material found in an organism. **Genomics** is the study of all of the nucleotide sequences, including structural genes, regulatory sequences, and noncoding DNA segments, in the chromosomes of an organism. This field of study would not be possible without our ability to sequence DNA and map genes.

The Human Genome Project was originally conceived by the National Institutes of Health and James Watson served as the first director of the project. The first goal of the project was to sequence the entire genome of our species. The initial NIH team led by Watson extended to include a host of international teams, leading to the International Human Genome Sequencing Consortium (IHGSC). Amidst concerns

TABLE 1.9

Biotechnology Milestones

Timeline		
Prior to 1750:		Plants used for food
		Animals used for food and to do work
		Plants domesticated, selectively bred for desired characteristics
		Microorganisms used to make cheese, beverages, and bread by fermentation
1797:	Edward Jenner	Used living microorganisms to protect people from disease
1750–1850:		Increased cultivation of leguminous crops and crop rotations to increase yield and land use
1850s:		Horse drawn harrows, seed drills, corn planters, horse hoes, 2-row cultivators, hay mowers, and rakes
		Industrially processed animal feed and inorganic fertilizer
1859:	Charles Darwin	Hypothesized that animal and plant populations adapt over time to best fit the environment
1864:	Louis Pasteur	Proved existence of microorganisms
		Showed that all living things are produced by other living things
1865:	Gregor Mendel	Investigated how traits are passed from generation to generation— called them factors
1869:	Johann Miescher	Isolated DNA from the nuclei of white blood cells
1893:	Koch, Pasteur	Fermentation process patented
	Lister	Institutes Diphtheria antitoxin isolated
1902:	Walter Sutton	Coined the term "gene"
		Proposed that chromosomes carry genes (factors which Mendel said could be passed from generation to generation)
1904:		Artificial "silks" developed
1910:	Thomas H. Morgan	Proved that genes are carried on chromosomes
1917:	Karl Ereky	"Biotechnology" term coined
1918:		Germans use acetone produced by plants to make bombs
		Yeast grown in large quantities for animal and glycerol
		Made activated sludge for sewage treatment process
1927:	Herman Mueller	Increased mutation rate in fruit flies by exposing them to X rays
1928:	Frederick Griffiths	Noticed that a rough kind of bacterium changed to a smooth type when unknown "transforming principle" from smooth type was present
1928:	Alexander Fleming	Discovered antibiotic properties of certain molds
1920–1930:		Plant hybridization
1938:		Proteins and DNA studied by X-ray crystallography
		Term "molecular biology" coined
1941:	George Beadle and Edward Tatum	Proposed "one gene, one enzyme" hypothesis
1943–1953:	Linus Pauling	Described sickle cell anemia, calling it a molecular disease
		Cortisone made in large amounts
		DNA is identified as the genetic material
1944:	Oswald Avery	Performed transformation experiment with Griffith's bacterium
1945:	Max Delbruck	Organized course to study a type of bacterial virus that consists of a protein coat containing DNA
Mid-1940s:		Penicillin produced

TABLE 1.9

Biotechnology Milestones (*continued*)

Timeline		
1950:	Erwin Chargaff	Determined that there is always a ratio of 1:1 adenine to thymine in DNA of many different organisms
		Artificial insemination of livestock
1952:	Alfred Hershey and Martha Chase	Used radioactive labeling to determine that it is the DNA, not protein, which carries the instructions for assembling new phages
1953:	James Watson and Francis Crick	Determined the double helix structure of DNA
1956:	Fred Sanger	Sequenced insulin (protein) from pork
1957:	Francis Crick and George Gamov	Explained how DNA functions to make protein
1958:	Arthur Kornberg	Discovered DNA polymerase
1960:		Isolation of messenger RNA
1965:		Classification of the plasmids
1966:	Marshall Nirenberg and Severo Ochoa	Determined that a sequence of three nucleotide bases determine each of 20 amino acids
1970:		Isolation of reverse transcriptase
1971:		Discovery of restriction enzymes
1972:	Paul Berg	Cut sections of viral DNA and bacterial DNA with same restriction enzyme
		Spliced viral DNA to the bacterial DNA
1973:	Stanley Cohen	Produced first recombinant DNA organism
	Herbert Boyer	Beginning of genetic engineering
1975:		Asilomar Conference on Recombinant DNA
		Moratorium on recombinant DNA techniques
1976:		National Institutes of Health guidelines developed for study of recombinant DNA
1977:		First practical application of genetic engineering
		Human growth hormone produced by bacterial cells
1978:	Genentech, Inc.	Genetic engineering techniques used to produce human insulin in *E. coli*
		First biotech company on New York Stock Exchange
	Stanford University	First successful transplantation of mammalian gene
		Discoverers of restriction enzymes receive Nobel Prize in medicine
1979:	Genentech, Inc.	Produced human growth hormone and two kinds of interferon DNA from malignant cells transformed a strain of cultured mouse cells, a new tool for analyzing cancer genes
1980:		U.S. Supreme Court decided that man-made microbes could be patented
1983:	Genetech, Inc.	Licensed Eli Lily to make insulin
		First transfer of foreign gene in plants
1985:		Plants can be patented
1986:		First field trials of DNA recombinant plants resistant to insects, viruses, and bacteria
1988:		First living mammal was patented
1993:		Flavr savr tomatoes sold to public

Data from: Ann Murphy and Judi Perrella.

that the IHGSC was mired in bureaucracy and red tape, Watson's colleague Craig Venter pioneered cheaper and faster methods to identify genes and eventually branched out from the project to found his own company, Celera Genomics. From this point onward, two separate teams, one private and the other publicly funded, raced to sequence the human genome. In 2001 each team published its own working draft that included about 90% of the genome. In 2003, the IHGSC team published a complete draft of the over three billion base pairs in the sequence. From this draft, the next goal is to characterize each of our approximately 25,000 to 35,000 genes found on all 24 chromosomes by determining each gene sequence and location. In addition to genes, genetic markers located in noncoding regions of our DNA must also be identified. The final objective is to have a complete map of the entire human genome, including both coding and noncoding regions. This wealth of information is already proving useful for the biotechnology industry and will only become more valuable as our understanding of the genome increases.

1000 Genomes Project

As its name implies, the 1000 Genomes Project is sequencing the genomes of a large and diverse sample of humans. In actuality, closer to 2,500 genomes will be sequenced by the end of the project. Launched in 2008, the project builds upon the database of knowledge obtained through the Human Genome Project and focuses on identifying genetic variations within our species. In particular, 1000 Genomes Project researchers are searching for variations with frequencies of at least 1% within the human population. These variations are known as **haplotypes** and appear to be correlated with individuals of different geographic origins. Also similar to the Human Genome Project, the 1000 Genomes Project makes all data available to the public within searchable databases.

Genome 10K Project

The Genome 10K Project was founded in 2009 by a collective calling themselves the Genome 10K Community of Scientists (G10KCOS). The purpose of the project is to collect tissue samples from 10,000 vertebrate species in order to sequence the DNA. The 10,000 number equates to approximately one DNA sequence per vertebrate genus. The project is unique as its primary objective is unrelated to unlocking the mysteries of human health. Conservation is the goal, as the database would act as a genomic "zoo." In the event any of the sampled organisms became extinct, we would still have their genetic information available. The potential to reintroduce genes via recombination or even the extinct species itself via cloning would be possible.

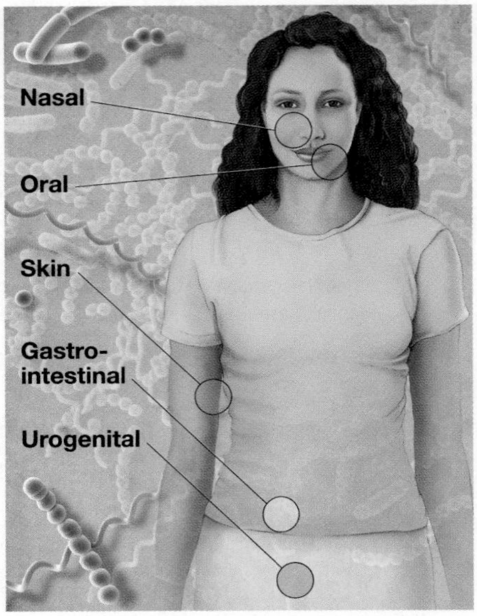

FIGURE 1.39 Areas of the human body being sampled by the Human Microbiome Project are known habitats of symbionts associated with our species. Courtesy of NIH Medical Arts and Printing

Human Microbiome Project

The human body is often described as a community in and of itself. It is composed of numerous populations of species other than our own, in the form of microbial hitchhikers. We have played host to microbes throughout the history of our evolution; in many cases we engage in mutualistic relationships where both species benefit. The reciprocal nature of our relationship with endosymbionts is an example of coevolution. As they change, we react; as we change, they react. Spearheaded by the National Institutes of Health, the Human Microbiome Project is sampling microbial populations that inhabit the human body. DNA sequencing of the representative species will create genome databases for the microbes with whom we have coevolved (**FIGURE 1.39**). The microbes are being sampled from the following areas of the human body:

- Urogenital tract
- Gastrointestinal tract
- Skin
- Oral cavities
- Nasal passages

Currently over 800 strains of microorganisms have been or are in the process of being sequenced. The genetic information obtained from our miniscule residents will illuminate their roles in human health and disease.

Genographic Project

The Genographic Project was initiated in 2005 by the National Geographic Society. Nearly 600,000 individuals in over 140 countries have provided their DNA as part of the project. The project has successfully supported the Out of Africa Hypothesis on the origin of *Homo sapiens*, which ascertains our species first emerged in Africa and later migrated beyond the continent to populate Earth. The alternative hypothesis previously considered, known as the Candelabra Hypothesis, which states *Homo sapiens* evolved independently in Africa, Europe, and Asia, has been negated. Building upon that foundation, the project is now weaving a tapestry of genetic profiles across the globe in order to determine the patterns of human migration out of Africa. The timing and order in which we came to inhabit the different continents is evident in our DNA. As stated by the project's team lead, geneticist Spencer Wells, "The greatest history book ever written is the one hidden in our DNA."

Not only is the Genographic Project unveiling the early stages of human history, it also invites public participation. It is one of several organizations that will analyze your DNA, though it is unique for being nonprofit. Volunteers contribute DNA samples which are then sequenced and analyzed for **genetic markers**, a class of noncoding DNA regions of known location that vary among individuals. Participants have the option of making their information public (yet anonymous), thereby adding to the viability of the overall sample. They in turn receive a detailed report on their genetic "lineage(s)," which includes their geographic origin(s) and whether they possess any DNA from our hominin cousins, the Neanderthals and Denisovans (**FIGURE 1.40**).

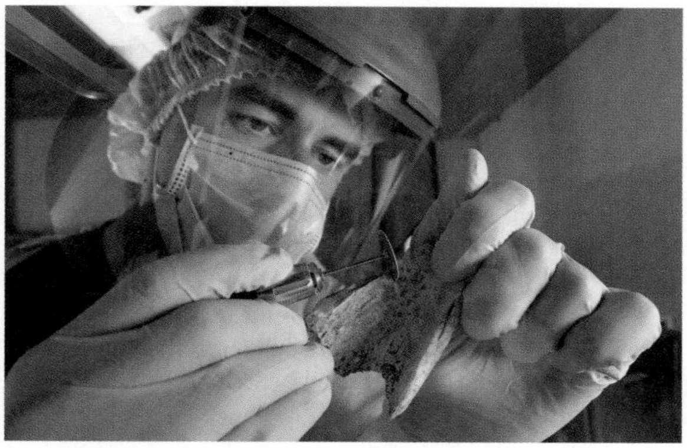

FIGURE 1.40 The Genographic Project is unraveling the mysteries of early human history. Here a technician prepares a Neanderthal bone for DNA extraction. © Volker Steger/Science Source

SUMMARY

- Biotechnology involves industrial production of goods and services by processes using biological organisms and systems. Molecular biotechnology specializes in the manipulation of these organisms and systems at the molecular level, including genetic engineering and its products.
- Molecular biotechnology relies heavily upon knowledge gained through scientific experimentation, drawing upon biological subdisciplines such as genetics, molecular biology, microbiology, and biochemistry. This knowledge is then applied in technology designed for practical utilitarian purposes and ultimately commercial enterprise.
- Gregor Johann Mendel is known as the Father of Genetics for the contribution of his two laws of inheritance. The Law of Segregation states that all individuals possess two factors for each trait; the two factors segregate during the formation of gametes and result in gametes with only one factor for each trait; and the process of fertilization unites factors from each parent, resulting in offspring with two factors for each trait. The Law of Independent Assortment states the two factors for a given trait segregate independently from other sets of two factors which exist for separate traits, thus allowing all possible combinations of factors in the resulting gametes. Mendel's laws served as the foundation for theories including evolution and chromosomal inheritance.
- Natural selection is the gradual change of organisms over time that leads to adaptations to the environment. The process was recognized independently and conceived as a mechanism for evolution by Charles Darwin and Alfred Russel Wallace. Darwin referred to natural selection as "descent with modification" while Wallace used the expression "survival of the fittest."
- The modern synthesis is the currently accepted paradigm of evolutionary biology which bridges gaps between microevolution, occurring at the genetic level, and macroevolution, occurring at the species level. Molecular events which contribute to evolution can operate at both the population level and the level of an individual organism. The modern synthesis has created a hierarchical understanding of evolution with one mode of action having a cascade effect upon the next level.

- Theodor Boveri and Walter Sutton performed experiments which supported the chromosomal theory of inheritance. Subsequent research by Thomas Morgan and his colleagues further asserted that chromosomes are comprised of the DNA molecule and consist of discrete units of heredity known as genes.
- Oswald T. Avery, Colin MacLeod, and Maclyn McCarty determined bacterial transformation was possible with DNA. Alfred Hershey and Martha Chase confirmed DNA as the genetic material by proving it is transmitted by bacteriophages in the process of its life cycle.
- The four bases found in DNA include thymine, adenine, guanine, and cytosine. Erwin Chargaff determined adenine always pairs with thymine and guanine always pairs with cytosine, thus establishing Chargaff's Rules of complementary base pairing. Adenine is complementary to thymine while guanine is complementary to cytosine. The complementary base pairs comprise the "rungs" of the DNA molecule, which when unwound from its double helical structure appears shaped as a ladder.
- The genetic code lists every possible codon, or set of three nucleotides, found in messenger RNA and identifies the amino acid encoded by each. Each codon is the result of transcription, while the encoded amino acid is the result of translation. These two steps make up the process of protein synthesis, which is also known as gene expression. The central dogma of molecular biology describes the flow of genetic information, first from a gene to messenger RNA (transcription), and then onto a final protein product (translation).
- Recombinant DNA technology involves the creation of novel gene sequences, potentially combining genetic material from separate organisms. Restriction endonucleases are enzymes that cut DNA at predictable locations. Genetic material may then be spliced into the cut fragments, thus resulting in recombinant DNA.
- The goal of the Asilomar conference was to discuss the risks involved with recombinant DNA technology and to set forth guidelines on how the technology should and should not be applied.
- DNA sequencing is the process of determining the exact order of nucleotide bases within a DNA molecule. The polymerase chain reaction amplifies a small fragment of DNA into millions of identical copies.
- PCR and DNA sequencing allows for genetic mapping by categorizing the known DNA sequences into chromosomal fragments: genes, or coding regions, and the remaining DNA, or noncoding regions. Compiling a map of an entire organism's genetic information is a science known as genomics. The related field of proteomics determines the proteins encoded for within a given genome. The two pursuits of genomics and proteomics are the basis for several notable biotechnology projects.

KEY TERMS

Bacterial transformation
Bacteriophage
Biotechnology
Central dogma of molecular biology
Chargaff's Rules
Chromosome Theory of Inheritance
Codon
Degenerate
DNA sequencing
Gene
Genetic code
Genetic mapping
Genetic markers
Genome
Genomics
Haplotype
Mendelian genetics
Modern synthesis
Morphology
Mutation
Natural selection
Nucleic acids
Parsimony
Plasmid
Polymerase chain reaction (PCR)
Proteomics
Purines
Pyrimidines
Recombinant DNA technology
Restriction endonuclease
Transgenic

DISCUSSION QUESTIONS

1. What was the significance of the Asilomar Conference? How did it steer the course of molecular biotechnology?
2. Research and describe a genomics project other than those presented in this chapter. What are the project's objectives and milestones? Who is participating and how are they funded?
3. What issues do you see with the overlap of biotechnology research and commercial industry? What benefits does this relationship pose? What risks?

REFERENCES

Chargaff, Erwin. *Heraclitean Fire: Sketches from a Life Before Nature*. Baltimore: Paul and Company Publishing Consortium, 1978.

Cochran, Gregory and Henry Harpending. *The 10,000 Year Explosion: How Civilization Accelerated Human Evolution.* New York: Basic Books, 2010.

Crick, Francis. *What Mad Pursuit: A Personal View of Scientific Discovery.* New York: Basic Boooks, 1988.

Crick, Francis, J.S. Griffith, and L.E. Orgel. "Codes Without Commas." *Proceedings of the National Academy of Sciences of the United States of America* 43, (1957): 416–421.

Crow, Ernest W. and James F. Crow. "100 Years Ago: Walter Sutton and the Chromosome Theory of Heredity." *Genetics* 160, no. 1 (2002): 1–4.

Dahm, Ralf. "Discovering DNA: Friedrich Miescher and the Early Years of Nucleic Acid Research." *Human Genetics* 122, no. 6 (2008): 565–581.

DNA Learning Center. "DNA from the Beginning." Cold Spring Harbor Laboratory. Cold Spring Harbor, New York, 2002.

The Genographic Project. "About the Project." The National Geographic Society, Washington DC, 2013.

Harwood, Jonathan. "A History of Heredity: From Mendel to Genetic Engineering." Centre for the History of Science, Technology, and Medicine, University of Manchester, Manchester, United Kingdom, 2002.

Huxley, Julian. *Evolution: The Modern Synthesis*. London: Allen and Unwin, 1942.

James, Clive. "2012 ISAAA Report on Global Status of Biotech/GM Crops." International Service for the Acquisition of Agri-biotech Applications, Manila, Philippines, 2012.

Jha, Alok. "Genomes Project Publishes Inventory of Human Genetic Variation." *The Guardian*, London, United Kingdom, October 31, 2012.

Jumba, Miriam. *Genetically Modified Organisms: The Mystery Unraveled.* Bloomington, Indiana: Trafford Publishing, 2010.

National Agricultural Statistics Service. "Genetically Engineered Varieties of Corn, Upland Cotton, and Soybeans, by State and for the United States, 2000-12." United States Department of Agriculture, Washington DC, 2012.

Pauling, Linus. "Linus Pauling Oral History Interview." American Philosophical Society: March 1, 1971.

Watson, James. *The Double Helix: A Personal Account of the Discovery of the Structure of DNA.* New York: The New American Library, 1968.

Watson, James and Francis Crick. "A Structure for Deoxyribose Nucleic Acid." *Nature* 171, no. 4356 (1953): 737–738.

Watson, James and Francis Crick. "Genetical Implications of the Structure of Deoxyribonucleic Acid." *Nature* 171, no. 4361 (1953): 964–967.

Wetterstrand, K.A. "DNA Sequencing Costs: Data from the NHGRI Genome Sequencing Program (GSP)." Last modified February 11, 2013. Available at http://www.genome.gov/sequencingcosts.

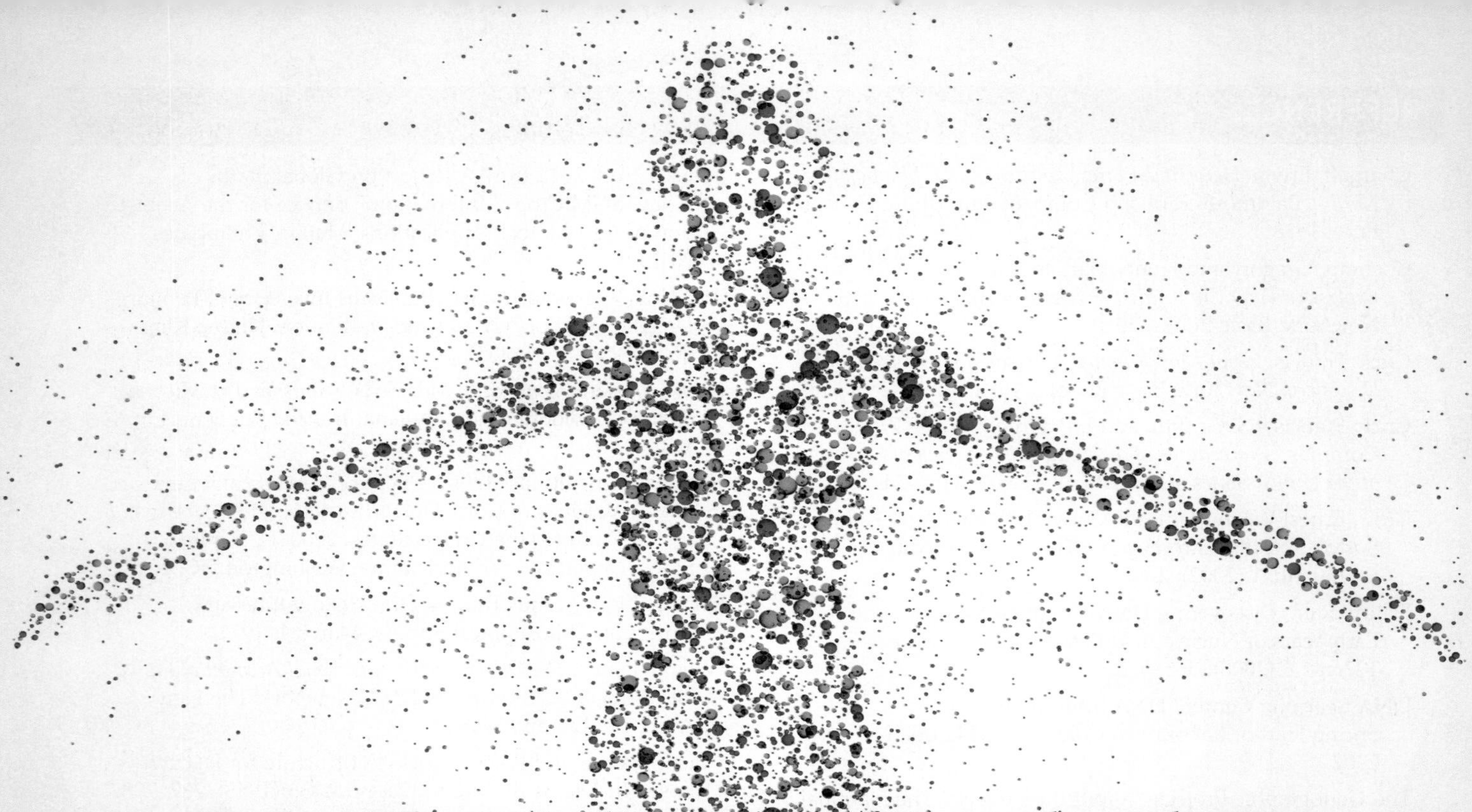

2 The Molecular Biotechnology Industry Today

LEARNING OBJECTIVES

Upon reading this chapter, you should be able to:

- Describe the interdisciplinary nature of the molecular biotechnology industry.
- Characterize the advances in medicine due to molecular biotechnology and differentiate among biopharma, genetic counseling, and gene therapy.
- Define "genetically modified organism" and explain the basic methods used in their production.
- Provide reasoning for the establishment of seed banks.
- Recall the uses of aquaculture including its role in bioprospecting.
- Recognize the role of industrial biotechnology in consumer products.
- Discuss green biotechnology while explaining bioremediation, biofuels, and conservation methods aided by molecular biotechnology.
- Describe DNA fingerprinting and its role in forensics.
- Identify examples of biodefense.
- Explain the utility of evo devo and provide examples of its potential.
- Detail common activities and objectives of molecular biotechnology professionals.
- Compare and contrast the careers available in the molecular biotechnology industry and associated fields.

The tools available to the molecular biotechnology industry are essentially universal. The consistency of the genetic code amongst the diverse forms of life on our planet translates to our ability to pick and choose traits, modify genetic sequences, and deconstruct the aberrations that have caused trouble in the past. The technological tools of the industry have allowed for a variety of applications. This makes **molecular biotechnology** an interdisciplinary field, making use of the wealth of knowledge previously discussed and drawing from biology, chemistry, mathematics, computer science, engineering, philosophy, and ethics.

Applying Molecular Biotechnology to Modern Lifestyles

The goals of this field have in large part determined the course of each segment within the industry. Scientists who choose this field as a career may seek to rid the world of its diseases or eliminate pollution from the environment. Or perhaps they are interested in increasing the world's food supply and ridding the world of starvation. They may wish to improve industrial processes in order to make them safer or easier to complete, more cost effective, or more environmentally friendly. Molecular biotechnologists may also focus on the origins of life, the evolutionary relationships among organisms, and how organisms utilize gene expression in the processes of growth and development. The major segments of the molecular biotechnology industry and how they influence modern living are now outlined in brief, with subsequent chapters providing greater detail.

At the Doctor, In Your Medicine: Life Science and Health Care

The Western world has its host of health problems associated with industrialization, such as heart disease, obesity, and cancer. Add this growing list of "first world diseases" to the known inherited diseases and there is an abundance of research material waiting to be studied. A large majority of today's molecular biotechnology industry is devoted to health care applications. Biopharma (bioengineered pharmaceuticals), gene therapy, and genetic counseling are all important areas of research and development.

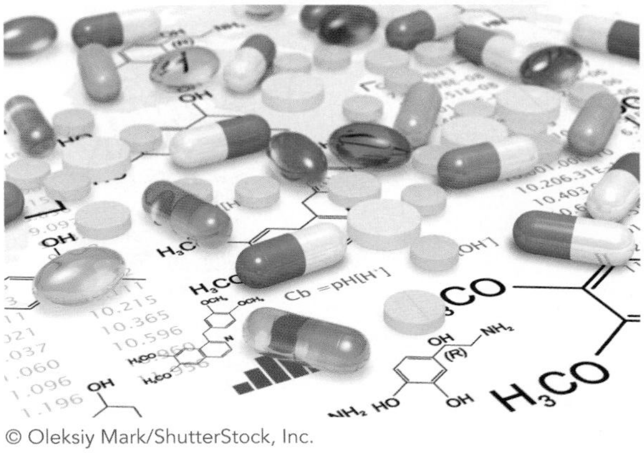

Biopharma

The development of pharmaceuticals via genetic engineering, known as **biopharma**, is a primary goal. When dysfunctional genes or gene products are discovered to be the cause of a particular disease, the delivery of a functional gene or gene product may be all that is needed to "medicate" the patient. Since the first biopharma product was introduced in 1982 (see **BOX 2.1**), literally thousands of drugs produced with molecular biotechnology have been researched and hundreds have been approved for market (**FIGURE 2.1**). While biopharma aims to fight against the major diseases affecting humanity, over half of the drugs investigated relate to cancer treatment.

Gene Therapy

Gene therapy is an extension of biopharma technology. **Gene therapy** involves techniques that correct defective genes, or alter their expression, in order to treat or even potentially cure disease. Gene therapy is in its relative infancy but already shows promise in repairing disease-causing alleles. It is encouraging that there is so much still to be discovered. The processes of human cell culture and high throughput screening, to be described in detail, are improving our ability to systematically narrow down potential therapeutics and ensure their compatibility within the human body.

Genetic Counseling

Genetic counseling is also an aspect of the health-care industry's involvement in molecular biotechnology. **Genetic counseling** involves the identification of disease in an individual's family history and determining the appropriate measures one might take to avoid passing on the disease to his or her offspring. While previously genetic counseling relied heavily upon creating family **pedigrees** to trace the inheritance patterns of disease, today genetic counseling more often involves genetic profiling (**FIGURE 2.2A** and **B**). **Genetic profiling** identifies genetic markers in an individual, some of which are genetic sequences for known disease-causing alleles and mutations associated with susceptibility to disease (**FIGURE 2.3**).

On Your Plate: Agriculture and Food Production

Humans have been selectively breeding plants and animals for thousands of years. We have used microbes such as bacteria and yeast in the production and preservation of food. Although it has been met with its

Box 2.1 Biopharma Begins

Patented in 1982, Humulin (short for "human insulin") was the first genetically engineered human therapeutic approved by the U.S. Food and Drug Administration and thus the first biopharma product on the market. The pharmaceutical was developed by Genentech in the late 1970s but licensed for production through Eli Lilly and Company in the year of its patent. The pharmaceutical giant is credited with guiding Humulin through the FDA approval process, as Genentech was unfamiliar with the regulatory terrain.

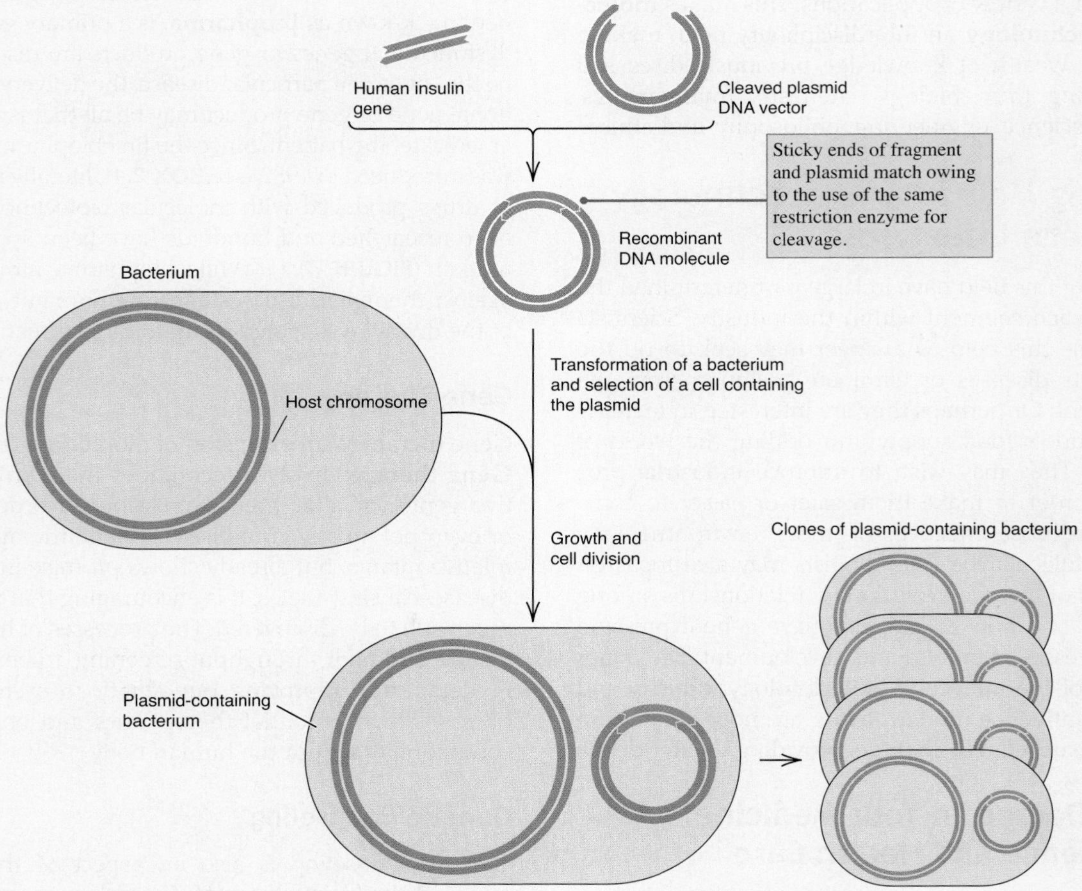

FIGURE 1 Recombinant DNA and cloning techniques utilized in the process of Humulin production.

Humulin replaces insulin in diabetes patients who are unable to produce the protein hormone themselves. Creating this drug was simply a matter of identifying the human insulin gene and inserting it into a bacterial host, in this case *E. coli*. Once the *E. coli* was transformed, the human insulin gene could be expressed in endless supply, thus producing a steady stream of human insulin for pharmaceutical consumption.

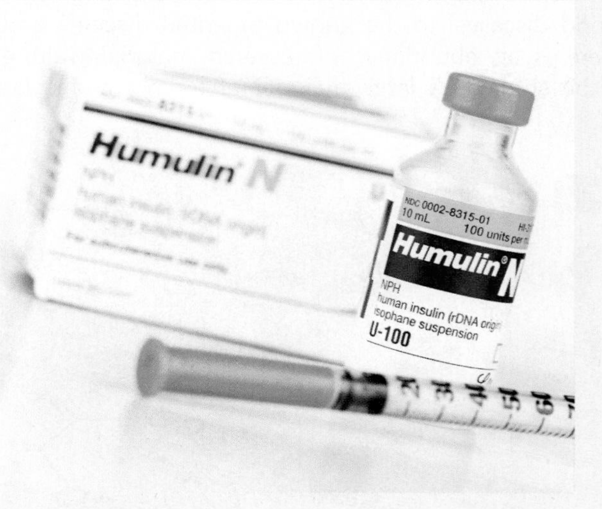

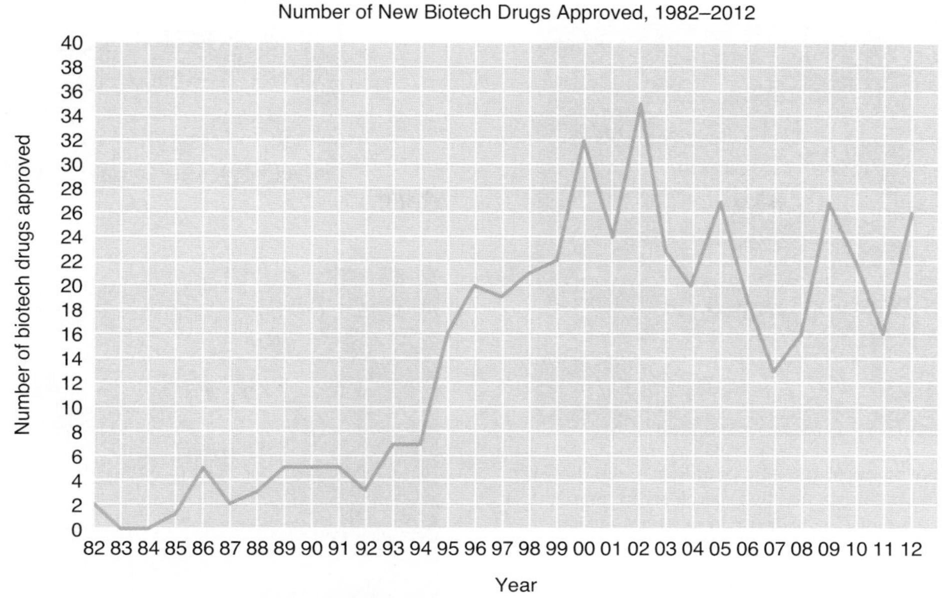

FIGURE 2.1 The number of drugs produced through biotechnology approved for market has increased steadily since Humulin.

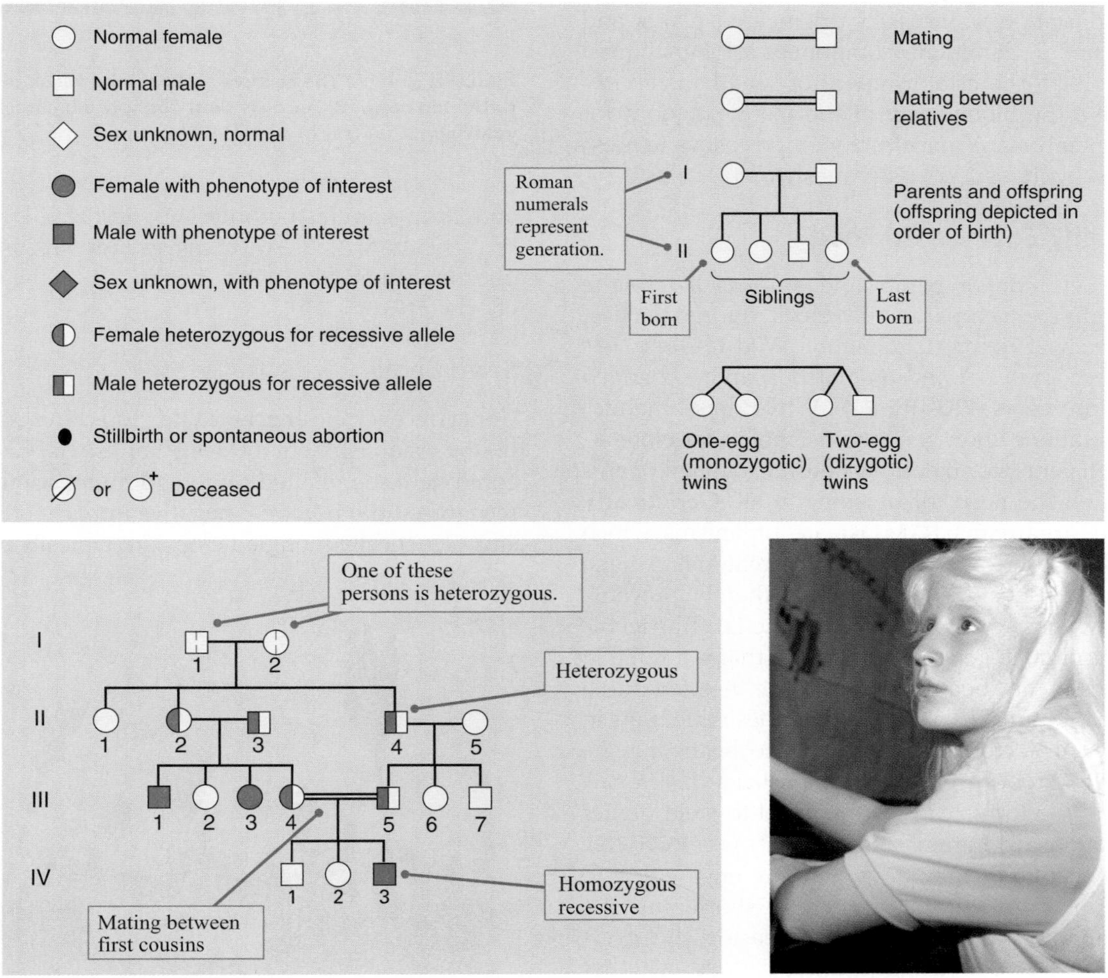

FIGURE 2.2 (A) The standard icons used in the creation of genetic pedigree charts. (B) A pedigree for albinism. Note the incidence of inbreeding and how it affects the frequency of recessive alleles. What kind of inheritance pattern does albinism exhibit?

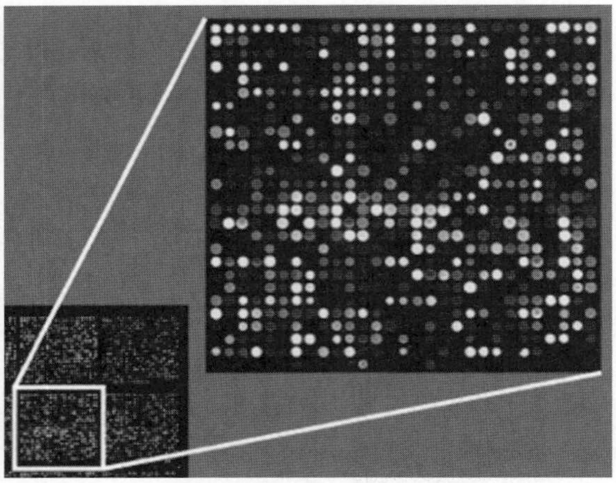

FIGURE 2.3 **The outcome of genetic profiling.** Each glowing DNA microarray spot indicates a positive match for a genetic marker. Courtesy of Jeffrey P. Townsend and Duccio Cavalieri, Yale University

share of controversy, agricultural biotechnology may be viewed as a further progression of our history of food modification. Genetically engineered plants and animals are a dominant component of agricultural biotechnology. The establishment of seed banks seeks to preserve the biodiversity of the plant kingdom for future generations. Aquaculture methods alleviate pressure faced by the world's natural fisheries.

Genetically Engineered Food

Artificial selection of plants and animals led to the domestication of crops and livestock during the last 10,000 years of human agriculture. Wild teosinte was bred to become corn and animal husbandry changed aurochs into cows (**FIGURE 2.4**). While these changes took generations upon generations, molecular biotechnology can generate an entirely new line of plant or animal without the need for breeding at all. **Genetically modified organisms (GMOs)** are organisms which contain knockout or inserted genetic material. A gene **knockout** renders a gene inoperable, thereby preventing the synthesis of its protein product(s). Transgenic organisms contain genetic information that is foreign in origin. Genes may be knocked out of or inserted into a food crop or animal in order to confer desirable traits in the organism. A crop such as corn or soybeans may be modified to produce herbicide or pesticide resistance. Fruits and vegetables can be enhanced to yield higher concentrations of vitamins and nutrients (see **BOX 2.2**). Plants and animals may be genetically engineered to increase their growth rates, thereby shortening the time to market and potentially increasing the overall food supply. Genetically modified foods, whether whole or as an ingredient, are the subject of debate and controversy. Although they are deemed safe by the U.S. Environmental Protection Agency (EPA), Food and Drug Administration (FDA), and U.S. Department of Agriculture (USDA), many groups and individuals question their long-term health and environmental effects (**FIGURE 2.5**).

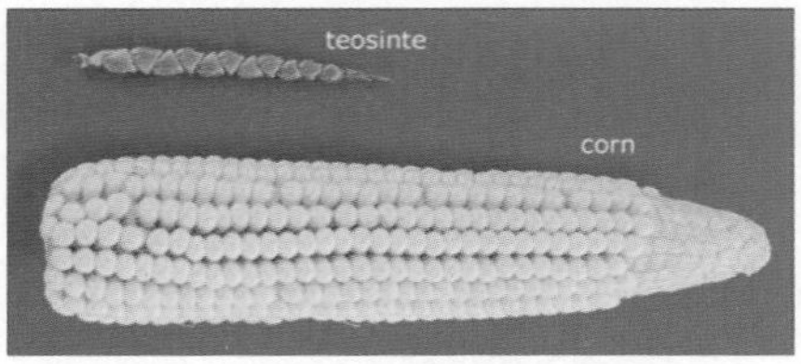

(A)

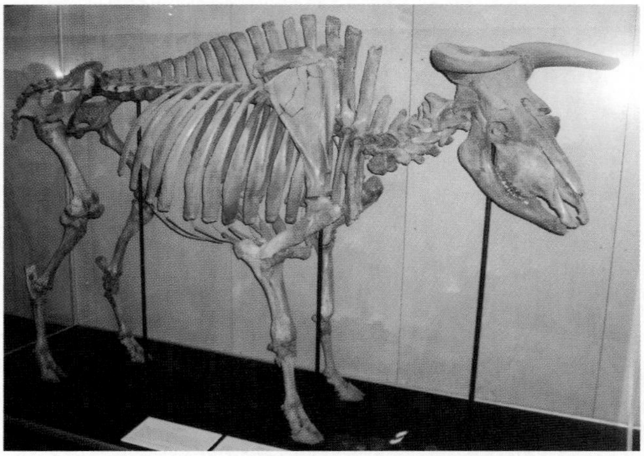

(B)

FIGURE 2.4 **(A) The relative sizes of wild teosinte and cultivated corn. (B) Aurochs died out approximately 400 years ago.** Courtesy of John Doebley; Courtesy of Monk, Inc.

Seed Banks

Concerns over the effects of GM crops have contributed to the establishment of seed banks. As GMO crops continue to be planted more and more, scientists have recognized the potential negative impacts on biodiversity. Genetically modified crops are typically planted as

FIGURE 2.5 **A Greenpeace protest in front of the Berlin parliament building in 2003 depicts a killer GMO corn field.** © Tobias Schwarz/Reuters/Landov.

Box 2.2 "Pharming: Food as Medicine"

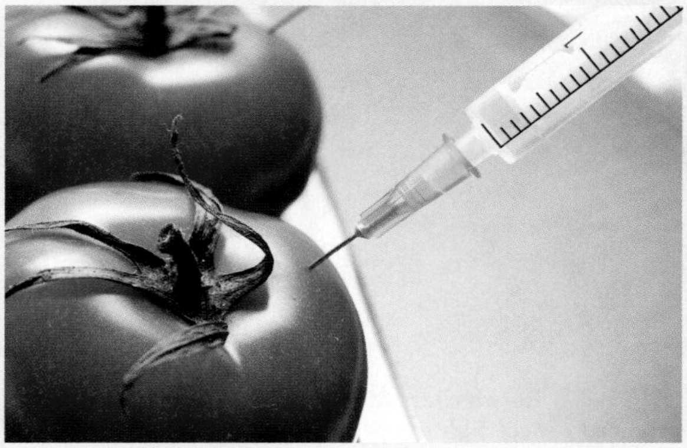

© NiDerLander/iStockphoto.com

Another example of the interdisciplinary nature of molecular biotechnology can be seen in particular genetically modified foods used to deliver drugs and vaccines. Using GMOs as living bioreactors to produce therapeutic proteins is known as "pharming." Pharming entails both the development of a genetically engineered medication or vaccine and the utilization of transgenic plants and animals as edible delivery vehicles. Transgenic goats, sheep, and cattle have been successfully created to yield human therapeutics in their milk. For example, transgenic sheep are used in the United Kingdom to produce the protein alpha-antitrypsin. Normally this protein is influential in maintaining the integrity of connective tissues. In its absence, tissues degrade, such as in the liver where it can lead to cirrhosis. The milk is then consumed as a treatment for liver disease and hemorrhages.

monocultures, meaning a single variety is planted in the entire acreage available (**FIGURE 2.6**). As monocultures contain just one genetically engineered variety, this essentially constitutes a field of clones. Every plant put in the ground is genetically identical. The danger of monoculture practices can be illustrated by the Irish Potato Famine of the 1840s. If Ireland's potato plantings had been more diverse, it would have been less susceptible to blight.

As transgenic crop varieties become more commonplace, the number of unique cultivars growing in the ground and being sold in the store has decreased. In an effort to preserve the genetic information found in the diverse plant species historically used as food, seed banks aim to collect, catalog, and store as many genetic varieties as possible. In fact, routine practice now includes determining the genome sequence of stored seeds. In the event a field of clones falls prey to drought or disease, or if biodiversity decreased by any other means, the seed banks would act as a reserve by which species could be reintroduced. Thus, seed banks act as a modern day Noah's Ark for plant life (**TABLE 2.1**).

Note that a cultivar could be reintroduced by traditional propagation methods as a whole plant, or select genes could be utilized for creating new genetically modified crops. For example, U.S. corn yields were cut in half during the 1970s due to widespread fungal infection. Researchers identified a fungal resistance gene in wild corn preserved in seed banks, and were able to modify the existing corn to express the resistant trait.

FIGURE 2.6 An example of monoculture. © Zoran Orcik/iStock/Thinkstock

TABLE 2.1

The Reasons Behind Seed Banks

Cause of Concern	Negative Impact Alleviated by Seed Banks
Natural Disaster	Existing species may be wiped out.
Disease	Existing species are susceptible and may not adapt fast enough.
Anthropogenic Causes	War and other human-made disasters can destroy existing biodiversity.
Climate Change	Changing patterns in weather, pollination events, and exposure to pests and disease, all of which may be the results of climate change, could alter the viability of existing species.
Timely Research	General trends in biodiversity decline may eliminate species with beneficial properties, such as medicinal compounds.

Data from United Nations. International Treaty on Plant Genetic Resources for Food and Agriculture. Food and Agricultural Organization. Rome 2001

Aquaculture

Much like animal husbandry and selective breeding of crops, aquaculture is not a new technique. **Aquaculture** involves the production of fish or shellfish for human consumption in a controlled environment. Aquaculture accounts for nearly 50% of total fish consumption worldwide, making it a vital source of seafood for humans. As wild fish populations are pressured by commercial fishing enterprises, aquaculture sources will prove more and more valuable to fulfill this need.

Molecular biotechnology plays a role in aquaculture by providing the means to genetically modify seafood species. Through recombinant DNA technology, aquaculture researchers have created disease-resistant strains of oysters. Just as livestock are treated with antibiotics due to their confined living arrangements, vaccines for aquaculture operations have become necessary. Such vaccines have successfully resulted from applications of molecular biotechnology. Genetically modified salmon are another product of aquaculture biotechnology; GM varieties that overproduce growth hormones exhibit a much shorter growing period and allow more salmon to go to market in less time (**FIGURE 2.7**).

Another aspect of aquaculture is **bioprospecting**. The concept is similar to seed banks, only in this case aquatic species are the focus. Aquatic life represents some of the oldest organisms in the evolutionary progression life on this planet. The species found within the freshwater and marine ecosystems, whether unicellular or multicellular, have survived under extremely harsh conditions. This would create the potential for a wide variety of useful genes and gene products that evolved as adaptations for survival. By harvesting and classifying the various aquatic organisms of the world, molecular biotechnologists can create a bank of genetic "tools" for future use. One example of bioprospecting is the marine sponge, which is being explored as a potential source of anticancer compounds (**FIGURE 2.8**).

FIGURE 2.7 A 2010 photo of an 18-month-old GM salmon modified with growth hormone (background) compared to a non-GM salmon of the same age (foreground). © MCT/Landov

FIGURE 2.8 Researchers Murray Monro and John Blunt discovered cancer-fighting properties in this sea sponge off the coast of New Zealand. © Fairfax Media/The Press.

Around the House, At the Superstore: Industrial Biotechnology

This branch of molecular biotechnology is much like a catch-all in that it encompasses techniques and processes from a variety of industries. Those aspects of the field that do not fit in nicely would be found here. When we consider the vast quantity of tangible goods produced in the factories of the world, each manufacturing process might benefit from molecular biotechnology in some way. For example, enzymes produced through biotechnology can act as catalysts in order to speed up many processes. Genetically engineered microbes may be introduced into manufacturing systems as decontaminants to eliminate waste streams. Proteins obtained through molecular biotechnology techniques can act as ingredients to improve different aspects of a product.

Green Biotechnology: Environmental Science and Conservation

Since the early part of the twentieth century, humanity has engaged in a continuous conversation on how to become more environmentally friendly. It is a reaction to the negative side effects of industrialization: air, water, and soil pollution, dwindling of fossil fuels, and mounting evidence in support of global warming and climate change. Ways to minimize our impact upon the natural ecosystems of the world have been outlined, debated, and tabled time and again. The advent of molecular biotechnology is changing our approach to old questions, such as how to reduce pollution and the development of alternative fuels. As biodiversity depletion persists, molecular biotechnology may allow us to preserve threatened and endangered species, or even resurrect those which have become extinct.

Bioremediation

Bioremediation uses molecular biotechnology to degrade environmental pollutants. Industrial processes lead to the release of harmful chemicals that persist in the natural environment. Left to naturally occurring chemical cycles, these pollutants continue to accumulate because of their extremely slow turnaround. Oil spills act as an extreme example of both monumental environmental damage and the bioremediation techniques employed to alleviate the massive pollution. Beginning with the 1989 Exxon Valdez oil spill off the coast of Alaska in Prince William Sound, scientists have let loose oil-degrading strains of bacteria. The bacteria act as a living sponge, metabolizing the chemicals found within the oil (**FIGURE 2.9**). They thereby break down large molecules into smaller components that can be absorbed by the environment much more easily. However, oil may still be trapped beneath the superficial layers of soil. Long term impacts of the Valdez spill are still apparent over a quarter century later. This technology was improved with the Deep Water Horizon oil spill in 2010, when the oil-eating bacteria were delivered via nanoparticles over the Gulf of Mexico. It is still too soon to tell how effective these treatments were to the Gulf environment. Impact studies continue.

Biofuels

Alternative energy sources continue to be on the forefront of environmental and economic policies. The needs for energy independence, cheaper energy with less price volatility, and cleaner energy have culminated in our desire for technologically advanced alternatives. The decomposition and fermentation of various organisms are being investigated as potential biofuel sources. **Biofuels** are sources of energy derived from biological carbon fixation. Corn and its byproducts, agricultural waste, prairie and switch grasses, and other sources of concentrated cellulose are being investigated as potential fuel sources.

Conservation

The preservation of biodiversity certainly plays a role in seed banks and bioprospecting, as previously presented. Both of these enterprises seek to collect and maintain threatened genetic profiles before they become extinct. An additional area of conservation is seen in projects like the Genome 10K Project, which is assembling a genomic zoo of 10,000 vertebrate species.

Some would even propose reviving extinct species through cloning, as is the case for the Long Now Foundation's Revive and Restore Project. Headed by Stewart Brand, known for his *Whole Earth Catalog*, Revive and Restore maintains a list of potentially revivable extinct species. Whether an extinct species is well-suited for the project depends on criteria such as how desirable and practical the organism is considered to be, and how easily it may be reintroduced into the wild.

Currently, the passenger pigeon acts as the model candidate. The passenger pigeon was iconic until its extinction in the early twentieth century, and its reintroduction would be feasible and practical. DNA from this wild pigeon is readily available from museum specimens and the species has close extant relatives that might ease reintroduction (**FIGURE 2.10**).

FIGURE 2.9 A bioremediation project along an oil-spilled shoreline. Courtesy of the Exxon Valdez Oil Spill Trustee Council/NOAA

FIGURE 2.10 Revive and Restore researcher Ben Novak examines the last living carrier pigeon, Martha. She died in captivity in 1914. © Ryan Phelan/The Long Now Foundation

Protecting Citizens and Nations: Forensics and Biodefense

Techniques of the molecular biotechnology industry are useful in the identification and tracking of organisms. Our ability to confirm the identity of individuals through genetic analysis has dramatically improved the level of confidence in positively determining perpetrators of crime. The technology may even be used to exonerate a wrongly convicted inmate. The tracking of genetic material through space and time can help epidemiologists describe the infection patterns of disease. This has the potential to prevent contamination and the outbreaks that follow. If the inspectors of our food, drug, and water supplies incorporate these tracking abilities, such a practice would constitute preemptive defense of our nation.

Forensics

The term **forensics** refers to science and technology used in the investigation of crimes. The field has exploded thanks to molecular biotechnology. The 1995 murder trial of O.J. Simpson brought this to the attention of the American public when 45 blood samples taken from crime scene evidence were checked against his DNA. Of the 45 samples, 30 were found to be a match with the suspect using restriction fragment length polymorphism (RFLP) analysis. RFLPs are varying lengths of the DNA molecule cut at specific locations with restriction enzymes. The specific length aids in identification. At the time, DNA sequencing was not a part of everyday vernacular and thus likely was not valued as highly as it is today. This may be one reason the jury did not commit Mr. Simpson of murder.

DNA fingerprinting is an additional forensic application. Also referred to as genetic profiling, this technology identifies genetic markers that are unique in each person. For example, genetic markers with variable numbers of tandem repeats (VNTRs) that vary from one person to the next. It is an additional layer of certainty when identifying a suspect of a crime, a victim of a crime or accident, or in cases of paternity.

Biodefense

Molecular biotechnology is being utilized to improve national security, an area known as **biodefense**. Biological contamination of food, water, or medicine has the potential to be accidental or intentional. In either event, our capability to identify the source of contamination can lead to improved security measures for the future. If intentional contamination took place, it would be considered bioterrorism. Biotechnologists have the potential to prevent bioterrorism by closely monitoring all known biological specimens posing such a threat (**FIGURE 2.11**). They can also actively develop treatments such as vaccines in the event a pathogen was released. An additional application in the area of biodefense is the use of biosensors to detect explosives, chemicals, and pathogens in places such as airports or international borders.

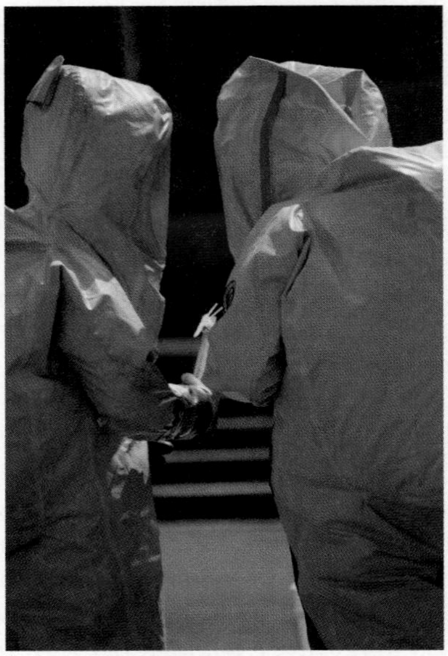

FIGURE 2.11 Biodefense researchers work with the deadliest known pathogens, requiring them to wear biohazard suits. © Photodisc

Evo Devo: The Development of Life and the Human Family Tree

The data obtained from various DNA sequencing projects can be used to understand both development and evolution. The order of events that transform a single-celled zygote to a full-sized individual entail the development of an organism. Comparisons of DNA sequences, genomes, and proteomes can all offer insight into the evolutionary relationships. We can use this information to understand the place of our species in the tree of life.

Evo Devo

Evo devo is a relatively new branch of molecular biotechnology. The term is an abbreviation of "evolutionary developmental biology." **Evo devo** represents the convergence of study within the fields of evolution and embryonic development. The myriad discoveries in genomics revealed that we share a number of genes with other animals. The fact that these genes were upheld throughout evolutionary eons indicates they share a common purpose across species. These universal genes act as a body-building tool kit that determines the specific timing and intensity of gene expression during development. The toolkit acts as a conductor to orchestrate the complex series of events that tell a developing embryo whether it will be bilateral

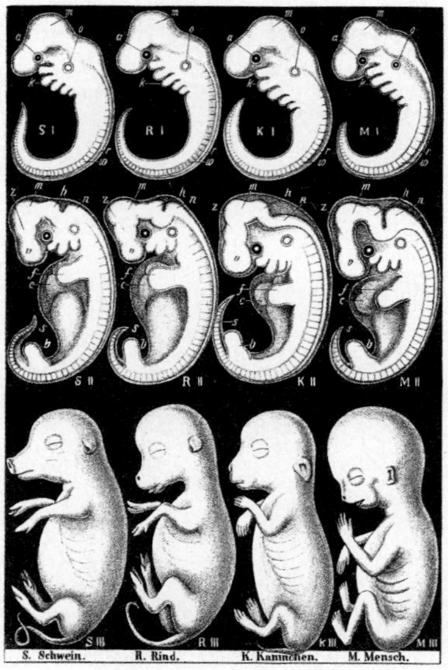

FIGURE 2.12 The comparative anatomy of mammalian embryos. © Photos.com/Thinkstock

or radially symmetrical, if it will have a tail, walk on two or four feet, or fly, and so on. By characterizing the exact pattern of orchestrations occurring during the development of different organisms, researchers can then recognize the differences and begin to establish an order of progression. This order directly correlates to the evolution of various creatures.

Comparative embryology is not a new pursuit; Charles Darwin engaged in comparing various embryos in attempts to determine their evolutionary relationships (**FIGURE 2.12**). What makes evo devo unique is that we now add a molecular understanding: how genes and environment interact to produce a new organism. It is analogous to the initial theory of natural selection as an evolutionary mechanism and the eventual modern synthesis. While Darwin and Alfred Russel Wallace were able to conceive the mechanism of natural selection by observing species at the organismal level, the modern synthesis nevertheless added depth to the concept of evolution. In the same vein, comparative embryology can draw conclusions based upon organismal observations, but evo devo is layering molecular events onto existing principles.

The Human Family Tree

The field of evolutionary biology has benefitted tremendously from molecular biotechnology. As the DNA of more and more species are sequenced, the data are used to determine the step-by-step progression of one species giving rise to another wherever possible. Specific segments of DNA, including particular genes, can be honed in on for small changes over time. Then a phylogenetic tree is created, which is a visual schematic of evolutionary relationships (**FIGURE 2.13**).

As previously discussed, projects such as the Genographic Project are using this information to learn more about our own species and where we fit in the grander scheme of life. Anthropologists are using DNA evidence in concert with archaeological evidence to piece together the story of our past. By sampling populations around the world, the project has constructed a map of human migration out of Africa, showing how our species came to populate the continents of Earth (**FIGURE 2.14**).

Molecular Biotechnology Industry Practices

The molecular biotechnology industry follows its interdisciplinary nature when it comes to practice as well. Research scientists often begin the flow of work when they identify a gene or gene product. This leads to the development of genetic engineering in some form. Moving from the initial discovery to a final product involves a number of steps.

Funding: Public and Private

There is no getting around the associated costs to conduct research. Money must come from somewhere, and the source may be public or private. Funding is awarded based on the viability of the product or information

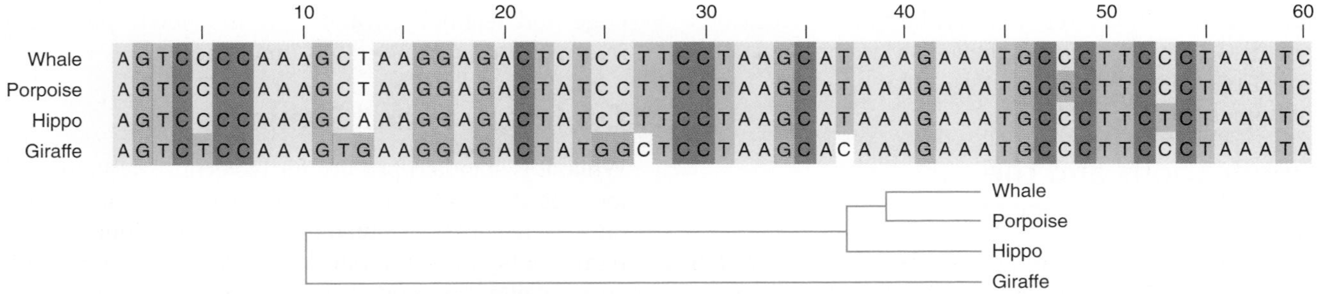

FIGURE 2.13 Evolutionary biologists used DNA sequence comparisons to construct a phylogenetic tree linking hippopotamuses and giraffes with their aquatic cousins, whales and porpoises.

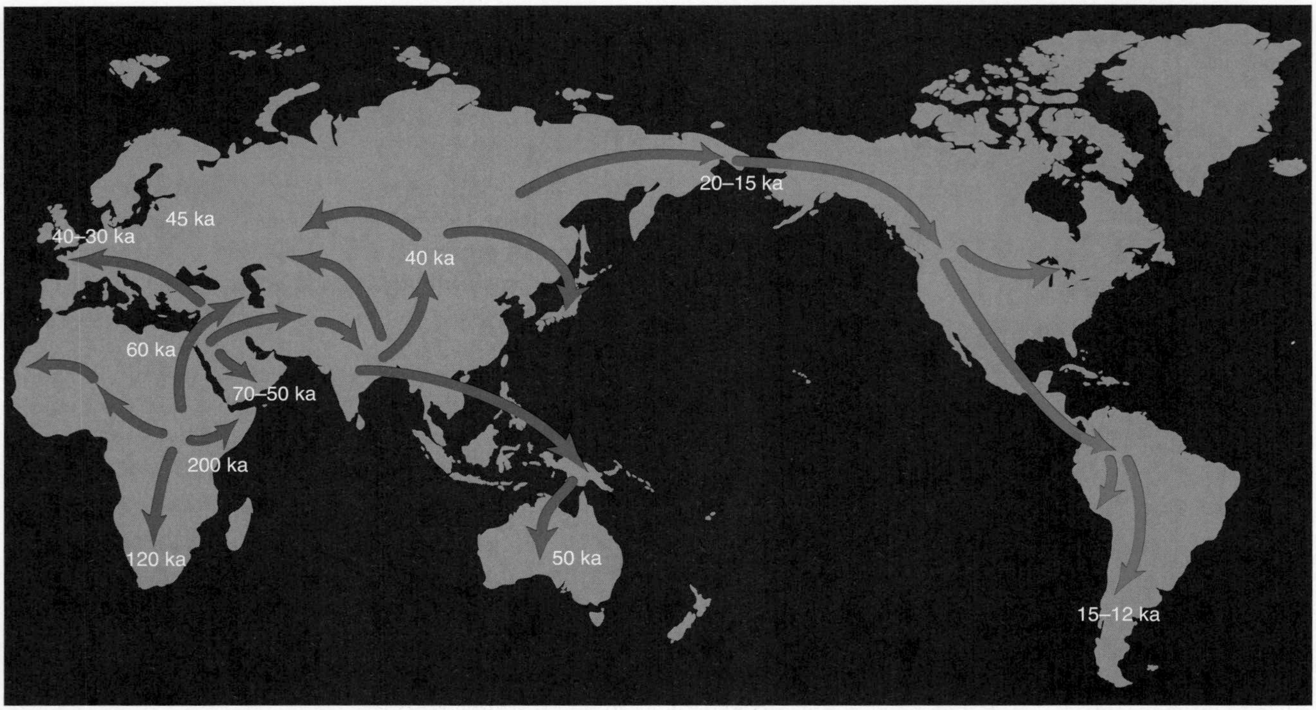

FIGURE 2.14 **The Genographic Project has pieced together the migratory patterns of *Homo sapiens* out of Africa.**

to be obtained through the research. The value of the research is also considered, both in terms of the educational value of the knowledge gained and in terms of the profitability of the project's output.

Research scientists may work in a publicly funded laboratory at state colleges and universities or state or federal agencies such as the National Institutes of Health or the Center for Disease Control. Public and private companies may also draw funding through grants from federal sources. Alternatively, private entities, such as venture capital firms or pharmaceutical corporations, may fund research for a laboratory as well.

The funding process starts with the submission of a research proposal, which a panel of scientists and biotechnology professionals then reviews. Funding decisions may also be subject to government regulation. For example, the 1996 Dickey-Wicker law limited federal funding in stem cell research. While funding decisions are based on the criteria previously mentioned, personal relationships and political influence have a way of influencing the outcome.

Publications and the Peer-Review Process

As with any scientific endeavor, part of the research process is to communicate your findings. The nature of science involves the collaboration of diverse interested parties. Communicating information is a necessary step that is carried out through professional journal articles. A wealth of journals pertain to the field of molecular biotechnology, including *Cell, Journal of Molecular Biology, Nature, Genetics, Molecular and Cellular Proteomics,* and the *American Journal of Public Health.*

Peer review is a process associated with journal article publication to ensure the accuracy and validity of a proposed paper. Once the author(s) submits a paper, a journal editor sends advanced copies to experts on the particular subject. The experts review the paper and return comments and suggestions to the editor, which the author is usually then privy to for the purposes of modifying the original work as needed.

Scientists are also on the receiving end of this communication. It is crucial to stay current with the cutting edge to know what events are taking place in your field, who amongst your colleagues is contributing knowledge and technology, and how the newly published information can be used in future studies.

Conferences and Seminars

While it is standard practice for biotechnologists to read journals, it is also beneficial for them to meet face to face on a regular basis. Conferences draw large numbers of participants in vast exhibition halls. Members of industry and academia convene to discuss the latest developments in their areas of expertise. Conferences include visiting booths of different vendors to learn more about the latest products and technologies and sitting in on talks and symposia led by scientists and industry executives. Conferences are usually several-day-long events that require travel and associated costs, such as entry fees.

Seminars are more intimate in nature and tend to be much less commercial than conferences. They are smaller gatherings at colleges and universities typically held within a single department. Seminars may also be held at organizational or association meetings. They offer researchers, in some cases including student researchers, the opportunity to present information on topics related to their projects. Seminars are usually free and open to the public, unless membership is required.

Intellectual Property

Any knowledge obtained from biotechnology research could yield intellectual property that may be patentable. Once patented, the owner now holds rights to the information and can incur licensing fees toward any researcher seeking to use it. Any additional technologies discovered or invented would then be subject to royalties due to the patent holder(s).

Patenting becomes tricky when the intellectual property originates from living organisms. It raises ethical implications when something that could be construed as public domain becomes private property. Regardless, in 2011 a federal appeals court ruled that human genes can be patented. In 2013 the Supreme Court overturned this ruling; at present, human genes are not patentable. Until this recent verdict, the molecules found in each of our cells could essentially have part ownership—what's yours was not yours. Unfortunately, if this pipeline to profit was eliminated, many players in the biotechnology industry might lose their incentive to conduct additional research.

Careers in Molecular Biotechnology

As we have seen, the molecular biotechnology industry spans a growing number of endeavors. By virtue of its interdisciplinary nature, the number of careers pertaining to the field is likewise increasing. There are comparatively few areas of occupation that do not involve some aspect of the science, technology, engineering, and mathematics (STEM) required of molecular biotechnology. Collectively, the public and private sector occupations related to STEM disciplines involve these four elements. A career in molecular biotechnology is included in the rapidly growing STEM industries.

When considering a career in molecular biotechnology, there are a number of roles to choose from. Not surprisingly, many require advanced education beyond the bachelor's degree level. Medical and Doctor of Philosophy degrees are common among biotech professionals. A higher degree obtained correlates to a higher earning potential. Jobs are concentrated around technology centers such as colleges, universities, and research parks.

FOCUS ON CAREERS

In this chapter, an overview of the various careers available pertaining to molecular biotechnology is presented. Throughout the remainder of this text, be sure to notice the *Focus on Careers* features. These features will relate to the specific topics being presented in each chapter and will provide a more detailed look into the type of work being done within each field utilizing molecular biotechnology. *Focus on Careers* includes up-to-date, relevant information such as educational requirements, day-to-day activities, and performance expectations. Through *Focus on Careers*, you will be exposed to real life professionals and institutes of excellence along with the strides they are making to innovate science, technology, and society.

Careers in the Laboratory

Laboratory Managers

Laboratory managers act as overseers of the day-to-day operations performed within the lab setting. They are responsible for hiring decisions, budget management, and regulatory compliance. When a research project is initiated, the laboratory manager searches for the necessary personnel and orients new employees to the good laboratory practices (GLPs) and standard operating procedures (SOPs) of the particular project. It is an ongoing process to maintain a sufficiently trained and functional team, which the laboratory manager must ensure takes place. This position may require someone to conduct trainings on laboratory safety and the laws pertaining to that lab, or facilitate the presentation of such trainings. The laboratory manager must also keep the lab running smoothly by keeping materials and supplies well stocked, maintaining the working order of all tools and instruments, and providing a regulatory compliant environment. As a budget manager and the buyer for the laboratory, this person often cultivates relationships with vendor representatives and maintenance servicers. They may attend trade conferences for such purposes as well. The same is true when performing duties as a regulatory compliance officer. A laboratory manager is in regular contact with the agencies of government which regulate his or her lab and may need to attend meetings or conferences to stay current with regulations. The position may also include keeping permits current and filing regular paperwork with government agencies as necessary.

Principal Investigators, Senior Research Scientists, and Medical Scientists

Principal Investigators, Senior Research Scientists, and Medical Scientists are all terms that refer to the lead role of a research team. They may work in a college, university, or hospital setting, or find employment in government agencies or private companies. The team lead is responsible for steering the course of research and determining the objectives of each project. Other responsibilities of the team lead include:

- Design of the project and its experiments
- Delegation of tasks to the research team
- Participation in grant proposals and patent applications
- Report on the progress of each project
- Troubleshoot unexpected results and modify procedures as needed

Principal Investigators, Senior Research Scientists, and Medical Scientists nearly always have a PhD, MD, or both. They tend to be highly specialized in one or perhaps a handful of areas within the field of molecular biotechnology. As the number of biotechnology enterprises continues to grow, the level of specialization is expected to increase as well.

Laboratory Technicians and Technologists

Laboratory technicians and technologists act as the support team to the principal investigators of a research project. They perform the day-to-day work of experimentation and laboratory management. Depending on the exact nature of the laboratory, duties of laboratory technicians and technologists vary widely. They can be involved with equipment maintenance, preparation of materials, carrying out experiments, and assembling data. One may become specialized in certain tasks with a distinct division of labor assigned to each member of the support team.

Due to the growth of this industry, the availability of jobs with less educational requirements is increasing. However, these tend to be the lowest paying biotechnology-related jobs as they may require a bachelor's degree, or even an associate's of science degree in some cases. It bears repeating that when it comes to careers in molecular biotechnology, the more education the better. Two- and four-year degrees limit opportunities for advancement and minimize income potential in this highly technical and specific field.

Crime Lab Technicians

Forensic investigation now relies heavily upon biotechnology techniques and crime lab technicians perform this work. It is a rewarding role for individuals interested in both science and criminal justice. Crime lab technicians use their knowledge of biotechnology to analyze biological evidence such as hair, skin, and blood samples. They report their findings to members of law enforcement and may also testify as experts in court trials.

Quality Control Analysts

Once a biotechnology product is researched and developed, the manufacturing division steps in to create the final product. Quality control (QC) analysts act as part of the manufacturing team by testing the product within a laboratory, both during the process of manufacture and upon its completion. QC analysts also test the quality and efficacy of each ingredient involved in the manufacturing process. They collect and analyze data from their tests and present reports of their findings. If any specifications are not met, QC analysts will alert management and work to find a solution. In such an event, they are then responsible for updating operating procedures and communicating the information to the appropriate personnel.

Non-Laboratory Careers

Many aspects of the biotechnology industry take place outside of the laboratory. While the science and technology is initiated and refined in the laboratory, the resulting products will go nowhere without the assistance of a downstream pipeline. Professionals outside of the laboratory contribute to the necessary elements of completion both prior to release and aftermarket of their product. Educators of biotechnological pursuit would likewise be considered professionals outside of the laboratory, as would the technicians and engineers that provide the tools and instruments needed in the laboratory itself. Careers outside of the laboratory are among the highest paying of all STEM employment.

Educational Careers

Professors, instructors, and teachers can specialize in biotechnology education. These educators must be well-versed in the various disciplines that contribute to the study of biotechnology, such as molecular biology, biochemistry, genetics, and immunology. They provide the knowledge necessary to educate the upcoming biotechnology workforce and can act as mentors for budding biotechnologists. Therefore, they must stay abreast of current events in the field and may have outside interests or relationships with industry. An educational career may be divided between laboratory and non-laboratory duties, as university professors may also perform research.

Management Professionals

Positions in management exist in several realms of molecular biotechnology. They represent the highest paying STEM career of any, with an average salary of over $130,000 per year. All branches of molecular biotechnology require individuals to manage operations. A management professional could perform a leadership role in research and development, project management, clinical trials, manufacturing, or sales, among other areas.

Biostatisticians

The role of a biostatistician straddles the fields of biology, mathematics, and computer science, making it a highly technical career. Analyzing the massive amounts of data generated in DNA sequencing requires information technology and highly specialized computer programs. Biostatisticians extract useful information from the billions of nucleotide base pairs routinely generated in the world of molecular biotechnology. This type of job usually requires a PhD, but with sufficient computer and math skills, it is possible to obtain employment at the bachelors or masters level.

Clinical Research Associates

Clinical research projects determine the effectiveness of experimental medications in human subjects. The clinical research associate designs, executes, and manages these research projects. Individuals with a medical background are usually preferred for these roles, such as registered nurses with experience dealing with patients and FDA regulations. Knowledge of the regulations governing clinical trials is a necessity for this career path.

Sales and Marketing Professionals

Around 400,000 sales and marketing professionals are employed in scientific and pharmaceutical organizations across the United States. Among the highest paying careers in the biotechnology industry, sales and marketing roles are attractive to individuals with a scientific background but who do not wish to work in a laboratory.

SUMMARY

- The tools of biotechnology have allowed for a variety of applications. Molecular biotechnology is an interdisciplinary field that incorporates a broad base of scientific disciplines. It draws upon biology, chemistry, mathematics, computer science, engineering, philosophy, and ethics and influences a number of fields including medicine, agriculture, justice and defense, environmental science, and anthropology.
- A large majority of today's molecular biotechnology industry is devoted to healthcare applications. Biopharma is the name given to bioengineered pharmaceuticals. Gene therapy is a form of medical treatment that removes or replaces defective gene(s) in order to treat or even cure injury or disease. Genetic counseling involves the identification of disease in an individual's family history and determining the appropriate measures one might take to avoid passing on disease to his or her offspring. It may include the use of pedigrees and/or genetic profiling.
- Genetically modified organisms (GMOs) are organisms that contain knockout or inserted genetic material. Genes may be knocked out of or inserted into a food crop or animal in order to confer desirable traits in the organism. A gene knockout renders a gene inoperable, thereby preventing the synthesis of its protein product(s). GMOs may also possess inserted genetic information from a separate species.
- Seed banks serve the purpose of collecting and preserving a diverse assortment of plant seeds. The establishment of seed banks maintains biodiversity for future prospecting, which may be useful in the event of famine or crop disease. This threat is particularly acute as monoculture becomes more prevalent in agriculture worldwide.
- Aquaculture produces fish and shellfish for human consumption under controlled and sustainable conditions. In addition to serving as a food source, aquaculture also provides products from aquatic organisms such as pearls, adhesives, and pharmaceutical compounds. Bioprospecting includes harvesting and classifying the various aquatic organisms of the world, allowing molecular biotechnologists to create a bank of genetic "tools" for future use.
- Industrial biotechnology is a broad class of applications involved in manufacturing processes and production of consumer goods. Examples include enzymes created by GMOs used to enhance the productivity of manufacturing processes as well as vitamins and drug compounds produced through microbial fermentation.
- Green biotechnology seeks to reduce pollution, develop alternative fuels, and conserve biodiversity. Bioremediation utilizes living organisms to decompose chemical pollutants and other human-made waste. Biofuels such as corn ethanol

and biodiesel are being investigated as potential fuel sources. Conservation is aided by biotechnology through DNA and tissue preservation and enhanced breeding methods.

- DNA fingerprinting has innovated the field of forensics by adding the additional layer of genetic evidence to crime scene investigation and cases of identity such as paternity. Also known as genetic profiling, DNA fingerprinting is a nearly indisputable method used by forensic analysts as it provides an insurmountable level of certainty.
- Biodefense is the application of molecular biotechnology to defending peoples and places of a nation. Examples of biodefense include detecting and tracing biological contaminants and/or weapons, developing vaccines against biological warfare, and the creation of biosensors.
- Evo devo combines the study of evolution and embryological development. Evo devo has the potential to contribute knowledge of evolutionary relationships among organisms. The field may also lead to advances in medicine as we develop a greater understanding of the molecular and cellular mechanisms which make us who we are as a species.
- Molecular biotechnology professionals are often involved in obtaining funding for their projects, performing research and development under applicable laws and regulations, and communicating their findings in peer reviewed journals, meetings, and conferences. The molecular biotechnology industry spans a growing number of endeavors. By virtue of its interdisciplinary nature, the amount of careers pertaining to the field is likewise increasing. Positions may be set in a laboratory or non-laboratory setting and are typically divided into institutional and private sector occupations.

KEY TERMS

Aquaculture
Biodefense
Biofuels
Biopharma
Bioprospecting
Bioremediation
DNA fingerprinting
Evo devo
Forensics
Gene therapy
Genetic counseling
Genetic profiling
Genetically modified organism
Knockout
Molecular biotechnology
Pedigree

DISCUSSION QUESTIONS

1. What area of molecular biotechnology do you foresee having the greatest impacts on society and why?
2. Select a career path in molecular biotechnology that interests you most. Research current prospects for this career. Parameters to search include average salaries, necessary education and experience, available job postings, and companies that employ individuals in this career.
3. Search the U.S. Patent and Trademark database for recent patents in molecular biotechnology. Describe a patent awarded in the past year. Include the characteristics which granted the technology or product a patentable status.

REFERENCES

Cover, Ben, John I. Jones, and Audrey Watson. "Science, Technology, Engineering, and Mathematics (STEM) Occupations: A Visual Essay." *Monthly Labor Review*, May (2011): 3–15.

Elbehri, Aziz. "Biopharming and the Food System: Examining the Potential Benefits and Risks." *AgBioForum* 8, no. 1 (2005): 18–25.

Emory University. "Medicine and GMO." Last modified February 2007. Available at http://www.scienceandsociety.emory.edu/GMO/GMO%20and%20Medicine.htm

Frierman-Hunt, Gina and Julie Solberg. *Careers in Biotechnology: A Counselor's Guide to the Best Jobs in the United States, 3rd ed*. Sacramento, California: Chancellors Office California Community Colleges, 2008.

Lange, Tom and Philip Vannatter. *Evidence Dismissed*. New York: Pocket Books, 1997.

United Nations. *International Treaty on Plant Genetic Resources for Food and Agriculture*. Food and Agricultural Organization. Rome, 2001.

United States Department of Energy. "Gene Therapy." Last modified August 24, 2011. Available at http://www.ornl.gov/sci/techresources/Human_Genome/medicine/genetherapy.shtml

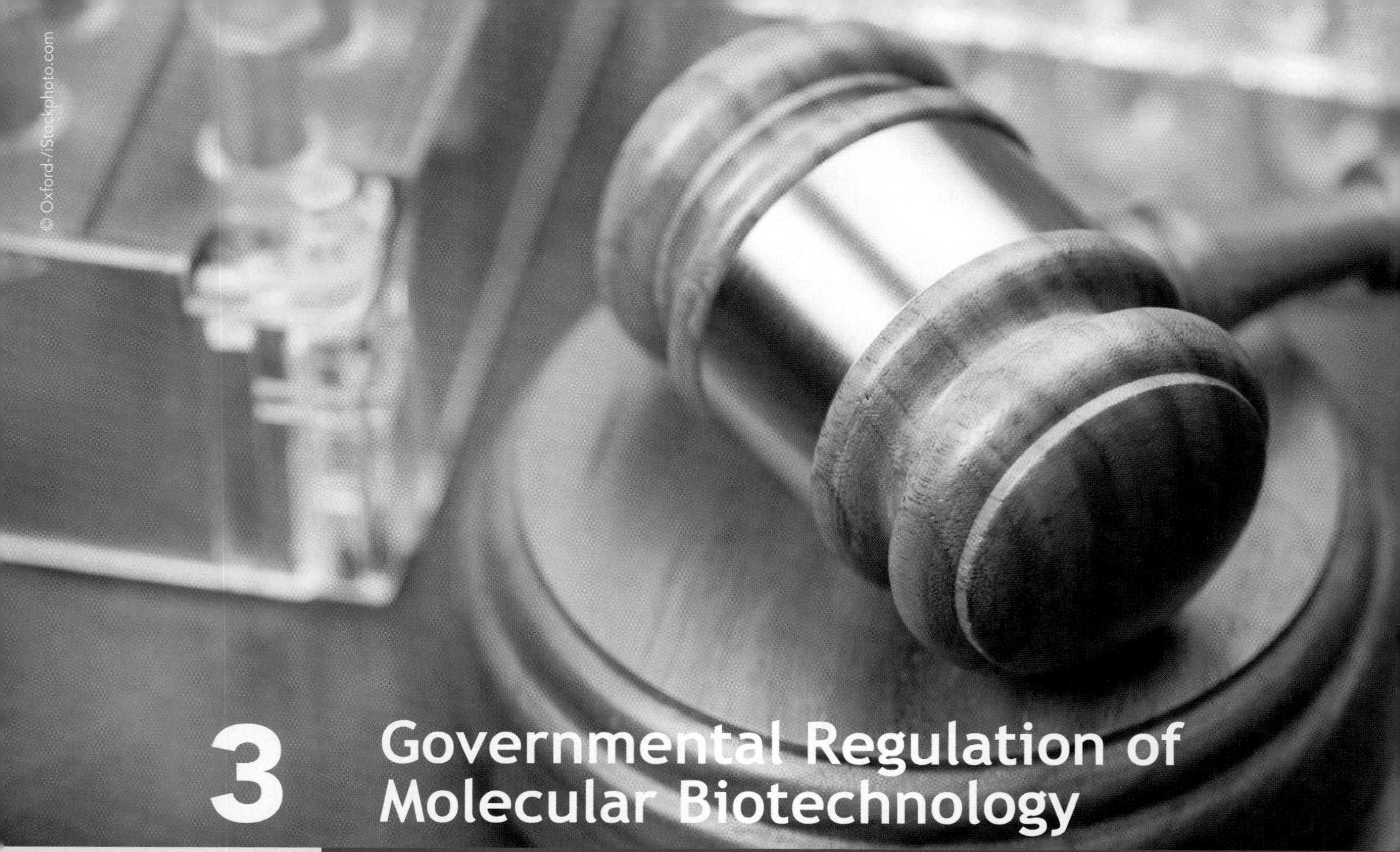

3 Governmental Regulation of Molecular Biotechnology

LEARNING OBJECTIVES

Upon reading this chapter, you should be able to:

- List the federal agencies charged with regulating biotechnology.
- Identify the goals of regulating biotechnology.
- Summarize the role of the Environmental Protection Agency in the regulation of biotechnology.
- Discuss the involvement of the Food and Drug Administration in regulating bioengineered products.
- Explain how the United States Department of Agriculture regulates the biotechnology industry.
- Characterize the relationship among the National Institutes of Health, the Coordinated Framework, and the biotechnology industry in terms of the Asilomar Conference, the Recombinant DNA Advisory Committee, and Institutional Biosafety Committees.
- Describe the process of bringing a genetically modified organism to market, from inception to commercialization.
- Indicate the criteria for labeling a product containing genetically modified organisms.
- Analyze the drawbacks and benefits of labeling products containing genetically modified organisms.

As an industry, molecular biotechnology has experienced a rather brief history when compared to other major scientific disciplines. Yet in the four decades or so of its existence, we struggle as a society to keep up with its rapid pace of innovation. The legal and regulatory environments as they exist today are more of a reaction to biotechnology than they are a guiding light. Most of the statutes governing the industry were created long ago and have been superimposed on decidedly new processes and products. Research and development costs are extremely high, but the products obtained from biotechnology provide us with clear benefits which grow in number day by day. Yet they also pose risks to human health and the delicate balance of our natural ecosystems. Much of the concern our society faces today derives from the same dilemma faced by early biotechnologists in the 1970s. Is the pursuit of scientific knowledge the goal of this field, or are profits the primary objective?

Since its inception, biotechnology has faced two opposing forces. On the one hand, the burgeoning industry may be considered halted by regulatory terrain. The ability to patent biotechnology discoveries means there is private ownership of information that may be necessary to move forward with research. Opponents of privatization would argue that patent fees can be cost prohibitive in accelerating scientific discovery. On the other hand, the need to earn profits can also be seen as a motivational force. Proponents of industry would argue that the potential to make a profit is what drives technological innovation (**BOX 3.1**). Without profit potential, a research project may never be funded.

Whichever side appeals to you, the reality is biotechnology research has been intertwined with industry since the late 1970s. Discovery fuels profit and profit fuels discovery. Even the purists uninterested in commercial gain often must answer to private entities because corporations are increasingly the ones paying the bills. The lines between science and technology and risks versus rewards have blurred beyond recognition, other than, perhaps, as an esoteric discussion in textbooks. Somewhere in the middle lies the regulation of biotechnology.

Regulatory Oversight: The Federal Agencies

The Asilomar Conference of 1975 brought together 150 scientists from 13 different countries to discuss the fate of recombinant DNA technology. The conference took place in the midst of a self-imposed moratorium on molecular biotechnology research activities, an unprecedented event. Molecular biologists and geneticists involved with the early stages of biotechnology wanted to establish universal guidelines for research in order to prevent unintended consequences. In addition to the scientific representation at Asilomar, attorneys and government officials attended as consulting authorities on existing laws and regulations that could be applied to biotechnology research. To encourage transparency and public awareness, 16 journalists attended the conference. Not only was the research moratorium historic, so were the cautionary measures taken by the scientific community in order to preemptively establish defined governing rules. Previously, regulations were typically constructed in reaction to scientific innovations, not prior to their adoption. Anticipating possible hazards before they were known to exist was a unique proposition.

Even after the Asilomar Conference, concerns that bioengineers are "playing God" or altering the natural course of evolution have led to a coordinated regulatory environment. In addition to these philosophical issues, there are more concrete concerns over the impacts of biotechnology on human health and the environment. However, existing statutes that apply to molecular biotechnology are not specific to bioengineering processes and products. While there has been widespread activity when it comes to patenting biotechnology discoveries and innovations, with few exceptions the federal government has chosen to monitor the industry with existing regulations. Thus, from the viewpoint of regulatory agencies, the industry is equivalent to any other and therefore operates on an essentially level playing field.

Several federal agencies provide rules and guidance to biotechnology researchers and industry, including the Environmental Protection Agency (EPA), the Food and Drug Administration (FDA), and the United States Department of Agriculture (USDA). Current regulations aim to effectively evaluate potential products in a safe, timely, and cost efficient manner, while keeping an open dialogue between industry and the public in order to address health, safety, and environmental concerns (**BOX 3.2**).

Environmental Protection Agency

The EPA was established in 1970 in response to unchecked industrial pollution. The EPA has the authority to regulate and monitor any activities that might cause pollution of the air, land, and/or water. The agency sets limits on tolerable levels of pollutant released into the environment and the maximum acceptable concentrations of toxic substances in food and water. The EPA also regulates hazardous waste disposal and the use of pesticides and herbicides. Because there are transgenic plants that synthesize such chemical compounds, they fall under EPA scrutiny. Like the FDA (see below), the EPA assesses risks posed by genetically modified organisms (GMOs), including environmental impact and allergenicity. Three congressional acts entail the EPA's involvement with regulating molecular biotechnology:

Box 3.1 Who Owns Your Genes?

Historically, only inventions, designs, and novel ideas were patented. The United States Patent and Trademark Office (PTO) will consider patent approval for anything meeting the following requirements:

- The product or the process used to create the product is **novel**, meaning the product or process is not currently protected with an existing patent.
- The product or process must contain a non-obvious step indicative of independent invention. A discovery made with obvious steps, typically those that are currently in use, is ineligible.
- The invention must be exhibit **utility**, meaning the product or process is useful to humanity in some way.

Today, unveiling information was often enough to apply for a patent because the steps taken in its discovery were ruled to be non-obvious. For example, proteins and their antibodies as well as isolated nucleic acid sequences were patentable until quite recently, including sequences of DNA, RNA, and complimentary DNA determined from messenger RNA sequences. Even though these molecules and sections thereof occur naturally, the processes to isolate and characterize them are not. That is the key element rendering them patentable. Due to the sheer volume of patent applications for partial gene sequences, a standard for utility was also established. It became clear to the PTO that industry had a use for partial sequences. Therefore, these discoveries are likewise eligible for patent protection, even if the function of the derivative gene is unknown. There is a catch: the actual use of a full or partial sequence must be identified in the patent application; a potential use is not sufficient. Nearly 20% of the human genome is now patent-protected.

Some interesting cases have made it to our court systems that highlight the new territory we are covering. The U.S. Supreme Court ruled on the case of John Moore in 1991. Moore was hospitalized in 1976 for hairy cell leukemia treatment at the University of California at Los Angeles. In the process of his treatment, Moore's spleen was removed and he exhibited a rapid progression into remission. Because hairy cell leukemia is rare and nearly always fatal, his spleen tissue was intensely studied for unique properties. The spleen was later found to contain an unusual blood gene that encourages growth of white blood cells. The implications of medical value were not ignored and UCLA patented "Mo's gene" in 1984 and licensed his gene sequence to the Sandoz Pharmaceutical Corporation. Using Mo's gene, Sandoz was able to develop an anticancer drug that boosts the immune system in order to fight off disease. The federal Supreme Court upheld the 1990 ruling of the California Supreme Court that argued allowing Moore to share in profits obtained from his cell lines would open a floodgate of similar litigation. A patient does not own any biological specimens taken from his or her body; once removed they become the property of the medical facility. The court sided with industry, but did recognize "a physician who is seeking a patient's consent for a medical procedure must... disclose personal interest unrelated to the patient's health, whether research or economic, that may affect his medical judgment." Moore passed away in 2001, having spent over a decade fighting for individual rights of gene ownership.

Another similar case was recently profiled in the best-selling book entitled, "The Immortal Life of Henrietta Lacks." Henrietta Lacks fell prey to cervical cancer in 1951 at the age of 30. Her tissues, taken without consent, were used in the development of the "HeLa" cell line. HeLa cells became the first commercially-available biological materials. They have proved immensely useful in medical research as they are apparently immortal. The cultures taken from her over 60 years ago are thriving in laboratories around the world. Notably, HeLa cells were used by Jonas Salk in creation of the Polio vaccine. While Ms. Lacks did not live to see the truth, her family fought unsuccessfully for any claim to profits obtained from HeLa cells.

The recent Supreme Court case, *Association for Molecular Pathology vs. Myriad Genetics*, has modified existing patenting protocols and now forbids patents on naturally occurring gene sequences. However, the intent of the ruling remains subject to interpretation and debate continues on the matter.

the Federal Insecticide, Fungicide, and Rodenticide Act, the National Environmental Policy Act, and the Toxic Substances Control Act (**TABLE 3.1**).

Food and Drug Administration

The FDA is a division of the Department of Health and Human Services. The FDA is responsible for ensuring the safety and purity of food and food additives, drugs, and cosmetics. The FDA regulates their manufacture by testing the final products for contamination. Additionally, the two congressional acts that are administered by the FDA also apply to bioengineered products (**TABLE 3.2**). Moreover, the Federal Food, Drug, and Cosmetic Act and the Public Health Service Act have been periodically amended since their original enactments in order to reflect advancements in technology.

Manufacturers of products regulated by the FDA must also submit several applications necessary for approval to sell the product to consumers. When the product is

Box 3.2 The Coordinated Framework

During the infancy of molecular biotechnology, regulatory oversight was unclear. Early scientific research focused on obtaining knowledge related to human health, disease, and heredity. The National Institutes of Health was the primary force in establishing standards and guidelines for research projects via the Recombinant DNA Advisory Committee. Initially, the NIH advised biocontainment levels for recombinant DNA research as follows:

1. Recommended: Preference for host organisms to be transformed included microorganisms with the least likelihood of transferring their DNA to other species and the least ability to proliferate outside of the laboratory environment if accidentally released.
2. Recommended: If research activity was to involve a known pathogen, self-contained laboratory facilities with negative pressure filtration systems were recommended in order to minimize contamination.
3. Recommended: Research with nonpathogenic microorganisms should utilize self-contained laboratories with high quality air filtration systems, though negative pressure was not a necessary element.
4. Required: "Deliberate release" of any recombinant organism was strictly prohibited.

Over time, as the industry matured, the initial guidelines above were revised to reflect the experiences of biotechnology researchers. For example, the bacterium *Escherichia coli* K-12 became the preferred host microorganism due to its inability to proliferate outside of the laboratory environment. This eliminated concerns of accidental release and its potential consequences. Additionally, no evidence was found that a non-pathogenic gene in its original form would become pathogenic once inserted into its host organism, thus relaxing item 1.

Existing biocontainment measures routinely performed by microbiologists were tested, accepted as suitable safety standards, and adopted by the industry. A consensus was reached that microorganisms with recombinant DNA do not pose additional risks beyond those previously ascertained in comparable non-recombinant strains. Items 2 and 3 were considered redundant in light of this information.

Perhaps the most obvious and dramatic divergence from the original guidelines is with regard to item 4. The deliberate release of GMOs produced via recombinant DNA technology has become a routine step in final commercial approval of new GMO products when they involve agriculture. Field testing is a necessary element in the evaluation of these products in order to ensure their safety to the environment and human health. How was the only required element of the original guidelines pushed to the wayside? Through careful selection, the RAC eventually permitted small scale field tests. Upon analysis of their results, the RAC determined transgenic microorganisms carry out identical bioactivity whether in the laboratory or in nature. They remain confined to the test area and do not persist in the environment beyond a few months. This quelled the fears of significant environmental impact and item 4 is now a thing of the past.

As the field of biotechnology shifted from pure science to a commercial industry, the NIH began to reduce its oversight role and relinquished many aspects of its involvement to the agencies described above. The interagency environment sparked concerns over potential redundancies and inefficiencies which were addressed through a policy known as the Coordinated Framework for the Regulation of Biotechnology, often abbreviated to the **Coordinated Framework**.

Coordinated Framework

The Coordinated Framework was initially launched by the Biotechnology Science Coordinating Committee (BSCC) of the Office of Science and Technology Policy in 1986. The Coordinated Framework was designed to streamline procedures between the three agencies governing biotechnology regulation and clearly delineate responsibilities among them. It also aimed to codify a common language within the agencies to prevent ambiguity or miscommunication. Most notably the Coordinated Framework sought to regulate bioengineered products under existing statutes, which meant final product quality, safety, and efficacy were to be evaluated (instead of manufacturing processes).

Nevertheless conflicts among agencies emerged and in certain cases existing regulations were insufficient to address novel technologies and products. To alleviate these problems, the BSCC was dismantled and replaced with the Biotechnology Research Subcommittee (BRS) of the Committee on Health and Life Sciences. The BRS was created by the Federal Coordinating Council on Science, Engineering, and Technology in 1990.

The functions of the BRS are similar to the BSCC, but membership is larger and drawn from a broader representation of government departments and agencies. While the intent of the BRS is to act upon the drawbacks of the BSCC, it has also fallen short of its purpose. There are still significant issues related to molecular biotechnology that are not addressed by the EPA, FDA, or USDA. Many of the statutes regulated by these agencies originated in an effort to guide and monitor the chemical industry. The unique nature of biological products (they respond to their environment and have the ability to reproduce) was not anticipated by the congressional acts superimposed upon the biotechnology industry. The rapid progress of molecular biotechnology only amplifies the insufficient nature of the Coordinated Framework. It is a challenge that must be overcome, and quickly.

TABLE 3.1

Acts Regulated by the Environmental Protection Agency

Name of Act	Year of Inception	Mandates
Federal Insecticide, Fungicide, and Rodenticide Act (FIFRA)	1947	Requires registration of any pesticide with the EPA that includes data on its composition, effects on human health and other non-target organisms, and its environmental fate.
National Environmental Policy Act (NEPA)	1970	Requires preparation of an environmental impact statement with any major federal project that might adversely affect the human environment.
Toxic Substances Control Act (TSCA)	1976	Authorizes the EPA to monitor and control usage of chemicals in research and commerce in addition to evaluating their potential impact on human health and the environment.

Data from United States Environmental Protection Agency Website. "Laws and Executive Orders." Accessed February 25, 2014. Available at http://www2.epa.gov/laws-regulations/laws-and-executive-orders

pharmaceutical (human or veterinary) this includes the New Drug Application, the Product License Application, and the New Animal Drug Application. Edible products are compared to existing non-GMO varieties where applicable. As long as they are nutritionally equivalent and do not contain any unusual allergens, they are not treated any differently. From the FDA's perspective, the average consumer would have no way of knowing whether they are eating a bioengineered food.

United States Department of Agriculture

The USDA has existed in various forms since 1839, when the Department of State created an Agricultural Division. President Abraham Lincoln established the Department of Agriculture in 1862, although it was not elevated to a cabinet-level department as it is known today until 1889. The main role of the USDA is to promote and protect the agriculture and forestry industries. It oversees and funds research within these industries and regulates the approval of commercial products. It also inspects and ensures the safety and quality of meat, poultry, dairy, and eggs. The USDA encompasses three divisions that monitor the research and release of GMOs (**TABLE 3.3**).

In addition to the separate divisions within the USDA which govern biotechnology regulation, there are also two congressional acts upheld by the department pertaining to the industry: Federal Plant Protection Act and the Virus-Serum-Toxin Act (FPPA; **TABLE 3.4**). Previous Acts have been superseded by the FPPA, including the Plant Quarantine Act of 1912, the Federal Plant Pest Act of 1957, and the Federal Noxious Weed Act of 1974. The FPPA was enacted in order to streamline the myriad regulations affecting agriculture, in addition to modernizing the statutes in consideration of molecular biotechnology. Genetically modified plants are considered potentially invasive, qualifying them as plant pests under the FPPA.

Beyond Regulation: National Institutes of Health Guidelines

The Asilomar Conference launched the involvement of a newly formed entity, the Recombinant DNA Advisory Committee, known as the RAC. The National Institutes of Health (NIH) was responsible for its creation and

TABLE 3.2

Acts Regulated by the Food and Drug Administration

Name of Act	Year of Inception	Mandates
Federal Food, Drug, and Cosmetic Act (FFDCA)	1938	Oversees manufacturing processes in order to ensure quality-control standards. Genetically engineered food products such as those containing recombinant DNA are subject to this Act.
Public Health Service Act (PHSA)	1944	Ensures manufacturers of food and food additives, cosmetics, and drugs are appropriately licensed.

Data from United States Food and Drug Administration Website. "Regulatory Information." Accessed February 25, 2014. Available at http://www.fda.gov/RegulatoryInformation/default.htm

TABLE 3.3

United States Department of Agriculture Divisions Overseeing Biotechnology

Division	Authority
Animal and Plant Health Inspection Service (APHIS)	Reviews proposals for the release of GMOs into the environment and prepares environmental impact statements as required by NEPA. Evaluates the safety of field testing genetically engineered crops on an individual basis.
Committee on Biotechnology in Agriculture (CBA)	Coordinates regulation of biotechnology within the USDA and acts as a liaison with National Institutes of Health research projects pertaining to agriculture.
Food Safety Inspection Service (FSIS)	Regulates the research, labeling, and sale of genetically engineered meat products.

Data from United States Department of Agriculture Websites. "About FSIS." Accessed February 25, 2014. Available at http://www.fsis.usda.gov/wps/portal/informational/aboutfsis and "Advisory Committee on Biotechnology and 21st Century Agriculture (AC21)." Accessed February 25, 2014. Available at http://www.usda.gov/wps/portal/usda/usdahome?contentid=AC21Main.xml

continues to manage its oversight. The RAC was initially tasked with three main objectives:

1. Identify and evaluate potential risks associated with recombinant DNA technology.
2. Develop procedures to minimize the threat of biological contamination via spread of recombinant DNA into unintended populations.
3. Establish guidelines for recombinant DNA researchers that address items 1 and 2.

The ensuing guidelines published by the RAC were directly influenced by the Asilomar Conference proceedings. From 1975 until 1982 the RAC has continuously released and revised its guidelines, to the point they are now a shadow of their initial recommendations. As previously discussed, the proposed guidelines set forth at the Asilomar Conference were much more stringent than those of today (see *The Coordinated Framework* in Box 3.2). As it now stands, the RAC has limited authority over NIH-funded research projects, but none over private sector research. Privately funded projects are subject to voluntary compliance only. Instead of strict RAC involvement, NIH-funded projects are now subject to oversight by **institutional biosafety committees (IBCs)**. IBCs are in-house advisory panels that steer the research and development process. They establish project milestones and procedures, provide legal advice and ethical guidance, and weigh potential risks versus benefits. The RAC requires at least five individuals in each IBC, composed of the following:

1. At least two members with no affiliation with the organization or institution
2. A plant expert
3. An animal expert
4. A biosafety officer

Therefore, the NIH provides the guidelines, the RAC funnels the information into a national framework, and the IBC provides local oversight (**FIGURE 3.1**).

The NIH guidelines are routinely updated in response to current events. For example, a very recent update took effect in February 2013. Strains of the H5N1 avian influenza virus used in research are now subject to additional biocontainment measures in order to ensure public safety (**FIGURE 3.2**). There is mounting evidence these strains can exhibit airborne transmission to mammals. There is real concern for human outbreaks should the virus be released, whether accidentally

TABLE 3.4

Acts Regulated by the United States Department of Agriculture

Name of Act	Year of Inception	Mandates
Federal Plant Protection Act (FPPA)	2000	Oversees field testing of genetically engineered crops and plan pests. Regulates the introduction of GMOs into and within the United States that may pose threats to existing flora.
Virus-Serum-Toxin Act	1913 (amended 1985)	Controls the import, export, production, and distribution of biological veterinary products. Coordinates field testing with APHIS where applicable.

Data from United States Department of Agriculture Website. "Biotechnology Regulatory Services: Regulations." Accessed February 25, 2014. Available at http://www.aphis.usda.gov/wps/portal/aphis/ourfocus/biotechnology?1dmy&urile=wcm%3apath%3a%2FAPHIS_Content_Library%2FSA_Our_Focus%2FSA_Biotechnology%2FSA_Regulations

through insufficient handling or purposefully if it fell into the wrong hands. As an additional change in 2013, the guidelines now encompass not just recombinant DNA but also synthetic nucleic acids that were designed from scratch in laboratories.

Regulation of Genetically Modified Organisms

All three federal agencies—the EPA, FDA, and USDA—play a role in regulating genetically modified organisms and any product with ingredients derived from them, including food, drugs and biologics, and chemicals (**TABLES 3.5** and **3.6**). A **biologic** is any medical

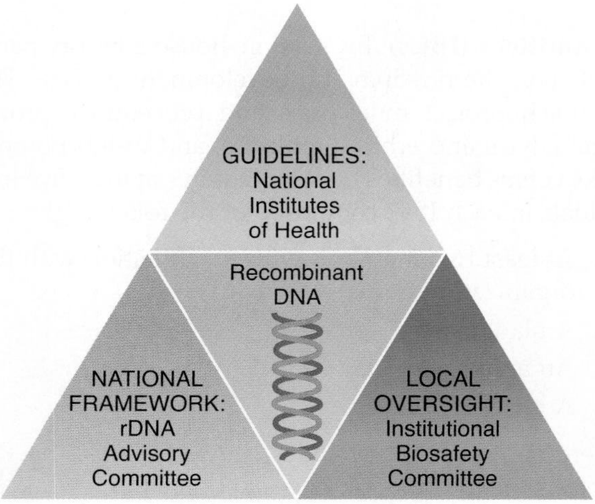

FIGURE 3.1 NIH-funded research projects have a three-tiered approach: the NIH provides guidelines, the RAC constructs a national framework, and the IBC oversees individual projects. Data from Offices of Biotechnology Activities. "The NIH Guidelines and IBC Responsibilities." Presentation at IBC Basics: An Introduction to the NIH Guidelines and the Oversight of Recombinant DNA Research, San Diego, CA. March 12, 2005

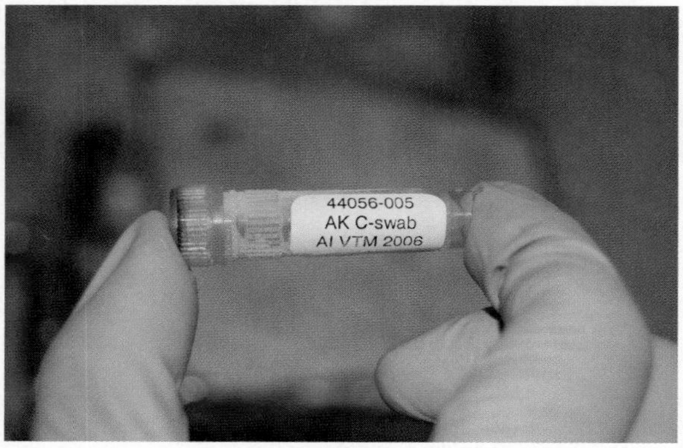

FIGURE 3.2 Restrictions on avian flu samples such as this vial collected by U.S. Fisheries and Wildlife researchers have tightened recently. Courtesy of Don Becker/U.S. Fisheries and Wildlife Service

TABLE 3.5

Regulation of Genetically Modified Organisms by Classification

Genetically Modified Organism	Agency	Act
Plants		
Plant pests	USDA	PPA
Plant-incorporated protectorants	EPA	FIFRA
Plants producing toxins	EPA	TSCA
Animals		
Animals	FDA	FFDCA
Animals producing toxins	EPA	TSCA
Microorganisms		
Microorganisms	EPA	TSCA
Plant pests	USDA	PPA

Data from Fish, Andrew C. and Larisa Rudenko. "Guide to U.S. Regulation of Genetically Modified Food and Agricultural Biotechnology Products." Pew Initiative on Food and Biotechnology. Washington, DC. 2001

TABLE 3.6

Regulation of Commercial Products Derived from Genetically Modified Organisms

Genetically Modified Product	Agency	Act
Food for Human Consumption		
Meat, Poultry, Eggs	USDA	FSIS
Produce	FDA	FFDCA
Food Additives	FDA	FFDCA
Dietary Supplements	FDA	FFDCA
Food for Animal Consumption		
Animal Feed	FDA	FFDCA
Drugs and Biologics		
Human Drugs	FDA	FFDCA
Human Biologics	FDA	PHSA
Animal Drugs	FDA	FFDCA
Animal Biologics	USDA	VSTA
Other Products		
Pesticides	EPA	FIFRA
Cosmetics and Personal Care Items	FDA	FFDCA
Toxic Substances	EPA	TSCA

Data from Fish, Andrew C. and Larisa Rudenko. "Guide to U.S. Regulation of Genetically Modified Food and Agricultural Biotechnology Products." Pew Initiative on Food and Biotechnology. Washington, DC. 2001 preparation involving living organisms or their products, such as a vaccine. Before a genetically modified organism is approved for commercialization, a number of steps must take place. These may be divided into activities taking place during laboratory research, field testing (if applicable), and final product testing, which includes a number of public comment periods (**FIGURE 3.3**). **Stewardship**, the responsible management of all stages of development, from inception to commercialization, is critical.

Obtaining approval of a new GMO is expensive and time-consuming. Cost estimates from start to finish range between $20 and $30 million. The process can take upward of 5 to 8 years. In light of this reality, some detractors of the industry reason these hurdles are too great for small organizations and start-up biotech companies to overcome. In other words, the regulatory system in place is only suited to large corporations and well-funded research organizations. In any case, the entity proposing a new GMO product will weigh these

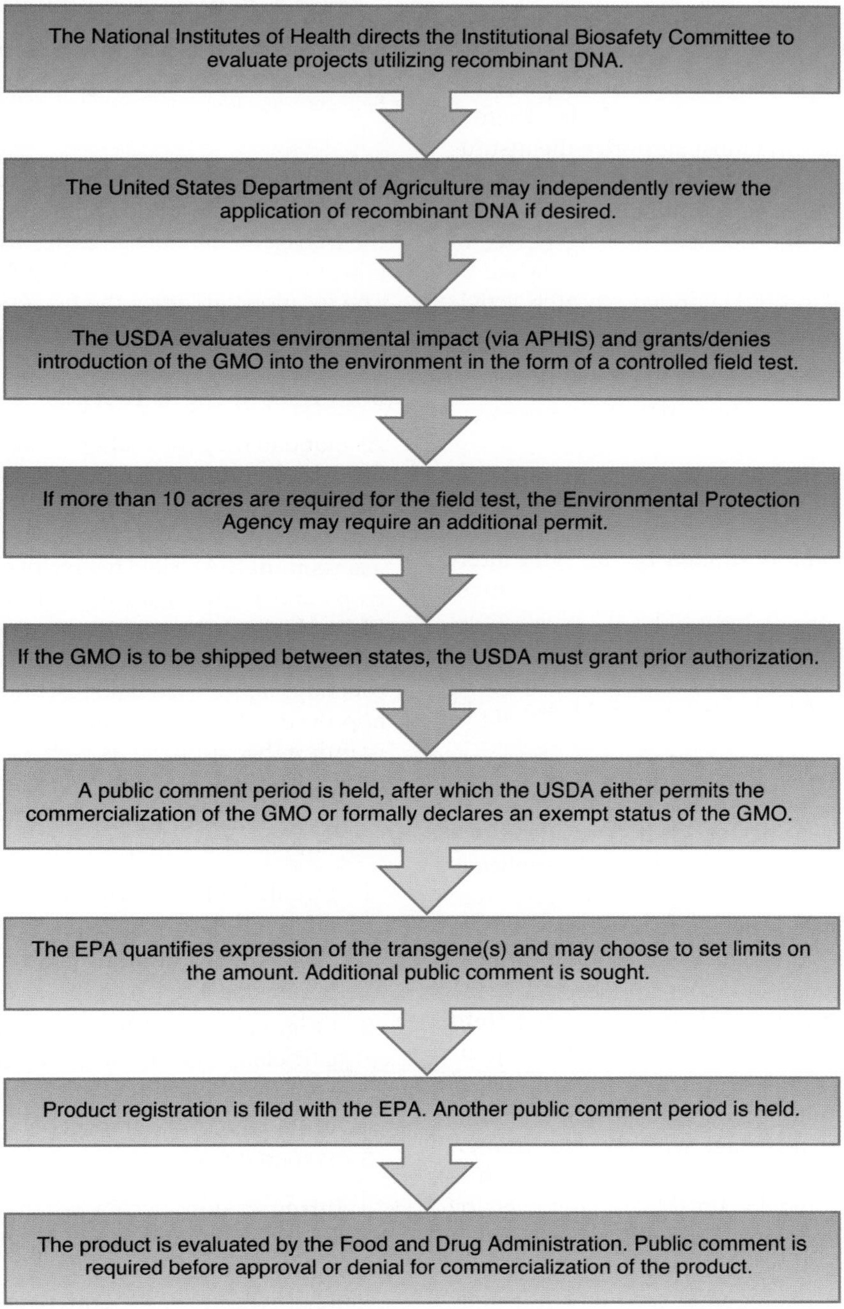

FIGURE 3.3 The steps to commercialization for a GMO product include laboratory research (blue), field testing (green), and final product testing (purple). Data from Fish, Andrew C. and Larisa Rudenko. "Guide to U.S. Regulation of Genetically Modified Food and Agricultural Biotechnology Products." Pew Initiative on Food and Biotechnology. Washington, DC. 2001

costs against the potential profits to be had. If the GMO is a member of the plant kingdom, commodity crops planted on a large scale, such as corn, cotton, and soybeans, are most commonly utilized as host organisms because the return on investment is likely to be greater. Specialty crops have a lower probability of exhibiting the law of diminishing returns as they are typically planted in relatively smaller quantities. Nevertheless, there are examples of these, too.

Laboratory Research

Any project making use of recombinant DNA technology, whether publicly or privately funded, may require approval of the USDA, at minimum. When a project is sponsored by the National Institutes of Health, there is additional scrutiny. The NIH must authorize the institutional biosafety committee so that it may consider and approve or deny the use of recombinant DNA in the project. In some cases the decision-making power of the IBC is intercepted by NIH guidelines established by the RAC. Prohibitions on NIH-funded research activities are limited to the following:

- Manipulation of genes that encode lethal toxins
- Transfer of drug resistant traits that do not occur naturally
- Deliberate release of subject organisms into the environment

Any research laboratory funded by the NIH interested in pursuing one of the above activities must request a procedural review from the RAC, which must then be approved by both the NIH and governing IBC. If release into the environment is sought, the regulations pertaining to field testing below would also apply.

Field Testing

Field testing is required for any GMO considered for release into the environment, including genetically modified crops and microorganisms used in agriculture or bioremediation. Once a transgenic organism grown in the laboratory is considered sufficiently promising for commercialization, it is introduced into the environment upon approval. This is known as a **field test**. Field tests are conducted under controlled conditions (**FIGURE 3.4**). Procedures must be put in place to minimize environmental impact both during and after a field test. A GMO in this stage will often go through small scale field testing (less than 10 acres), which is approved and monitored by the USDA, followed by large scale field testing (greater than 10 acres) that is regulated by the EPA.

A separate field test is required each time a genetically modified product is proposed for public use. This is true even if a previously approved transgene is utilized for more than one GMO product, because in the case of plants insertion of the transgene into the genome is random. The exact location within the plant genome where the transgene ultimately resides will vary with each new GMO, even if it is introduced into the same crop variety. The location can affect the expression of the transgene, so each case must be tested and evaluated separately.

FIGURE 3.4 APHIS personnel inspect genetically modified corn during a field test. Courtesy of R. Anson Eaglin/Biotechnology Regulatory Services/USDA-APHIS

Small Scale Field Testing

As mandated by the National Environmental Policy Act (NEPA), environmental impact studies are a necessary element when a project might negatively affect the natural landscape. This is known as an **environmental assessment** (EA). Field testing of genetically engineered crops and microorganisms, referred to by Animal and Plant Health Inspection Service (APHIS) as **regulated articles**, would apply. As part of this process, the USDA evaluates a proposed field test and grants permits when approved. More specifically, APHIS is responsible for granting this approval as well as monitoring all stages of field testing. The major considerations when APHIS evaluates a proposed field test include:

Purpose of the proposed field test

1. Detailed information on the host organism containing the transgene
2. GMO impact on non-target and beneficial organisms
3. Origin and characteristics of inserted gene(s): specific elements such as promoters, markers, gene(s), etc., and their order
4. Product(s) of the transgene(s)
5. Potential for introgression

Introgression occurs when the inserted gene(s) unintentionally jump to additional species, usually through outcrossing of a GMO crop with unrelated plants or wild relatives. **Outcrossing** takes place when gene pools merge, which occurs when closely related organisms exchange genetic material through repeated hybridization over several generations. The potential for

FIGURE 3.5 An APHIS inspector verifies the isolation distance of a genetically modified corn field test using a laser distance measuring device. Courtesy of R. Anson Eaglin/ Biotechnology Regulatory Services/USDA-APHIS

this type of gene contamination manifests in a variety of ways in addition to outcrossing. For example, pollinators such as insects, birds, and bats may transfer GMO material across long distances. If approved, the field test must include provisions to minimize the risks of such occurrences, such as minimum distances between test fields and neighboring fields, usually at least one mile (**FIGURE 3.5**). An additional 25-foot perimeter of fallow ground must surround the entire test site. Following a field test, the entire site is kept fallow for an entire season before any new plantings can occur. When a field test is approved by the USDA, the decision is known as a **finding of no significant impact (FONSI)**, meaning in the opinion of the USDA the field test is not expected to alter the environment.

It is possible to fast track a field test application through APHIS. This alternative system is known as **notification**. To be eligible for the notification option, certain criteria must be met:

1. The host organism must be an eligible species that has been previously field tested and thoroughly evaluated. These species include corn, potatoes, soybeans, tobacco, cotton, and tomatoes.
2. Transgene function must be known. The expressed protein cannot confer any plant disease.
3. The transgene must be inserted into the nucleus of the host organism, as this minimizes any possibility of introgression.
4. If the GMO is intended for consumption, the transgene cannot express any known toxins, cause disease, or produce any substance used in pharmaceutical applications.
5. A transgene obtained from a plant virus cannot have the potential to create a new virus.
6. The transgene cannot come from an animal virus, whether human or not.

When notification is successful, APHIS permits the field test within 30 days of application.

Large Scale Field Testing

In the event the field test covers more than 10 acres, the EPA has the option of requiring an additional permit, known as an **Experimental Use Permit (EUP)**. The EPA is primarily concerned with toxic effects of herbicide and pesticide products expressed by GMO crops, firstly on the environment and ultimately on human health. Therefore EPA consideration for large scale field testing focuses more on the protein products expressed by a GMO rather than the organism as a whole. To warrant issue of an EUP, the EPA will request and assess the following information:

1. Effects on human health: known implications of traditional and existing herbicide and/or pesticide products of similar origin
2. Ecological impact: factors including introgression, the potential for increased or decreased resistance to herbivores and disease, and effects on non-target organisms
3. Environmental fate of the herbicide and/or pesticide: the quantity released during and after the GMO's lifespan and how long the herbicide and/ or pesticide persists in the environment (soil, air, and water)
4. Biochemistry and ecology of the transgenic plant and inserted gene(s)
5. Expression levels of the transgene(s) and bioactivity of gene product(s)

To date more than 6,500 field tests have taken place within the United States. A field test for one GMO may take place at several locations. This is valuable to ascertain the viability of crops or microorganisms in different soils and climates. Also, field testing at more than one site improves statistical validity of data obtained. When multiple sites are involved in a field test, it may necessitate even greater USDA involvement. The agency must grant permission if the test subject will be transported across state borders. Over 18,000 test sites have been used for the U.S. field trials that have been carried out thus far. This chapter's *Focus on Careers* features the role of APHIS inspectors in field testing of genetically modified organisms.

Final Product Testing

It is rare for a product to make it through laboratory and field testing and not receive approval for commercial

FOCUS ON CAREERS
APHIS Inspectors

The Animal and Plant Health Inspection Service (APHIS) division of the USDA was established in 1972. APHIS serves to promote and protect agriculture of the United States, carry out wildlife management activities and administer the Animal Welfare Act, and most importantly to this discussion, regulate genetically modified organisms. Inspectors hired by APHIS are tasked with monitoring the various stages of small and large scale field testing in addition to overseeing the safe import and export of agricultural products.

APHIS inspectors may serve in domestic or foreign settings. They bear the official title of "Plant Inspection Technician." Foreign positions are often involved in the Agriculture Inspection Quarantine program (AQI), with the goal of limiting the spread of agricultural pests and diseases across borders. Domestic positions are likewise involved in quarantine programs, but with a focus on containing GMOs during field testing. As these are federal positions, they are offered in pay grades, which are strict categories of compensation levels based upon the applicant's knowledge, skills, and abilities. Plant inspection technicians may be hired at three different pay grades. At minimum, potential employees must hold a bachelor's degree that included at least 16 semester hours in courses such as biology, plant pathology, entomology, zoology, botany, forestry, chemistry, agriculture, or physics. Desirable qualities of the applicant include the ability to distinguish among agricultural products, their origins, and potential pests and diseases that may affect different species. Plant inspection technicians should be able to collect and interpret data, diplomatically cooperate with stakeholders, and engage in planning and oversight of projects. Besides this technical knowledge, APHIS looks for employees with excellent written and verbal communication skills, familiarity with GPS mapping techniques, and the ability to interpret and apply applicable regulations.

Full-time, paid "Student Trainee" positions are also available within APHIS in a number of specialties, including biology. The trainee positions are designed to introduce young career-minded individuals to the regulatory agency and are term-limited appointments. Positions with APHIS, as with all federal job opportunities, are advertised on the website usajobs.gov.

use. Following a field test, the USDA is also responsible for final approval to market the GMO for sale to the public. Before the agency makes its decision, a 60-day public comment period is held to garner input from outside sources. The EA and FONSI must be made available for review. Should a member of the public decide to challenge approval of the GMO, the USDA is required to delay its decision until the challenge is vetted. The USDA may deny approval, approve commercialization, or declare the product is exempt from regulation. If exempt, the organism is no longer considered a regulated article by APHIS and has achieved **non-regulated status**. Under non-regulated status, the GMO producer and any farmer wishing to plant the GMO no longer needs permits from the USDA each time the GMO is planted. The transgenic crop is considered identical to its non-GMO counterpart, with the exception of the inserted gene(s). In the event of a non-regulated status request, a final EA must be conducted by the USDA utilizing data from the large scale field testing overseen by the EPA and the public comments obtained. In any case, when the GMO is approved for commercial production and sale, at any time the USDA has the authority to remove the product from the market if necessary. A recent House of Representatives resolution has temporarily altered some aspects of USDA involvement (**BOX 3.3**).

The EPA also weighs in on final approval of a new GMO and opens its own period of public comment. Like the USDA, the EPA can also grant non-regulated status, which allows the GMO to be produced, distributed, and sold without any additional permits. The agency will test the levels of transgene expression. It is within the EPA's authority to limit the amount of gene product produced in each individual GMO, particularly if it may alter the ecosystem. For example, there may be concern that a plant expressing an herbicide will confer resistance to weedy plants. In such a case, the expression

Box 3.3 Rider 735 Changes the Regulatory Landscape...For Now

On March 26, 2013, President Obama signed a continuing resolution spending bill known as H.R. 933 into law. Also known as the Consolidated and Further Continuing Appropriations Act of 2013, this bill was created in order to fund the federal government through the end of that fiscal year. As is common with spending bills, ancillary items collectively referred to as "riders" are buried within, and in this particular bill Section 735 was one such rider to be passed. Officially dubbed the "Farmer Assurance Provision," Section 735 is a few paragraphs of the 575-page bill, and contains provisions for the biotechnology industry that effectively amend the Farmland Protection Policy Act (FPPA). Specifically, Section 735 requires the U.S. Secretary of Agriculture "notwithstanding any other provision of law, upon request by a farmer, grower, farm operator, or producer, immediately grant temporary permit(s) or temporary deregulation," even if a federal court has ordered the planting or sale of seeds to be halted. In effect, the Secretary must ignore court rulings in favor of continued GMO plantings and sale of their seeds. Under the rider, the public is limited in seeking redress from the court system as judges would have virtually no legal authority to halt the process. The rider has been popularized in the media as the "Monsanto Protection Act" as it affords biotech companies temporary permits to plant and grow GMO crops while pending judicial review. Republican Senator Roy Blunt of Missouri openly acknowledges drafting the Section 735 language under the advisement of Monsanto, noting, "what it says is if you plant a crop that is legal to plant when you plant it, you get to harvest it."

Previously, any public comment period resulting in legal challenges, any court-ordered challenge to a USDA FONSI status, or any reexamination of commercial approval on behalf of the USDA due to health or environmental findings would suspend the permitting process, halt, or even rescind final commercial approval until the legal case was settled. Section 735 has lifted this requirement, at least temporarily. It is set to expire within six months. Proponents of the rider, primarily biotechnology industry interests, argue it is necessary to increase production of scarce commodities, such as sugar beets, while opponents of Section 735 suggest it is simply intended to deregulate genetically modified crops. The temporary nature of the rider means the ultimate fate of this issue is still uncertain.

must be modified in order to meet the demands of the agency. The EPA also administers registration for bioengineered products when they contain herbicides, pesticides, toxins, and/or allergens, even when they obtain non-regulated status. It is the responsibility of the product's developers to apply for EPA registration where applicable. Registration requires evidence there are no "unreasonable adverse effects" caused by the product. Public comment is sought once again prior to product registration. Lastly, the EPA maintains oversight over genetically modified animals that are not intended as food sources, such as insects modified to control agricultural pests. Upon final commercial approval, the EPA may still remove the product from the market at any time if any problems were to arise.

Finally, the FDA will characterize any proposed GMO that falls under its jurisdiction. This includes not only the transgenic crops and microorganisms discussed thus far, but also any genetically modified animal, whether developed for human consumption or otherwise. All transgenic animals and their subsequent offspring, regardless of their intended use, are considered "new animal drugs" by the FDA. The FDA is legally mandated to validate the safety of a new GMO by demonstrating it poses "no unreasonable adverse effects on the environment or human health when used according to label directions." The specific characteristics of each GMO determine how the FDA assesses its safety. It will depend on the host crop, origin of the donor gene(s), the construction of the transgene(s), and the gene product(s) of the GMO. If the GMO expresses a pharmaceutical product, the FDA also evaluates it as a drug, which opens the door to additional scrutiny (**BOX 3.4**). While there is no standard battery of tests for all GMOs under consideration, novel products created through biotechnology may be evaluated for the following:

1. Chemical composition
2. Allergenicity and toxicity
3. Novel metabolites
4. Nutritional profile
5. Production, processing, and disposal methods*
6. Biocontainment measures*
7. Gene source
8. History of use

(*applies to transgenic animals only)

Gene source and history of use important factors when determining the safety of bioengineered products under consideration for market approval. For example, the FDA is interested in knowing the odds of the transgene entering into unintended organisms. If the transgene is currently used in existing products, the precedent has been set and approval may be simplified. Public comment is sought once more during FDA evaluation and considered along with product characteristics prior to final approval for commercialization. Like all other agencies, the FDA has the authority to remove the product from the market if it is later deemed unsafe.

Box 3.4 FDA Drug Regulations

In addition to oversight by the Coordinated Regulatory Framework, all molecular biotechnology products designed for pharmaceutical applications are subject to a drug approval process regulated by the Food and Drug Administration. There are also instances of genetically modified organisms unintended for use as drugs that must follow the same approval process, in particular transgenic animals. The FDA mandates several phases of drug testing prior to approval.

- Phase I Testing: Safety
  A group of healthy volunteers (around 20 to 80 individuals) take the medicine. Side effects are noted and dosage levels are established. As long as there are no severely detrimental side effects, the process moves on to phase II.
- Phase II Testing: Efficacy
  The medication is tested on a group of volunteer patients who have the disease for which the product is intended to treat. This group is larger, composed of 100 to 300 individuals. If results are favorable, phase III is initiated.
- Phase III Testing: Benefits Comparison
  An even larger group of patients, between 1,000 and 3,000, participates in a double blind study of the medication. Double blind is a type of study format that ensures neither the patients nor the researchers involved are aware of who is taking a placebo and who is taking the drug being investigated. As long as the drug appears statistically superior to the placebo, **new drug application (NDA)** may be sought.
- Phase IV Testing: Continuous After-Market Monitoring
  Once NDA is obtained and the medication is sold on the market, the FDA continues to monitor the drug indefinitely for its long-term safety. Biologics must also be approved by the FDA, including those produced through biotechnology methods. Rather than an NDA, the company seeking approval of a biologic must file for a **Biological License Agreement (BLA)**.

Whether a company seeks an NDA or BLA, the FDA will inspect the testing laboratories and manufacturing facilities in order to verify **good laboratory practice (GLP)** and **good manufacturing practice (GMP)**, respectively. GLP requires pharmaceutical laboratories to follow standard operating procedures (SOPs), which are established written protocols. The laboratories must also be adequate for the work being performed, give proper care to any research animals, carry out approved toxicity tests, and record detailed notes and data sets. GMPs are similar but apply to manufacturing facilities. They require written protocols, prescribed quality testing throughout the manufacturing process, and ensure the safety of the final product.

Labeling

Organisms modified through recombinant DNA technology that are approved for commercialization are indeed newly created living entities and, therefore, what some might call "unnatural." But as far as the federal government is concerned, the final products are just the same as any other living thing. The statutes governing GMOs were not created specifically to address these novel organisms; rather, they were already in place to regulate existing products. The case of labeling GMOs underscores the viewpoint that they are equivalent to non-GMOs.

The FDA regulates the labeling of genetically modified organisms in food products. This includes both whole foods, such as produce, and processed foods with GMO ingredients. The FDA considers genetically engineered food to be equivalent to traditional varieties in most cases. Under current rules established in 1992, GMOs do not have to be labeled, unless it meets one of the following criteria:

1. The GMO contains an allergen that is not normally present in the non-GMO variety.
2. The GMO has significantly different nutritional value than that of the non-GMO variety.

This is not the case in over 60 countries including those of the European Union, where all GM food products must be labeled (**FIGURE 3.6**). Even foods derived from GM products must be labeled; for example, corn oil extracted from GM corn would require a label.

In 2012, the state of California proposed an amendment to their state constitution requiring GMOs to be labeled. Known as Proposition 37 (Prop 37), the amendment narrowly failed to pass voter approval after an onslaught of media hype. Residents of California were exposed to countless advertisements from both sides of the issue, although opponents of GM labeling significantly outnumbered the proponents (**TABLE 3.7**). Opponents of the amendment spent an estimated $46 million to defeat the measure, including biotechnology organizations and chemical companies. By comparison, proponents of Prop 37 campaigned with approximately $9 million. Proponents were represented by organic food companies and individuals. Whether this discrepancy in funding was a direct cause of voter outcome is a matter of debate (**FIGURE 3.7**).

The California ballot initiative highlights the opposing viewpoints related to GM labeling. Consumer advocacy groups and individuals concerned over potential health

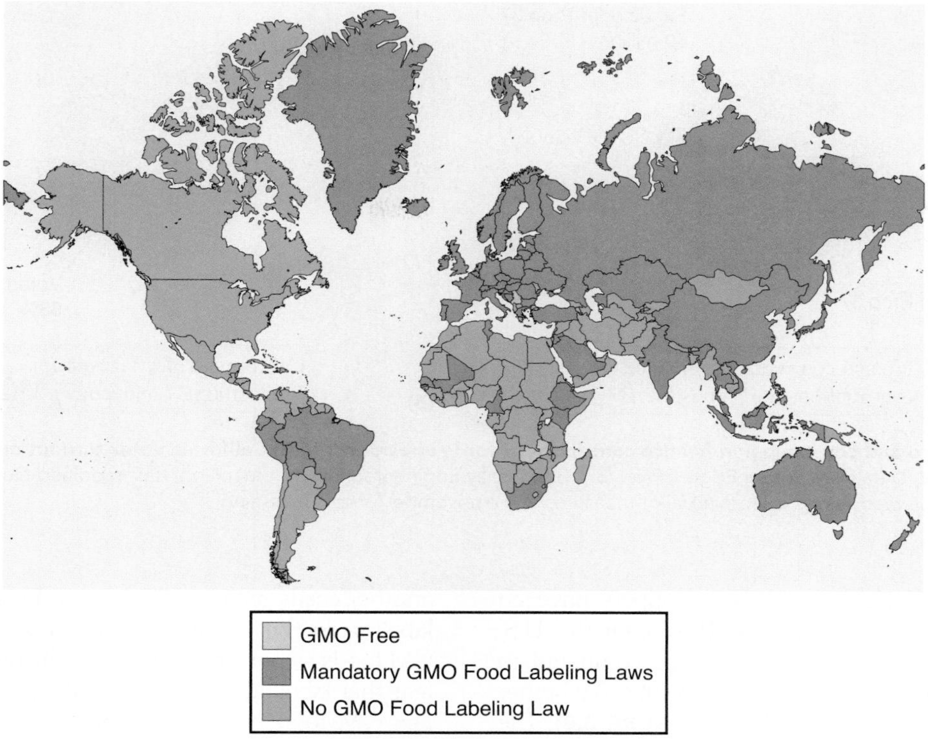

FIGURE 3.6 A worldwide map of countries that require labels on GMO products. Data from: Center for Food Safety

TABLE 3.7

Top 10 Pro and Con Funding Sources of Proposition 37, California 2012*

	Pro		Con	
Rank	**Contributor**	**Total $**	**Contributor**	**Total $**
1	Mercola.com Health Resources, LLC	$1,199,000	Monsanto Company	$8,112,867
2	Kent Whealy	$1,000,000	E.I. DuPont de Nemours & Co.	$5,400,000
3	Nature's Path Foods USA, Inc.	$660,709	PepsiCo, Inc.	$2,485,400
4	Dr. Bronner's Magic Soaps All-One-God-Faith Inc.	$620,883	Grocery Manufacturers Association	$2,002,000
5	Organic Consumers Fund	$605,667	Kraft Foods Global, Inc.	$2,000,500
6	Ali Partovi	$288,975	Bayer CropScience	$2,000,000
7	Mark Squire	$258,000	Dow Agrosciences LLC	$2,000,000
8	Weha Farm, Inc. DBA Lundberg Family Farms	$251,500	BASF Plant Science	$2,000,000
9	Amy's Kitchen	$200,000	Syngenta Corporation	$2,000,000
10	The Stillonger Trust, Mark Squire Trustee	$190,000	Coca-Cola Company	$1,700,500

*Though Prop 37 was on the California ballot, funding may come from any state.
Data from Voter's Edge. "Prop. 37: Genetically Engineered Foods." MapLight, last modified November 6, 2012, http://votersedge.org/california/ballot-measures/2012/november/prop-37/funding

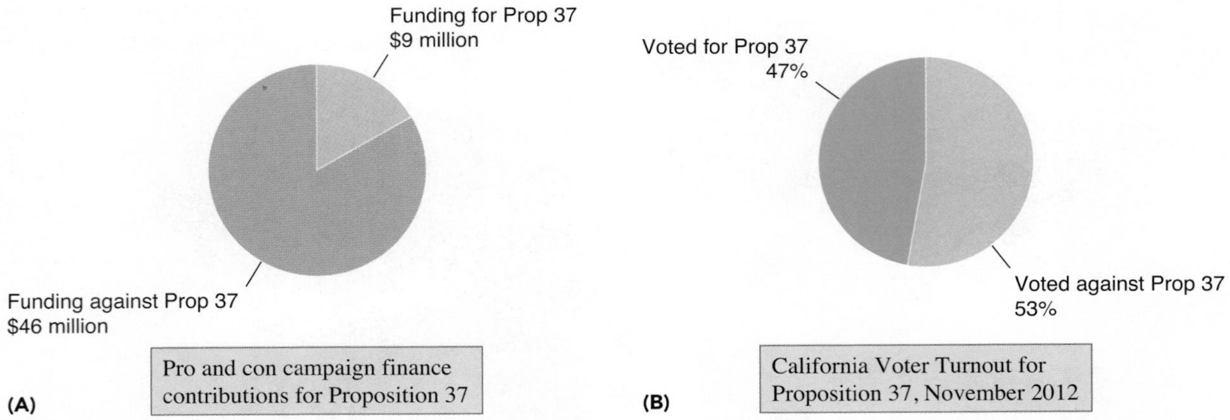

FIGURE 3.7 (A) Pro and con campaign finance contributions for Proposition 37. (B) California voter turnout on Proposition 37, November 2012. Data from Voter's Edge. "Prop. 37: Genetically Engineered Foods." MapLight, last modified November 6, 2012. Available at http://votersedge.org/california/ballot-measures/2012/november/prop-37/funding

effects recommend labeling so that shoppers have an informed choice at the grocery store. They argue that U.S. market consumers are at a disadvantage compared to countries that provide a choice when it comes to whether to purchase and consume GM foods (**FIGURE 3.8**). The food industry and biotech companies responsible for creating GM food products oppose labeling for two primary reasons. First, required labeling would increase costs, as processed food products would have to be tested for GM constituent concentrations. Packaging would have to be redesigned to include the required labeling information, further adding to costs. Labeling also opens the door to potential lawsuits in the event a food item was mislabeled. Such a situation could lead to recalls as well, another costly measure. The second reason they oppose labeling is the negative connotations associated with GM foods as perceived by the public (**FIGURE 3.9**). They fear that labeling would essentially put a black mark on any GM food product and devastate the marketability of these products. This concern is valid, considering many food producers now conspicuously label non-GM foods with the suggestion they are somehow healthier or safer to eat, further confusing the issue (**FIGURE 3.10**). Consumer advocates believe the very apprehension exhibited by food companies to label GM products is enough to prove these products are inherently different. From the standpoint of advocacy groups, if the food industry doesn't want to label GM products, it must indicate they have something to hide.

Although narrowly voted down, Proposition 37 has not settled the labeling debate. Similar labeling laws have been proposed in over 30 states, representing the majority of states in the country (**FIGURE 3.11**). Retailer Whole Foods Market announced that by 2018 their U.S. and Canadian stores will voluntarily label all GM products,

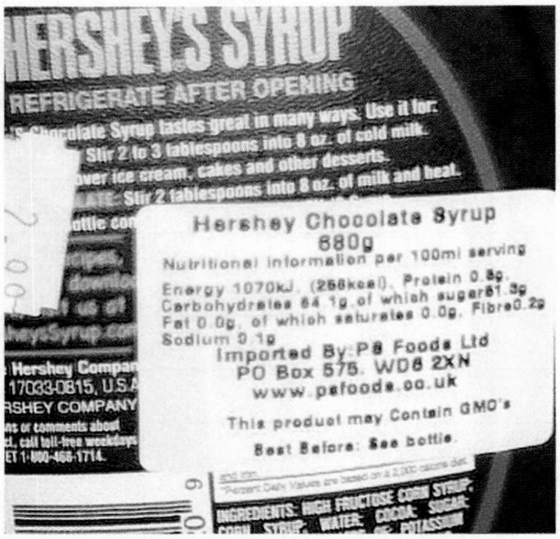

FIGURE 3.8 "This product may contain GMOs." Because of GMO labeling requirements, foods produced in the United States and exported elsewhere may need additional labels, such as this one on a Hershey's chocolate syrup bottle for sale in the United Kingdom. Due to the high fructose corn syrup content, the product likely contains GM corn derivatives. © Heather Gilderdale

FIGURE 3.9 A 2012 protest in support of Proposition 37 in California demonstrates the public's negative perception of GMOs. © Reuters/Stephen Lam/Landov

invoking the consumers' "right to know." According to a Whole Foods press release on the matter, this decision was in response to customer requests. It is apparent that U.S. consumers prefer the labeling of GM foods. A 2010 Reuters poll revealed 93% of individuals polled believe GM foods should be labeled as such. Notably, the likelihood an individual supported GM food labeling decreased with both age and education level (**FIGURE 3.12**).

As is the case with food products, healthcare products manufactured with biotechnology are deemed equivalent to traditional products. They fall under the same jurisdiction as all other drugs and cosmetics overseen by the FDA. Thus, it is not the manufacturing process that is scrutinized; it is the safety of the final product going to market that concerns regulatory agencies. With the number of states proposing new labeling requirements, it will be interesting to see the changes taking place in retail environments.

FIGURE 3.10 Some food producers highlight the absence of GMO ingredients in their products. The inclusion of third party verification seals, as seen here, is a recent trend. The USDA Organic seal indicates the products lack GMOs.

FIGURE 3.11 Over 30 states have initiated proposed amendments that would require GMO labeling. Data from Swanson, Abbie Fentress. "GMO labeling laws on deck in the Midwest." KBIA Mid-Missouri Public Radio, March 18, 2013. kbia.org/post/gmo-labeling-laws-deck-midwest

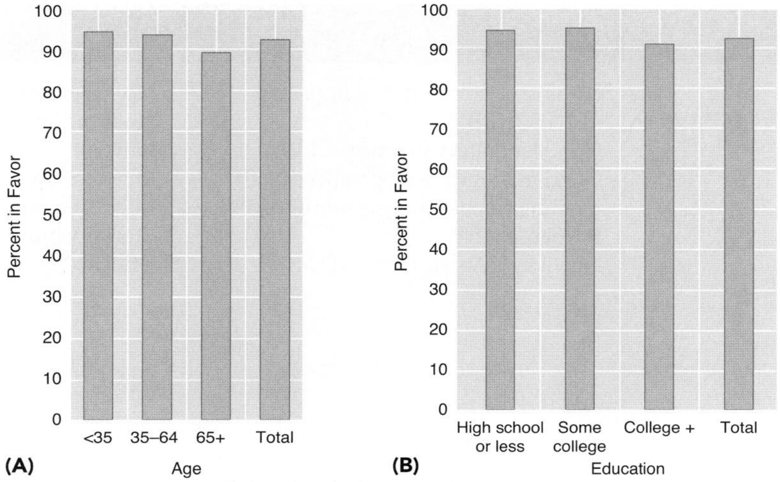

FIGURE 3.12 (A) The number of supporters for GM food labeling requirements decreases with age. (B) The number of supporters for GM food labeling requirements decreases with education level. Data from Thomson Reuters. "National Survey of Healthcare Consumers: Genetically Engineered Food." October 2010. Retrieved from http://www.factsforhealthcare.com/pressroom/NPR_report_GeneticEngineeredFood.pdf

SUMMARY

- Three federal agencies provide rules and guidance to biotechnology researchers and industry, including the Environmental Protection Agency (EPA), the Food and Drug Administration (FDA), and the United States Department of Agriculture (USDA).
- Federal regulations aim to effectively evaluate potential products in a safe, timely, and cost efficient manner, while keeping an open dialogue between industry and the public in order to address health, safety, and environmental concerns.
- The EPA assesses risks posed by genetically modified organisms, including environmental impact and allergenicity. The EPA enforces three congressional acts applicable to the biotechnology industry: the Federal Insecticide, Fungicide, and Rodenticide Act, the National Environmental Policy Act, and the Toxic Substances Control Act.
- The FDA is responsible for ensuring the safety and purity of food and food additives, drugs, and cosmetics. Two congressional acts are administered by the FDA and all apply to bioengineered products: the Federal Food, Drug, and Cosmetic Act and the Public Health Service Act. The FDA also requires the New Drug Application, the Product License Application, and the New Animal Drug Application when the product is pharmaceutical (human or veterinary).
- The USDA encompasses three divisions that monitor the research and release of genetically modified organisms: the Animal and Plant Health Inspection Service (APHIS), the Committee on Biotechnology in Agriculture (CBA), and the Food Safety Inspection Service (FSIS). Two congressional acts upheld by the department pertain to the biotechnology industry: Federal Plant Protection Act and the Virus-Serum-Toxin Act.
- Initial attempts to regulate the biotechnology industry were self-imposed and took shape at the Asilomar Conference. The National Institutes of Health (NIH) became involved through the formation of the Recombinant DNA Advisory Committee, which served to identify and evaluate risks and to develop safeguards in order to minimize these risks. The RAC continues to operate in the oversight of NIH-funded projects, while private entities voluntarily engage in Institutional Biosafety Committees that serve similar purposes. The NIH has steadily relinquished regulatory authority with the advent of the Coordinated Framework, which shifted oversight responsibilities to the FDA, USDA, and EPA.
- The biotechnology engages in stewardship to ensure the responsible management of all steps involved to commercialize a genetically modified organism (GMO). The process begins with laboratory research which is followed by small and large scale field testing if the GMO is a plant. Final product testing is then conducted. Under certain circumstances, the final product may require labeling to indicate it contains GMOs.
- A final product containing GMOs must be labeled in the event either of the following criteria is met: if the GMO contains an allergen that is not normally present in the non-GMO variety and/or if the GMO has significantly different nutritional value than that of the non-GMO variety.
- Required GMO labeling is beneficial as it allows shoppers to have an informed choice when making purchasing decisions. It may also be considered a benefit because it would put the United States on par with other developed countries that do require GMO labeling. One drawback of required labeling is it would likely increase costs, as processed food products would have to be tested for GM constituent concentrations. Packaging would have to be redesigned to include the required labeling information, further adding to costs. Another drawback is the potential for lawsuits and/or recalls in the event a product was mislabeled. The negative connotations perceived by the public are an additional drawback to required labeling of GMOs.

KEY TERMS

Biologic
Biological License Agreement (BLA)
Coordinated Framework
Environmental assessment (EA)
Experimental Use Permit (EUP)
Field test
Finding of no significant impact (FONSI)
Good laboratory practice (GLP)
Good manufacturing practice (GMP)
Institutional biosafety committee (IBC)
Introgression
New Drug Application (NDA)
Non-regulated status
Notification
Novel
Outcrossing
Regulated article
Stewardship
Utility

DISCUSSION QUESTIONS

1. Describe the Coordinated Framework. Do you believe its structure is appropriately applied to molecular biotechnology? Is it effective to use existing regulations, or do you believe laws should be enacted specifically for the industry, its processes and practices, and/or final products?
2. What is the status of GMO labeling in your state? Visit "Just Label It!" (http://www.justlabelit.org/state-initiatives/) for an interactive map. Click on each state to see details on any initiatives related to GMO labeling.
3. At the time this book was written, Section 735, the "Monsanto Protection Act," was just signed into law. Research the status of this bill rider. Is it still in effect? Has it been signed permanently into law, or was it abandoned? Have any similar riders or laws been introduced and/or passed?

REFERENCES

Animal and Plant Health Inspection Service. "Biotechnology." United States Department of Agriculture. Available at http://www.aphis.usda.gov/biotechnology/index.shtml

Department of Health and Human Services, National Institutes of Health. "Guidelines for Research Involving Recombinant DNA Molecules." *Federal Register* 47 (21 April 1982): 17180–17198.

Fish, Andrew C., and Larisa Rudenko. "Guide to U.S. Regulation of Genetically Modified Food and Agricultural Biotechnology Products." *Pew Initiative on Food and Biotechnology*. Washington, DC. 2001.

Knowles, David. "Opponents of Genetically Modified Organisms in Food, or GMOs, Rail Against Provision That Would Limit the Courts' Ability to Stop Food Producer Monsanto from Growing Crops Later Deemed Potentially Hazardous." *The New York Daily News*, March 25, 2013.

Lowery, Kate and Libba Letton. "Whole Foods Market Commits to Full GMO Transparency." (ND). Available at http://media.wholefoodsmarket.com/news/whole-foods-market-commits-to-full-gmo-transparency

McClellan, Dennis. "John Moore, 56; Sued to Share Profits from His Cells." *Los Angeles Times* (Los Angeles, CA), Oct. 13, 2001.

Office of Biotechnology Activities. "The NIH Guidelines and IBC Responsibilities." Presentation at IBC Basics: An Introduction to the NIH Guidelines and the Oversight of Recombinant DNA Research, San Diego, California, March 12, 2005.

Office of Science and Technology Policy. "Coordinated Framework for Regulation of Biotechnology." *Federal Register* 51 (26 June 1986): 23302.

Sesana, Laura. "Frankenveggies: Monsanto Protection Act Passes Senate." *The Washington Times*, March 25, 2013.

Sheets, Connor Adams. "Furor Growing Against Obama Over 'Monsanto Protection Act.'" *International Business Times*, March 27, 2013.

Skloot, Rebecca. *The Immortal Life of Henrietta Lacks*. New York: Broadway Books, 2010.

Swanson, Abbie Fentress. "GMO Labeling Laws on Deck in the Midwest." *KBIA Mid-Missouri Public Radio*, March 18, 2013. Available at http://www.kbia.org/post/gmo-labeling-laws-deck-midwest

Tarver-Wahlquist, Sarah. "Who Requires Labels?" *Green American*, April/May 2012.

Thelen, Jennifer. "FDA Regulation of Food and Drug Biotechnology." Harvard Law School LEDA, Cambridge, Massachusetts, 2010.

Thomson Reuters. "National Survey of Healthcare Consumers: Genetically Engineered Food." October 2010. Retrieved from http://www.justlabelit.org/wp-content/uploads/2011/09/NPR_report_GeneticEngineeredFood-1.pdf

Voter's Edge. "Prop. 37: Genetically Engineered Foods." *MapLight*, last modified November 6, 2012, http://votersedge.org/california/ballot-measures/2012/november/prop-37/funding.

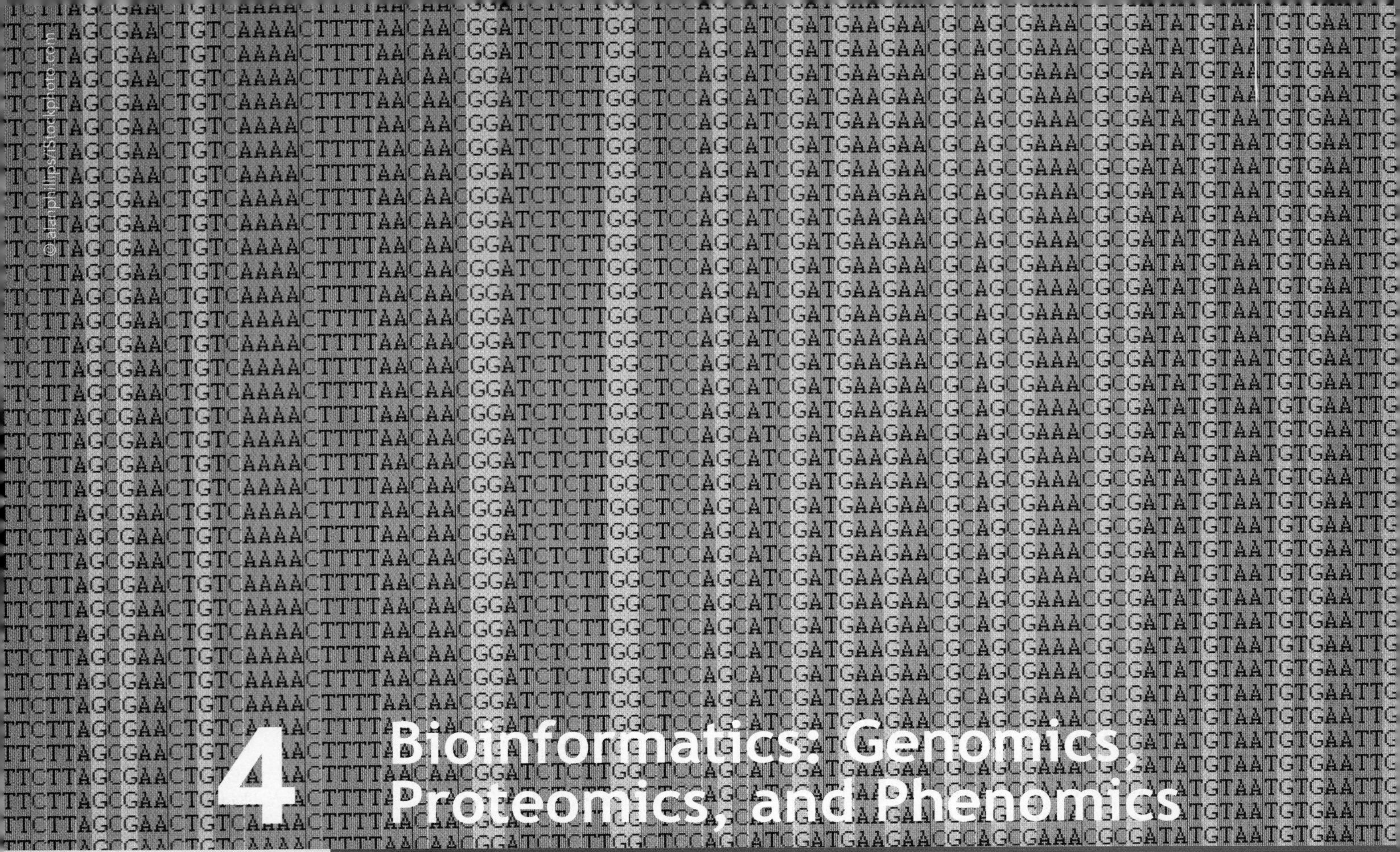

4 Bioinformatics: Genomics, Proteomics, and Phenomics

LEARNING OBJECTIVES

Upon reading this chapter, you should be able to:

- Relate bioinformatics, computational biology, and data mining.
- Provide examples of databases used in bioinformatics.
- Describe genomics and its intended outcomes.
- Identify the tools utilized in genomics.
- Discuss the revelations of comparative genomics.
- Compare and contrast proteomics with phenomics.
- Explain alternative splicing and its importance in computational biology.

Fortuitously, the Biotechnology Age predicted by Karl Ereky has coincided with the Information Age. We would not be where we are today if it were not for the computing power we have at our fingertips. It is highly improbable we would ever have obtained the amount of information we now have readily available if it weren't for the processing power of today's computers. And even if we were able to get our hands on the enormous and exponentially growing amount of data, how else but through microchips and binary code would we be able to comb through it, statistically analyze it, and even manipulate it by design? For example, imagine trying to find a single nucleotide polymorphism that distinguishes two individuals without an appropriate software program. To accomplish this task by hand would be an insurmountable task, much like the proverbial needle in the haystack. Only now the needle is a tiny portion of a molecule invisible to the naked eye, and the haystack is billions and billions of base pairs. How could we ever arrive where we wanted to go, and more importantly, how would we even know where we wanted to go in the first place? This chapter discusses the computational tools used in molecular biotechnology, an area known as bioinformatics. We also explore the major applications of bioinformatics: genomics and proteomics.

Bioinformatics

The problem of collecting and interpreting genetic data is aptly described by Spencer Wells, director of the Genographic Project and Harvard professor of genetics. The revolution taking place as computers were introduced as a primary tool for analyzing DNA is something he witnessed firsthand:

> When I'd started graduate school in genetics in the 1980s, the limiting factor in any research project had been simply generating enough data to test a hypothesis. Now data flowed like water from a fire hose, and the hard part was interpreting it and generating the many hypotheses that could explain the statistical patterns.

The field of bioinformatics is the direct response to the roadblocks experienced by Wells and his contemporaries. **Bioinformatics** is the application of biology, computer science, and information technology in order to organize, store, retrieve, and analyze data obtained from molecular analyses, including nucleic acids and proteins. A specialized field within bioinformatics, known as **computational biology**, uses computers to evaluate and interpret biological data. Prediction of gene or protein structure and function is also possible by analyzing previously characterized genes and proteins through the process of **data mining**. Bioinformatics pursues the following objectives:

1. Develop computational tools that aid in the management of biological data and provide searchable access to information.
2. Interpret and analyze biological data to reveal practical knowledge. The activities and discoveries within genomics and proteomics (below) are the direct result of this objective.
3. Actively create new software programs and algorithms for specialized data analysis. These are particularly useful for creating biological model systems and predicting outcomes such as protein interaction or mutation rates.

To date a wide variety of computational tools have been developed to perform such tasks as retrieval of DNA or protein sequences, classification of genes and proteins, determination of evolutionary relationships amongst species, annotation of the genome, and identification of genetic markers. Specific examples of these tools are discussed below, while careers related to bioinformatics are presented in the *Focus on Careers* box.

The Databases

Since the inception of bioinformatics, which was spurred greatly by needs of the Human Genome Project, a number of publicly available databases are now available. There are DNA databases that provide nucleotide sequences as well as protein databases containing amino acid sequences. Widespread Internet access means the "flowing water," as Wells called it, is just a few mouse clicks away for any interested party. Representatives of databases are outlined next, though this should not in any way suggest a researcher is limited to those found here. Niche research projects often collect and publish data, and it is within the realm of possibility for any researcher to develop customized database software for these purposes. Development of specialized algorithmic software programs is commonly part of this option.

In the United States

GenBank is maintained by the National Center for Biotechnology Information (NCBI), a division of the National Institutes of Health. GenBank accepts DNA sequence data from researchers and compiles the information into searchable databases. The Life Sciences Search Engine *Entrez* is a tool linked to GenBank. It enables a single search into the various GenBank databases, essentially linking all of the data into a one-stop shop. The database is searchable through several queries, including a DNA or protein sequence, species of interest, gene function, or genetic marker.

FOCUS ON CAREERS
Biostatisticians

Biostatistics is a field in which the data obtained from bioinformatics is statistically analyzed in order to define meaning and relevance. Biostatisticians often work as members of research teams focusing on medicine, epidemiology, and public health. The ideal biostatistician has equal competency in biological science and applied statistics, in addition to the advanced computer skills required to either access and utilize or in other cases design and implement bioinformatics software programs.

The primary role of a biostatistician is to actively participate in the scientific method, including formulating a working hypothesis, designing experiments, analyzing data, and drawing conclusions. They are critical in determining the sampling methods and data collection techniques to be used in a research project. Furthermore, they play a pivotal role in selecting and utilizing the most appropriate statistical analysis for particular data sets.

Currently, there is a shortage of biostatisticians and there are excellent opportunities for the right individual. Positions exist in government, industry, and academia in such settings as academic research teams, clinical trials in hospitals, and pharmaceutical companies. Most formal education in this field is at the postgraduate level. To become a biostatistician, the most direct path is to obtain a master's degree in biostatistics, although a doctorate degree would be preferable. Several universities offer these specialized degrees in addition to graduate certificates in either biostatistics or bioinformatics.

NCBI also maintains *Map Viewer*. This tool displays gene maps of different species through either a "whole-genome view" or a "chromosome or map view." Sequence data are also available here upon selecting a genomic region of interest. Their *Amino Acid Explorer* provides detailed information on amino acid structure and their chemical properties, making it particularly useful in the field of proteomics.

BLAST is another computational tool offered through NCBI. The Basic Local Alignment Search Tool is an algorithm which allows for gene sequence comparisons. This tool is particularly useful in the field of genomics where comparison studies are a significant endeavor. The BLAST website allows researchers to search within species genomes including human, mouse, fruit fly, and several others. BLAST also offers searches for genetic markers, nucleotide sequences (both DNA and their transcripts), and gene expression profiles.

International Databases

Databases maintained outside of the United States are also available. The *European Nucleotide Archive* (ENA) is spearheaded by the European Molecular Biology Laboratory. The *DNA DataBank of Japan* (DDBJ) is another major database maintained outside of the United States by the National Institute of Genetics of Japan. Both the ENA and the DDBJ collaborate regularly with NCBI under an umbrella organization, the *International Nucleotide Sequence Database Collaboration.* For nearly two decades this database has integrated information from all three sources into one location in order to streamline searches.

Genomics

The tools of bioinformatics have enabled an explosion of research projects relating to the study of genomes. Genomics is the study of the entire DNA content of a species (its genome), or in the case of RNA viruses, all of its ribonucleic acid content. In fact, the first complete nucleotide sequence was obtained from an RNA virus, the bacteriophage MS2, in 1976 at the University of Ghent. The first DNA genome sequence was of the bacteriophage Φ-X174. The 1990s saw complete genomes sequenced from all three domains of life: bacteria, in the form of *Haemophilus influenzae*, archaea (*Methanococcus jannaschii*), and eukarya. The first eukaryotic organism with a completed genome was the single-celled yeast *Saccharomyces cerevisiae.* Multicellular eukaryotes followed.

After sequencing viruses, bacteria, and yeasts, the efforts of the Human Genome Project determined the entire sequence of all 3.2 billion base pairs found within our 23 pairs of chromosomes. Further projects have revealed the genome content of species of all major taxa, ranging from microorganisms to fungi to plants and even our extinct relative *Homo neanderthalensis*, more commonly known as the Neanderthal.

The biodiversity of our planet grants many opportunities for specialization in this field, and the inclusion of extinct species under particular circumstances only makes the potential to retrieve genetic information greater. The intended outcome of each research project typically has one or more of these elements:

1. Sequence the genome of the species.
2. Annotate the genome and construct a gene map of the species.
3. Utilize the obtained data for genomic comparison studies.

Consider how difficult or outright impossible this would be without bioinformatics. Even with the fruition of computer technology, the sharp decline in DNA sequencing costs played a key role in the mass adoption of genomics as a viable branch of molecular biotechnology.

Genome Sequencing

Sequencing efforts began with relatively simple organisms such as viruses and bacteria. The single chromosomes found in bacteria contain far fewer nucleotides and prokaryotic genes are less complex in their design (see Appendix 1 for a closer look at prokaryotic and eukaryotic genome organization). Likewise, viruses often contain short fragments of DNA or RNA much smaller than the chromosomes seen in eukaryotic genomes. Scientists hypothesized that simpler organisms would have fewer genes than higher organisms. It was expected that prokaryotes and viruses would only possess the genes necessary for reproduction and survival. Yet upon sequencing these types of genomes scientists discovered there are also genes of unknown function (**FIGURE 4.1**).

It was soon realized that sequencing alone would have its limitations. Once the Human Genome Project was initiated, new methods to interpret the nucleotide data flowing from their machines were sought. The development of annotation and mapping became necessary as a means to organize and interpret the sequencing data.

Annotation and Mapping

Sequencing a genome merely results in a long string of As, Ts, Cs, and Gs. The information is not particularly useful until the sequence is broken down into readable pieces. If you were handed a written passage with a long chain of letters from the alphabet, you would have to identify individual words in order to read it. **Annotation** acts in the same manner by distinguishing one meaningful sequence from another. The process can identify regions of the genome such as those detailed below.

- 5′ and 3′ untranslated regions of a gene flank the assortment of exons and introns on either side.
- **Exons** are coding regions of the genome that express protein products, either individually or in concert. One exon may suffice to code for a single protein, though in combination with other genes it

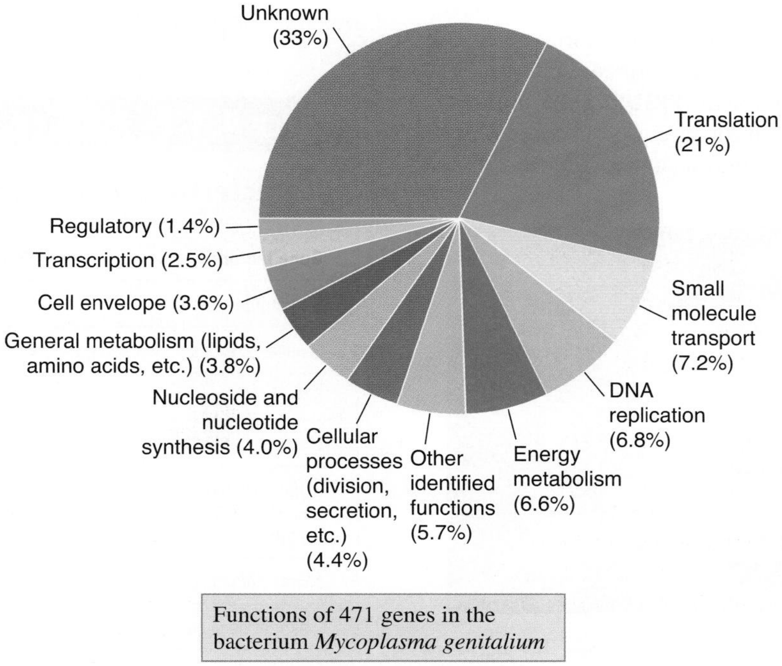

FIGURE 4.1 Types of genes found in the bacterium *Mycoplasma genitalium* classified by utility. Note that a third of the genome is of unknown function. Data from C.M. Fraser, et al., *Science* 270 (1995): 397–404

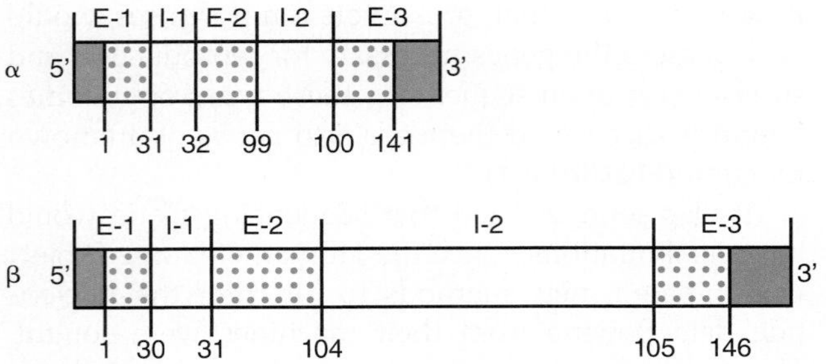

FIGURE 4.2 The human genes alpha globin (top) and beta globin (bottom) display the presence of introns (letter I) and exons (letter E). The numbers seen below each gene represent the number of amino acids expressed in each exon.

may produce an entirely different protein altogether. Only about 1.25% of the human genome actually encodes proteins. Exons may be interrupted within a DNA sequence by the presence of one or more introns (**FIGURE 4.2**).

- **Introns** are noncoding regions of the genome interspersed amongst exons. When first observed by the Japanese-American geneticist Susumu Ohno, he coined the term "junk DNA" to describe them. Introns are now known to play a role in gene regulation, though this may not be their only function. Introns account for about 20–25% of the human genome. Exons and introns are discussed further below as they relate to alternative splicing.
- **Transposable elements,** or transposons, are regions of the genome that are able to detach at one location and reattach in another. Transposable elements enable the dynamic nature of the genome in the sense it grants the ability for DNA sequences to change. When inserted into a gene sequence, a transposable element can lead to drastic results for the organism (**FIGURES 4.3** and **4.4**). They are a mechanism for phenotypic change in the short term and evolution in the long term.

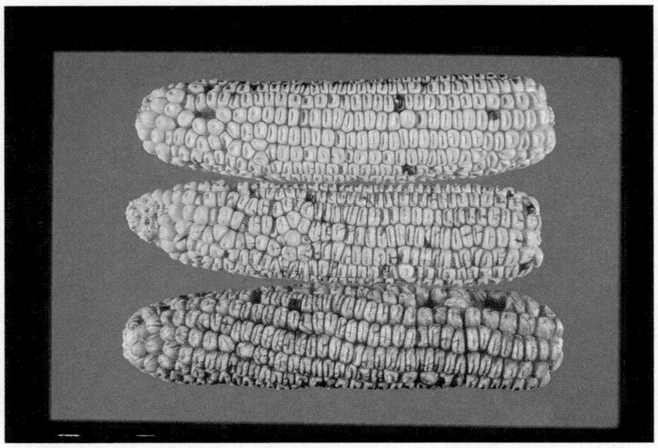

FIGURE 4.3 The color variations seen in these corn kernels are the result of transposable element insertion. Courtesy of Jerry L. Kermicle, Professor Emeritus, University of Wisconsin at Madison

- Functional RNA molecules which are not translated into proteins, such as transfer RNA involved in translation and small nuclear RNA involved in alternative splicing (see below).

Annotations of different genes may be compared; the International Human Genome Sequencing Consortium has statistically analyzed transcripts of human genes (**TABLE 4.1**). On average, a human gene contains 27,000 base pairs.

Craig Venter, the member of the Human Genome Project who branched out to form Celera Genomics, was inspired to do so once he developed an important annotation tool. Venter created a method of using **expressed sequence tags (ESTs)**, which are transcript sequences obtained through expression of complementary DNA (cDNA), in order to speed up the process. **Complementary DNA** is a sequence of engineered DNA that contains exons only. Existing cDNA libraries were utilized to speed up the availability of ESTs and identify previously unidentified human genes. Because cDNA are actually fragments of the genome, it was found that known cDNA sequences often contained redundant segments. This was helpful in matching the sequences through computational biology so that the volumes of data could be whittled down to unique sequences only (**FIGURE 4.5**). Upon study of the unique sequences, human genes could be divided into distinct classes by their functional role in the body (**FIGURE 4.6**).

Annotation data can also be utilized in the creation of genetic maps. Genetic mapping constructs a diagrammatic interpretation of the DNA sequence. A gene map is the visual landscape of a gene outlining its various loci, or physical locations of genes (**FIGURE 4.7**). If

TABLE 4.1

Characteristics of Human Transcripts

Gene feature	Median	Mean
Size of internal exon	122 bp	145 bp
Number of exons	7	8.8
Size of introns	1023 bp	3356 bp
5′ untranslated region	240 bp	300 bp
3′ untranslated region	400 bp	770 bp
Length of coding sequence	1101 bp	1341 bp
Number of amino acids (aa)	367	447
Extent of genome occupied	14 kb	27 kb

Source: Data from International Human Genome Sequencing Consortium, *Nature* 409 (2001): 860–921.

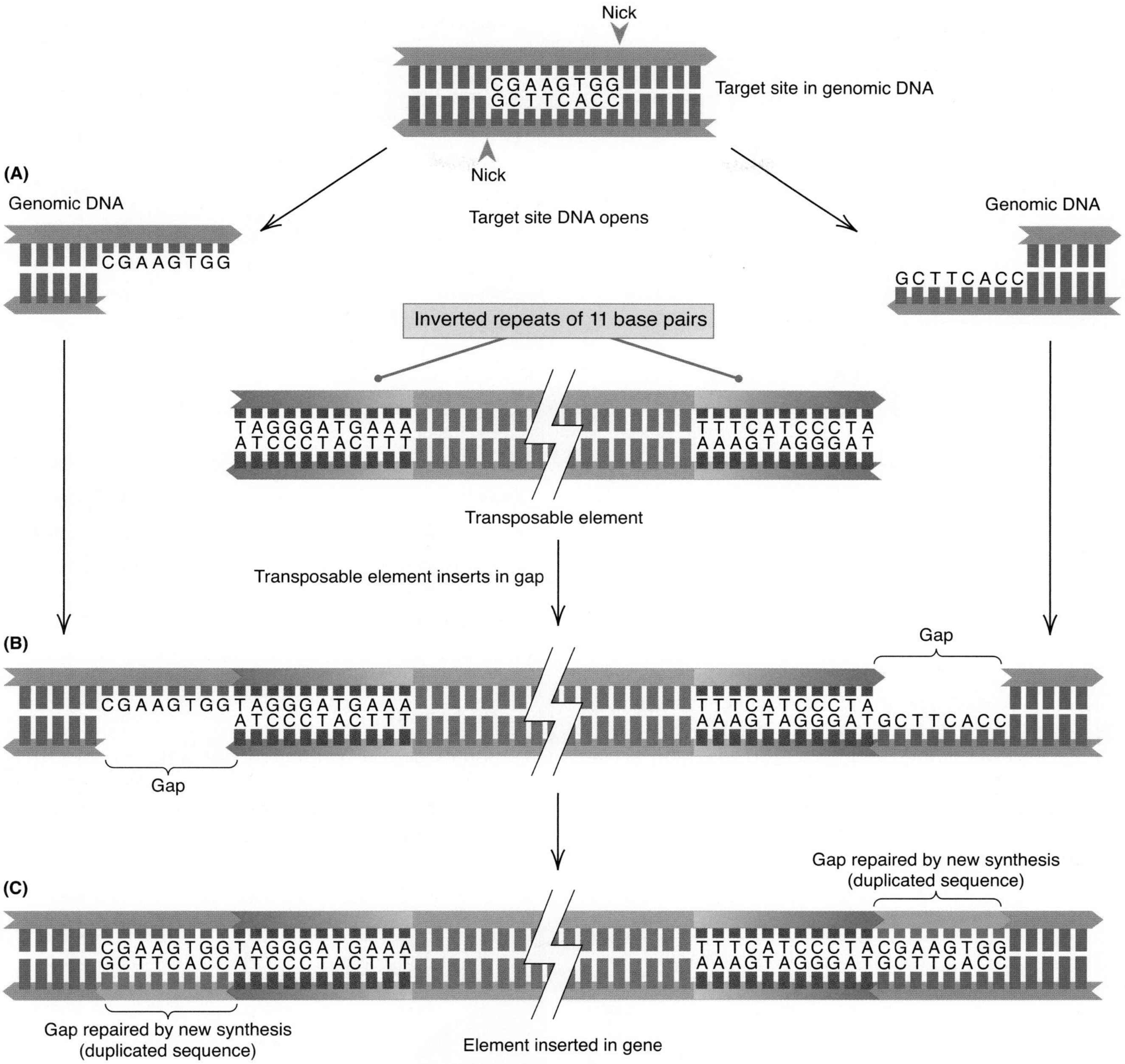

FIGURE 4.4 The "cut and paste" mechanism of transposable elements. (A) The target site is cleaved by the transposase enzyme. (B) The transposable element is inserted between the cleavage sites. (C) Gaps are filled by DNA polymerase via complementary base pairing.

a gene map were thought of as a map of a city, the loci would be the building addresses. Then each city would represent a unique species. Knowing gene locations may be directly applied to its study and possible manipulation via gene therapy.

Comparative Genomics

Through the study of genomics several insights have come to the surface. In some cases unexpected characteristics are observed. Interesting trends have likewise emerged. Comparison studies of different species have enabled the discovery of these insights.

DNA Molecule and Genome Size

As was expected, genome sizes of different taxa are a function of their relative simplicity or complexity. The same is true of DNA molecule size (**TABLE 4.2**). In general, the higher up the evolutionary tree an

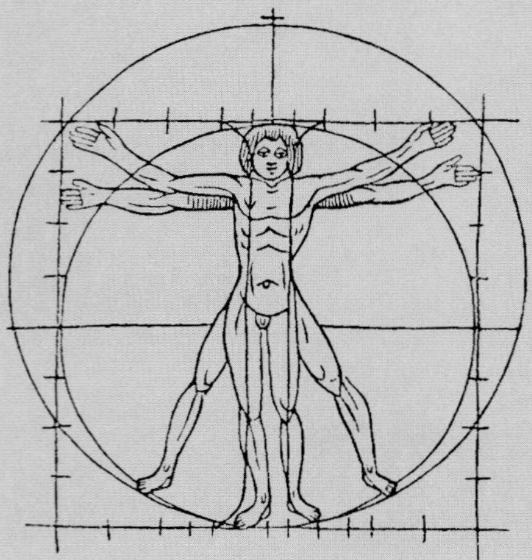

Source of cDNA	Sequences	Source of cDNA	Sequences
Skin	3,043 (629)	Brain	67,679 (3,195)
Blood cells	23,505 (2,194)	Eye	1,932 (574)
Bone	5,736 (904)	Salivary gland	186 (17)
Thyroid	2,381 (584)	Esophagus	194 (76)
Parathyroid	197 (46)	Adipose tissue	2,412 (581)
Endothelial cells	5,736 (1,031)	Heart	9,400 (1,195)
Liver	37,807 (2,091)	Peritoneum	884 (163)
Gallbladder	3,754 (768)	Spleen	7,924 (924)
Colon	4,832 (879)	Adrenal gland	3,427 (658)
Smooth muscle	297 (127)	Pancreas	5,534 (1,094)
Small intestine	1,009 (297)	Kidney	3,213 (712)
Prostate	7,971 (1,203)	Epididymus	1,716 (370)
Testis	7,117 (1,232)	Skeletal muscle	4,693 (735)
Ovary	3,848 (504)	Synovial membrane	3,889 (813)
Uterus	6,392 (1,059)	Embryo	19,291 (1,989)
Placenta	12,148 (1,290)	Thymus	2,412 (261)

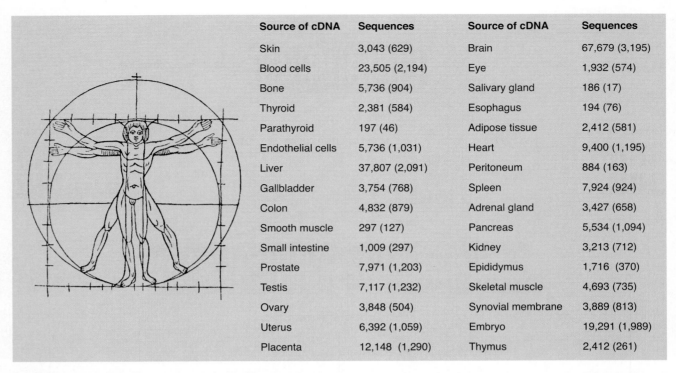

FIGURE 4.5 Known numbers of unique cDNA sequences by tissue or organ type. Data from M.D. Adams, et al., *Nature* (6547 Suppl.) 377 (1995): 3–174

organism is, the larger the genome will be (**TABLE 4.3**). However, upon closer inspection of eukaryotic kingdoms (Protista, Fungi, Animalia, and Plantae), it was discovered there is great variation in genome size within their various phyla (**FIGURE 4.8**). This suggests genome size may not be directly correlated to the complexity of the organism.

Gene Number

Genomes can be compared not only on their total size in nucleotide volume, but the number of genes is another aspect to consider. Another presumption of comparative genomics was a correlation between gene number and organism complexity. It was thought

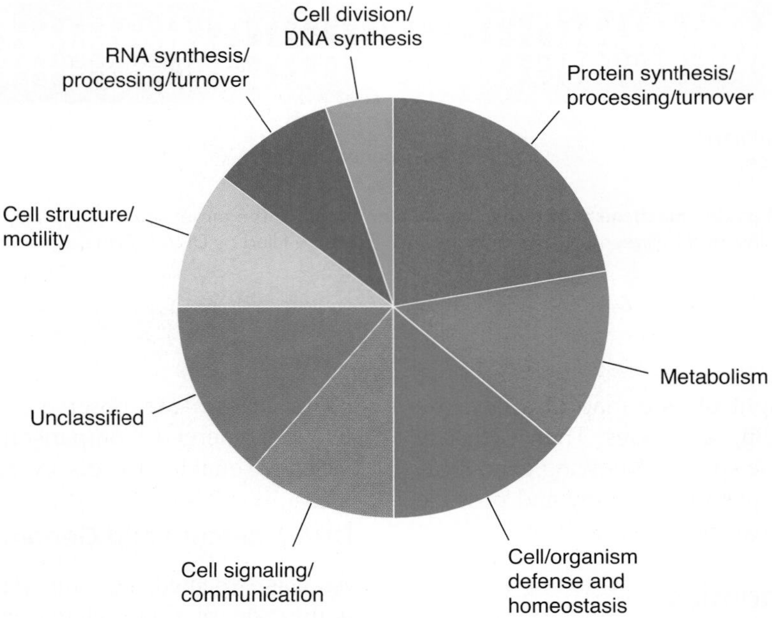

FIGURE 4.6 Classes of cDNA in the human body by function, based upon over 13,000 random and different sequences. Data courtesy of Craig Venter and the Institute for Genomic Research

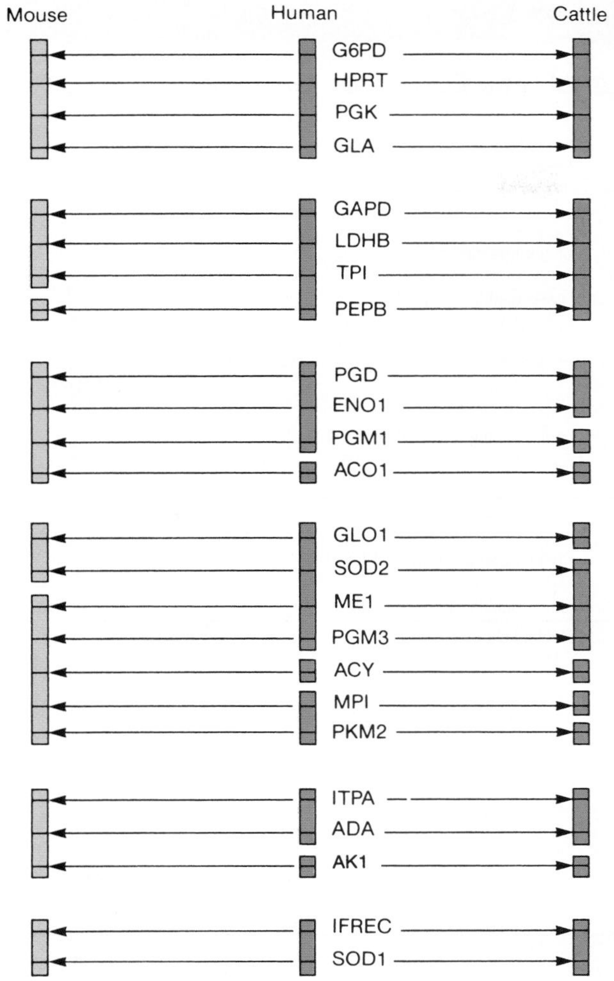

FIGURE 4.7 A linkage map of four chromosomes in the Drosophila melanogaster genome.

that more genes would be found in the more highly evolved multicellular species. The folly of human gene predictions serves as proof of the inaccuracy of this notion. Recall that most genes are essentially recipes to make proteins. The Human Genome Project anticipated discovery of over 100,000 human genes, because the human body is capable of synthesizing approximately that number of proteins. Instead, the Project found we possess only about a quarter that many. It was also found that genes are not evenly distributed throughout the genome; rather, they are concentrated in seemingly random regions. Gene-dense regions of the genome are rich in guanine (G) and cytosine (C) bases, while gene-poor regions are rich in adenine (A) and thymine (T). This distribution creates visible banding patterns in our chromosomes, as seen in the human karyotype (**FIGURE 4.9**). GC regions appear as dark bands and AT regions are much lighter in color. Chromosome 1 has the most genes and the Y chromosome, found only in males, has the least number of genes present. Study of additional taxa demonstrates no rhyme or reason to the number of genes in each genome (**TABLE 4.4**).

TABLE 4.2

Sizes of Various DNA Molecules

Source of DNA	Size in Base Pairs (bp)
Plasmid pBR322*	4,361
Simian virus 40 (SV40)	5,200
Phage T7*	39,937
Phage λ*	48,502
F plasmid*	99,159
Vaccinia virus strain WR	194,711
Fowlpox virus	266,145
Mycoplasma genitalium	580,073
Yeast chromosome IV	1,531,929
Escherichia coli	4,639,221
Human chromosome 1	245,522,847

Note: Phages (viruses that infect bacteria) and plasmids marked with an asterisk have *E. coli* as a host. *Mycoplasma genitalium* is the smallest known free-living bacterium. For yeast and humans the molecular mass of the largest DNA molecule in the organism is given.

Genetic Markers

The specific differences among members of a species, or between separate species, may be identified through comparisons of genetic markers. Genetic markers are also known as DNA polymorphisms, as they represent variations in genetic make-up. A proprietary "GeneChip®" was developed for use in the Human Genome Project to identify genetic markers, resulting in the observation of various types (**FIGURE 4.10**). Most of these markers involve repetitive sequences that account for at least 50% of the total human genome. Genetic markers are used in a variety of applications, such as evolutionary and population genetics and identification of individuals.

Single Nucleotide Polymorphisms

A **single nucleotide polymorphism (SNP)** is a difference of one nucleotide pair between two different DNA sequences. SNPs are located with the use of DNA microarrays. DNA microarrays are glass slides peppered with millions of short single-stranded DNA sequences. A DNA sample of interest will hybridize with a dot on the microarray when there is a complementary match

TABLE 4.3

Genome Size of Some Representative Viral, Bacterial, and Eukaryotic Genomes

Genome	Approximate genome size in thousands of nucleotides	Form
Viruses		
MS2	4	Single-stranded RNA
Human immunodeficiency virus (HIV)	9	Single-stranded RNA
Colorado tick fever virus	29	Linear double-stranded RNA
SV40	5	Circular double-stranded DNA
ϕX174	5	Circular single-stranded DNA; double-stranded replicative form
λ	50	Linear double-stranded DNA
Herpes simplex	152	Linear double-stranded DNA
T2,T4,T6	165	Linear double-stranded DNA
Smallpox	267	Linear double-stranded DNA
Prokaryotes		
Methanococcus jannaschii	1,600	Circular double-stranded DNA
Escherichia coli K12	4,600	Circular double-stranded DNA
Borrelia burgdorferi	910	Linear double-stranded DNA
Eukaryotes		Haploid chromosome number
Saccharomyces cerevisiae (yeast)	13,000	16
Caenorhabditis elegans (nematode)	97,000	6
Arabidopsis thaliana (wall cress)	100,000	5
Drosophila melanogaster (fruit fly)	180,000	4
Takifugu rubripes (fish)	400,000	22
Homo sapiens (human being)	3,000,000	23
Zea mays (maize)	4,500,000	10
Amphiuma means (salamander)	90,000,000	14

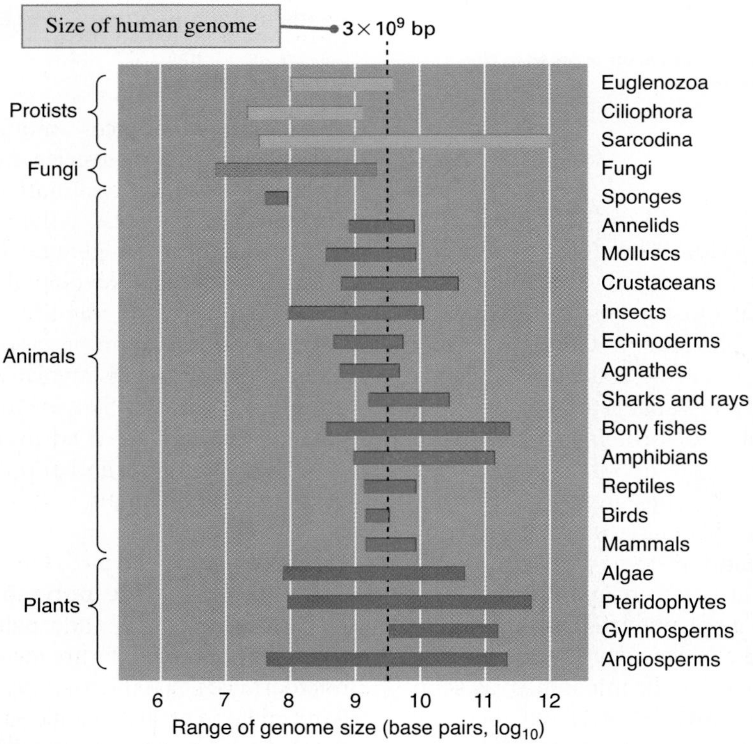

FIGURE 4.8 Genome size ranges within the phyla of all four eukaryotic kingdoms suggest that size is not a function of complexity.

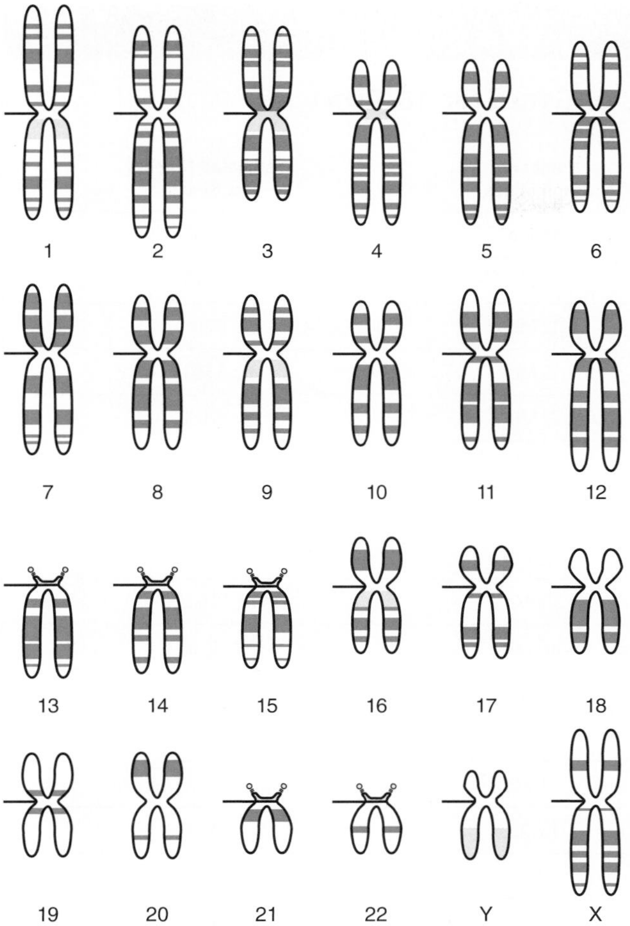

FIGURE 4.9 A graphical representation of the human karyotype displays all 24 chromosomes in the human genome. The banding patterns represent GC regions (dark bands) and AT regions (light/colorless bands).

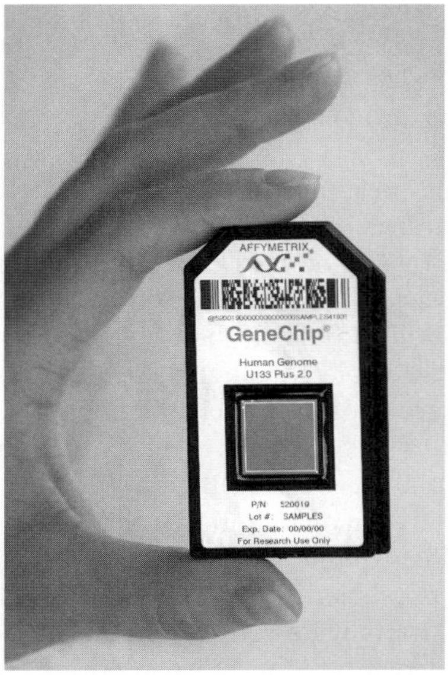

FIGURE 4.10 The GeneChip® employed by the Human Genome Project. Note its small size. Image courtesy of Affymetrix

FIGURE 4.11 A DNA microarray will visualize positive matches for genetic markers such as SNPs and CNVs through use of fluorescent dyes. © Andre Nantel/iStock/Thinkstock

between the donor and the sequence originally applied to the microarray types. Presence of fluorescent dyes act as indicators for a positive hybridization match (**FIGURE 4.11**).

Copy-Number Variation

When a large segment of DNA appears in the genome as multiple copies, it is known as **copy-number variation (CNV)**. This type of polymorphism applies to repeated segments ranging from 1,000 to 1,000,000 base pairs in length. Like SNPs, they are identified through DNA microarrays.

Tandem Repeat Polymorphisms and Dispersed Repetitive Sequences

Tandem repeat polymorphisms are similar to copy-number variations, except the repeated segments are much smaller in length and the repeats occur one after the other within DNA (in tandem). *Simple sequence repeats (SSRs)* are comprised of two to nine nucleotides repeating again and again (**TABLE 4.5**) and *variable number of tandem repeats (VNTRs)* contain anywhere from 10 to 60 nucleotides in tandem repetition. Because VNTRs are longer than SSRs, they are easier to detect and are frequently used in DNA fingerprinting.

Dispersed repetitive sequences are much like tandem repeat polymorphisms, except the repeats occur individually in various locations throughout the genome. They are also subdivided by size just like tandem repeat polymorphisms. Small repeats are seen in Short INterspersed Elements or SINEs. Long INterspersed Elements are referred to as LINEs. One class of LINE represents nearly 15% of the human genome and, thus, these genetic markers are a significant component in distinguishing members of our species. Most if not all dispersed repetitive sequences have the ability to relocate around the genome, much like transposable elements. They are actively investigated as there are over 30 dispersed repetitive sequences

TABLE 4.4

Numbers of Nucleotides and Estimated Numbers of Genes for 18 Taxa

Species	Number of Nucleotides	Estimated Number of Genes
Yersinia pestis (bacterium that causes bubonic and pneumonic plague)	4.5 Mb	3,956
Escherichia coli (bacterium)	5 Mb	3,470
Saccharomyces cerevisiae (yeast)	12 Mb	6,604
Cryptococcus neoformans (yeast)	20 Mb	6,500
Caenorhabditis elegans (nematode)	97 Mb	19,000
Arabidopsis thaliana (flowering plant from the mustard family)	115 Mb	25,498
Drosophila melanogaster (fruit fly)	137 Mb	14,100
Ciona intestinalis (ascidian)	160 Mb	15,852
Anopheles gambiae (mosquito)	278 Mb	13,683
Takifugu rubripes (Fugu)	365 Mb	c. 30,000
Tetraodon nigroviridis (freshwater pufferfish)	370 Mb	27,918
Oryza sativa (rice)	466 Mb	c. 50,000
Bombyx mori (silkworm)	530 Mb	
Danio rerio (zebrafish)	1.7 Gb	30,000
Mus musculus (house mouse)	2.4 Gb	c. 30,000 (24,174)
Rattus norvegius (Norway rat)	2.7 Gb	c. 30,000 (21,166)
Homo sapiens (humans)	2.9 Gb	c. 30,000 (26,966)
Pan troglodytes (common chimpanzee)	2.9 Gb	c. 30,000

known to cause disease. Though they cause harm, their multiple copies may assist in adaptation and evolution as they grant our genome additional plasticity under selective pressures.

Restriction Fragment Length Polymorphisms

A **restriction fragment length polymorphism (RFLP)** occurs when an SNP is present at a restriction site. The variation in restriction site sequences can lead to different cleavage patterns in DNA where they would otherwise not occur (**FIGURE 4.12**). Their presence in the genome can be detected through the Southern blotting technique, which involves digestion of the DNA being compared with restriction enzymes followed by electrophoresis analysis (**FIGURE 4.13**). The end result is the detection of distinct alleles that have been formed by RFLPs (**FIGURE 4.14**).

Ethical, Legal, and Social Issues

The pioneers of genomics also had to deal with bioethical concerns. Communication of new information has always been part and parcel of the scientific method. As scientists, members of the Human Genome Project desired to publish their data and make genetic sequences available to others. However, because the sequences are a part of each member of our species, privacy issues had to be contended with, as did concerns over malicious use of the data.

With the best of intentions, the Human Genome Project painstakingly sequenced our 3.2 billion base pairs and published a first draft in 2000, and two completed drafts were released in 2001. President Bill Clinton hailed it as "the most important, most wondrous map ever produced by humankind." To this day bioethics rears its head; researchers, industry, and the

TABLE 4.5

Some Simple Sequence Repeats in the Human Genome

SSR repeat unit	Number of SSRs in the human genome
5′-AC-3′	80,330
5′-AT-3′	56,260
5′-AG-3′	23,780
5′-GC-3′	290
5′-AAT-3′	11,890
5′-AAC-3′	7,540
5′-AGG-3′	4,350
5′-AAG-3′	4,060
5′-ATG-3′	2,030
5′-CGG-3′	1,740
5′-ACC-3′	1,160
5′-AGC-3′	870
5′-ACT-3′	580

Data from International Human Genome Sequencing Consortium, *Nature* 409(2001): 860–921.

public must weigh the risks and benefits of each new technology derived from this and similar projects.

Proteomics

Owing to the genetic code deciphered by Marshall Nirenberg, Har Gobind Khorana, and their research teams, as the field of genomics carves out gene sequences of various species it is possible to translate that information into amino acid sequences of corresponding proteins. Proteomics aims to perform this task by exploring the production, location, interaction, and metabolism of proteins. Another major goal is identification of all proteins expressed in an organism, otherwise known as the **proteome**. Furthermore, due to cell differentiation the various cells of a multicellular organism will differ in the proteins they synthesize. In other words, specialized cells within a single organism express different proteins; therefore the proteome of each cell within a single organism will likewise differ. The proteome of a multicellular organism must consider every cell type and their resulting protein expression.

All of this information can be used in the creation of synthetic proteins, biological mechanisms, and metabolic pathways. Protein molecules are the major change agents within an organism, performing functions such as:

- Controlling the flow of traffic in and out of cell membranes as protein channels
- Driving its chemical reactions and metabolic pathways as enzymes
- Amplifying its immune system as antibodies
- Providing structure as collagen, elastin, and keratin
- Aiding in movement as actin and myosin

To name a few. The performance of proteins in an organism can drastically affect its health. Take, for example, sickle cell anemia. The malformed hemoglobin proteins manifested in sickle-shaped erythrocytes contribute to a devastating, life-shortening illness (**FIGURE 4.15**). Lack of a protein can likewise result in disease, as is the case with phenylketonuria. Patients with this disease lack the hepatic enzyme phenylalanine hydroxylase, rendering them unable to metabolize the amino acid phenylalanine into tyrosine. They must remove phenylalanine from their diets or else suffer

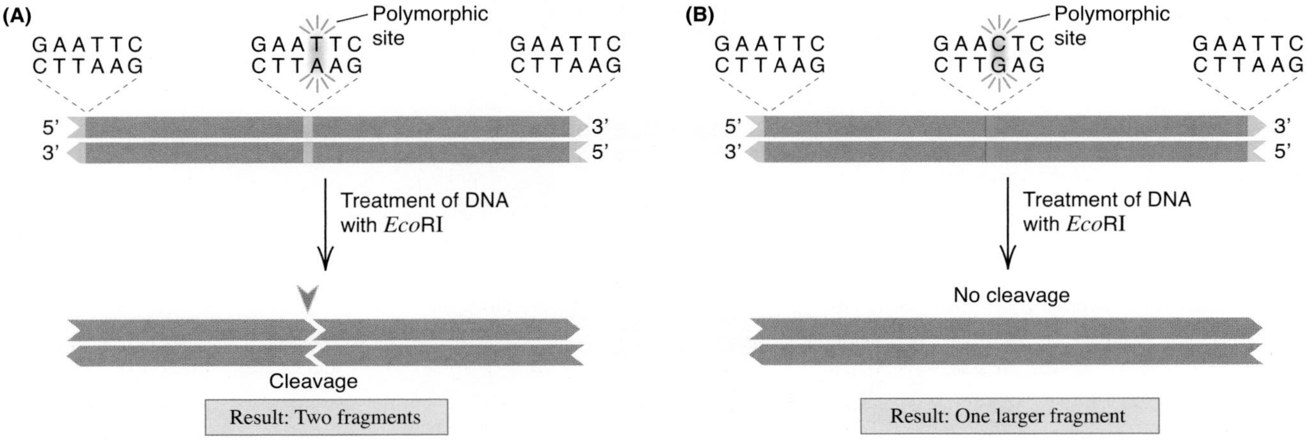

FIGURE 4.12 A single nucleotide polymorphism located within a restriction site can affect the digestion of DNA with restriction enzymes.

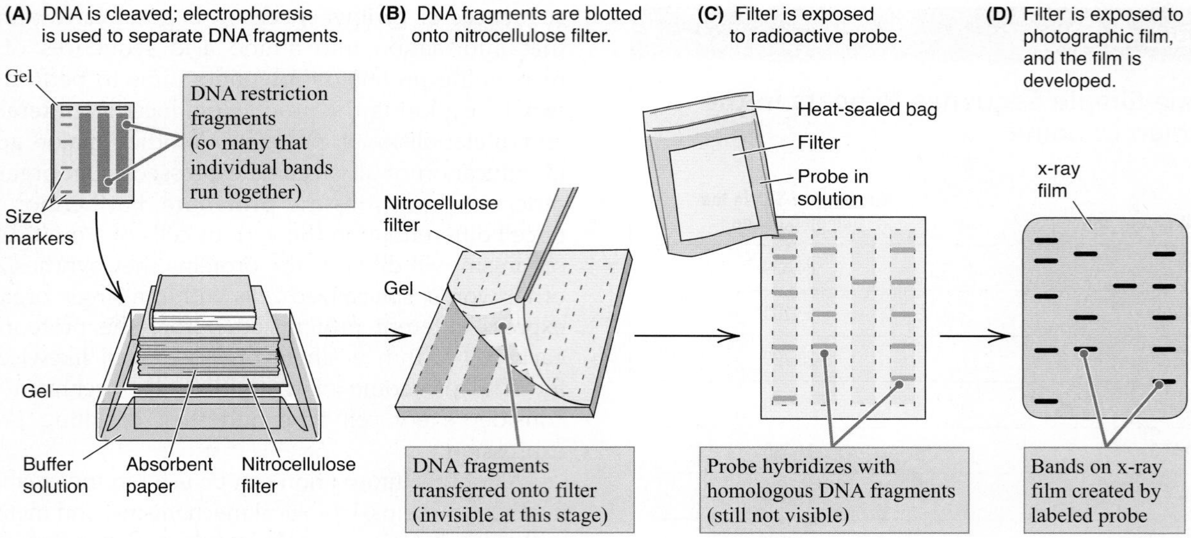

FIGURE 4.13 **The steps of the Southern blot technique.**

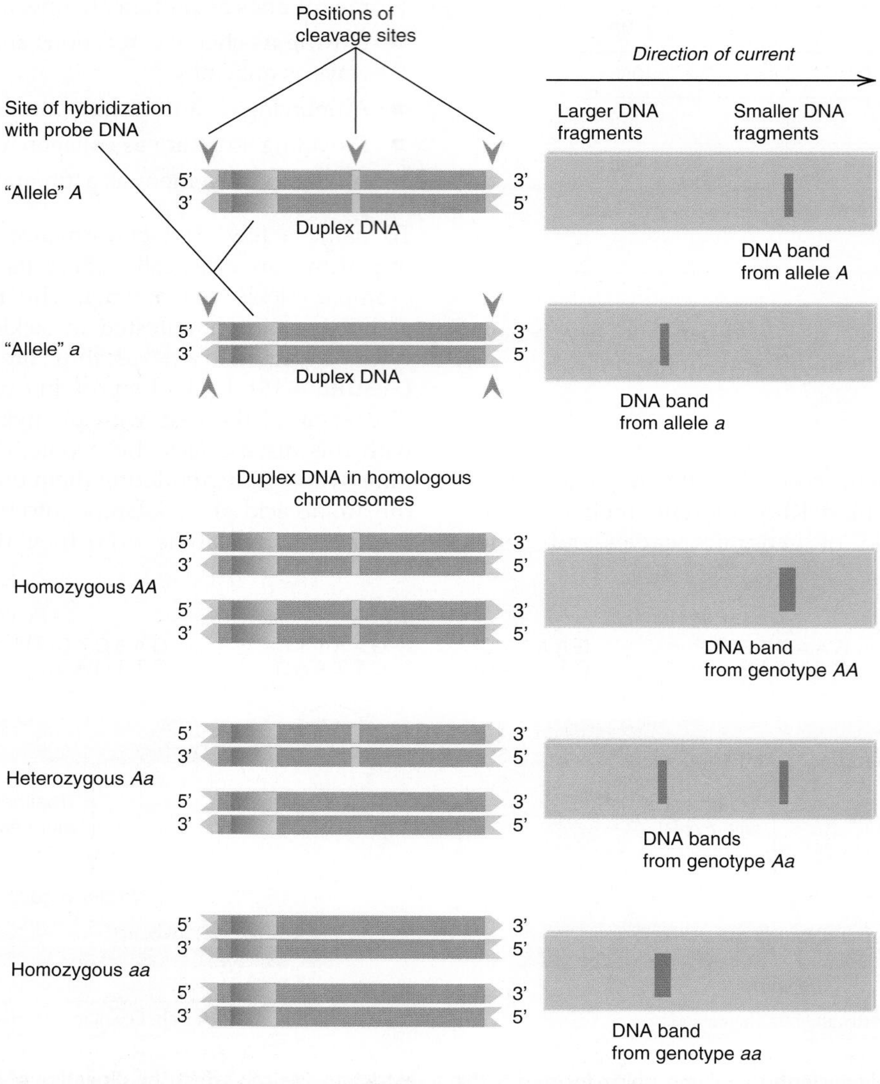

FIGURE 4.14 **The presence or absence of a restriction fragment length polymorphism can result in distinct alleles of a gene.**

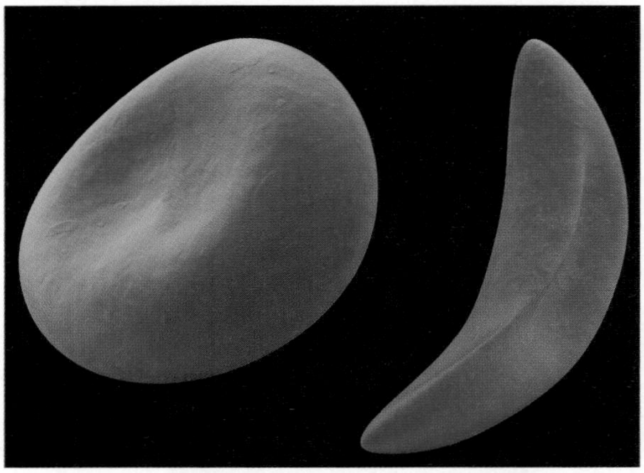

FIGURE 4.15 Malformed hemoglobin proteins result in sickle-shaped red blood cells, a signature trait of sickle cell anemia. © Sebastian Kaulitzki/ShutterStock, Inc.

symptoms such as seizures or even mental retardation (**FIGURE 4.16**).

Thus, understanding the structure and function of an organism's proteomics has enormous promise in medical applications as well as other areas of molecular biotechnology. The creation of genetically modified organisms is usually concerned with the protein products of the introduced gene(s). The structure exists in layers and in combination confers a final three-dimensional shape upon the molecule. In most instances, it is the shape of a protein that ultimately determines its function, or lack thereof. Therefore, the structural layers of a protein interact to decide just how the protein behaves within the organism.

FIGURE 4.16 Food items containing phenylalanine provide warning labels to phenylketonurics, patients with phenylketonuria who cannot metabolize the amino acid.

Structure of Proteins

In the process of gene expression, the steps of transcription and translation facilitate the central dogma of molecular biology. The central dogma relays the transformative chain reaction of events beginning with DNA (and its replication) which is transcribed into RNA which is then translated into a protein. Interestingly, proteins act upon nucleic acids (DNA and RNA) as well as other proteins, and many steps of protein synthesis rely upon enzymatic (read: protein) action, much like the chicken-or-the-egg quandary (**FIGURE 4.17**). A sequence of nucleotides in our DNA encodes for a particular gene (or, as we shall see, more than one gene). Recall the nucleotides contain one of four of the nitrogen-containing bases adenine (A), guanine (G), cytosine (C), and thymine (T). Chargaff's Rules state that A pairs with T and G pairs with C (**FIGURE 4.18**). The composition of bases varies with each species (**TABLE 4.6**).

The sequence is first transcribed into a messenger RNA (mRNA) molecule, comprised of a nucleotide sequence complementary to the gene sequence. Only in messenger RNA, the nucleotide uracil (U) is utilized instead of thymine (**FIGURE 4.19**). Therefore, if

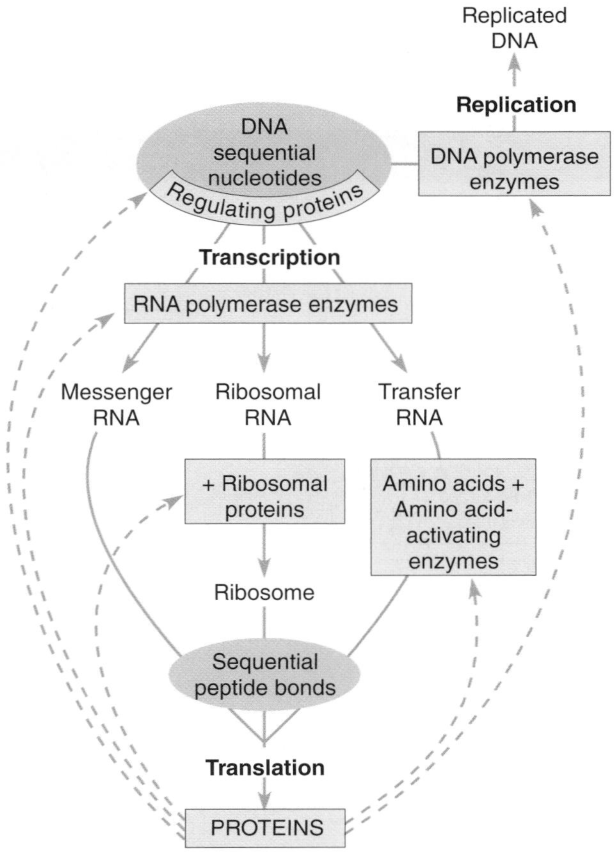

FIGURE 4.17 A schematic of the central dogma of molecular biology highlights the interdependence of nucleic acids and proteins.

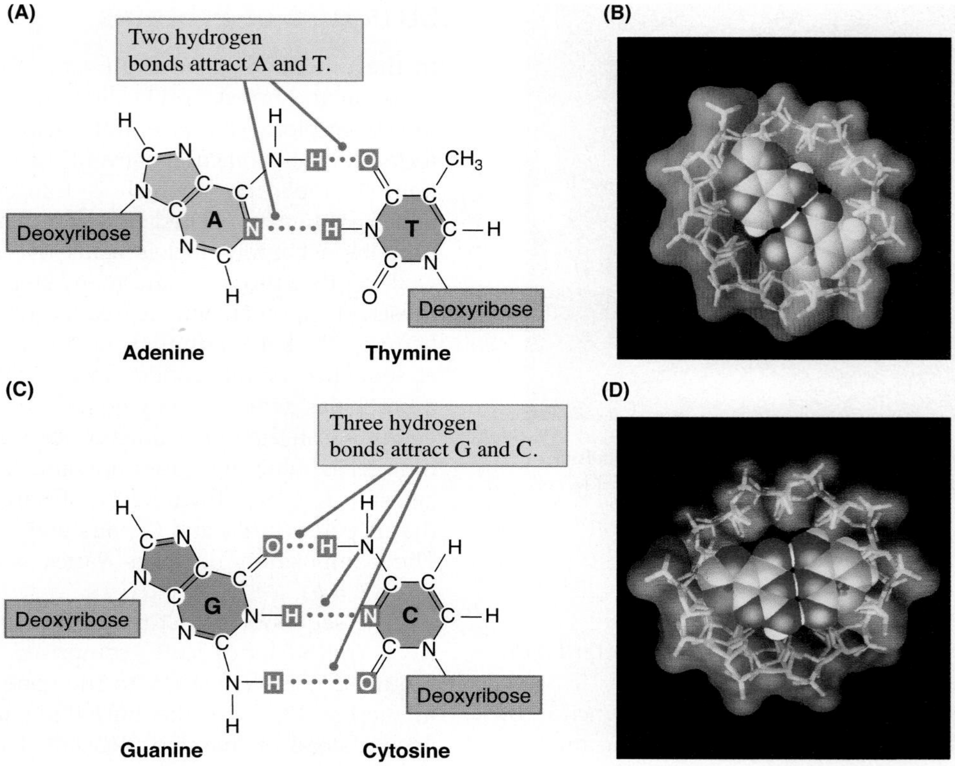

FIGURE 4.18 The chemical structures and pairings of the four nitrogen-containing bases of DNA. On the left: (A) and (C) show the hydrogen bonds between adenine and thymine (two bonds) and cytosine and guanine (three bonds), respectively. On the right: (B) and (D) are space-filling models with hydrogen bonds represented as white disks. The stick figures along their perimeters represent the sugar-phosphate backbones of the DNA molecule.

TABLE 4.6

Base Composition of DNA from Different Organisms

Organism	Base (and percentage of total bases)				Base composition (percent G + C)
	Adenine	Thymine	Guanine	Cytosine	
Bacteriophage T7	26.0	26.0	24.0	24.0	48.0
Bacteria					
Clostridium perfringens	36.9	36.3	14.0	12.8	26.8
Streptococcus pneumoniae	30.2	29.5	21.6	18.7	40.3
Escherichia coli	24.7	23.6	26.0	25.7	51.7
Sarcina lutea	13.4	12.4	37.1	37.1	74.2
Fungi					
Saccharomyces cerevisiae	31.7	32.6	18.3	17.4	35.7
Neurospora crassa	23.0	22.3	27.1	27.6	54.7
Higher plants					
Wheat	27.3	27.2	22.7	22.8*	45.5
Maize	26.8	27.2	22.8	23.2*	46.0
Animals					
Drosophila melanagaster	30.8	29.4	19.6	20.2	39.8
Pig	29.4	29.6	20.5	20.5	41.0
Salmon	29.7	29.1	20.8	20.4	41.2
Human being	29.8	31.8	20.2	18.2	38.4

*Includes one-fourth 5-methylcytosine, a modified form of cytosine found in most plants more complex than algae and in many animals

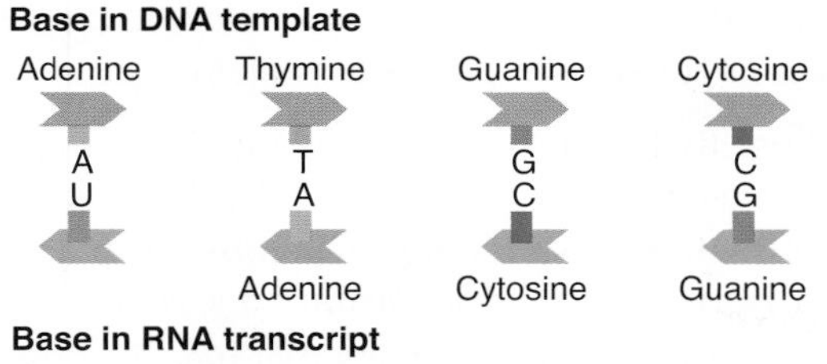

FIGURE 4.19 Base pairing between DNA and RNA involves the addition of uracil as a complement to adenine, unlike in DNA which would utilize thymine.

ACGTTAGGCACT were a gene sequence discovered in the field of genomics, the complementary transcript would be UGCAAUCCGUGA (the sequence is shortened for simplicity; an actual gene includes an average of 3,000 bases).

The second step in gene expression is translation of the mRNA into a polypeptide sequence, a future protein. Each set of three mRNA nucleotides (a codon) codes for one amino acid in the resulting polypeptide sequence (**FIGURE 4.20**). (Although it is important to note that not all genes code for proteins; some merely encode RNA and would go through the transcription process only.)

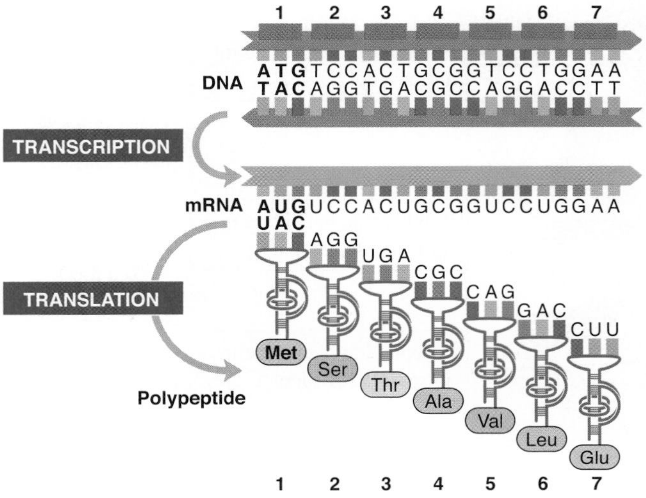

FIGURE 4.20 The steps of protein synthesis result in gene expression and construct the amino acid sequence of the resulting polypeptide.

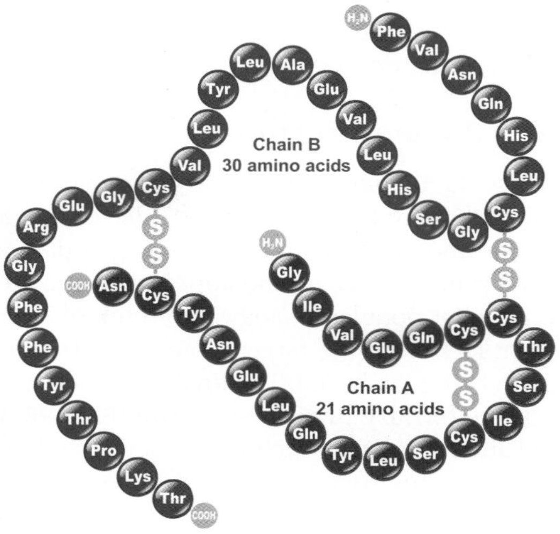

FIGURE 4.21 The primary structure of human insulin contains two chains of amino acids. Chain A contains 21 amino acids and chain B contains 30. © spline_x/ShutterStock, Inc.

Primary Protein Structure

The polypeptide's amino acid sequence obtained from translation is known as the **primary protein structure (FIGURE 4.21)**. If we were to take the example transcript above and determine the amino acid for each codon, we would come up with a primary protein structure of Cysteine-Asparagine-Proline-STOP (A STOP codon tells the cell to cease constructing the protein, as it has reached the end of the transcript). Though two amino acids may contain the same exact amino acids, it is the order of these subunits that distinguishes one amino acid from another.

Secondary Protein Structure

The sequence of amino acids will play a role in how each of these subunits might have the chance to interact with one another. One amino acid in the polypeptide chain will inevitably bond with another, setting forth a cascade of interactions which cause the polypeptide chain to fold into a unique arrangement known as the **secondary protein structure**. Secondary structural levels often exhibit three types of sub-structures: the alpha helix, the beta sheet, and random coils. The alpha helix appears spiral shaped (helical) while the beta sheet is a flattened pattern. Random coils are self-explanatory; they do not exhibit a repetitive pattern as do the other types of secondary structure (**FIGURE 4.22**). The entirety of these three sub-structures comprise the secondary

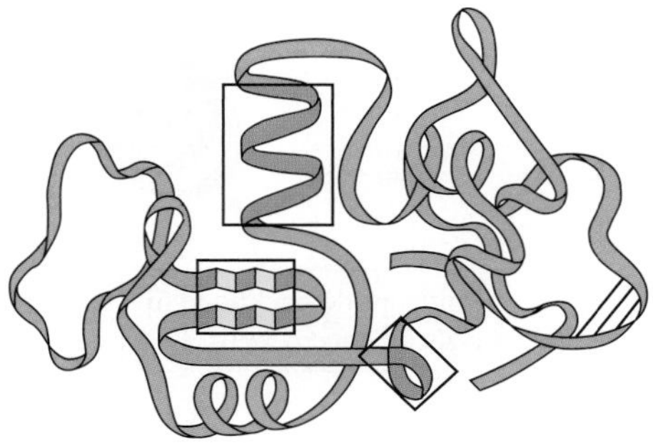

FIGURE 4.22 The secondary structure of a protein can take on several forms, including beta sheets, alpha helices, and random coils. Adapted from B.E. Tropp. *Biochemistry: Concepts and Applications*, First edition. Brooks/Cole Publishing Company, 1997

structure. Two proteins containing the exact same amino acids in different sequence will have distinct secondary structures.

Tertiary Protein Structure

Following its folding into a secondary structure, a protein will then take on a final three-dimensional shape. Enzymes are often referred to as globular, while structural proteins take on the appearance of a pleated sheet. Another popular analogy is to think of a protein as a giant lock, with the target molecule(s) the protein acts upon being the key(s). The final three-dimensional shape is the **tertiary protein structure** (**FIGURE 4.23**). The cell attains tertiary structure in a newly synthesized protein by linking together various side chains of the molecule via electrical charge, hydrogen bonding, or disulfide bonding (**FIGURE 4.24A**). Specialized molecules known as **folding moderators** often aid in these linkages. As the final shape of the protein is crucial to its functionality, it can safely be inferred that the correct action of the folding moderator molecules is equally important.

One type of folding moderator acts to speed up folding; these are called *folding catalysts*. Others have the ability to change the tertiary protein structure, so that more than one three-dimensional shape may arise from the same amino acid sequence. These folding moderators are known as *chaperones* (**FIGURE 4.24C** and **D**). There are five classes of chaperones:

- Those that enable desired folding are *folding chaperones*.
- Those that assist folding chaperones by holding the protein in place are *holding chaperones*.
- Those that reshape proteins that have lost their desired folding are *disaggregating chaperones*.
- Those that aid in protein secretion through the plasma membrane in order to exit the cell are *secretory chaperones*.
- Those that assist folding of very complex, slow-forming proteins are *chaperonins*.

Folding moderators are valuable to molecular biotechnology due to their relative nonspecificity. For example, a particular chaperone may act upon any number of proteins in the process of folding. Bioengineers involved in synthesizing proteins are able to create artificial environments similar to those of a living cell in order to properly fold the protein of interest.

Understanding how proteins are shaped is also useful in exploring the nature of misshapen proteins that cause harm to an organism. Two major classes of misfolded proteins are known to cause conditions in human health that we might like to modify or remedy. **Amyloids** are insoluble fibers from old proteins

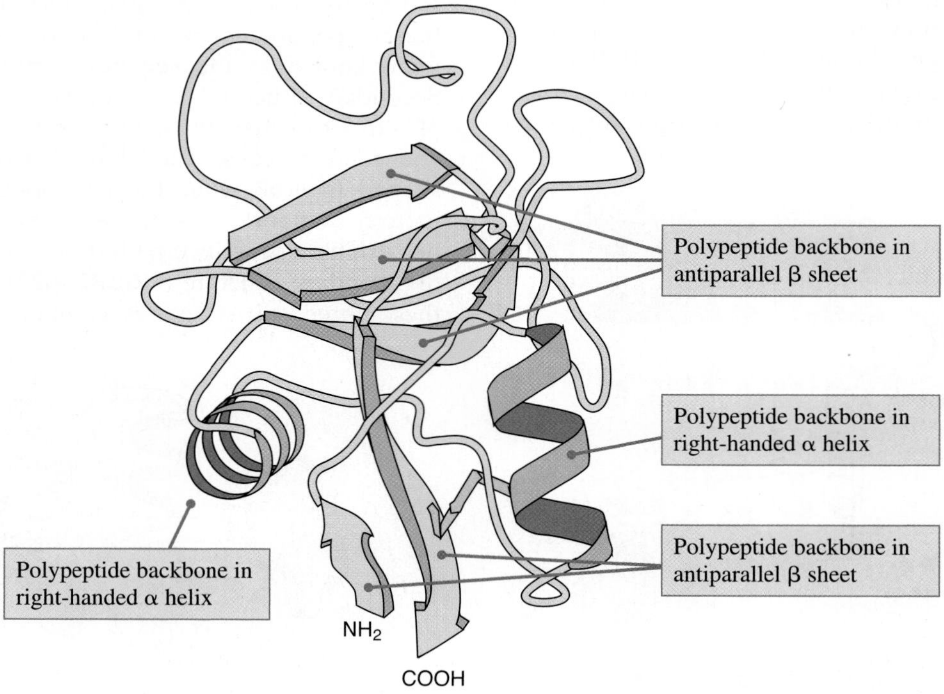

FIGURE 4.23 A "ribbon" diagram of a mannose-binding protein shows tertiary protein structure, the final three-dimensional shape of the molecule. Note that two forms of secondary protein structure (alpha helices and beta sheets) are present in this example. Adapted from W.I. Weiss, K. Drickamer, and W.A. Hendrickson, *Nature* 360 (1992): 127–134

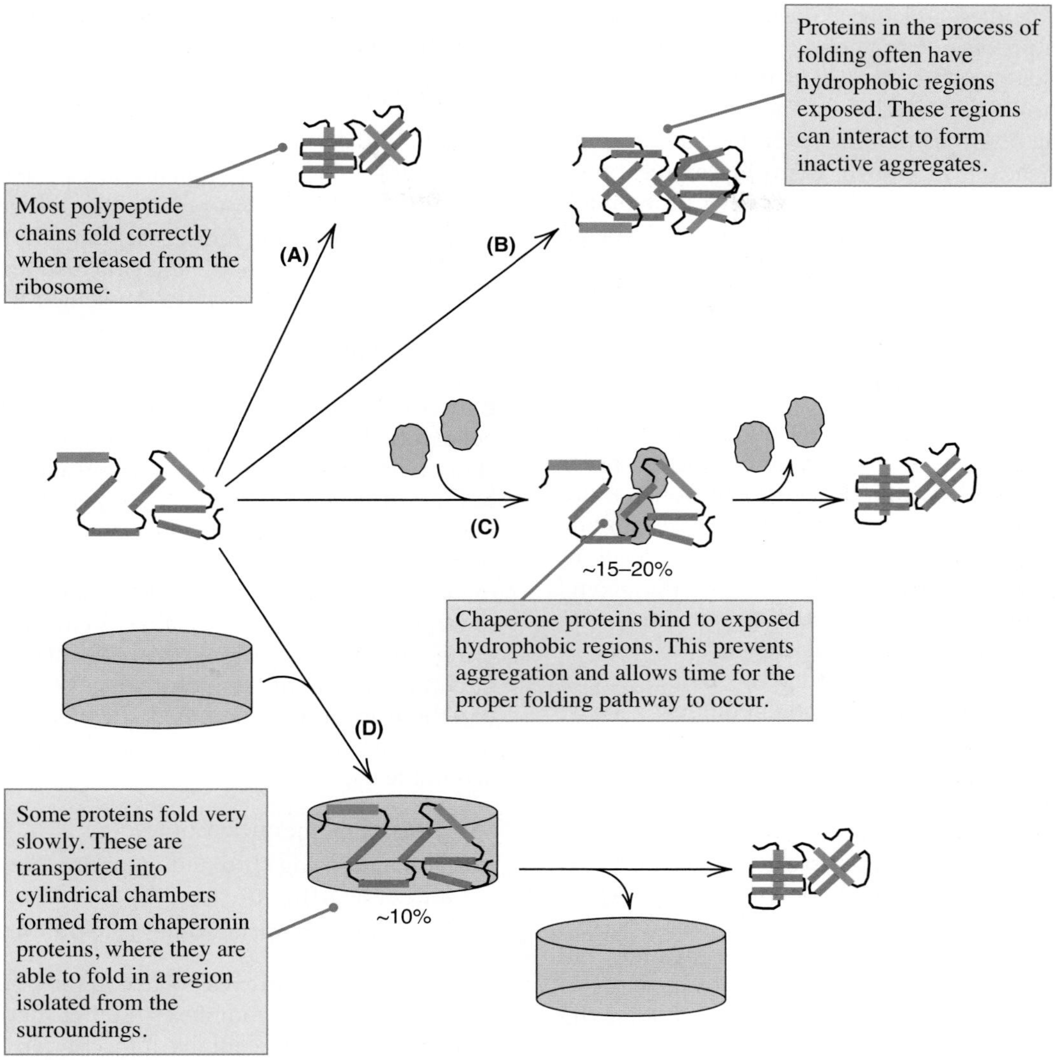

FIGURE 4.24 Folding pathways observed in proteins: (A) folding occurs independently, (B) misfolding results in aggregates known as inclusion bodies, (C) chaperone classes assist in folding, or (D) chaperonins aid in slow, complex folding.

that accumulate as we age. Excessive amyloids have been linked to illnesses including Alzheimer's and Parkinson's disease, cataracts, and type II diabetes mellitus. **Prions** are misshapen protein fragments that can infect an organism and cause disease. These pathogenic agents trigger healthy proteins in brain tissue to take on altered forms, resulting in spongy tissue containing dysfunctional proteins. Prion diseases include mad cow disease (bovine spongiform encephalopathy), scrapie (a similar disease affecting sheep), Creutzfeldt-Jakob disease (CJD), and Kuru (**FIGURE 4.25**). The latter two diseases affect humans.

In addition to losing functionality, if a protein is not folded properly it may end up aggregating with others of similar fate. These insoluble masses of misshapen proteins are known as **inclusion bodies** (**FIGURE 4.24B**). The organism is able to rid itself

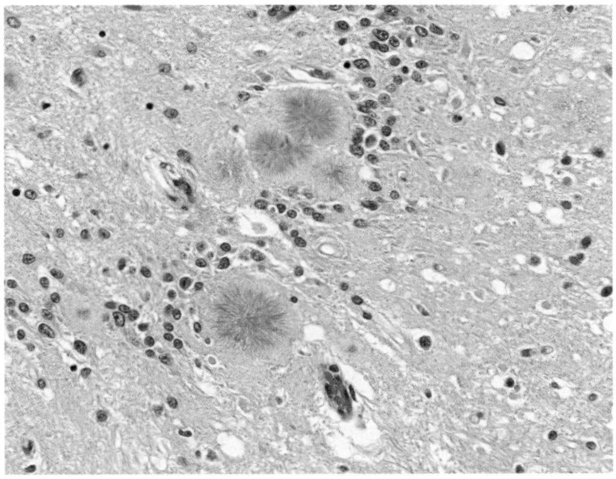

FIGURE 4.25 The prion-caused Creutzfeldt-Jakob disease leads to spongy brain tissue in humans, as seen in this micrograph. Courtesy of Teresa Hammett/CDC

of inclusion bodies and other dysfunctional proteins through the process of proteolysis. **Proteolysis** is the breakdown of proteins into their amino acid monomers, which may be recycled for the synthesis of new proteins. The enzymes that aid in proteolysis are proteolytic enzymes, also known as **proteases** (yes, proteins break down proteins). In the field of proteomics, the study of proteolysis and the proteasomes involved is useful in determining how we might engineer the event if desired, and also how to prevent it if that is the preferred outcome.

Quaternary Protein Structure

A number of proteins do not act alone; rather, they form protein complexes of several molecules in order to play a specific role within the organism. When more than one polypeptide chain folds into a three-dimensional protein complex it has reached the level of **quaternary protein structure**. Several important complexes exhibiting quaternary structure are of interest to molecular biotechnologists, as they are involved in regulating eukaryotic gene expression; these include transcription complexes, chromatin remodeling complexes, and silencing complexes. Not only can proteins lyse other proteins, they can also play a role in building them.

Transcription complexes are assembled in the process of recruitment in order to activate gene expression (**FIGURE 4.26**). Transcription complexes may be designed *in vitro* by researchers to study their bioactivity. A complex that binds to an enhancer region of the genome will promote gene expression. To construct a transcription complex under laboratory conditions, the minimal elements required include:

- *Transcriptional activator proteins (TAPs)* are proteins that bind to a region upstream of the gene (usually the enhancer region) and stimulate transcription.
- *Transcription factors* are proteins that promote the interaction of members of the transcription complex. A class of these proteins, the basal transcription factors, are seemingly universal and have been conserved in various organisms during the process of evolution. This suggests they have a fundamental role in facilitating transcription and can be utilized in a wide variety of research studies. Transcription factors assemble at the promoter region of a gene and attach to transcriptional activators, forming a "hairpin loop" in the DNA. Some transcription factors are quaternary protein complexes themselves, such as TFIID (**FIGURE 4.27**). TFIID contains the TATA-box binding protein (TBP) that binds to the promoter region of the gene that contains the "TATA box." The TATA box is an example of a **consensus motif** (a highly conserved nucleotide sequence among various genes) that, when bound with TBP, aids in transcription initiation (**FIGURE 4.28**).
- *Pol II Holoenzyme* is a quaternary protein complex containing RNA polymerase and additional proteins including more basal transcription factors. RNA polymerase is the enzyme responsible for transcribing the gene into messenger RNA.

A bioengineer may construct a transcription complex in the hopes of expressing a gene of interest, yet transcription nevertheless does not initiate. Because eukaryotic DNA is packaged as chromatin, a chromosome may be so thoroughly condensed that the gene of interest is inaccessible to the transcription complex. DNA binding sites must be available for the transcription activator protein. In these cases, addition of a chromatin remodeling complex (CRC) may be necessary in order to make the gene available. A **chromatin remodeling complex** restructures the molecular arrangement of chromatin. Depending on the specific class of CRC, and there are several identified, the mechanism of rearrangement will vary. CRCs make DNA binding sites accessible by:

- Disassembling nucleosomes.
- De-condensing chromatin (spacing nucleosomes further apart).
- Repositioning nucleosomes (**FIGURE 4.29**).

Silencing complexes behave in the opposite manner as transcription complexes. Rather than promoting gene expression, they inhibit transcription from taking place. Instead of binding to an enhancer region of the genome as with transcription complexes, the silencing complex will bind to a silencer sequence of DNA in order to prevent transcription. Silencing complexes are highly active during the development of an organism and therefore are useful in understanding cell fate and differentiation, such as in stem cell research (this will be discussed in greater detail later on). This is also true of transcription complexes and CRCs, making all of these quaternary protein structures valuable to molecular biotechnology.

The Phenome

While genomics results in the genome that characterizes the DNA of an organism, proteomics reveals the entire inventory of proteins expressed by a cell or organism, which is known as the proteome. The **phenome** is the complement of phenotypes that result from said genome and proteome. The terminology correlates to the concepts of genotypes and phenotypes

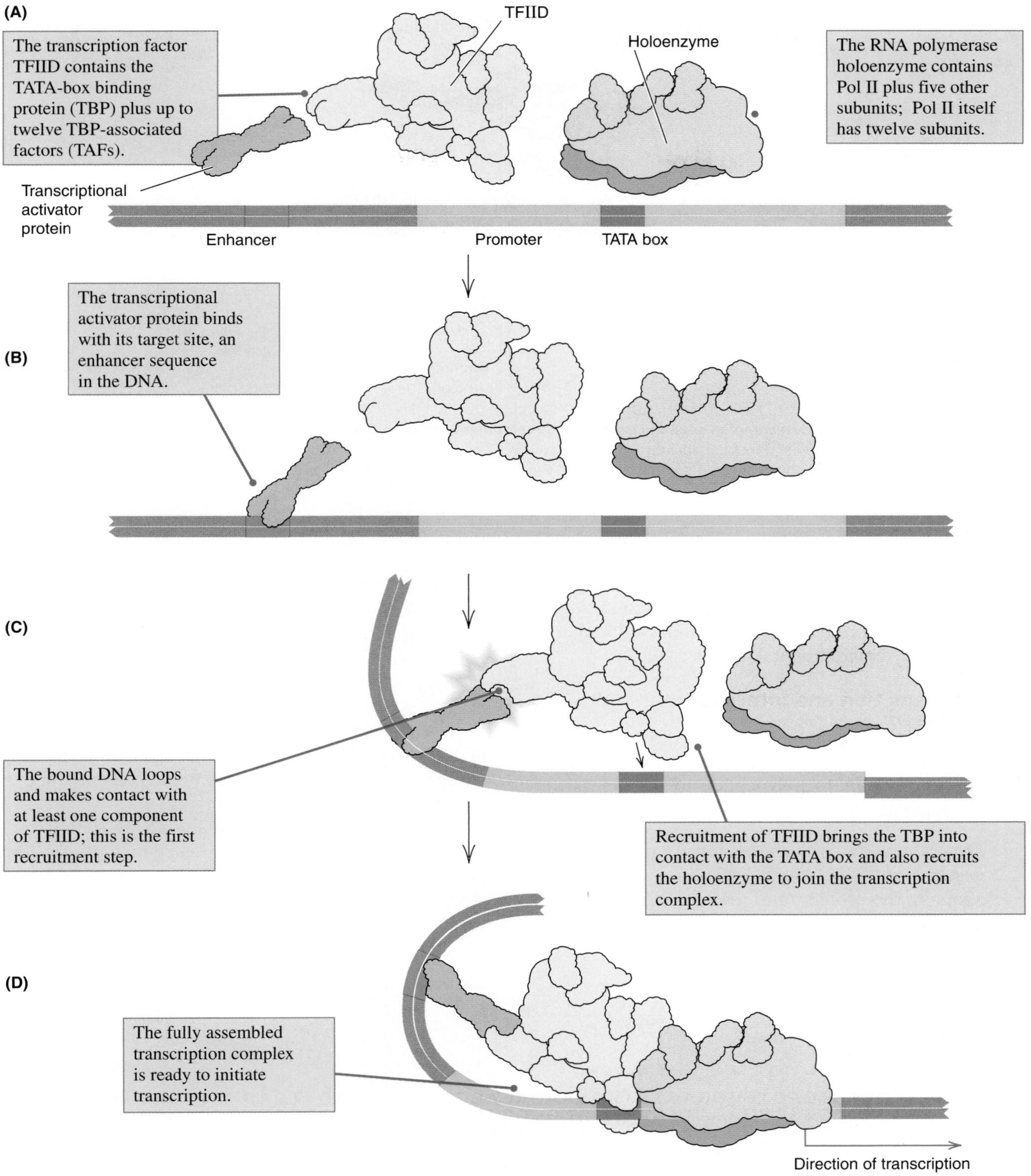

FIGURE 4.26 **Recruitment of the transcription complex activates transcription.**

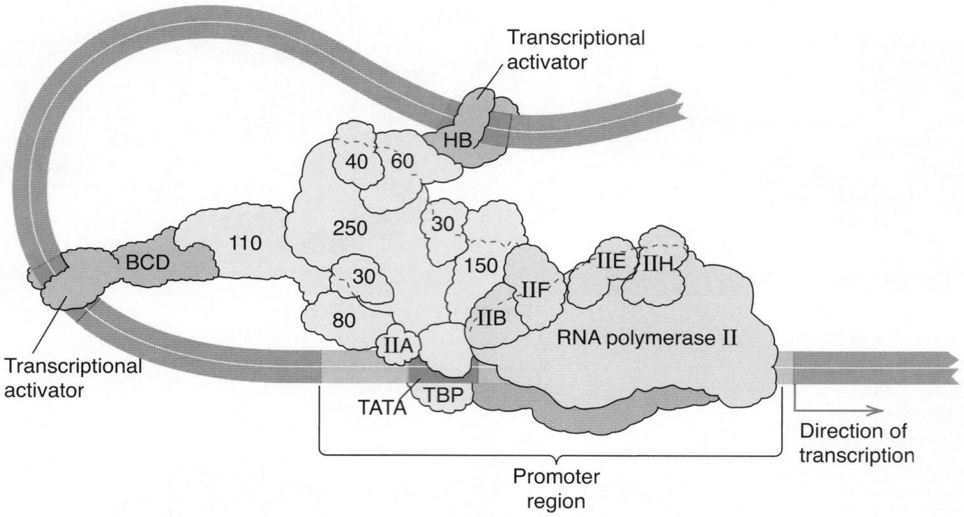

FIGURE 4.27 A transcription complex seen in *Drosophila* development binds to DNA and forms a hairpin loop. The transcriptional activators, seen in dark blue, include bicoid protein (BCD) and hunchback protein (HB). The numbered subunits comprise the TFIID transcription factor complex (seen in light blue) while the Pol II holoenzyme is represented by the brown subunits.

in classical genetics, where the genotype is the genetic make-up of an individual and the phenotype is the physical trait observed in the individual. In addition to the genome and proteome, the phenome is ultimately determined by environmental factors as well.

Protein Function and Interaction

Studies of the phenome have revealed errors in protein function can arise at a number of steps along the path from DNA to RNA to polypeptide to protein as the central dogma entails.

1. Translational error alters the primary protein sequence, affecting the ultimate three-dimensional shape of the protein (and therefore its functionality).
2. Folding moderators incorrectly shape the polypeptide into its three-dimensional shape.
3. Secretory chaperones fail to release the synthesized protein, so that even if it is appropriately shaped it is unable to reach its target and perform its physiological role.

Details about protein aberrations can reveal root causes of illnesses related to dysfunctional or missing proteins. Disaggregating chaperones that refold proteins are one investigatory route being taken.

Phenome investigation has the ability to link particular proteins of interest to their genetic locus (the physical chromosome location). It is now possible to study illnesses by determining the associated protein malfunction and then pinpointing its location in

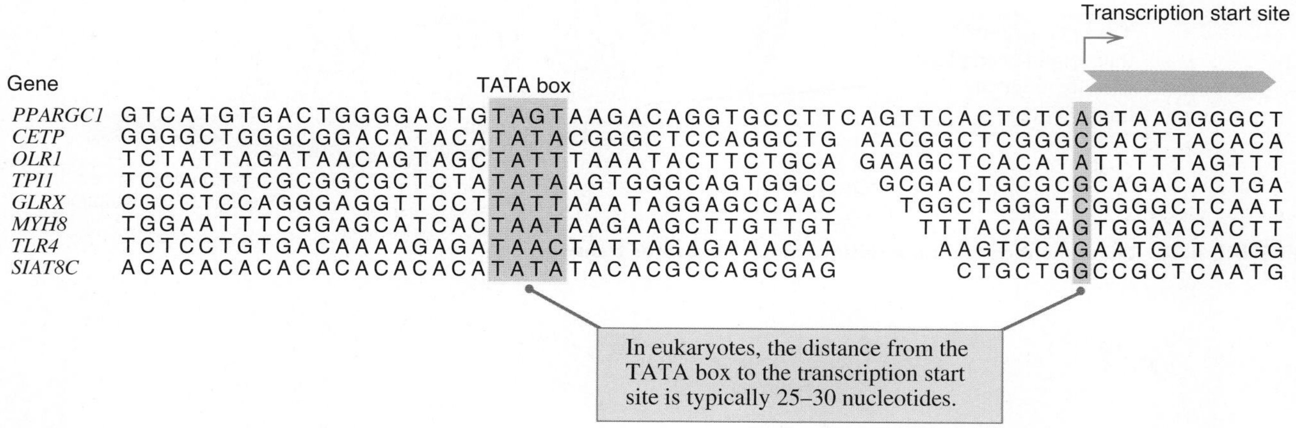

FIGURE 4.28 The TATA box regions of some human genes, illustrating the variations seen on the TATA box motif of TATAAT.

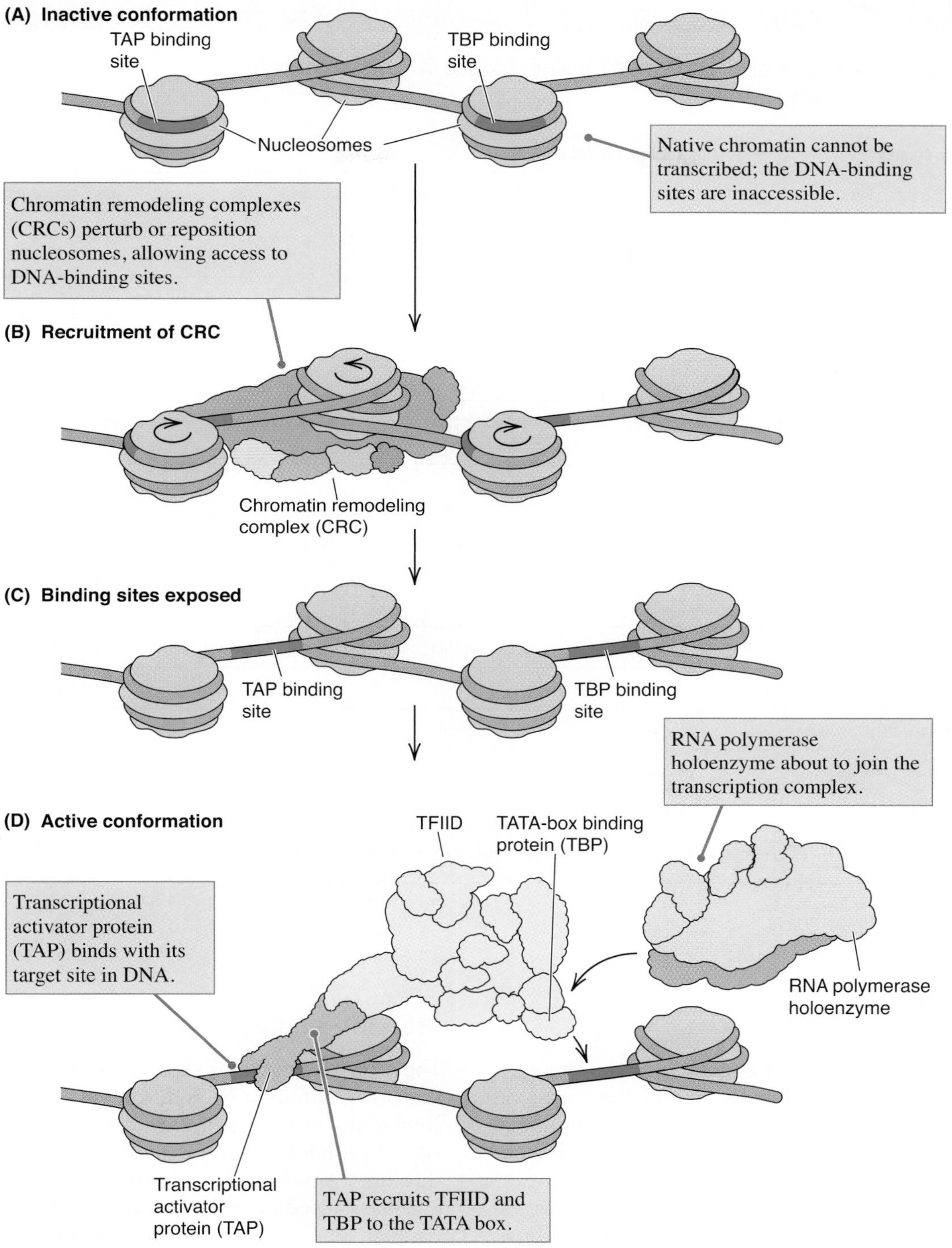

FIGURE 4.29 One way a chromatin remodeling complex can rearrange DNA in order to activate transcription is by repositioning nucleosomes. (A) DNA binding sites TAP and TBP are locked up within nucleosomes. (B) The CRC repositions the nucleosomes. (C) The TAP and TBP binding sites become accessible. (D) The transcription complex may now bind appropriately and initiate gene expression.

TABLE 4.7

Some Gene Products and Diseases Linked to the X Chromosome in Humans and Other Mammals

X-Linked in Humans	X-Linked in Other Mammals
α-galactosidase deficiency	More than 25 species, including chimpanzee, gorilla, sheep, cattle, pig, rabbit, hamster, mouse, cat, dog, kangaroo
Anhidrotic ectodermal dysplasia	Cattle, dog, mouse
Bruton-type agammaglobulinemia	Mouse, cattle, horse
Copper transport deficiency	Mouse, hamster
Duchenne/Becker muscular dystrophy	Mouse, dog
Glucose-6-phosphate dehydrogenase	More than 30 species, including chimpanzee, gorilla, sheep, cattle, pig, horse, donkey, hare, hamster, cat, mouse, kangaroo, opossum
Hemophilia A (factor VIII deficiency)	Dog, cat, horse
Hemophilia B (factor IX deficiency)	Dog, cat, mouse
Hypoxanthine-guanine phosphoribosyl	More than 25 species, including horse, hamster, dog, cat, mouse, cattle, gibbon, pig, rabbit, kangaroo
Ornithine transcarbamylase deficiency	Mouse, rat
Phosphoglycerate kinase	More than 30 species, including chimpanzee, gorilla, cattle, horse, hamster, mouse, kangaroo, opossum
Steroid sulfatase deficiency (ichthyosis)	Mouse, wood lemming
Testicular feminization syndrome	Cattle, dog, mouse, rat, chimpanzee
Vitamin D-resistant rickets	Mouse
Xg blood cell antigen	Gibbon

Adapted from Strickberger, M. W. 1985. *Genetics*, 3d ed. Macmillan, New York, with modifications and additions. Further listings can be found in J. R. Miller (1990).

order to determine what genetic defects have led to the disease (**TABLE 4.7**). Of course, it is important to remember that not all phenotypes are the result of a single gene. Many more are affected by multiple genes and the interactions amongst them and their environment.

Protein–protein interactions are another area of study in proteomics. We have seen examples of how proteins influence the activities of other proteins, such as in gene expression, protein folding, and proteolysis. Researchers implement a transcriptional activator protein known as GAL4 obtained from budding yeast cells in order to study protein–protein interactions through a technique called two-hybrid analysis (**FIGURE 4.30**). The GAL4 protein has two domains, or regions, which can each be utilized in the analysis. A reporter gene is introduced that enables the detection of transcription when the GAL4 protein and the protein in question do indeed interact.

Protein Families

The proteome contains super-groups of protein classes known as families. Protein families are identified by similarities in their amino acid sequences. The diversity of proteins is largely due to duplication of genes throughout evolutionary history, followed by nucleotide changes caused by mutation events. Therefore, through comparisons of amino acid sequences within the proteome, researchers can infer evolutionary relationships and construct phylogenetic trees depicting the history of protein families, much like a pedigree depicts the genetic history of a human family.

While the Human Genome Project has shown that our species expresses upward of 100,000 proteins, when these are subdivided into their corresponding families the number of distinct protein types diminishes to 10,000. There is a clear trend as organisms evolve where certain protein families are lost over time. Yet even so, multicellular organisms conserve a

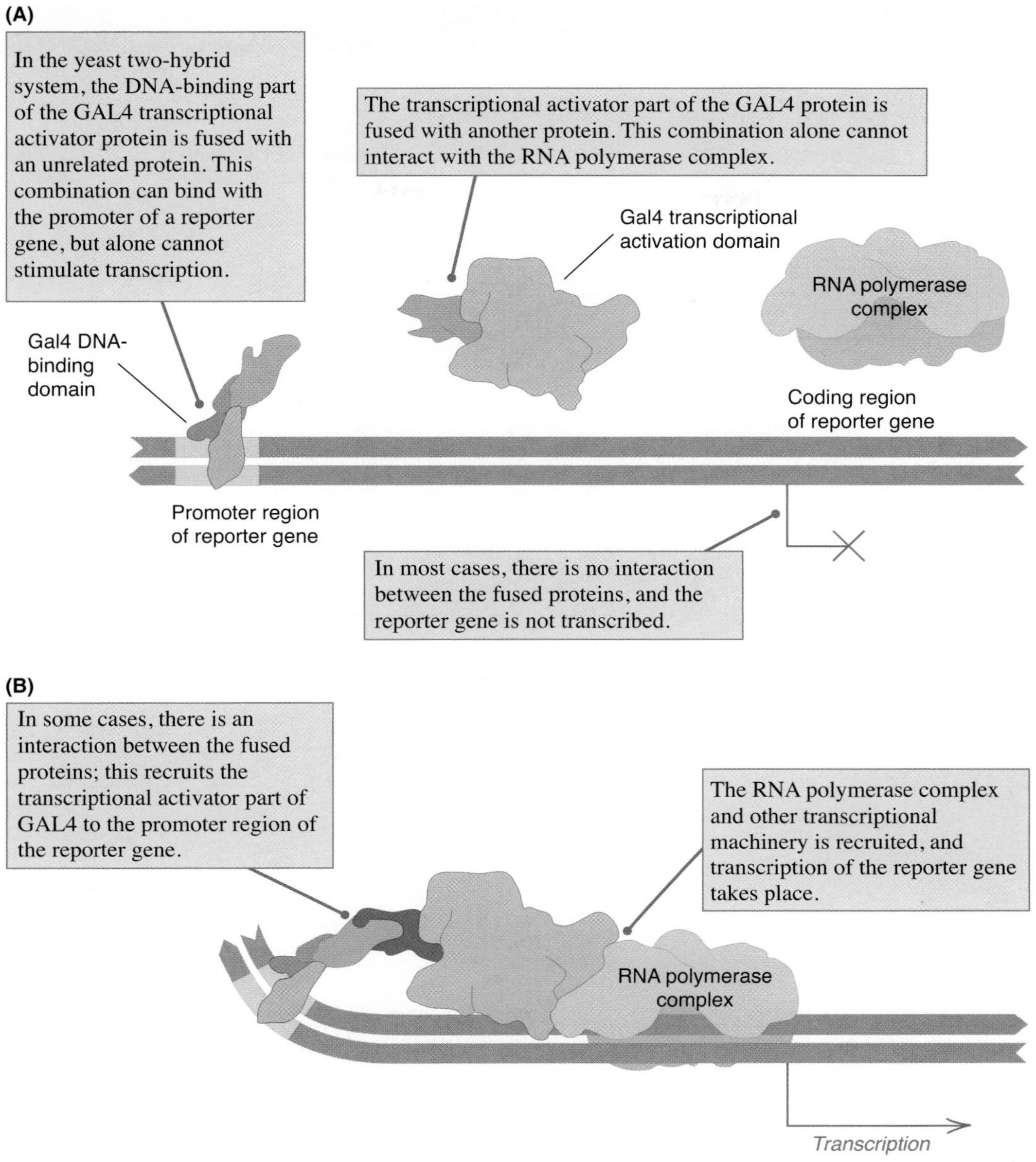

FIGURE 4.30 The two-hybrid analysis technique provides a method for determining protein-protein interactions, with either (A) a negative result (no interaction) or (B) a positive result (interaction).

large portion of protein families; among multicellular taxa about 5,000 to 10,000 families are shared. When compared to single-celled eukaryotes like yeast, there are still around 3,000 protein families in common with multicellular organisms. Going back further to a distant relative such as single-celled bacteria, 1,000 protein families, or approximately 10%, are found in common with multicellular species such as ours (**TABLE 4.8**)

Alternative Splicing

A surprise to members of the Human Genome Project was the number of genes present compared to the number of proteins we are known to produce. The diversity of proteins (somewhere on the order of 60,000 to 100,000) compared to the smaller number of genes (around 25,000 to 35,000) is the result of alternative splicing. **Alternative splicing** is the process of rearranging a primary mRNA transcript to produce a completely different nucleotide sequence. The process removes introns from the primary mRNA transcript and combines one or more exons in different orders (**FIGURE 4.31**). It allows one gene to encode for more than one protein.

The discovery of alternative splicing negated the "one gene–one protein" model previously ascribed to

TABLE 4.8

Comparisons of Genes and Protein

Organism	Genome size, Mb[a] (approximate)	Number of genes (approximate)	Number of distinct proteins in proteome[b] (approximate)	Shared protein families
Hemophilus influenzae (causes bacterial meningitis)	1.9	1700	1400	
Saccharomyces cerevisiae (budding yeast)	13	6000	4400	3000
Caenorhabditis elegans (soil nematode)	100	20,000	9500	5000
Drosophila melanogaster (fruit fly)	120[c]	16,000	8000	7000[d]
Mus musculus (laboratory mouse)	2500	25,000	10,000	9900[f]
Homo sapiens (human being)	2900[e]	25,000	10,000	

[a] Millions of base pairs.
[b] Excludes "families" of proteins with similar sequences (and hence related functions).
[c] Excludes 60 Mb of specialized DNA ("heterochromatin") that has a very low content of genes.
[d] Based on similarity with sequences in messenger RNA (mRNA).
[e] For convenience, this estimate is rounded to 3000 Mb elsewhere in this book.
[f] Based on the observation that only about 1% of mouse genes lack a similar gene in the human genome, and *vice versa*.

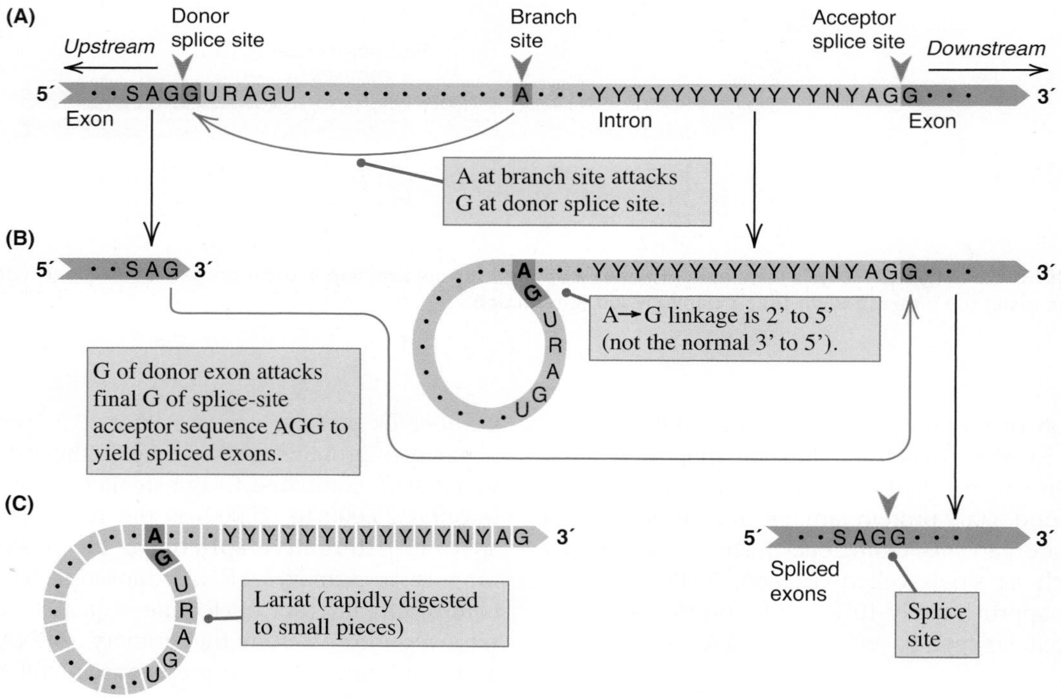

FIGURE 4.31 **The removal of one intron from a primary messenger RNA transcript involves (A) a splice at the first donor exon site, (B) a splice at the second donor exon site along with excision of the intron "lariat," and (C) the joining of the two exons at their splice sites.**

our understanding of molecular biology. Although some human genes do not contain introns, there are others where the introns account for 95% of the entire gene region. The average number of exons in each human gene is eight. At least one-third of human genes undergo alternative splicing. The number of mature mRNA strands transcribed from one gene can vary from two to seven in our species. The gene encoding for the alpha chain of the insulin receptor protein, found in humans and other mammals, undergoes alternative splicing depending on where the cell resides within the body (**FIGURE 4.32**). This demonstrates how alternative splicing also plays a role in gene expression and the determination of cell types.

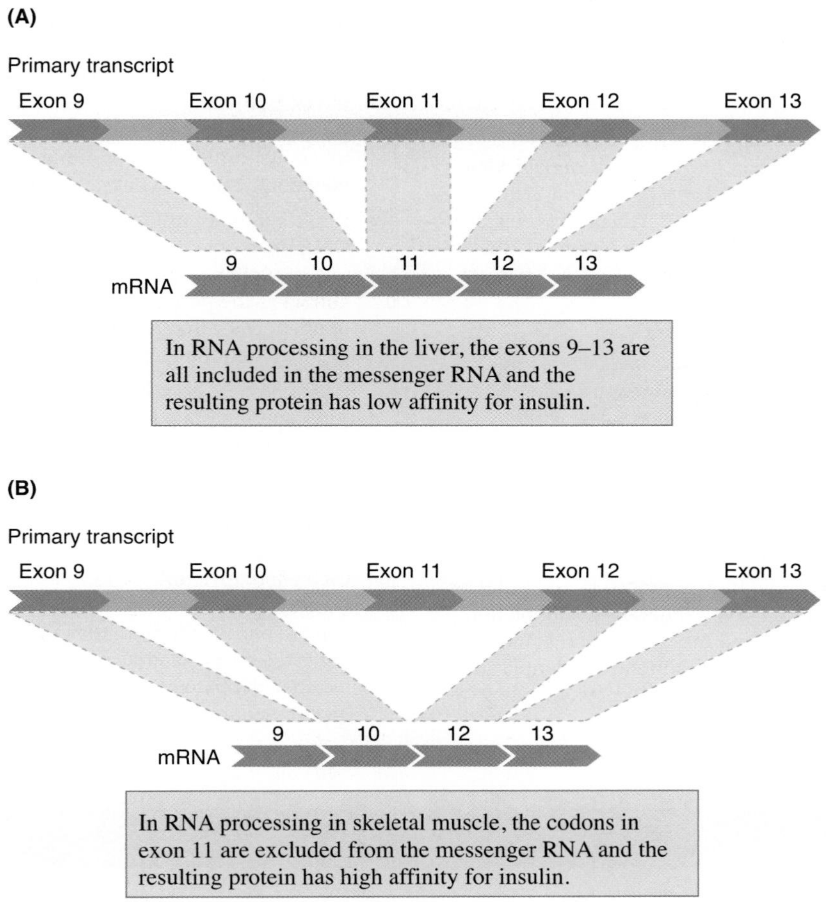

FIGURE 4.32 The gene encoding for the alpha chain of the insulin receptor protein can be spliced for different purposes in the liver (A) or in bone (B) in humans and other mammals.

SUMMARY

- Bioinformatics combines biology, computer science, and information technology in order to organize, store, retrieve, and analyze data obtained from molecular analyses, including nucleic acids and proteins. Computational biology uses computers to evaluate and interpret biological data. Prediction of gene or protein structure and function is also possible by analyzing previously characterized genes and proteins through the process of data mining. The main goals of bioinformatics are to develop computational tools, to interpret and analyze biological data, and to create new software programs and algorithms for specialized data analysis.
- Examples of bioinformatics databases maintained in the United States include *GenBank, Entrez, MapViewer, Amino Acid Explorer,* and *BLAST* (Basic Local Alignment Search Tool). Outside the United States, the databases *European Nucleotide Archive, DNA DataBank of Japan,* and the *International Nucleotide Sequence Database Collaboration* are popular.
- Genomics is the study of the entire DNA content of a species (its genome). The intended outcomes of genomic research projects include sequencing, annotation, and genetic mapping of a genome, in addition to using the obtained data in order to compare genomes of separate organisms and/or species.
- Genomics utilizes three main tools: sequencing, gene mapping, and annotation. Sequencing is the process of determining the exact order of nucleotide bases within a DNA molecule. Gene mapping identifies the location and order of individual genes in relation to each other, while annotation interprets both the sequence and map of a genome in order to provide meaningful information. Annotation identifies the introns and exons of each gene along with the noncoding DNA regions such as genetic markers.
- Comparative genomics has revealed genome sizes of different taxa are a function of their relative simplicity or complexity. In general, the higher up the evolutionary tree an organism is, the larger the genome will be. However, this does not correlate to the number of genes and their encoded proteins; there appears to be no relationship between the complexity of an organism and its number of genes or the size of its proteome. Gene-dense regions of the genome are rich in guanine and cytosine bases, while gene-poor regions are rich in adenine and thymine. Comparative genomics has also detected noncoding DNA: genetic markers, also known as DNA polymorphisms, which represent variations in genetic make-up. Some genetic markers are single nucleotide polymorphisms, copy-number variations, tandem repeat polymorphisms, and dispersed repetitive sequences.
- While genomics results in the genome that characterizes the DNA of an organism, proteomics reveals the entire inventory of proteins expressed by a cell or organism, which is known as the proteome. The phenome is the complement of phenotypes that result from said genome and proteome. Compared to the genome and proteome, the phenome is more likely to be altered by environmental influences.
- Alternative splicing is the process of rearranging a primary messenger RNA transcript to produce a completely different nucleotide sequence. The process removes introns from the primary mRNA transcript and combines one or more exons in different orders. This means more than one mRNA molecule may be transcribed from a single gene. Furthermore, each alternative mRNA transcript will be translated into a different protein. Therefore, alternative splicing allows one gene to encode for more than one protein. Alternative splicing is the best explanation for the discrepancies seen in gene number compared to proteins expressed in many organisms, including humans.

KEY TERMS

Alternative splicing
Amyloids
Annotation
Bioinformatics
Chromatin remodeling complex
Complementary DNA
Computational biology
Consensus motif
Copy-number variation (CNV)
Data mining
Dispersed repetitive sequence
Exons
Expressed sequence tag (EST)
Folding moderators
Inclusion bodies
Introns
Phenome
Primary protein structure
Prions
Proteases
Proteolysis
Proteome
Quaternary protein structure
Restriction fragment length polymorphism (RFLP)
Secondary protein structure
Silencing complex
Single nucleotide polymorphism (SNP)
Tandem repeat polymorphism
Tertiary protein structure
Transcription complex
Transposable elements

DISCUSSION QUESTIONS

1. How are genomics, proteomics, and phenomics related? Construct a flow chart or concept map that connects these three fields of molecular biotechnology. What applications do they have? What are some examples of knowledge obtained from these fields?
2. What do you consider to be fair use of computational data? What do you consider to be unfair use? For example, should insurers, employers, adoption agencies, and/or the military be able to use this information, and if so, how?
3. Visit GenBank (https://www.ncbi.nlm.nih.gov/genbank/). What types of information are you able to retrieve? Keep in mind the more specific you can be in your searches, the more pertinent the results will be. Describe why bioinformatics databases are necessary in molecular biotechnology.

REFERENCES

Bergman, N.H., editor. *Comparative Genomics: Volume 1*. Totowa, New Jersey: Human Press, 2007.

Buckingham, S. "Bioinformatics: Programmed for Success." *Nature* 425, (2003): 209–215.

Koonin, E.V. and M.Y. Galperin. *Sequence-Evolution-Function: Computational Approaches in Comparative Genomics*. Boston: Kluwer Academic, 2003.

Lawler, P.A. and R.M. Schaefer, editors. *American Political Rhetoric: A Reader*, 5th ed. Lanham, Maryland: Rowman & Littlefield, 2005.

Moody, G. *Digital Code of Life: How Bioinformatics Is Revolutionizing Science, Medicine, and Business.* New Jersey: Wiley, 2004.

National Center for Biotechnology (NCBI). *BLAST Help. (ND)*. Available at http://blast.ncbi.nlm.nih.gov/Blast.cgi?CMD=Web&PAGE_TYPE=Blastdocs

National Center for Biotechnology Information (NCBI), National Institutes of Health (NIH). (ND). *All Resources*. Available at https://www.ncbi.nlm.nih.gov/guide/all/

Office of Biological and Environmental Research, Human Genome Program. "The Science Behind the Human Genome Project" United States Department of Energy Human Genome Project Information, last modified August 13, 2013. Available at http://www.ornl.gov/sci/techresources/Human_Genome/project/info.shtml

Venter, J.C., et al. "The sequence of the human genome." *Science* 291 (2001): 1304–1351.

Watson, J. D., et al. *Recombinant DNA; Genes and Genomes*. New York: W.H. Freeman and Company, 2007.

Wells, Spencer. *Pandora's Seed: The Unforeseen Cost of Civilization*. New York: Random House, 2010.

Wong, Kate. "Vanished Humans: Our Inner Neandertal." *Scientific American* 22 (2012): 82–83.

5 Industrial Biotechnology

LEARNING OBJECTIVES

Upon reading this chapter, you should be able to:

- Discriminate industrial biotechnology from other sectors of the biotechnology industry.
- Identify examples of industrial enzymes and discuss the roles thereof.
- Explain the need for research enzymes and provide examples thereof.
- Discuss examples of biopolymers and their uses.
- Compare and contrast primary metabolites with secondary metabolites.
- Provide examples of primary and secondary metabolites.
- Compare and contrast the forms of fermentation involved in upstream processing.
- List the steps of downstream processing.

The advent of molecular biotechnology techniques has influenced a number of industrial processes and products. Organisms ranging from unicellular bacteria, algae, and fungi to multicellular plants and even animals are utilized as living factories that produce a variety of important substances. We are now able to manipulate gene expression of these organisms in order to synthesize molecules of commercial value. Thanks to the fields of genomics and proteomics, our knowledge of gene locations and functions enables biotechnologists to customize a suite of genes within an organism designed for a specific purpose. Additionally, it is possible to alter metabolic pathways of organisms to increase the efficiency and output of such biosynthesis.

The final products represent a variety of utilities. Certain molecules are useful in their own right as catalysts for industrial processes and as tools in laboratory research. Other molecules are prized as ingredients in a final product or even serve as the final product themselves. The general classes of molecules created and utilized in industrial biotechnology are presented below. The basic steps in creating these molecules are also described.

Commercial Products of Industrial Biotechnology

Industrial biotechnology has introduced several major classes of molecular products. Examples of enzymes, biopolymers, and metabolites are currently in use. Additional representatives of these classes are under investigation as potential substitutes for existing products and ingredients. It is the perspective of industrial biotechnology that biological systems are well suited to improve upon existing processes and products. Traditionally synthesized molecules usually require large amounts of energy and chemicals, and the resulting waste is likewise typically high. The substitution of molecules obtained through chemical synthesis with a biologically derived substance promises to reduce costs, energy expenditure, and downstream pollutants while potentially improving yield simultaneously. Considering their patentable nature, novel applications of biologically derived substances are also an area of intense study within the field of industrial biotechnology.

Enzymes

Enzymes are protein molecules that act as catalysts in chemical reactions. They speed up the reaction time by lowering the energy of activation, the amount of energy required for the reaction to proceed (**FIGURE 5.1**). Enzymes are by no means new to industrial products, but the production methods incorporated have changed with the introduction of biotechnological techniques.

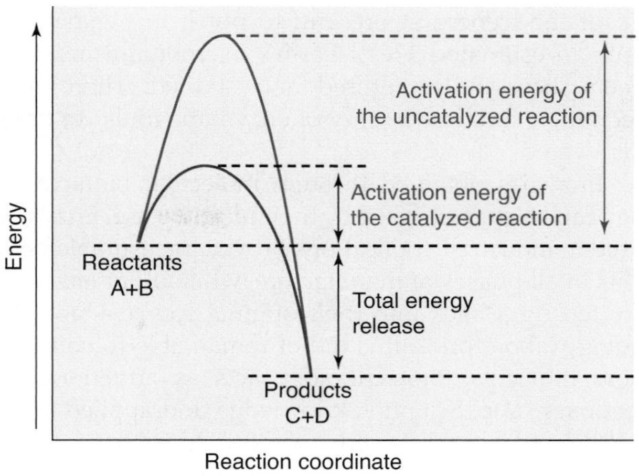

FIGURE 5.1 Progression of an uncatalyzed chemical reaction (red) and a catalyzed chemical reaction (blue) demonstrates the lowered energy of activation due to enzymatic action.

For example, the enzyme rennin used in cheese making was traditionally extracted from the stomachs of milk-producing animals, such as calves, goats, or sheep. The source of rennin would depend upon the geography and culture of the cheese maker, and in some cases the rennin even came from horses, camels, zebras, or certain plants such as thistle species. In the 1980s, scientists genetically modified both bacterial and fungal cells through recombinant DNA technology. The rennin gene was inserted and expressed in order to produce a cost-efficient, reliable, and consistent source of this commercially valuable enzyme. The recombinant version of rennin goes by the name **chymosin** and in 1990 became the first genetically modified food ingredient approved by the Food and Drug Administration (FDA).

Through the processes of evolution, every enzyme is well adapted to its environment. Therefore, any enzyme of interest has optimal conditions that foster maximum enzymatic action. Optima include pH, temperature, and concentration parameters. Locating an enzyme that will function best for a particular purpose requires the consideration of environmental conditions. If a manufacturing process operates at high temperatures, it is necessary to incorporate an enzyme with a corresponding optimal temperature. The same is true of pH, the relative acidity or alkalinity of an environment. If optimal conditions are not met, the enzyme will denature (become misshapen), losing its ability to function. The field of bioprospecting has increased our understanding of naturally occurring enzymes. As researchers discover new species living in all types of environments, the field of proteomics amplifies our knowledge of the enzymes they synthesize—and the optimal conditions of these enzymes. Target organisms may then become genetic donors, offering us libraries of their proteinaceous recipes for enzymes of value. We are likely only seeing the tip of the iceberg at present; to put it in perspective, only an estimated 1% of Earth's microorganisms have been successfully cultured and studied. There is a seemingly limitless bounty of enzymatic tools yet to be discovered.

Enzymes obtained through molecular biotechnology can be characterized by their ultimate use. Enzymes used in industrial applications provide measurable benefits in all phases of manufacture, while other enzymes are fed right back into their original source—biotechnology laboratories. It is rather remarkable to consider how molecular biotechnology acts as an enormous feedback loop. Scientific knowledge and applied technologies that are discovered and developed in the laboratory for commercial use can likewise improve upon the research process itself.

Industrial Enzymes

Industrial enzymes are used to catalyze reactions involved in manufacturing processes and final products. The introduction of enzymes into an engineered system adds several benefits. The enzyme may facilitate the steps of manufacture, by allowing it to proceed at a faster rate or by reducing the amount of mechanical effort and energy required. Chymosin is one such enzyme, as it catalyzes the coagulation of milk solids into curds (think of cottage cheese), separating the future cheese from the liquid portion of milk, known as whey (**FIGURE 5.2**). In some cases, the enzyme may simply serve as an ingredient which awards a unique property to the final product. Some examples of industrial enzymes are listed below.

Cellulase

Bacteria such as *E. coli* and *Streptomyces spp.* and fungi including *Trichoderma reesei* and *Aspergillus spp.* are used as biofactories in the production of the enzyme cellulase (**FIGURE 5.3**). As its name implies, this enzyme catalyzes the degradation of cellulose, a structural polysaccharide found in plant cell walls, into its glucose monomers. One use for cellulase is in the production of pet foods and livestock feed. When ingested, the enzyme eases the digestion of cellulose, the insoluble fiber within the food products. Cellulase is also used to treat plant matter in preparation for fermentation into biofuels.

FIGURE 5.2 The production of cheese involves the coagulation of milk aided by the addition of chymosin, which separates the curds (solid) from the whey (liquid). The curds will be pressed into molds and aged in order to become cheese.© Olaf Speier/ShutterStock, Inc.

FIGURE 5.3 Transgenic *Streptomyces* bacteria in culture. The bacteria have been genetically modified to produce the cellulase enzyme. © National Renewable Energy Laboratory/U.S. Department of Energy/Science Source

Another application for cellulase is within the textile industry. Cellulase is used extensively to soften plant-based fabrics, for example cotton. Cotton fabrics are subjected to treatment with cellulase enzymes just long enough to partially digest the tough cellulose fibers. Think about the last time you shopped for jeans and the ranges of softness (or firmness) available in the denim material. Clever marketing on the label may cite stonewashing as the method, but the variety of textures we enjoy as consumers is actually due to the action of cellulase.

Recent studies investigating cellulase performance have paired the enzyme with nanotechnology in order to provide additional benefits to textiles. When added to cellulase-treated cotton, nanoparticulate clay improves dye adherence and color brightness of fabrics. This treatment has the added benefit of rendering textiles antifungal and antibacterial.

Subtilisin and Lipase

Subtilisin and lipase are industrial enzymes providing functional properties to a final commercial product. While they have additional uses, the enzymes subtilisin and lipase are both ingredients in laundry detergents. They improve the cleaning performance by aiding in the digestion of food particles and other fabric stains.

Subtilisin belongs to a class of enzymes known as proteases. **Proteases** catalyze the breakdown of protein molecules into their amino acid monomers. They act by initiating proteolysis, which hydrolyzes the peptide bonds within the polypeptide chain. Thus, adding subtilisin to laundry detergent triggers the dissolution of protein-based stains.

Subtilisin is named after its source organism, the bacterium *Bacillus subtilis*. Proteases such as subtilisin are also useful in brewing beer. Protein solids can accumulate during the production process and create a cloudy appearance. The proteins are hydrolyzed with the enzymes, resulting in a beer with greater clarity.

Lipase acts in a similar manner to subtilisin, only the target molecules are lipids, or fats. Lipase belongs to a class of enzymes known as the **esterases**, which cleave ester compounds, such as lipids, into their constituents of alcohol and fatty acid. Derived from the bacterium *Pseudomonas alcaligenes*, lipase is essential in the removal of oil and grease stains. Otherwise, a combination of high temperature and high alkalinity is required to eliminate lipid-based stains, which is quite energy-intensive and often damaging to fabrics. Lipase enzymes are also often used in cheese making. The partial breakdown of naturally-occurring fats via lipase action creates a more complex flavor profile in the final cheese product.

Researchers have overcome the initial high costs of producing lipase through the intersection of genomics and recombinant DNA technology. Early experiments involving the insertion of the lipase gene (termed *lipA*) into a variety of organisms yielded prohibitively low expression levels of the lipase enzyme, making the production costs too high. Investigators discovered a second gene, dubbed *lipB*, that acts as a helper to *lipA*. When both genes were inserted into a host organism, the expression of lipase increased an average of 35 times greater than the source organism (**FIGURE 5.4**).

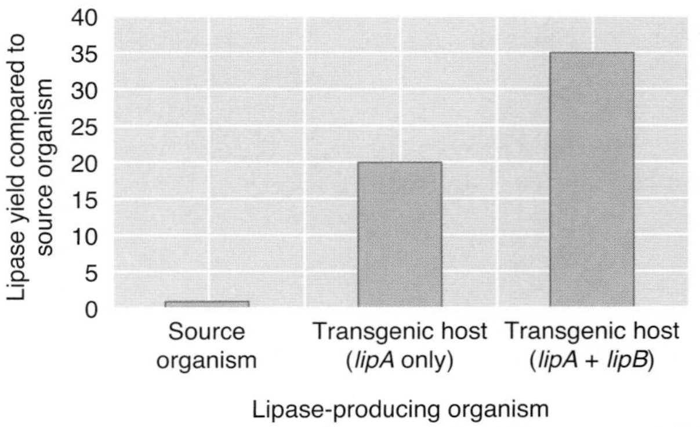

FIGURE 5.4 Insertion of the structural *lipA* gene into a variety of hosts averaged a 20-fold increase in lipase expression over the source organism, whereas the insertion of both *lipA* and *lipB* averaged a 35-fold increase. Data from Gupta, R. et al. "Bacterial lipases: An overview of production, purification, and biochemical properties." *Applied Microbiology and Biotechnology* 64, no. 6 (2004): 763–781

Like cellulase, lipase is also being investigated in the synthesis of biofuels. Lipase will hydrolyze vegetable oils in preparation for metabolic conversion into fatty acid methyl ester (FAME). In this instance, the fungus *Rhizopus oryzae*, which expresses lipase, is acting as the organismal factory. This application of industrial biotechnology is promising as a recycling stream for sources of waste oil.

Lactase

In addition to chymosin, other enzymes are useful in food production. Lactase is used as an ingredient in dairy products for lactose-intolerant individuals. While these products are often labeled "lactose-free," this is not strictly true. As lactose-intolerant individuals are unable to synthesize their own lactase enzymes in order to digest lactose, it is added to dairy products as a digestive aid. Thus, the end result is the same, and it is as if the products really were lactose-free.

A member of the β-galactosidase class of enzymes, the lactase enzyme degrades the lactose sugar found in milk, a disaccharide, into its monosaccharides, galactose and glucose (**FIGURE 5.5**). Without the supplemental enzyme, lactose-intolerant individuals must choose between a dairy-free diet or face the discomfort of indigestion after consuming a traditional dairy product (free of lactase).

Amylase

Amylases are a class of enzymes that hydrolyze starch into its glucose monomers. A major use of amylases in industry today is to facilitate the production of high

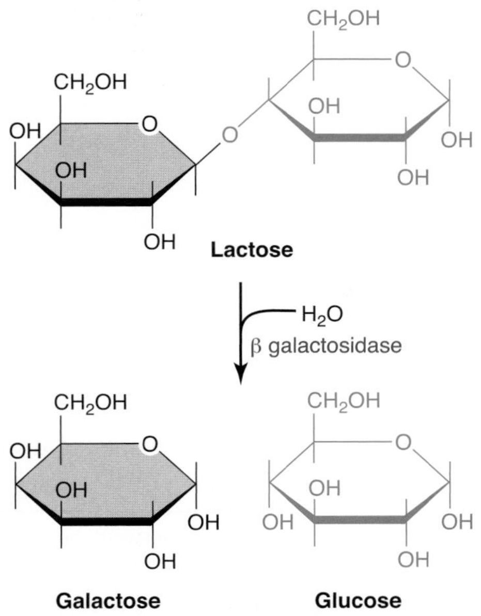

FIGURE 5.5 The β-galactosidase enzyme lactase catalyzes the hydrolysis of the disaccharide lactose into its monomer constituents, galactose and glucose.

fructose corn syrup (HFCS). Federal subsidies to corn producers have resulted in a long-standing surplus of feed corn (used for livestock), which industry has used as raw material for a variety of commercial ingredients and products. One example is utilizing the high levels of starch found in feed corn, which can be hydrolyzed into its sugary substructures of glucose and converted into HFCS with the help of three amylase types (**FIGURE 5.6**). The production of HFCS represents one of the most successful and widespread applications of industrial enzymes, primarily due to its unique economic circumstances. The availability of surplus corn and the high efficiency of the enzymatically-driven process makes HFCS significantly less expensive to produce than sucrose, more commonly known as table sugar. HFCS thus has a stronghold on the sweetener market (**FIGURE 5.7**). Amylases have a variety of additional uses (**TABLE 5.1**).

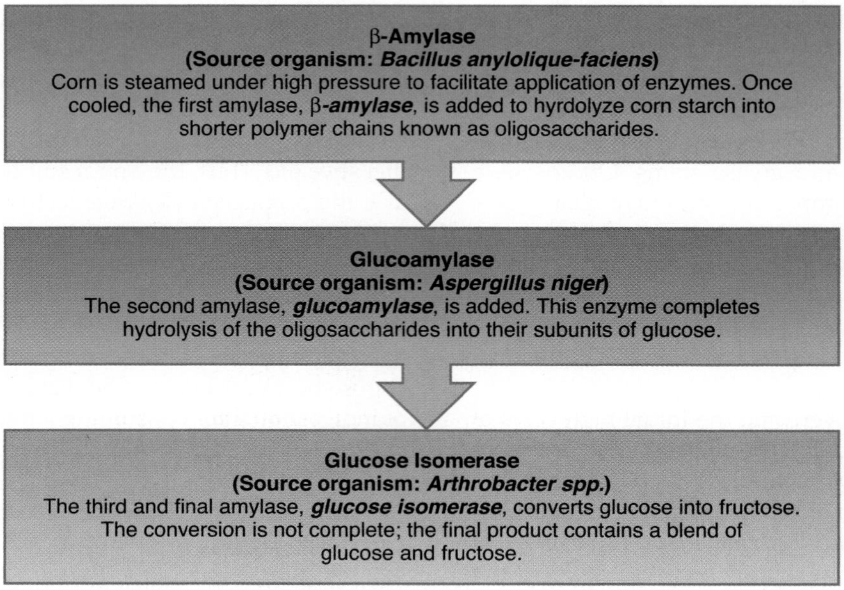

FIGURE 5.6 The production of high fructose corn syrup from corn starch requires the action of three amylase enzymes. Data from Litchfield, Ruth. *High Fructose Corn Syrup—How Sweet It Is*. Ames, IA: Iowa State University Extension and Outreach, 2008

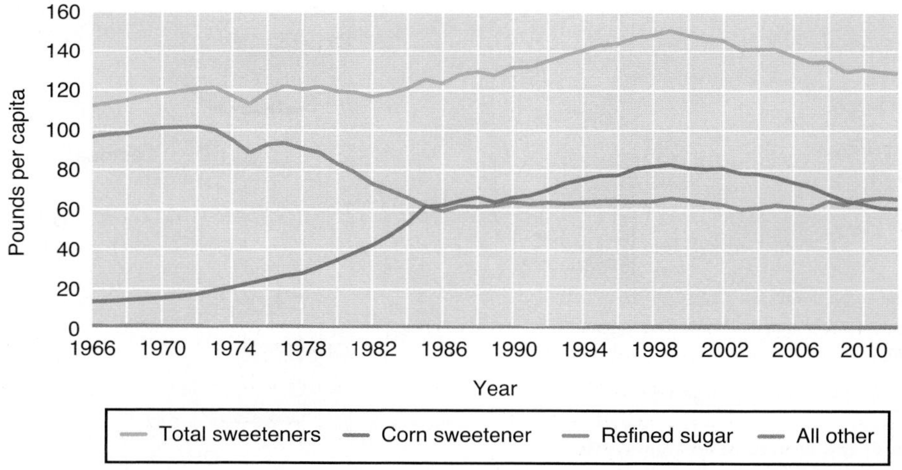

FIGURE 5.7 Sweetener consumption in the United States per capita, 1966–2010, as estimated by the USDA. High fructose corn syrup has overcome traditional table sugar. Data from USDA. "Sugar and Sweetener Outlook." Economic Research Service, United States Department of Agriculture. Washington, DC, 2013. Available at http://www.ers.usda.gov/data-products/sugar-and-sweeteners-yearbook-tables.aspx#25512

TABLE 5.1

Uses for Amylase Enzymes by Industry and Application

Industry	Application
Bread Making	Breaks down starches in flour sources; assists yeast in ethanol fermentation
Beer, Wine, and Spirits	Breaks down starches in materials used for alcohol production, such as barley, potatoes, and rice; assists yeast in ethanol fermentation and controls alcohol by volume (ABV)
Detergents	Used in laundry and dishwashing detergents to dissolve starch-based stains and food particles

Data from Aiyer, P.V. "Amylases and Their Applications." *African Journal of Biotechnology* 4, no. 13 (2005): 1525–1529

Research Enzymes

Some enzymes obtained from biotechnological discoveries stay in the laboratory and remain as research tools. These include restriction enzymes and the enzymes utilized in the polymerase chain reaction (PCR).

Restriction Enzymes

Restriction enzymes, also known as restriction endonucleases, cleave DNA at specific recognition sequences known as restriction sites (**TABLE 5.2**). Recombinant DNA technology and its myriad applications would not be possible without the discovery and commercialization of restriction enzymes. They are a vital tool in "cutting and pasting" DNA fragments for insertion of transgenes into genetically modified organisms. There are over 300 restriction enzymes currently available; as genomic and proteomic research continues the number will undoubtedly grow. Discovery typically entails the digestion of the same DNA molecule with different combinations of restriction enzymes, followed by gel electrophoresis to determine the resulting DNA fragment lengths. A map of recognition sites is then deduced based upon the fragment lengths (**FIGURE 5.8**).

Because restriction enzymes are found in a variety of organisms with a range of optimal growth conditions, it is much easier to insert the gene encoding a particular restriction endonuclease into a model species such as *E. coli*. This allows the uniform production of *E. coli* biofactories via cloning (**FIGURE 5.9**). This is preferred by biotechnologists, as otherwise one must directly culture source organisms in their different environmental requirements, a laborious and tedious process. Also, this method is preferred due to the rapid growth of *E. coli* and the bacterium's ability to easily overexpress the restriction enzyme under production.

PCR Enzymes

As previously discussed, the polymerase chain reaction is a method of obtaining multiple identical copies of

TABLE 5.2

Sequence Specificity of Some Restriction Endonucleases

Organism of Origin	Restriction Endonuclease	Recognition Sequence
Arthrobacter luteus	AluI	5′…AG↓CT … 3′
Anebaena variabilis	Ava I	5′...C↓PyCGPuG...3′[a]
Bacillus amyloliquefaciens H	Bam HI	5′...G↓GATCC...3′
Bacillus globigii	Bgl HI	5′...A↓GATCT...3′
Escherichia coli RY13	Eco RI	5′...G↓AATTC...3′
Escherichia coli J62 pLG74	Eco RV	5′...GAT↓ATC...3′
Haemophilus aegyptius	Hae II	5′... PuGCGC↓Py...3′
Haemophilus aegyptius	Hae III	5′...GG↓CC...3′
Haemophilus haemolyticus	Hha I	5′...GCG↓C...3′
Haemophilus influenzae Rd	Hind II	5′...GTPy↓PuAC...3′
Haemophilus influenzae Rd	Hind III	5′...A↓AGCTT…3′
Haemophilus parainfluenzae	Hpa I	5′...GTT↓AAC...3′
Haemophilus parainfluenzae	Hpa II	5′...C↓CGG...3′
Kliebsiella pneumoniae	Kpn I	5′...GGTAC↓C...3′
Moraxella bovis	Mbo I	5′... ↓GATC...3′
Nocardia otitidis-caviarum	Not I	5′...GC↓GGCCGC...3′
Providencia stuartii	Pst I	5′...CTGCA↓G...3′
Serratia marcescens	Sma I	5′...CCC↓GGG...3′
Streptomyces stanford	Sst I	5′...GAGCT↓C...3′
Xanthomonas malvacearum	XmaI	5′...C↓CCGGG...3′

[a]Py, pyrimidine; Pu, purine.

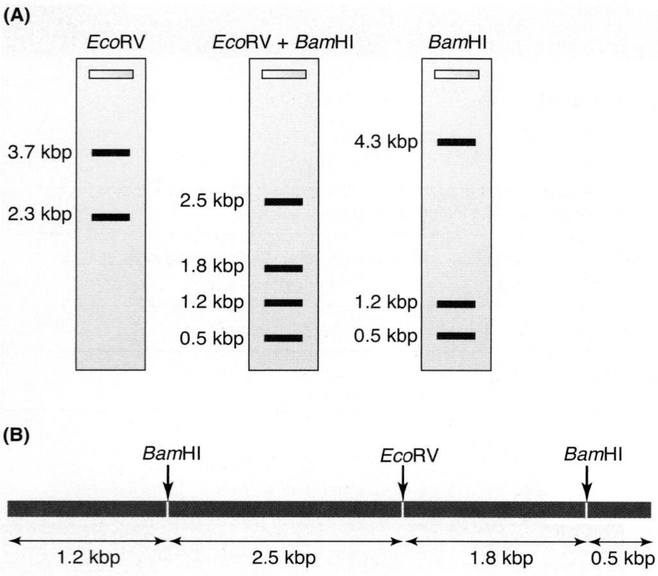

FIGURE 5.8 **Discovery of recognition sequences specific to each restriction enzyme involves (A) gel electrophoresis of identical DNA samples digested with various restriction enzyme combinations and (B) deduction of restriction site locations based upon fragment lengths obtained from step one.**

DNA. The steps of PCR are repeated to facilitate the exponential production of DNA clones:

1. DNA is denatured, whereby the two strands of the molecule are separated.
2. Primers are annealed to the single strands of DNA.
3. Elongation occurs via complementary base pairing.

Taq polymerase is a DNA polymerase enzyme obtained from *Thermus aquaticus*, a species belonging to Domain Archaea that inhabits hot springs (**FIGURE 5.10**). As a striking example of optimal temperature, this enzyme has adapted to its hot spring environment and therefore is ideal for the high temperatures required in the steps of PCR. Organisms that thrive in high temperature environments are known as **thermophiles** (literally "heat-loving"). The U.S. government grants special permits to researchers for the purposes of bioprospecting geysers in Yellowstone National Park, leading to the identification of additional

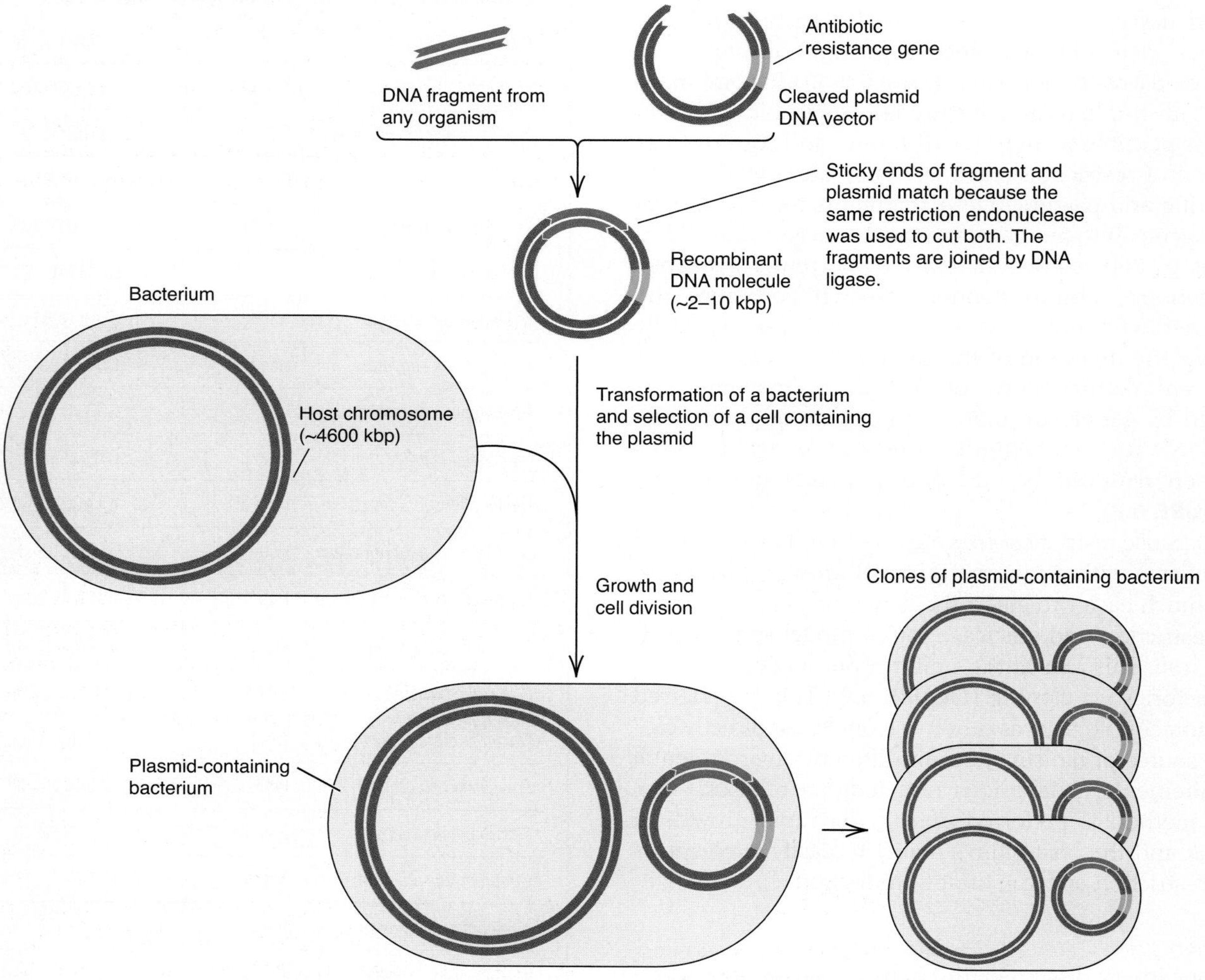

FIGURE 5.9 **The process of gene insertion and cloning utilized in production of restriction enzymes.**

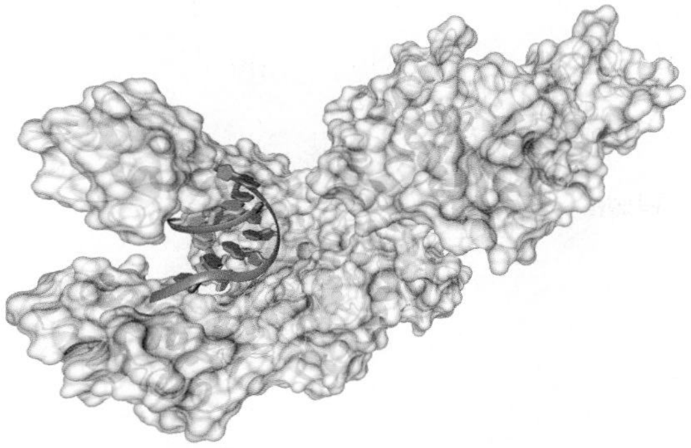

FIGURE 5.10 **A model of the Taq polymerase enzyme, obtained from the archaean *Thermus aquaticus*.** © molekuul.be/ShutterStock, Inc.

TABLE 5.3

PCR Enzymes by Thermophilic Source Organism

Polymerase Enzyme	Thermophile Source Organism
Taq	*Thermus aquaticus*
Pfu	*Pyrococcus furiosis*
Deep Vent	*Pyrococcus spp.* Strain GB-D
Vent	*Thermococcus litoralis*
Pfx	Thermococcus spp. Strain KOD

Data from Cline, Janice et al. "PCR Fidelity of Pfu DNA Polymerase and Other Thermostable DNA Polymerases." *Nucleic Acids Research* 24, no. 18 (1996): 3546–3551.

thermophiles (**FIGURE 5.11**). For example, *Pyrococcus furiosus* is now a source for Pfu DNA polymerase, which can be used as an alternative to Taq polymerase (**TABLE 5.3**).

Biopolymers

Biopolymers are large molecules comprised of multiple monomer subunits that are produced by transgenic organisms, such as microbes and plants. They are akin to traditional chemically-synthesized plastics and adhesives in function. Biopolymers provide several benefits to industry. They enable the production of in-demand molecules without the need for raw materials derived from fossil fuels. Rather, the insertion of polymer-encoded genes into living factories provides a self-replicating supply, effectively replacing an untenable model that harms the environment. Biopolymers therefore act as a sustainable source of plastics and adhesives that are readily renewable and pose no risk of depletion. Additionally, bioengineers are investigating ways to modify existing biopolymers for improved characteristics of both the production processes and the final product. Biopolymers promise to reduce costs, improve energy efficiency, and minimize downstream pollutants that are common to chemically synthesized polymers.

FIGURE 5.11 **Tom Brock, the microbiologist who discovered *Thermus aquaticus*, is shown collecting hot spring samples in Yellowstone National Park.** © Peter Menzel/Science Source

Bioplastics

Traditional plastics are used heavily for packaging, coatings, and films. As they are used for such purposes, desirable traits include strength and durability. These traits are useful while the plastic is carrying out its intended use, but poses problems when disposed. Commonly used plastics include polyvinyl chloride (PVC), polyethylene, polystyrene, and polypropylene, all of which are synthetic polymers produced from nonrenewable petroleum and/or natural gas. Because they are not biodegradable, it is a reasonable conclusion that every molecule of plastic that has ever been thrown in the refuse bin is still around somewhere on this planet, the only exception being the miniscule fraction of plastics either burned for energy or recycled each year. In 2008, nearly 86% of plastics consumed in the United States ended up in landfills. Perhaps even more disturbing, large islands of plastic debris float in the world's oceans, negatively affecting the marine ecosystem (**FIGURE 5.12**). Alternative sources of polymer materials are vital for industry as consumers demand more environmentally safe options. As fossil fuel resources diminish, the demand becomes even greater to find alternative sources of plastics.

FIGURE 5.12 An example of floating plastic debris observed near the port of Le Havre, France. © Alex Bartel/ Science Source

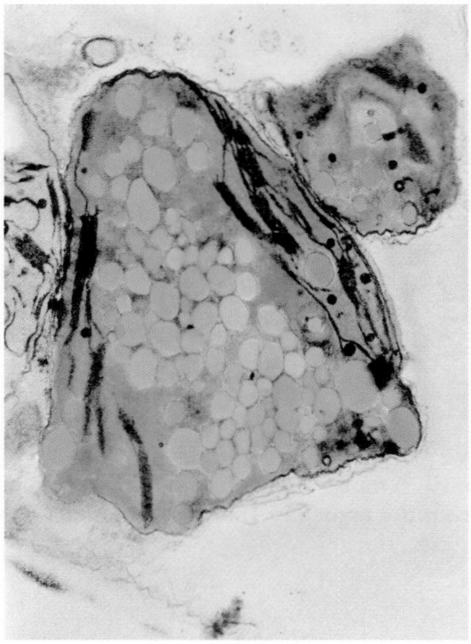

FIGURE 5.13 Bioengineered *Arabidopsis thaliana* exhibiting large inclusion bodies containing the biopolymer poly(β-hydroxybutyrate) (PHB). © Dr. Chris Somervile/Science Source

PHA

Poly(β-hydroxyalkanoate) (PHA) is a family of polymer molecules that are naturally synthesized by a number of microorganisms, including members of the genera *Pseudomonas, Streptomyces, Clostridium, Spirulina,* and *Vibria*. Depending on the species, anywhere from 30% to 80% of the organism's dry weight may be composed of PHA. Microbial organisms utilize PHA as an energy storage molecule. Synthesis of PHA is in response to an imbalance of environmental conditions; for example, when an excess of glucose is present, the microbes may sequester the energy available by producing PHA. The polymer is stored within the unicellular organisms as inclusion bodies (**FIGURE 5.13**). The most common molecule of the PHA family currently in commercial production is PHB [poly(β-hydroxybutyrate)], which has chemical and structural properties similar to polypropylene. While microbes such as *E. coli* are commonly used as PHA production factories, insertion of a PHB gene into plant species has demonstrated high yields as well. Transgenic *Arabidopsis thaliana* expressing a PHB gene produces up to 40% dry weight of the biopolymer.

PHA is promising in applications where long-term durability is unnecessary. Because bioplastic is biodegradable, it will be useful for disposable items. PHA can serve as carrier packaging for agricultural treatments (pesticides, fertilizers, herbicides, etc.), encapsulated medications, vitamins, and nutritional supplements, surgical items like sutures and staples, and personal hygiene products such as feminine napkins and diapers.

PLA

Polylactic acid (PLA) is a biopolymer produced from waste carbon sources. As with HFCS, PLA taps into the nation's corn surplus and converts sugars derived from corn starch into polylactic acid. The biopolymer is currently under production for use in packaging, clothing, and bedding (**FIGURE 5.14**). PLA has the advantage of being a biodegradable alternative to traditional plastics while also making use of a readily available carbon source. Outside of the United States, tapioca starch and sugarcane are raw materials for PLA production. Any instance of biopolymer synthesis that utilizes plant matter can be integrated into crop production. The biomass resulting from biopolymer extraction can be

FIGURE 5.14 A biodegradable plastic cup produced from PLA (polylactic acid).

recycled and applied to fields as fertilizer, providing another distinct advantage.

Xanthan Gum

The soil bacterium *Xanthomonas campestris* synthesizes the biopolymer xanthan gum, a material used extensively in commercial products as a thickener, stabilizer, and/or emulsifier. Products containing xanthan gum include food, cosmetics, and personal care items like toothpaste.

Xanthomonas campestris can make use of several carbon sources for energy capture, including glucose, sucrose, and starch. In consideration of cheap and plentiful carbon sources, biotechnologists have engineered a strain of *X. campestris* by inserting the *E. coli lacZY* genes. This enables the transgenic bacteria to metabolize lactose as an alternative carbon source, which is not possible in the wild type variety of the species. The engineered strain can feed off the large amounts of lactose-rich whey produced by the dairy industry, which would otherwise represent a waste product in cheese making.

Rubber

Rubber is unique among polymer materials in that it cannot be synthesized by chemical means. It represents a biopolymer used long before the advent of molecular biotechnology, as it is obtained directly from its only natural source, the latex-secreting Brazilian rubber tree (*Hevea brasiliensis*). Monoculture of these trees puts them at risk for pathogen infection and heightens the real potential for devastation of our sole source of rubber. Another concern is the fluctuating price of rubber, perpetuated by labor costs as latex must be extracted by hand (**FIGURE 5.15**). Genetic modification of other plant species, bacteria, and yeast through recombinant DNA technology are all under investigation as a means to diversify the availability of rubber. Biotechnology researchers have also tinkered with the chemical properties of rubber to make it hypoallergenic. This is a useful genetic manipulation due to increases in latex allergies.

Bioadhesives

Aquatic bioprospecting has led to the discovery of naturally occurring adhesive substances in bivalve organisms. The blue mussel (*Mytilus edulis*) is one such bivalve, as these mollusks utilize adhesive proteins called byssal threads in the process of aggregation, whereby the mussels attach themselves to one another, forming dense clumps on tidal zone surfaces (**FIGURE 5.16**). In ancient times, byssal threads, also known as "sea silk," were used to weave resilient fabric that was even used as currency. The adhesive properties of byssal threads are now the subject of biotechnology research.

Bioengineers inserted and expressed the gene encoding the adhesive protein into *Saccharomyces cerevisiae* yeast cells as a commercial means of production. This is more efficient to extract the adhesive than directly through its source organism as it involves a reliable source, requires less steps to harvest, and takes less time and energy, all while reducing costs. Additional studies are evaluating industrial waste streams as potential carbon sources for growing transgenic microbes that express adhesive proteins. Adoption of such methods would provide the additional benefit of redirecting waste into value-enhancing commercial processes.

Metabolites

Metabolites are molecules produced as natural byproducts and their intermediaries during intracellular reactions. **Primary metabolites** are normally produced during a microbial culture's growth phase, whereas secondary metabolites are typically synthesized late in the

FIGURE 5.15 A rubber tree plantation in Thailand displays the uniformity characterized by monoculture. The suspended bowls collect the tapped latex, which must be routinely monitored and collected by hand. © WathanyuSowong/ShutterStock, Inc.

FIGURE 5.16 Blue mussels (*Mytilus edulis*) in the process of aggregation that forms a clumped distribution pattern in the population. Byssal threads containing protein adhesives useful to industry enable aggregation. © AtWaG/iStockphoto.com

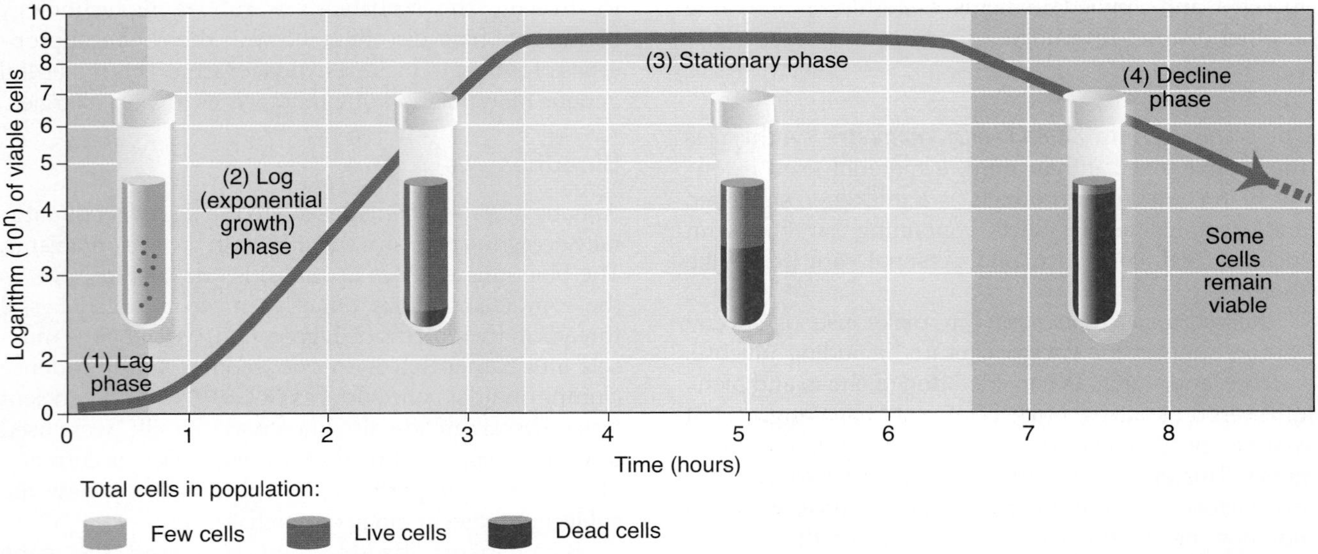

FIGURE 5.17 A typical growth curve of microbial culture exhibits an initial lag phase of slow growth, exponential growth during the log phase, followed by equilibrium (zero net population growth) during the stationary phase. The population then declines as cells die.

growth phase and into the stationary equilibrium phase (**FIGURE 5.17**). Primary metabolites include both intermediates and final products of metabolic pathways. They are essential for cell functions including growth, reproduction, and development. Primary metabolites are thusly produced in greatest quantities when the cell culture is experiencing exponential growth (i.e., high cell division rates). **Secondary metabolites** are not as well characterized; many have unknown functions. They are theorized to play a role in the ecological interactions displayed by the organism by conferring a selective advantage. As they are synthesized in the growth curve well after primary metabolites, they represent metabolic end products. Both classes of metabolites may be produced through biotechnology and possess members of commercial utility.

Nutritional Supplements

Nutritional supplements are a $27 billion per annum industry, with over half of Americans regularly taking them. Traditional methods of producing supplements such as vitamins and nutrients include chemical synthesis and microbial fermentation. Biotechnologists are discovering ways to manipulate metabolic pathways in order to bioengineer organisms that can produce supplements with improved benefits.

Vitamins

L-Ascorbic acid is a primary metabolite more commonly known as vitamin C. In traditional commercial production, it is derived from glucose that is first fermented and then chemically treated in a number of steps. By combining genes of different bacterial species, a bioengineered microbe eliminates any chemical treatment and confines the entire biosynthesis of vitamin C in a single organism. The genome of the bacterium *Erwinia herbicola* encodes a metabolic pathway that converts glucose into an intermediary of L-ascorbic acid, 2,5-diketo-D-gluconic acid. Fermentation of *E. herbicola* as is would yield similar results as step one of its traditional synthesis process. But rather than continuing with the chemical treatment steps, transgenic *E. herbicola* has been transformed to conclude vitamin C biosynthesis in its entirety. The enzyme 2,5-diketo-D-gluconic acid reductase is expressed in the bacterium *Corynebacterium spp*. The gene encoding the reductase was inserted into *E. herbicola* in order to construct a complete metabolic pathway for L-ascorbic acid within the bioengineered organism. The commercial scale-up of this technology will serve to diminish chemical waste and pollution while creating efficiencies in production leading to reduced costs.

Antioxidants

A variety of amino acids produced via recombinant DNA technology function as antioxidants. Commercially valuable amino acids may be produced by microbial fermentation. Prior to biotechnology, researchers would attempt to amplify metabolic pathways in order to improve production rates of these primary metabolites. The mutation of amino acid–producing microbes was induced through exposure to radiation or chemical mutagens. This was the proverbial shot in the dark, as the results were impossible to predict. Trial and error was the only means to create genetically modified microbes. With recombinant DNA, it is now possible to overexpress amino acids within microbes while maintaining fermentation as the production process. Antioxidant amino acids obtained via genetically modified biosynthesis include cysteine, tryptophan, and histidine.

Lycopene is a carotenoid pigment and nutritional supplement highly prized for its antioxidant properties. Studies have indicated lycopene lowers low density lipoprotein (LDL) cholesterol and the risk of heart disease and is being investigated as a form of cancer treatment. This antioxidant has been extracted from plant matter for the introduction into nutritional supplements. Insertion of the lycopene encoding–gene into bacteria and via in vitro plant tissue culture are successful methods of lycopene synthesis in biotechnology research. Engineering of lycopene into more bioactive forms is also an area of research.

Pigments and Dyes

Pigments and dyes are secondary metabolites. Natural dyes are often obtained in batch processes that make them inefficient and expensive. A multitude of synthetic dyes are on the market, though these are produced with petroleum-derived raw materials that are nonrenewable and sources of pollution. Engineering of metabolic pathways is an alternative method. The blue dye indigo has been extracted from true indigo (*Indigofera tinctoria*) and woad (*Isatis tinctoria*) plants for millennia and is more commonly synthesized by chemical means today. Production of indigo is cheap, though highly toxic. Bioengineered sources of indigo would improve occupational safety and prevent chemical pollution. *E. coli* bacteria are transformed to overproduce tryptophan, the amino acid that serves as raw material for the metabolic synthesis of indigo (**FIGURE 5.18**). The overproduction results in excessive amounts of indigo, improving yields and enabling dye extraction.

Industrial Chemicals

Biosynthesis of industrial chemicals through traditional microbial fermentation has tremendous potential for improvements by biotechnology. Primarily, bioengineering will improve upon the resulting quantity of derived substances. Inserting additional copies of the encoding gene for a given chemical of interest is one method. The simple concept is more genes yield more product. This is useful when the chemical is directly expressed by a gene or set of genes, but the process is more indirect when the desired molecule is a metabolite. In such cases, enzymatic genes may be more appropriate additions to a bioengineered pathway. The inclusion of such genetic information will encode and express the enzymes necessary to catalyze the targeted reaction, resulting in synthesis of desired metabolites.

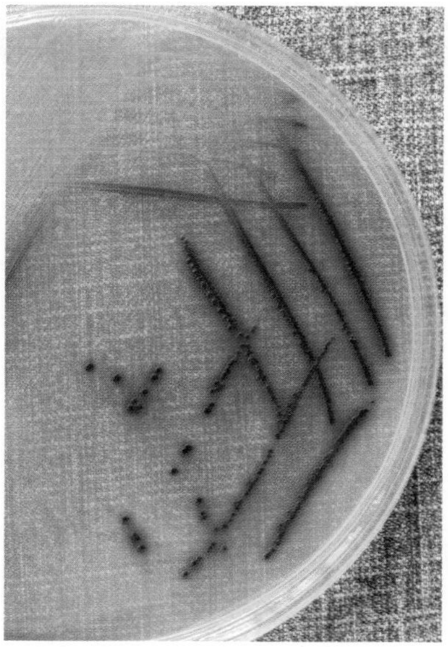

FIGURE 5.18 Culture of *E. coli* (foreground) transformed to overexpress tryptophan yields high indigo dye production. Denim dyed with indigo (background) is a source of comparison. © Dr. Jeremy Burgess/Science Source

Examples of industrial chemicals with improved biotechnological production methods are summarized in **TABLE 5.4**.

Commercial Processes of Industrial Biotechnology

Production of the various molecules described above involves two basic steps. **Upstream processing** entails the actual bioengineering of the genetically modified organism and allowing the biosynthesis of the molecule of interest to proceed. Several methods of biosynthesis exist for living factories, but they all require fermentation of some sort in order for a large population to grow. (A notable exception lies with plants and animals. Plant and animal cells may be cultured in the process described below, or the whole multicellular organism may be allowed to grow. This alternative is discussed separately.) High growth will yield the desired molecule in bulk. **Downstream processing** is the harvesting of the final product, which typically requires extraction, stabilization, purification, and preservation.

Upstream Processing

One of the first, and most important, steps of upstream processing is selecting the organism that will serve as the vehicle for biosynthesis. Bacteria, fungi, animal, and plant cells may be considered; each has advantages and disadvantages (**TABLE 5.5**). The examples above highlight some of the decisions researchers face when constructing a genetically modified organism or engineering a biochemical pathway. Once the chosen

TABLE 5.4

Industrial Chemicals Produced Through Molecular Biotechnology

Metabolite	Application	Source Organism
Acetic acid	Chemical synthesis Industrial solvent	*Acetobacter*
Acetone	Industrial solvents Nail polish remover Paint thinner Chemical synthesis Denaturation Pharmaceutical fillers Cosmetics and skincare	*Clostridium*
Acrylic acid	Chemical synthesis	*Bacillus*
Alcohols	Beer, wine, and spirits Biofuels Chemical synthesis	Multiple
Amino acids	Flavor enhancers Nutritional supplements Pharmaceuticals Cosmetics Food preservation Chemical synthesis	Multiple
Formic acid	Electroplating Leather treatments Textile dyeing	*Aspergillus*
Fumaric acid	Chemical synthesis Acidulant Dyeing	*Rhizopus*
Glycerol	Solvent Sweetener Explosives Cosmetics and skincare Antifreeze	*Saccharomyces*
Glycolic acid	Adhesives Cleaners	*Aspergillus*
Hyaluronic acid	Cosmetics Skincare products Pharmaceuticals	*Bacillus*
Melanin	Sunscreens UV-protective coatings Cosmetic pigments	*Streptomyces*
Methyl ethyl ketone	Solvent Explosives Resins	*Chlamydomonas*
Oxalic acid	Dyeing Bleaching Cleaners	*Aspergillus*
Propylene glycol	Antifreeze Solvent Resins Mold control agent	*Bacillus*
Succinic Acid	Flavoring agents Dyes Perfumes Coatings Pharmaceuticals	*Rhizopus*

species is engineered, the next step in upstream processing is to grow the cells in culture. Industrial biotechnology scales up growth parameters established during laboratory research in order to provide mass production. The challenges of upstream processing can be complicated and daunting, but the rewards make such struggles well worth the effort. Specific techniques used in upstream processing or the unique sequence of events may be patentable if they are novel and exhibit utility.

Bioreactors

Bioreactors are used to maximize growth by continuously monitoring and adjusting environmental conditions. **Bioreactors** are essentially large fermentation vats or tanks housing cells suspended in growth media broth (**FIGURE 5.19**). Fermentation can be defined in different ways. A microbiologist or biochemist would explain fermentation as the process of energy capture in the absence of oxygen, that is, under anaerobic conditions. For the industrial biotechnologist, **fermentation** refers to any process enabling cell growth for the purposes of biosynthesis via gene expression and/or metabolic pathways (whether aerobic or anaerobic). Bioreactors contain various sensors and probes that provide a constant stream of data regarding temperature, pH, oxygen availability, and other aspects of the environment within. Bioreactor fermentation results in a large population of transformed cells known as **biomass**.

Fermentation is accomplished within a bioreactor by either batch fermentation or continuous fermentation.

FIGURE 5.19 A bioreactor used in industrial biotechnology to produce biomass during upstream processing. © Maximilian Stock Ltd/Science Source

TABLE 5.5

Some Advantages and Disadvantages to Host Organisms Used in Upstream Processing

Organism	Advantages	Disadvantages
Bacteria	• Bacterial fermentation is well understood • Comparatively easy to genetically modify • Fast growth	• Prokaryotic cells limit protein modifications • Intracellular production requires extra downstream processing
Fungi	• Eukaryotic modifications possible • Sources of commercially valuable substances	• Intracellular production requires extra downstream processing • Cell walls require extra downstream processing
Plant	• Eukaryotic modifications possible • Sources of commercially valuable substances • Sexual reproduction (if whole organism used) • May simplify downstream processing • Fruit or vegetable as delivery vehicle	• Cell walls require extra downstream processing • Won't synthesize certain proteins • Eukaryotic modifications may be incompatible for humans
Animal	• Products compatible for human use • Eukaryotic modifications possible • Sexual reproduction (if whole organism used) • May simplify downstream processing • Eggs, meat, or milk as delivery vehicle	• High contamination risk • Complex nutritional requirements in culture • Slow growth

Regardless of the chosen method, certain basic steps are standard in upstream processing:

1. Sterilization of the bioreactor and all other fermentation equipment
2. Preparation and sterilization of the culture medium
3. Culture of the genetically modified organism
4. Inoculation of the culture into the growth medium
5. Cell growth and biosynthesis

During step five, environmental conditions must be meticulously controlled in order to provide optimal growth. Depending on the chosen host organism, the best conditions for growth will vary but are most often associated with the environment to which the organism is naturally adapted.

Designing the appropriate growth conditions requires consideration of several factors, including temperature, dissolved oxygen content, pH, and nutrition. Ideal temperature ranges are displayed in **FIGURE 5.20**.

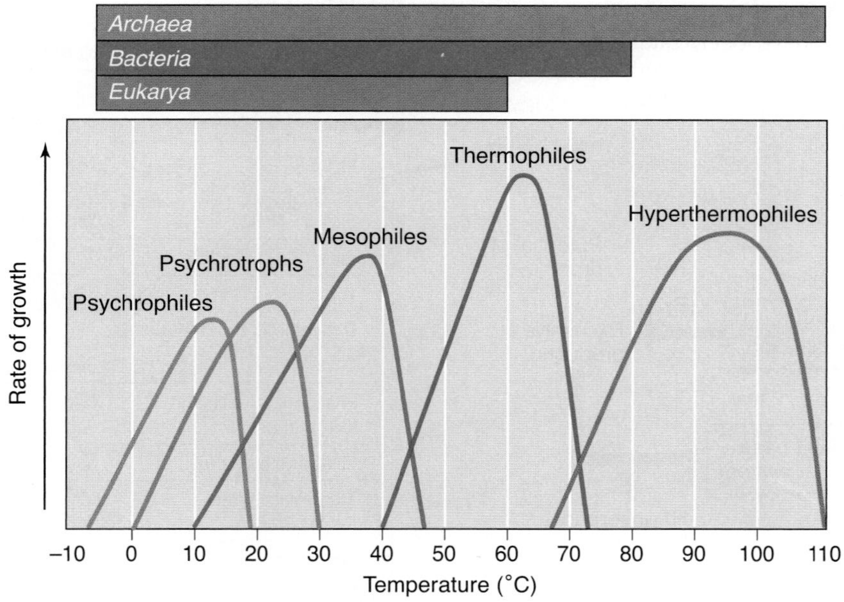

FIGURE 5.20 Ideal temperatures will vary upon the chosen host organism.

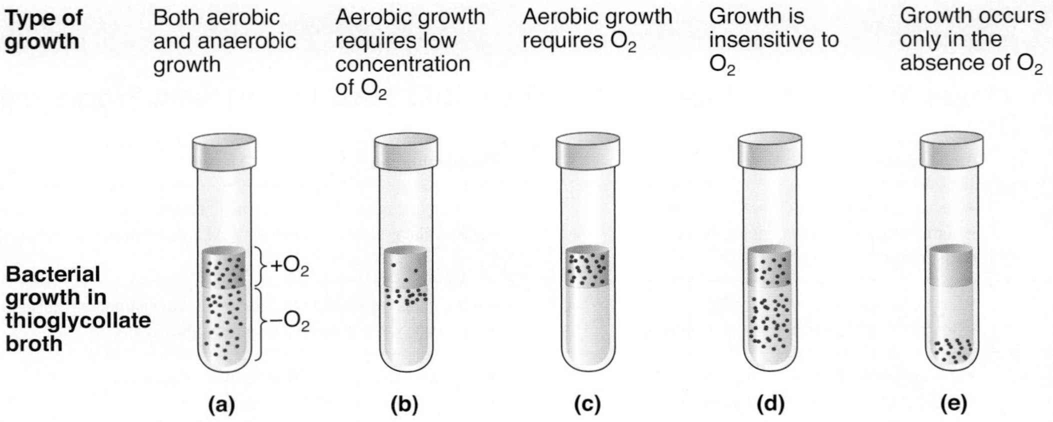

FIGURE 5.21 **Bacterial host organisms will vary in their oxygen requirements.**

If the host organism is bacterial, it may or may not require oxygen. If it requires oxygen, the amount must be controlled because some species thrive in high oxygen while others thrive with relatively low amounts (**FIGURE 5.21**). To maintain optimal pH levels, bioreactor systems are typically buffered to take up excess hydrogen and hydroxide ions as they are produced in metabolic processes (**FIGURE 5.22**). Another consideration is the nutritional requirements of the host organism as it grows. Culture media are available for different classes of organisms based upon their metabolic needs. Examples of available culture media are summarized in **TABLE 5.6**. Prepared media are not always utilized, however. For example, genetically modified *Xanthomonas campestris* used in xanthan gum production is grown on waste streams of whey. Thankfully the field of microbiology has contributed volumes of information regarding optimal environments for unicellular organisms (**FIGURE 5.23**). Prime conditions for animal and plant cultures have been established through biotechnology research, which has built upon existing knowledge obtained through the fields of zoology and botany, respectively.

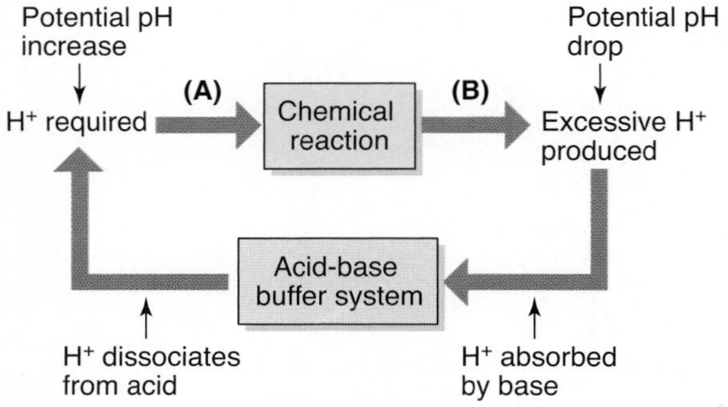

FIGURE 5.22 **Bioreactor systems are typically buffered to maintain a constant pH.** Ports on the bioreactor allow the addition of hydrogen and/or hydroxide ions as needed to adjust pH.

Batch Fermentation

Batch fermentation is just as the name would imply; biomass is cultured in batches. A schematic of a batch bioreactor is displayed in **FIGURE 5.24**. Each batch is allowed to grow to its desired population size and downstream processing begins. The five steps above are repeated to produce batch after batch. The specific time allotment for biomass growth in the batch method is a reflection of the target molecules to be harvested. For example, a secondary metabolite will require longer periods in the bioreactor because they are end products of metabolism. For reasons explained below,

TABLE 5.6

Composition of a Complex and Synthetic Growth Medium

Ingredient	Nutrient Supplied	Amount
A. Complex Agar Medium		
Peptone	Amino acids, peptides	5.0 g
Beef extract	Vitamins, minerals, other nutrients	3.0 g
Sodium chloride (NaCl)	Sodium and chloride ions	8.0 g
Agar		15.0 g
Water		1.0 liter
B. Synthetic Broth Medium		
Glucose	Simple sugar	5.0 g
Ammonium phosphate ($(NH_4)_2HPO_4$)	Nitrogen, phosphate	1.0 g
Sodium chloride (NaCl)	Sodium and chloride ions	5.0 g
Magnesium sulfate ($MgSO_4 \cdot 7H_2O$)	Magnesium ions, sulphur	0.2 g
Potassium phosphate (K_2HPO_4)	Potassium ions, phosphate	1.0 g
Water		1.0 liter

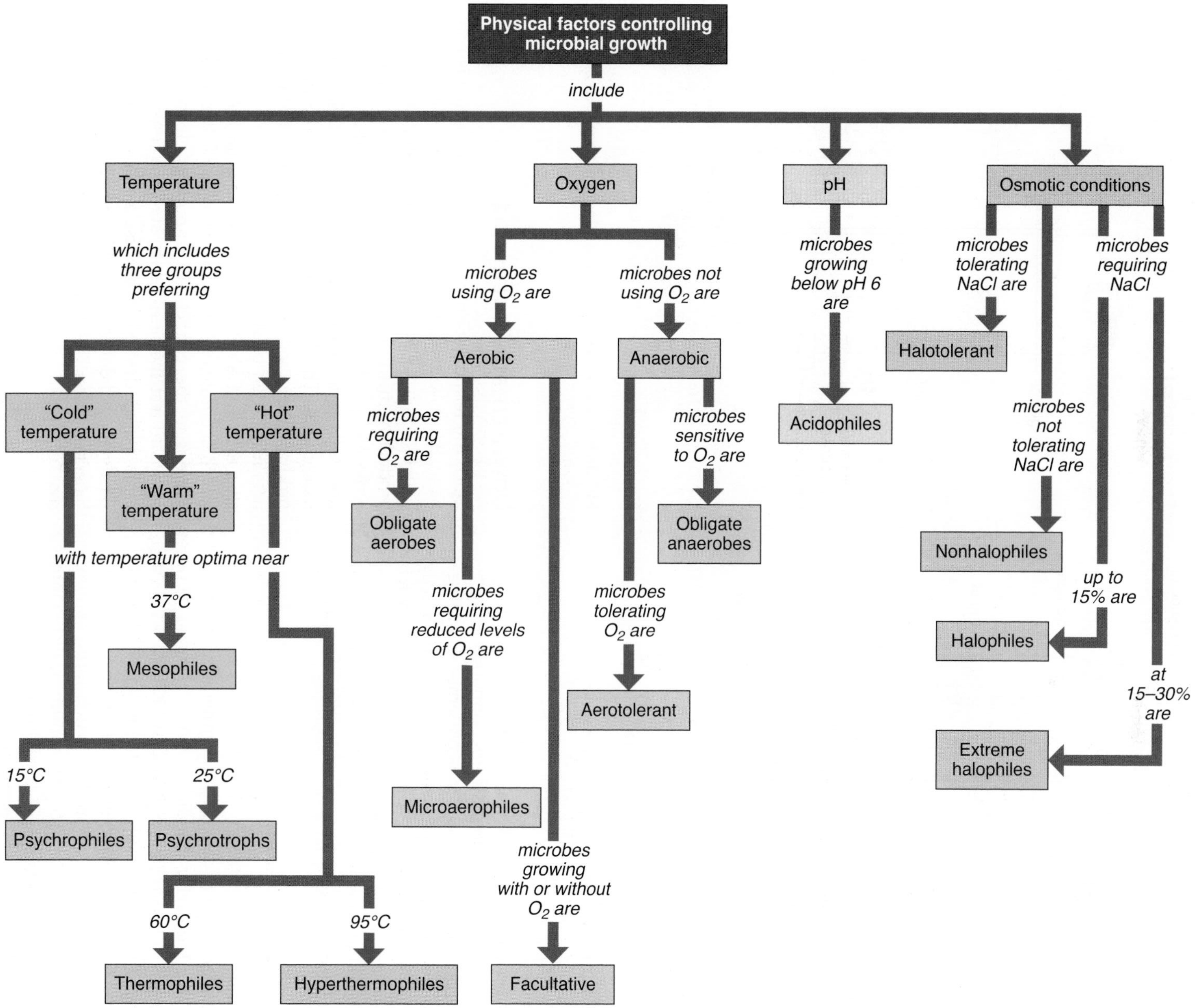

FIGURE 5.23 **A flow chart determines appropriate environmental growth conditions for different classes of microbes.**

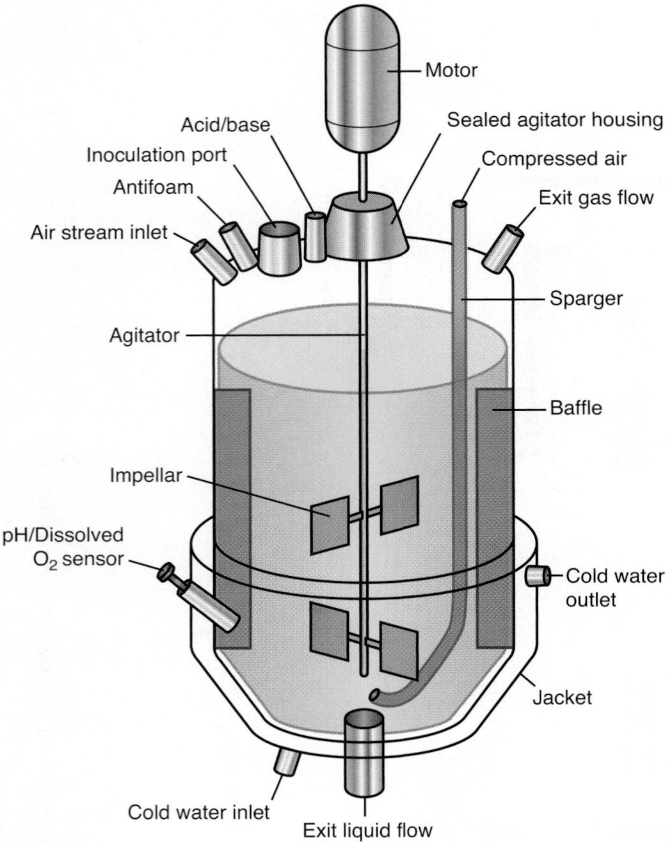

FIGURE 5.24 **Schematic of a batch bioreactor.**

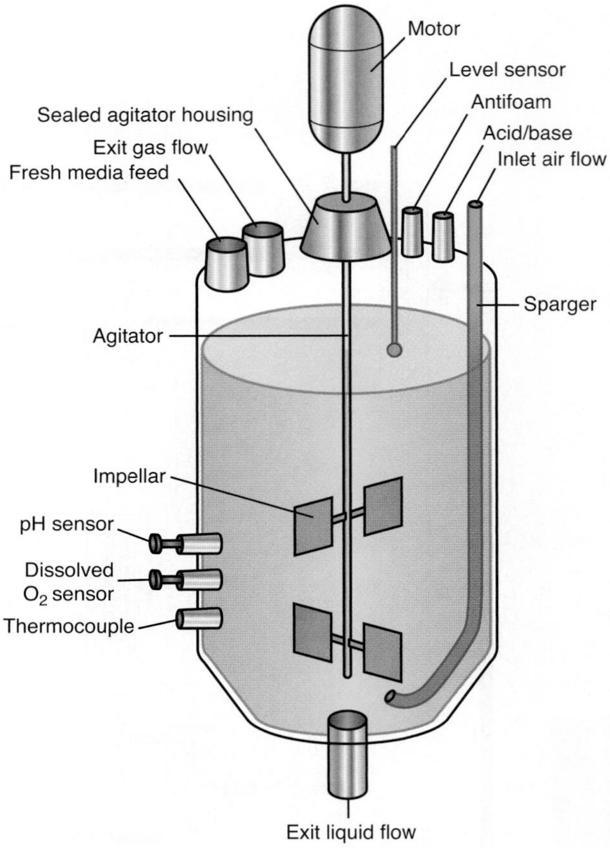

FIGURE 5.25 **Schematic of a continuous bioreactor.**

this fermentation method yields smaller quantities of biomass cells than the alternative (continuous fermentation). Thus, batch fermentation is preferred for synthesizing molecules where production rates are not directly correlated to biomass size. A specialized type of batch fermentation, known as fed-batch, feeds additional media into the bioreactor during phases of the growth curve so that nutritional resources are not exhausted.

Continuous Fermentation

In continuous fermentation, biomass is allowed to progress through sequential phases of growth by manipulating the bioreactor conditions. A schematic of a continuous bioreactor is displayed in **FIGURE 5.25**. At any given time, cells of different maturity levels will be present. Therefore, continuous fermentation results in a much larger biomass than does batch fermentation. Media or other carbon sources are continuously fed into the bioreactor to replenish nutritional requirements of the cultured cells. An equal quantity of product is removed from the bioreactor simultaneously in order to maintain the total volume within. Continuous fermentation is preferred for harvesting molecules synthesized as part of normal cell function, such as primary metabolites or proteins.

Downstream Processing

Once the fermentation of upstream processing is complete, the downstream processing of the product begins. The first step of downstream processing extracts the product from the cultured cells. In the event the molecule was produced and stored within the cell, extraction will require **cell lysis** (rupture) in order to release its contents. This is not necessary when the molecular product is secreted by the cells in culture. Protein products must then be stabilized in order to maintain their structure and ability to function. Purification follows, which separates the product from any other molecules present in the extract. An important element of this step is to test the purified substance to ensure its contents are as desired, a process known as **validation**. Lastly, the validated product is preserved in its purified, bioactive state. This final step prepares the product for market.

Extraction

Extraction is the first step of downstream processing necessary to isolate the product. The hundreds to thousands of liters held within a bioreactor may contain less than a percent of the target molecule. Cells are first separated from the growth medium by either filtration or centrifugation. When the molecule is a secreted product, the liquid medium will then be concentrated and downstream processing continues to its next steps. If the molecule is intracellular, the separated solids, containing the cells, are treated.

The cells must be lysed to remove their contents. Cell lysis should be turbulent enough to break the cell open, yet gentle enough to keep molecules intact. Rupturing cells is achieved in a variety of ways, each appropriate in different instances. One method is chemical treatment with solvents, detergents, or alkali. A biological approach involves the application of digestive enzymes. Cell lysis is also performed by physical means, including mechanical rupturing such as high-pressure treatments, grinding with small glass beads, and ultrasonication. Physical cell lysis may also occur in nonchemical procedures like freeze/thaw cycling, where ice crystals rupture the cell, and osmotic shock, where the cells are bathed in hypotonic solution, causing them to take on water and burst open.

Cellular structure will determine the best approach. Some bacterial cells possess a glycocalyx capsule, a thick layer of proteins and polysaccharides which must be penetrated (**FIGURE 5.26**). The cell walls of bacteria may be either gram-positive or gram-negative (**FIGURE 5.27**). Yeast cell walls contain a thick layer of polysaccharides, and tough cellulose fibers comprise plant cell walls. Animal cells, which lack walls altogether, require special handling. The different biochemistry of the cell walls demands different lysis treatments. The resulting solution is filtrated to remove cellular remains from the product-containing liquid, known as **lysate**. The product is now ready for the next step in downstream processing: stabilization.

Stabilization

The product may lose its integrity during downstream processing. To avoid this, stabilization of the product occurs. Consider proteins for example, which are denatured easily. Temperature and pH are two factors that must be controlled. Typically, stabilization takes place below room temperature in order to inhibit protease activity and encourage protein stability. Buffering agents are added to maintain pH at its optimum level. Downstream processing involves shearing, or mechanical agitation, which can cause the product to disintegrate. Foaming of the system is common with agitation as gases are introduced. Additives can be incorporated to minimize both shearing and foaming. An additional concern is protease enzymes that occur naturally in cells. They are released during cell lysis along with everything else, thus including protease inhibitors to the system is recommended. Any substance added in the process of stabilization must be removed before the final product is ready.

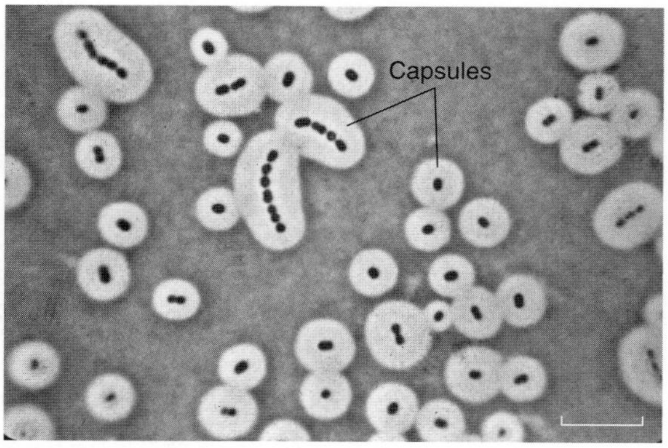

FIGURE 5.26 Members of the *Acitenobacter* species of bacteria possess a thick glycocalyx, or capsule. Courtesy of Elliot Juni, Department of Microbiology and Immunology, The University of Michigan

Purification

In the process of purification, the product of interest is isolated. Once the target molecule is concentrated, a validation step will guarantee its purity. To illustrate purification, a protein product will serve as our target molecule in this discussion.

Precipitation

Whether working from the stabilized medium (secreted products) or lysate (intracellular products), the product may be precipitated. **Precipitation** causes the proteins to aggregate and settle out of solution. Solvents or ammonium sulfate salts may be used for this purpose.

Filtration Methods

Filtration separates proteins from the liquid solution based upon their size. Centrifugation is one filtration method that compacts solids through high-speed spinning. Membrane filtration passes the solution through membranes with various pore sizes. Only those molecules small enough to pass through the pores are filtered out. Dialysis is a filtration method that uses the phenomena of osmosis and diffusion to concentrate water and molecules, respectively. This can separate target proteins from other chemical components in the solution. Diafiltration combines dialysis with membrane filtration.

Chromatography

Another purification method is chromatography, wherein proteins can be sorted and separated. Several chromatography techniques exist and vary by the characteristic used to distinguish proteins. This will be necessary to retrieve the target protein from myriad other proteins extracted from the culture.

Size-exclusion chromatography (SEC), also known as gel filtration chromatography, separates

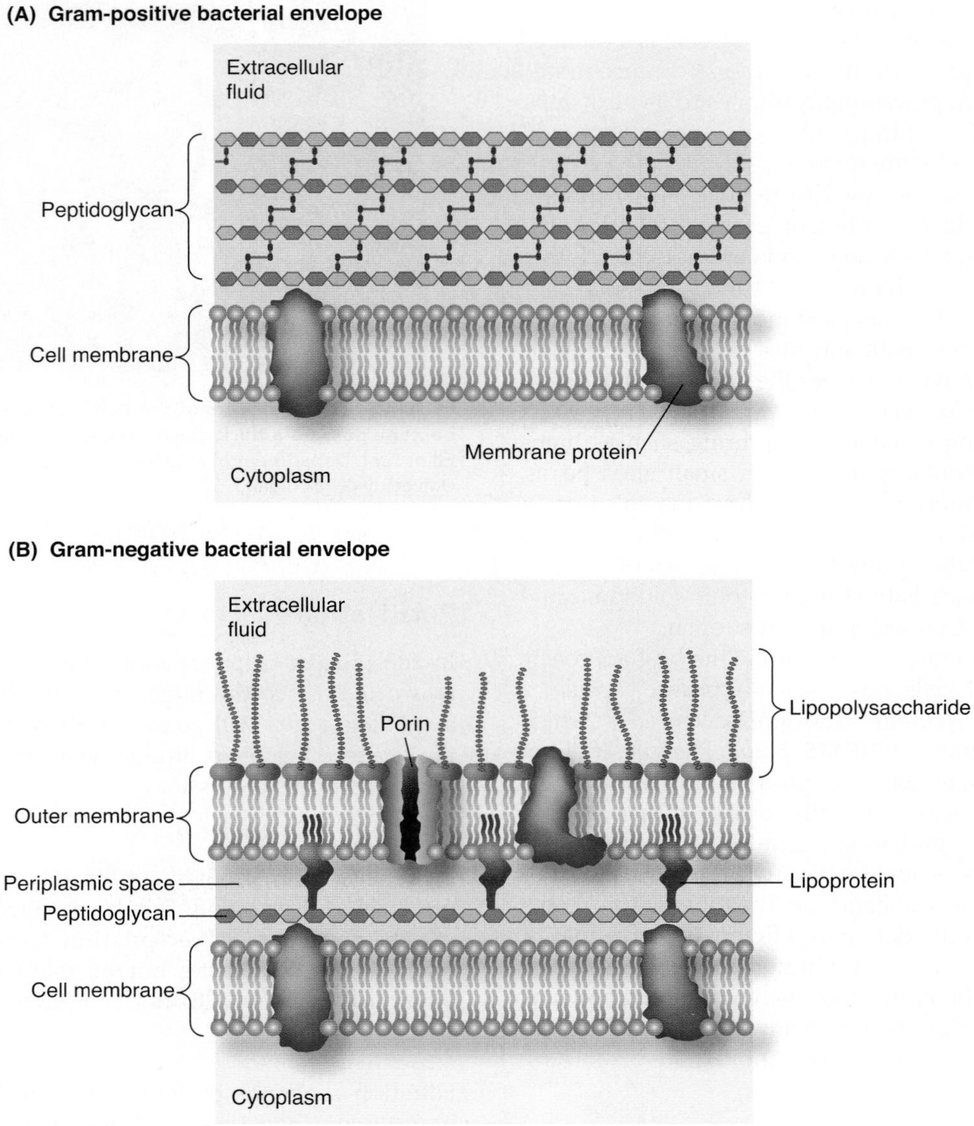

FIGURE 5.27 **Bacteria may possess either gram-positive (A) or gram-negative (B) cell walls.**

proteins based upon their molecular weight. Proteins are passed through a column of tiny gel beads punctured with small pores. Large proteins bleed through more quickly since they can't pass through the small spaces formed by the gel beads. Small proteins take longer as they navigate the narrow passageways of the column (**FIGURE 5.28**). SEC is best suited as a preliminary separation as it is rather rudimentary compared to other methods of chromatography.

Ion-exchange chromatography sorts proteins based on their electrical charges. In anion exchange chromatography, proteins are passed through a column of positively-charged resin-coated beads, with the result that all negatively-charged proteins will remain clinging to the beads and all positively-charged proteins will pass through (**FIGURE 5.29**). Cation exchange chromatography reverses the charges of the beads and binding proteins, but the concept remains the same.

In **affinity chromatography**, instead of beads coated in charged resin, they are coated with ligand molecules. **Ligands** are molecules of complementary shape to specific proteins. Target proteins adsorb to the ligands in a lock-and-key formation, allowing all other proteins to separate out. A desorption buffer solution then washes the bound proteins off of the beads.

Hydrophobic interaction chromatography (HIC) sorts proteins based on their interaction with water. Column beads are coated in hydrophobic molecules. Only hydrophobic proteins will bind, leaving hydrophilic proteins to pass through the column.

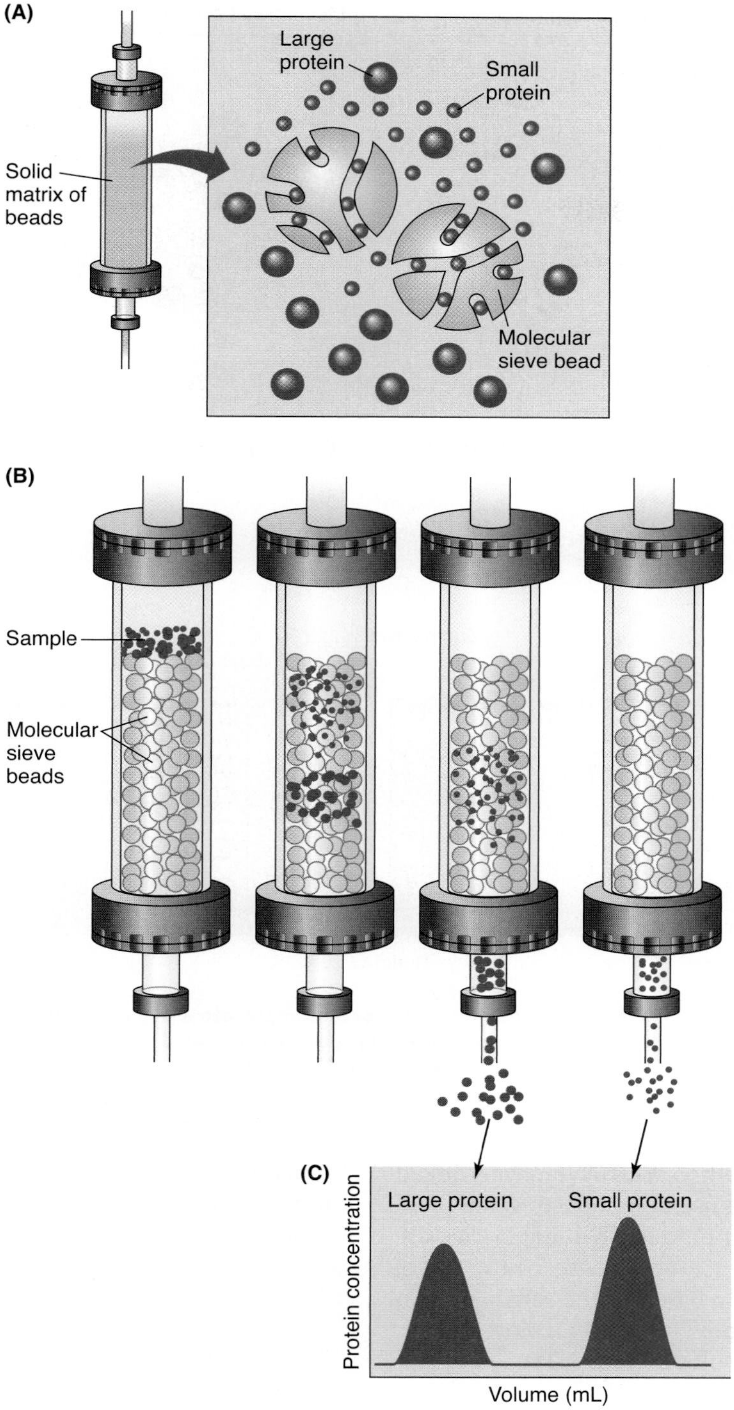

FIGURE 5.28 Size-exclusion chromatography separates proteins by size. (A) Narrow gaps are formed by the tiny gel beads within the column, trapping smaller proteins. (B) Larger proteins therefore pass through more quickly than smaller proteins.
Adapted from B.E. Tropp. *Biochemistry: Concepts and Applications*, First edition. Brooks/Cole Publishing Company, 1997

Validation

Validation has the dual roles of ensuring the presence of a desired protein product and determining its purity. Protein purity may be validated through **two-dimensional electrophoresis**, which separates the protein(s) resulting from purification by size and electrical charge. Isoelectric focusing is one dimension of the technique. This aspect separates proteins by electrical charge. Gel electrophoresis then separates proteins by size in the second dimension (**FIGURE 5.30**).

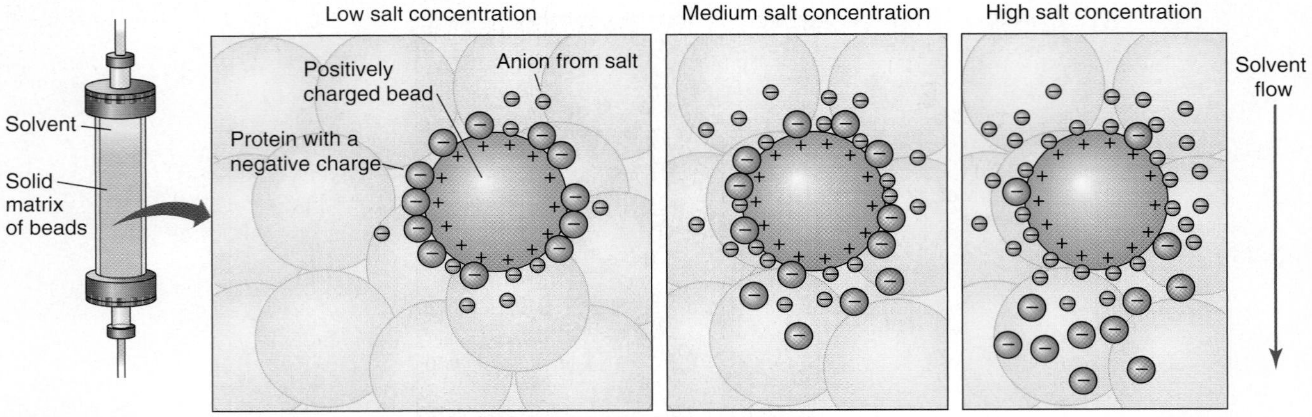

FIGURE 5.29 Anion exchange chromatography allows cationic proteins to pass through the column.

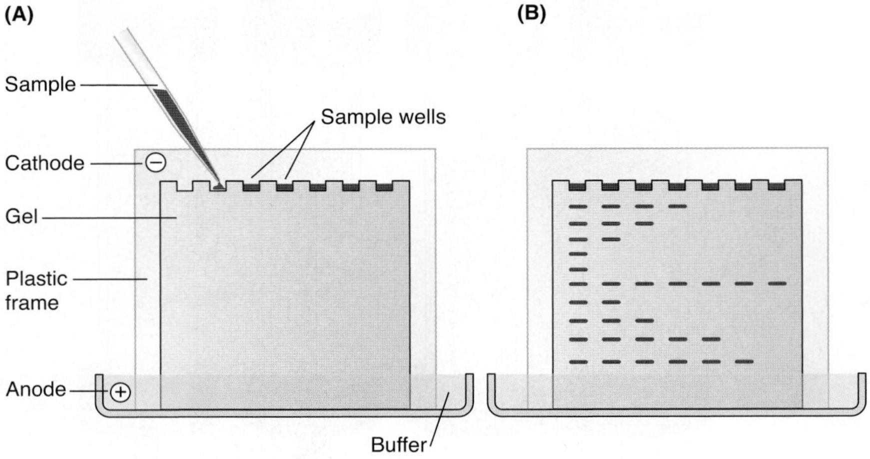

FIGURE 5.30 Two-dimensional electrophoresis introduces a protein sample into a polyacrylamide gel (A). After applying an electrical current, the resulting bands demonstrate how far each protein has traveled (B). The further down the gel a protein travels, the greater the negative charge and the smaller the size. Adapted from B.E. Tropp. *Biochemistry: Concepts and Applications*, First edition. Brooks/Cole Publishing Company, 1997

Another validation method, **ELISA** (enzyme-linked immunosorbent assay), is useful in detecting the presence of a desired protein product. Two antibodies are used to detect the presence or absence of the target protein. The first antibody binds to the target protein while the second antibody results in color change, indicative of a positive result (**FIGURE 5.31**).

Preservation

Preservation serves to keep the final product intact until its ultimate use. Preservation ensures the product does not degrade over time or lose its bioactivity. A common method in industrial biotechnology is lyophilization. In layman's terms, **lyophilization** is freeze drying. The purified protein solution is frozen, then a vacuum evaporates the water. The vacuum causes ice crystals to evaporate without ever thawing into liquid water first. Lyophilization maintains the structural and functional integrity of the isolated protein and allows for long-term room-temperature storage of the final product.

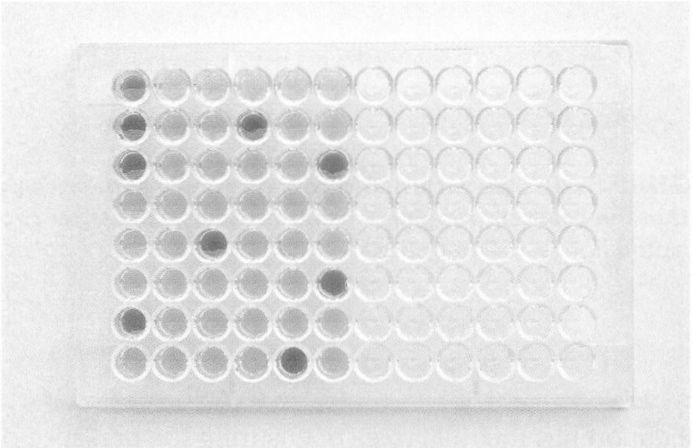

FIGURE 5.31 An ELISA plate. The color change (red) indicates a positive match. © jarun011/iStockphoto.com

SUMMARY

- Industrial biotechnology represents a myriad assortment of applications utilizing molecular biotechnology. Unlike other sectors of the biotechnology industry, this area does not have a single definitive goal or outcome. Organisms are utilized as living factories that produce a variety of important substances. Biotechnologists may insert desirable genes into an organism designed for a specific purpose. Additionally, it is possible to alter metabolic pathways of organisms to increase the efficiency and output of such biosynthesis. Certain molecules are used as catalysts for industrial processes and as tools in laboratory research. Other molecules act as part or all of a final product.
- Industrial enzymes are used to catalyze reactions involved in manufacturing. They include cellulases, proteases, esterases, lactase, and amylase. Cellulases catalyze the breakdown of cellulose fibers found in plant-based materials, while proteases and esterases hydrolyze proteins and lipids, respectively. Lactase is an important ingredient used to aid in the digestion of lactose (milk sugar). Amylases are used to break down starches into glucose monomers.
- Research enzymes are those discovered in biotechnology pursuits that have been adopted as tools in the laboratory. They include restriction enzymes and the enzymes utilized in the polymerase chain reaction (PCR).
- Biopolymers are large molecules produced by transgenic organisms and include bioplastics and bioadhesives. Biopolymers act as a sustainable source of plastics and adhesives. Biopolymers promise to reduce costs, improve energy efficiency, and minimize downstream pollutants that are common to chemically synthesized polymers. Bioplastics include PHA, PLA, xanthan gum, and rubber. Bioadhesives are derived from bivalve proteins.
- Primary and secondary metabolites are substances synthesized by microorganisms used for commercial purposes. Primary metabolites are produced during a microbial culture's growth phase, whereas secondary metabolites are synthesized late in the growth phase and into the stationary equilibrium phase.
- Primary metabolites include vitamins and antioxidants. Secondary metabolites include pigments and dyes. A number of industrial chemicals represent both classes of substances.
- Fermentation of microorganisms in bioreactors yields primary and secondary metabolites and enables commercial scale synthesis of these products. Batch fermentation involves the culture of biomass in batches. Batch fermentation is preferred for synthesizing molecules where production rates are not directly correlated to biomass size, such as secondary metabolites. Continuous fermentation allows biomass to progress through sequential phases of growth by manipulating the bioreactor conditions. Continuous fermentation is preferred for production of primary metabolites or proteins in addition to any other molecules synthesized as part of normal cell function.
- Downstream processing readies the products of upstream processing for their final sale. First, the product is extracted from the cultured cells. This may include rupture of the cells in order to release its contents. When the molecular product is secreted by the cells in culture, cell lysis is not necessary. The second step is to stabilize protein products in order to maintain their structure and ability to function. The next step is purification, which separates the product from any other molecules present in the extract. Purified products then undergo validation, which tests the purified substance to ensure its contents are as desired. Lastly, the validated product is preserved in its purified, bioactive state. This final step prepares the product for market.

KEY TERMS

Affinity chromatography
Amylases
Biomass
Biopolymers
Bioreactor
Cell lysis
Chymosin
Downstream processing
ELISA
Esterases
Fermentation
Hydrophobic interaction chromatography
Ion-exchange chromatography
Ligands
Lyophilization
Lysate
Precipitation
Primary metabolites
Proteases
Secondary metabolites
Size-exclusion chromatography
Thermophile
Two-dimensional electrophoresis
Upstream processing
Validation

DISCUSSION QUESTIONS

1. Research and describe a recent application of biotechnology to an industrial process or product.
2. How is batch fermentation different from continuous fermentation? What are the advantages and disadvantages for each process?
3. What factors must be considered when designing upstream processing? Downstream processing?

REFERENCES

Aiyer, Prasanna V. "Amylases and Their Applications." *African Journal of Biotechnology* 4, no. 13 (2005): 1525–1529.

Baig, Mirza Z. and Samita M. Dharmadhikari. "Optimization of Pre-Treatment and Enzymatic Hydrolysis of Cotton Stalk." *Journal of Pure and Applied Microbiology* 6, no. 3 (2012): 1437–1441.

Ban, Kazuhiro, et al. "Whole Cell Biocatalyst for Biodiesel Fuel Production Utilizing *Rhizopus oryzae* Cells Immobilized Within Biomass Support Particles." *Biochemical Engineering Journal* 8, no. 1 (2001): 39–43.

Belgacem, Mohamed N. and Alessandro Gandini, ed. *Monomers, Polymers, and Composites from Renewable Resources*. Philadelphia, Pennsylvania: Elsevier, 2008.

Cline, Janice, et al. "PCR Fidelity of *Pfu* DNA Polymerase and Other Thermostable DNA Polymerases." *Nucleic Acids Research* 24, no. 18 (1996): 3546–3551.

Demain, Arnold L., and Julian E. Davies, Ronald E. Atlas, eds. *Manual of Industrial Microbiology and Biotechnology*. Washington, DC: American Society for Microbiology Press, 1999.

Guo, Zheng and Xuebing Xu. "New Opportunity for Enzymatic Modification of Fats and Oils with Industrial Potentials." *Organic and Biomolecular Chemistry* 3, no. 14 (2005): 2615–2619.

Gupta, R., et al. "Bacterial Lipases: An Overview of Production, Purification, and Biochemical Properties." *Applied Microbiology and Biotechnology* 64, no. 6 (2004): 763–781.

Haley, Stephen, et al. "Sweetener Consumption in the United States: Distribution by Demographic and Product Characteristics." *Economic Research Service*, United States Department of Agriculture. Washington, DC, 2005.

Hemingway, Richard W., Anthony H. Connor, and Susan J. Branham, eds. *Adhesives from Renewable Resources*. Washington, DC: American Chemical Society Publications, 1989.

Lancini, Giancarlo and Rolando Lorenzetti. *Biotechnology of Antibiotics and Other Bioactive Microbial Metabolites*. New York: Plenum Press, 1993.

Litchfield, Ruth. *High Fructose Corn Syrup—How Sweet It Is*. Ames, IA: Iowa State University Extension and Outreach, 2008.

Lundberg, K. S., et al. "High-fidelity Amplification Using a Thermostable DNA Polymerase Isolated from *Pyrococcus furiosus*." *Gene* 108, no. 1 (1991): 1–6.

Maeder, Felicitas. "The Project Sea Silk: Rediscovering an Ancient Textile Material." *Archaeological Textiles Newsletter* 35 (2002): 8–11.

Marshall, R.O., et al. "Enzymatic Conversion of D-Glucose to D-Fructose." *Science* 125, no. 3249 (1957): 648–649.

Maryan, Ali Sadeghian, et al. "Aged-Look Vat Dyed Cotton with Anti-Bacterial/Anti-Fungal Properties by Treatment with Nano Clay and Enzymes." *Carbohydrate Polymers* 95, no. 1 (2013): 338–347.

Montalto, Massimo, et al. "Management and Treatment of Lactose Malabsorption." *World Journal* of *Gastroenterology* 12, no. 2 (2006): 187–191.

Nawrath, Christiane, and Yves Poirier. "Pathways for the Synthesis of Polyesters in Plants: Cutin, Suberin, and Polyhydroxyalkanoates." *Advances in Plant Biochemistry and Molecular Biology* 1 (2008): 201–239.

Park, Madison. "Half of Americans Use Supplements." *CNN*, Atlanta, Georgia, April 13, 2011.

Patel, A.H. "Plastics Recycling and the Need for Biopolymers." *International Society of Environmental Botanists* 9, no. 4 (2003).

Priya, P., et al. "Molecular Cloning and Characterization of the Rubber Elongation Factor Gene and Its Promoter Sequence from Rubber Tree (*Hevea brasiliensis*): A Gene Involved in Rubber Biosynthesis." *Plant Science* 171, no. 4 (2006): 470–480.

Silva, M. F., et al. "Production and Characterization of Xanthan Gum by *Xanthomonas campestris* Using Cheese Whey as Sole Carbon Source." *Journal of Food Engineering* 90, no. 1 (2009): 119–123.

Thompson, R.C., et al. "Lost at Sea: Where Is All the Plastic?" *Science* 304, no. 5672 (2004): 838.

Vandamme, E.G., ed. *Biotechnology of Vitamins, Pigments, and Growth Factors*. New York: Elsevier, 1989.

Vogel, Henry C., and Celeste L. Todaro. *Fermentation and Biochemical Engineering Handbook*. Westwood, NJ: Noyes Publications, 1997.

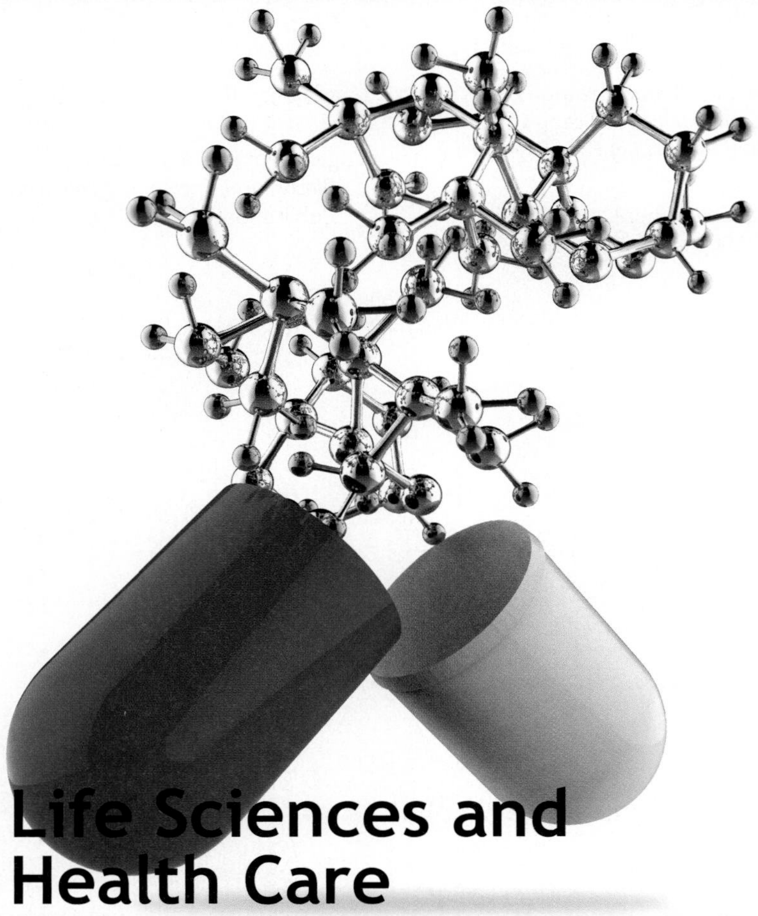

6 Life Sciences and Health Care

LEARNING OBJECTIVES

Upon reading this chapter, you should be able to:

- Describe the role of genetic counseling and the methods it employs.
- Compare and contrast ex vivo and in vivo gene therapy.
- Explain gene therapy and its various modes of delivery.
- Characterize biopharma's place in modern medicine.
- Discuss the promise of pharmacogenomics and nanomedicine.
- Summarize the available forms of immunotherapy.
- Relate the utility of xenotransplantation.
- Differentiate among advanced cancer treatments based upon medical biotechnology.
- Distinguish the types of stem cells and their various uses in regenerative medicine.
- Compare and contrast therapeutic cloning with reproductive cloning.

It can certainly be argued that molecular biotechnology has had its strongest influence in the area of medicine, perhaps with agriculture as its nearest rival. The Human Genome Project expanded our ability to identify genetic diseases over 20-fold. Where once we could test for about 100 genetic diseases, we are now capable of testing for over 2,000. There seems to be limitless applications to medicine and the future holds many opportunities for advancements in the promotion of human health. Indeed, many pharmaceuticals and medical treatments available today would have seemed like science fiction just a generation ago. We are approaching an era where personalized medicine seems possible. Yet on the other hand, medical biotechnology poses a number of bioethical concerns that have led to its position of controversy.

This chapter explores the major areas of research in medical biotechnology and presents some of the specific concerns regarding each application. Although it's important to remember this is an enormous field that could fill an entire textbook on its own (or a multivolume series), we visit a mere sampling in this discussion.

Genetic Counseling and Gene Therapy

Genetic counseling and gene therapy are intertwined subdisciplines within medical biotechnology. Genetic counseling aims to detect genetic disorders and can occur at all stages in the human life cycle:

- An embryo destined for in vitro fertilization
- A developing fetus in the womb
- A newborn (known as neonatal testing, required in all states to some degree)
- Juveniles and adults

Gene therapy offers a variety of techniques designed to treat genetic disorders. Approximately $400 million is spent each year to fund gene therapy research and over 100 clinical trials have taken place, yet no research project has resulted in a complete cure for any genetic disease. To date, gene therapy is an approved form of treatment for 13 genetic disorders in addition to cancer, AIDS, and heart disease.

Both genetic counseling and gene therapy are fields of intense research, making them costly enterprises. The price to patients may be a limiting factor in how widespread these areas of medical biotechnology ultimately become. The future will tell if these advances in medicine benefit the relatively few individuals comprising the upper crust (including industrialized nations), or if they will be accessible to the multitudes of underprivileged and poor who inhabit this planet (including less developed countries).

Detection of Genetic Disorders

Pharmacogenomics and gene therapy both demand prior genetic analysis of a patient. Without awareness of specific genetic disorders, genetic markers, and biomarkers, it is not possible to determine the appropriate course of treatment. Detection of genetic disorders and biomarkers is the work involved in **genetic counseling**. A genetic counselor will test an individual, create a genetic profile, and advise on the best course of action for the health of the patient. Counseling can also include planning advice if someone with a family history of illness is ready to conceive a child. For this reason, fetal genetic testing is also available. As a matter of fact, fetal testing thus far has become more established

FOCUS ON CAREERS
Genetic Counselors

Genetic counselors are trained to test for genetic disorders and advise patients on their healthcare and reproductive options. They often serve as patient advocates by connecting them with medical specialists and support groups. Individuals interested in this career should pursue a masters degree in genetics or more specifically in genetic counseling. It is recommended to seek out a degree program which is accredited by the Accreditation Council for Genetic Counseling (ACGC). Most positions require successful passage of a certification exam administered by the American Board of Genetic Counseling (ABGC). Other skills necessary include communication, understanding of available genetic tests, and sensitivity to the numerous bioethical concerns involved in genetic counseling. Internships and/or advocacy shadowing experiences are strongly encouraged to be a competitive applicant for genetic counseling positions. The National Society for Genetic Counselors (NSGC) is a professional organization for individuals in this field. NSGC offers online professional development, networking and career opportunities, and an annual education conference for its members. NSGC maintains a list of accredited institutions offering genetic counseling degree programs at http://gceducation.org/Pages/Accredited-Programs.aspx.

and routine than testing of adults. More information on genetic counseling as a career is presented in the *Focus on Careers* box.

Genetic disorders can manifest in several ways. Four basic types occur in the human genome.

- Single gene defects cause disease due to aberration of one gene. Examples include sickle cell anemia and Huntington disease. Single gene defects are the easiest to characterize and, therefore, they are the most understood (**TABLE 6.1**). Genomic studies estimate there may be more than 3,000 diseases caused by single gene defects.
- Polygenic defects are where several affected genes contribute to a disease. The interactions occurring between the multiple genes make polygenic defects more difficult to elucidate. Environmental factors can further complicate our ability to understand polygenic disorders. Examples include some forms of cancer, diabetes, and heart disease. With the notable exception of heart disease, medical biotechnology has not yet advanced to the point of applying gene therapy to polygenic defects due to their complex nature.
- Mitochondrial defects occur within extranuclear mitochondrial DNA, which is inherited solely from the maternal line. Examples include mitochondrial myopathy and Leigh syndrome (**TABLE 6.2**).
- Chromosome abnormalities include instances where an entire chromosome (or multiple chromosomes) is defective. A person may be missing all or part of a chromosome, have an extra copy, or pieces of separate chromosomes may have switched places. A DNA sequence within the chromosome may reverse direction, leading to a chromosomal abnormality known as an inversion. A well-known chromosomal abnormality is Down syndrome, which is caused by an extra copy of chromosome 21.

TABLE 6.1

Some Single Gene Defect Disorders

Chromosome	Associated Disease	Physiology
4	Huntington Disease	Neurodegeneration
5	Tay-Sachs Disease	Paralysis Mental retardation Blindness Respiratory ailments
7	Cystic Fibrosis	Mucous accumulation in lungs Respiratory infections Diminished function of pancreas, digestive system, and respiratory system
11	Sickle Cell Disease	Abnormal red blood cells Decreased oxygen circulation
19	Familial Hypercholesterolemia	High blood cholesterol
X	Duchenne Muscular Dystrophy	Muscle weakness and degeneration
X	Hemophilia A	Diminished blood clotting capability

The availability of genetic maps created in genomics studies facilitates identification of genetic disorders (**FIGURE 6.1**). In addition to genetic disorders, genetic markers and biomarkers can be analyzed when creating a genetic profile. Genetic markers and their detection were previously detailed; they encompass the DNA polymorphisms present in the human genome and detection tools such as DNA microarrays, gene chips, and restriction fragment length polymorphism (RFLP) analysis followed by Southern blotting. Genetic markers include single nucleotide polymorphisms and restriction fragment length polymorphisms, among several others (see the discussion on genomic comparisons). **Biomarkers** are substances that when present are the hallmarks of different diseases. They are usually proteins synthesized by diseased tissue(s). Elevated levels of a particular biomarker can help diagnose the disease with which it is associated. Blood or urine may be tested for this purpose. Specific methods for constructing a genetic profile are now outlined.

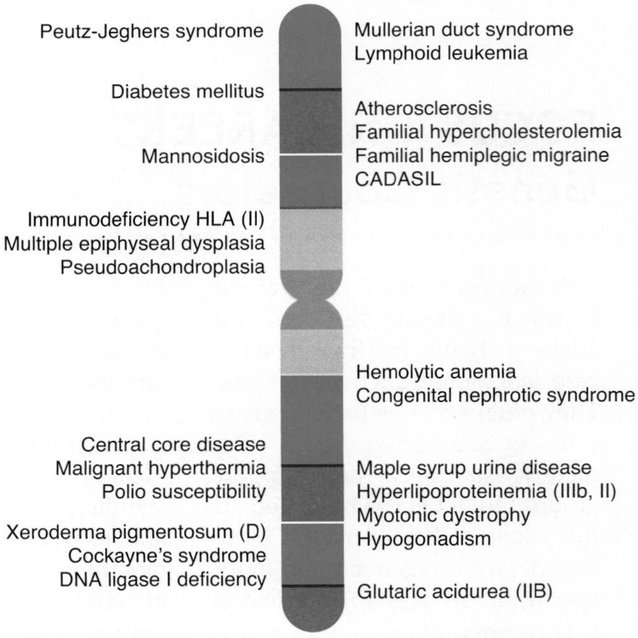

FIGURE 6.1 A map of chromosome 19 displays locations of genetic disorders. Reproduced from Lawrence Livermore National Laboratory/U.S. Department of Energy

TABLE 6.2

Phenotypes Associated with Some Mitochondrial Mutations

Nucleotide changed	Mitochondrial component affected	Phenotype[a]
3460	ND1 of Complex I[b]	LHON
11778	ND4 of Complex I	LHON
14484	ND6 of Complex I	LHON
8993	ATP6 of Complex V[b]	NARP
3243	$tRNA^{Leu(UUR)}$[c]	MELAS, PEO
3271	$tRNA^{Leu(UUR)}$	MELAS
3291	$tRNA^{Leu(UUR)}$	MELAS
3251	$tRNA^{Leu(UUR)}$	PEO
3256	$tRNA^{Leu(UUR)}$	PEO
5692	$tRNA^{Asn}$	PEO
5703	$tRNA^{Asn}$	PEO, myopathy
5814	$tRNA^{Cys}$	Encephalopathy
8344	$tRNA^{Lys}$	MERRF
8356	$tRNA^{Lys}$	MERRF
9997	$tRNA^{Gly}$	Cardiomyopathy
10006	$tRNA^{Gly}$	PEO
12246	$tRNA^{Ser(AGY)}$[c]	PEO
14709	$tRNA^{Glu}$	Myopathy
15923	$tRNA^{Thr}$	Fatal infantile multisystem disorder
15990	$tRNA^{Pro}$	Myopathy

[a]LHON Leber's hereditary optic neuropathy; NARP Neurogenic muscle weakness, ataxia, retinitis pigmentosa; MERRF Myoclonic epilepsy and ragged-red fiber syndrome; MELAS Mitochondrial myopathy, encephalopathy, lactic acidosis, stroke-like episodes; PEO Progressive external ophthalmoplegia.

[b]Complex I is NADH dehydrogenase. Complex V is ATP synthase.

[c]In $tRNA^{Leu(UUR)}$, the R stands for either A or G; in $tRNA^{Ser(AGY)}$, the Y stands for either T or C.

Amniocentesis and Karyotyping

Amniocentesis is a type of fetal genetic testing. It is typically performed when the fetus is around 16 weeks old. The procedure involves inserting a needle into the amniotic fluid surrounding the developing fetus and extracting a sample for analysis (**FIGURE 6.2**). The amniotic fluid contains cells shed from the fetus that are cultured under sterile conditions. Cells are then arrested and stained, causing the chromosomes to become visible. Chromosomes are sorted by size, centromere location, and banding patterns to create a karyotype of the fetus. The **karyotype** is a visual representation of the genome (**FIGURE 6.3**). Diagnosis of chromosomal abnormalities is possible by recognizing obvious errors, such as a trisomy (extra chromosome copy). Because the complement of sex chromosomes is visible in the karyotype, gender may also be determined (see Appendix 3 for more details on chromosomal abnormalities).

Chorionic Villus Sampling

Chorionic villus sampling is another type of fetal genetic testing. A major advantage of this method over amniocentesis is it can be performed much earlier in a pregnancy (at 8 to 10 weeks instead of 16). In this procedure, a small suction tube is inserted through the

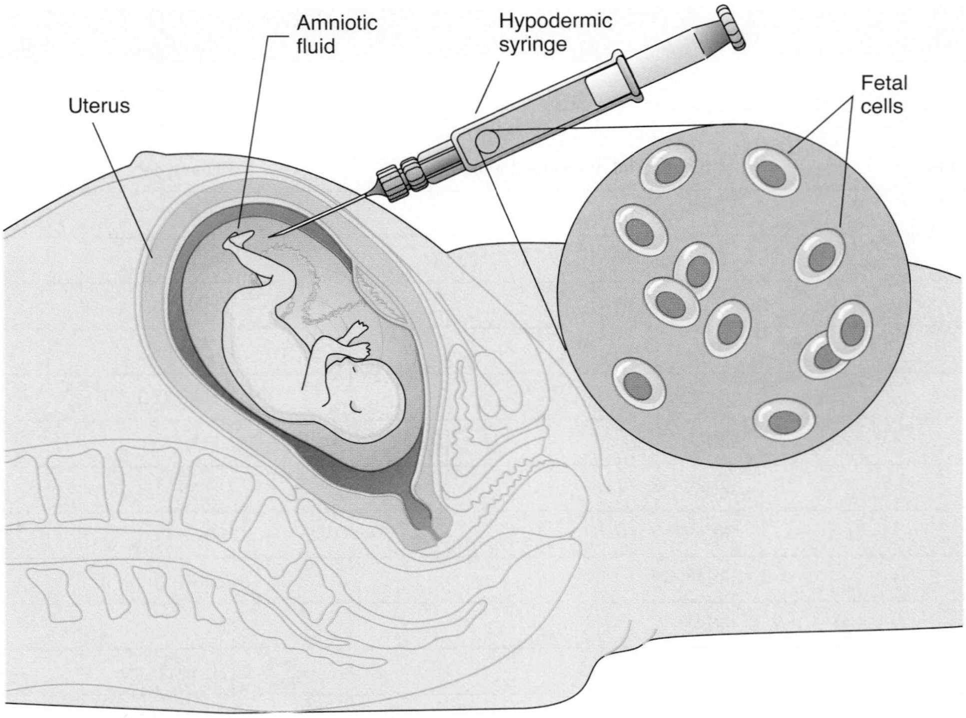

FIGURE 6.2 Amniocentesis procedure.

mother's cervix and chorionic villus cells are collected (**FIGURE 6.4**). The chorionic villus is tissue that aids in placental formation. The number of cells retrieved is much greater than with amniocentesis, so the sample does not need the extra step of culturing. This represents an additional advantage over amniocentesis. From this point in the procedure, the remaining steps are identical. Chorionic villus cells are arrested and stained in order to create and analyze the fetal karyotype.

Fluorescence In Situ Hybridization

Karyotyping can also be performed on adults. A method known as **fluorescence in situ hybridization (FISH)** may be performed on adult white blood cells for this purpose (**FIGURE 6.5**). FISH arrests and stains the cells and a karyotype is produced. The chromosomes are treated with fluorescent probes that attach to specific genetic markers. Thus, FISH is an enhanced karyotyping method that provides information not only on

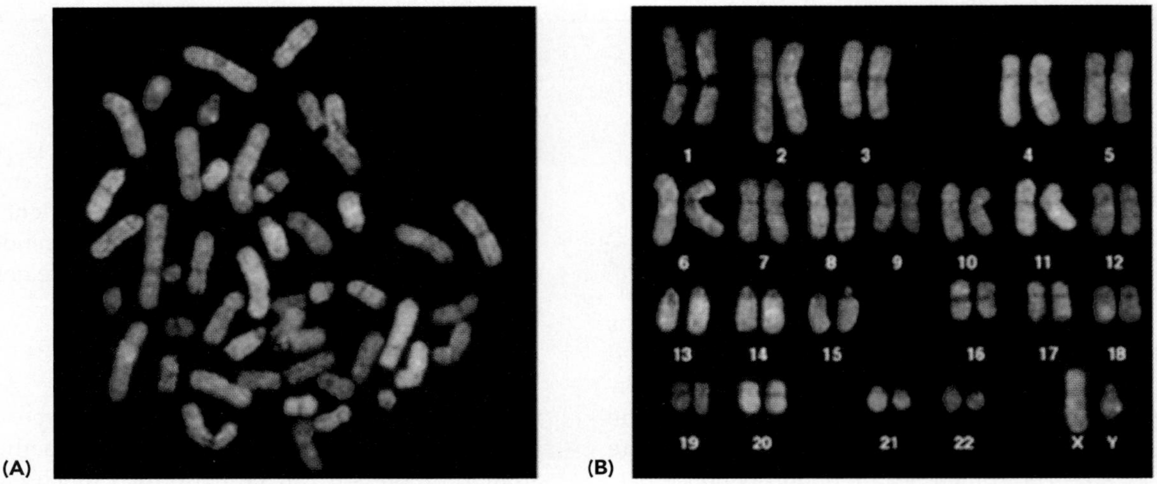

FIGURE 6.3 (A) Stained chromosomes collected from a dividing cell. (B) The same chromosomes aligned in pairs create a karyotype. Courtesy of Johannes Wienberg, Ludwig-Maximillians-University, and Thomas Ried, National Institutes of Health

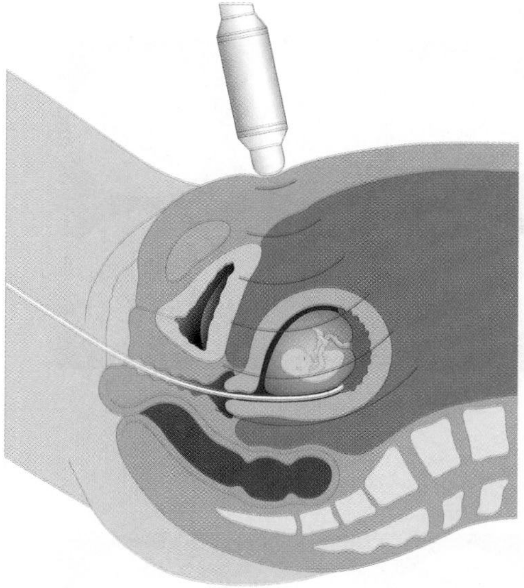

FIGURE 6.4 Chorionic villus sampling tests a developing embryo as early as eight weeks. © Dorling Kindersley RF/ Thinkstock

extra or missing chromosomes, but it can also identify defects present within the chromosomes. For example, the procedure may be used to detect chromosomal translocations, where pieces of two chromosomes have traded places. One known translocation, between chromosomes 9 and 22, causes a form of cancer known as chronic myelogenous leukemia (**FIGURE 6.6**). FISH can also be performed on fetal cells.

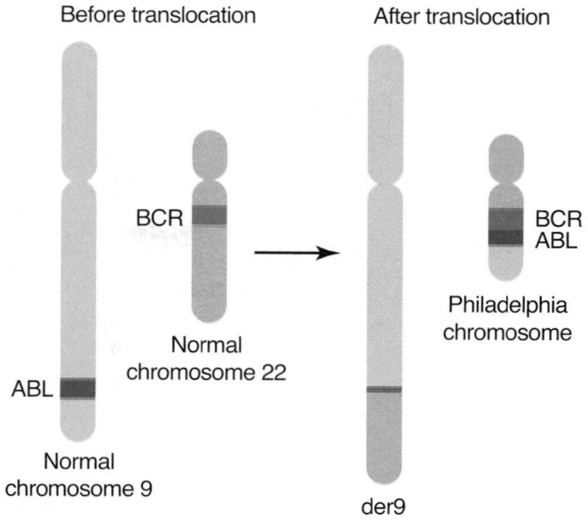

FIGURE 6.6 The Philadelphia chromosome is formed through a translocation between chromosomes 9 and 22. This chromosomal abnormality results in chronic myelogenous leukemia.

Protein Microarrays

We have seen the utility of DNA microarrays in identifying genetic markers in an individual (**FIGURE 6.7**). **Protein microarrays** are a more recent technique based upon similar principles. Instead of an array of DNA probes, the protein microarray is embedded with thousands of antigens associated with different pathogenic organisms. If a person is infected with one of these pathogens, antibodies produced by the immune system will be present in the sample to be tested. These antibodies will attach to their specific antigen on the protein microarray.

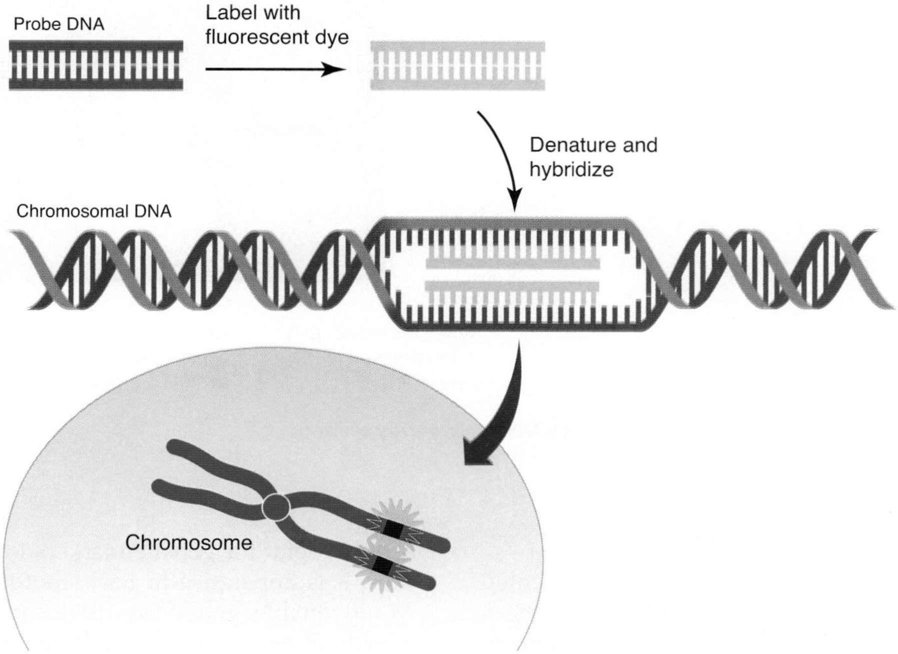

FIGURE 6.5 The fluorescent in situ hybridization technique. Adapted from an illustration by Darryl Leja, National Human Genome Research Institute (http://www.genome.gov)

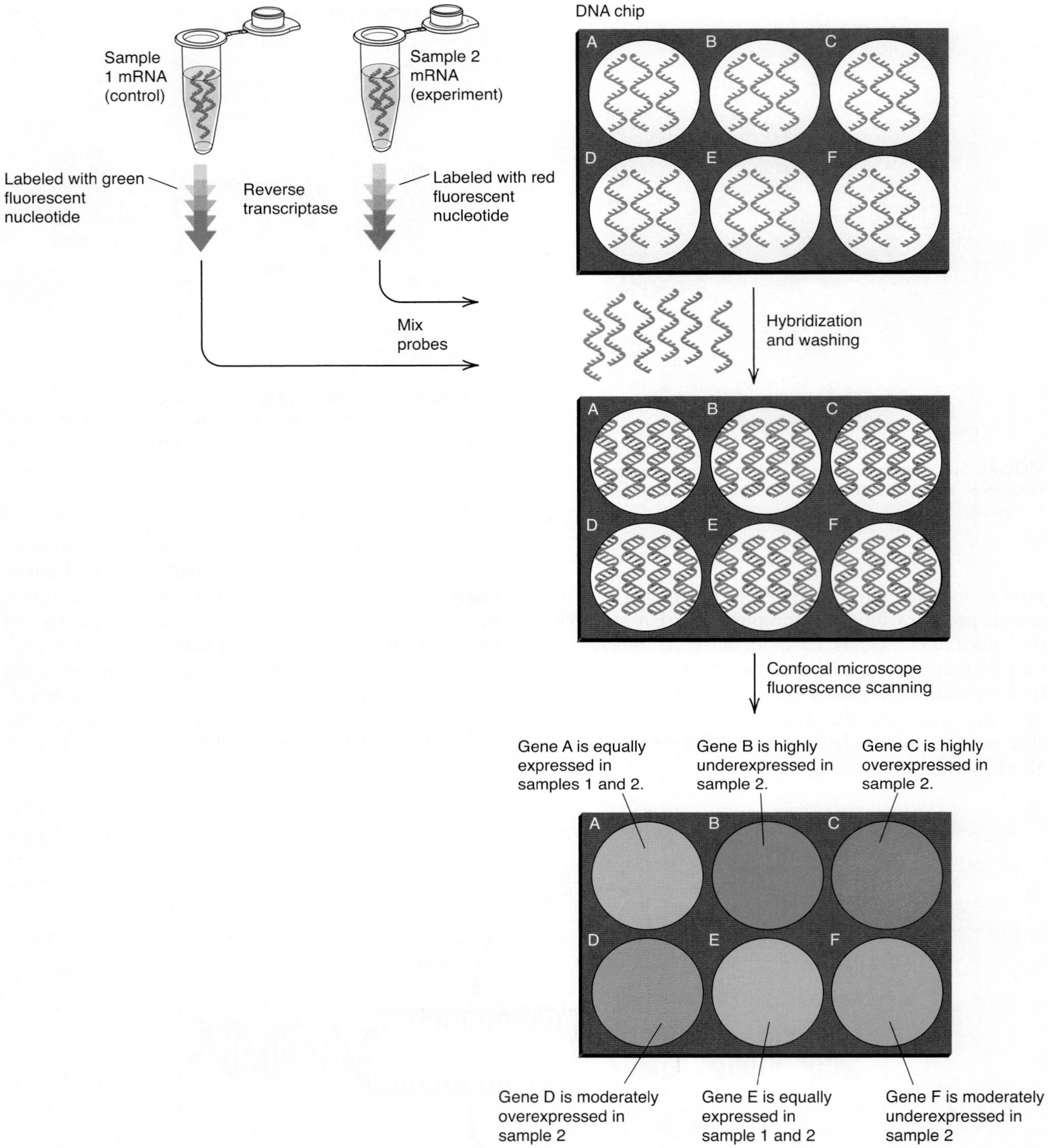

FIGURE 6.7 Identification of genetic markers through DNA microarray testing.

Allele-Specific Oligonucleotide Analysis

Our earlier exploration of genetic markers presented the technique of RFLP analysis followed by Southern blot (see the section on Genomics, above). A significant drawback of identifying restriction fragment length polymorphisms in an individual is that the technique is only useful for genetic markers located at restriction sites. It is imperative to have another approach which would enable geneticists to locate genetic markers outside of restriction sites. **Allele-specific oligonucleotide analysis (ASO)** solves this problem. Allele-specific oligonucleotides (single strands of DNA, in this

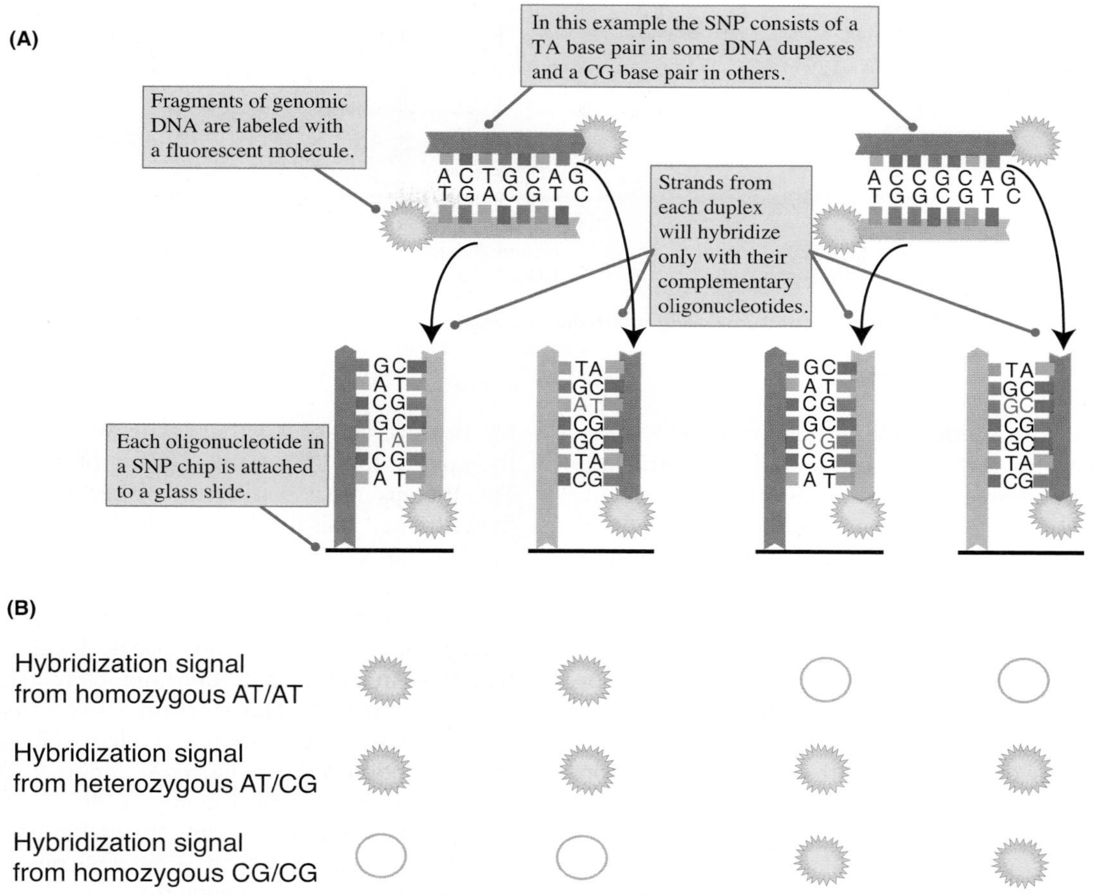

FIGURE 6.8 **Allele-specific oligonucleotide analysis detects the presence of healthy and/or mutant alleles.**

case about 20 nucleotides in length) complementary to target alleles are used as probes. White blood cells are collected, DNA is extracted and amplified by polymerase chain reaction (PCR), and two forms of ASOs are applied. One form of ASO will bind to a normal, healthy allele when present. The other form of ASO will bind to a defective, mutant allele when present (**FIGURE 6.8**).

Because PCR is part of ASO, this technique is well-suited in any instance where only a small sample of DNA is available. It is becoming more common as an element of preimplantation genetic testing performed prior to in vitro fertilization. A single cell can be taken from an embryo once it has reached the eight-cell stage without interfering in future development. All embryos produced for IVF are tested, though only genetically healthy embryos are used for implantation. Genetically defective embryos are either destroyed or donated for stem cell research (see below).

Epigenomics

Epigenomics literally means "above the genome." While genomics explores the genetic material of a given organism, epigenomics studies how the genome is regulated. In a popular analogy, genomics is comparable to the hardware of a computer, providing the genetic infrastructure of a living being. Epigenomics is akin to the software that controls how a computer performs based on its available hardware. Epigenomics therefore controls gene expression and tells a cell when to use a gene and how much. The epigenome is unique from the genome in part because it consists of methylation patterns within the genome. Epigenomics studies the timing and intensity of genetic expression (**FIGURE 6.9**).

Study of the epigenome is aiding researchers in understanding an individual's unique genetic profile. It is not only genetic markers, biomarkers, and chromosomal abnormalities that must be considered. Biopharma applications must also analyze how the epigenome influences them. One day it may be possible to alter an individual's epigenome in order to silence or activate a target gene to produce a desired impact on that person's state of health.

Gene Therapy

Detection of genetic disorders and creation of a genetic profile may lead to the recommendation of gene therapy. **Gene therapy** involves replacement of faulty genetic

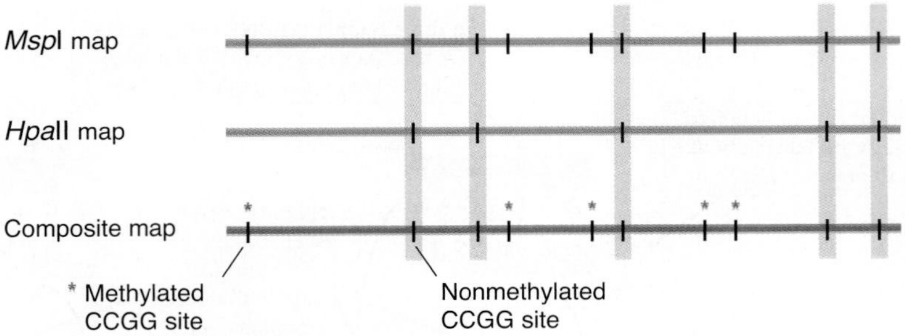

FIGURE 6.9 Detection of methylation patterns using restriction enzymes. MspI cleaves all CCGG sequences, while HpaII only cleaves nonmethylated CCGG sequences. Methylated sequences are thusly deduced.

sequences or the interference of their expression. Gene therapy may be performed on somatic or germ line cells; however, only somatic cells are routinely treated. If the genetic material of somatic, or body, cells is altered, all daughter cells resulting from cell division will exhibit the alteration as well, but the new genetic sequence is not passed on to offspring. This is not the case with genetic alteration of a germ line cell. Those cells ultimately become gametes (eggs in females or sperm in males) during the process of meiosis. Should a germ line cell that underwent gene therapy be involved in the process of fertilization, that genetic alteration would be inherited by the newly created individual. Thus, germ line cell gene therapy could pave the way for intentional genetic manipulation of future generations for desired traits, a field known as **eugenics**. Bioethical implications have resulted in a moratorium on germ line cell gene therapy (**BOX 6.1**). In addition to bioethical concerns, medical biotechnologists are faced with challenging technical issues:

- How can gene therapy be targeted to affected cells or tissues without affecting healthy areas of the body?
- Will results be permanent or will multiple (even indefinite) treatments be necessary to achieve desired effects?
- How can therapeutic gene expression be controlled? How can overexpression be prevented? How can a gene be switched on and kept on?
- Is there a way to introduce therapeutic genes to a specific region of the genome, or must we abide random integration that might disrupt functional genes?

Ex Vivo and In Vivo Gene Therapy

Gene therapy may be performed within the body or outside the body, known as in vivo and ex vivo gene therapy, respectively. In vivo (Latin for "in a living thing") gene therapy entails delivery of genetic information directly into affected cells or tissues. The biggest obstacle of in vivo techniques is targeting the affected region without wasting any of the (expensive) infusion on healthy areas. Another issue with in vivo gene therapy is the long-term efficacy of the treatment. The therapy must permanently alter gene expression in the treated cells for the treatment to persist throughout the life of the patient. Otherwise, effects of the treatment are only temporary.

Ex vivo (Latin for "out of a living thing") gene therapy involves removing diseased cells, applying treatment in culture, and returning the transfected cells back to the patient. **Transfection** is the introduction of genetic material into animal or plant cells, much like transformation alters a bacterial genome. One application of ex vivo gene therapy is with bone marrow transplants. Instead of finding a suitable bone marrow donor, which can take years, a patient's own bone marrow can undergo ex vivo gene therapy. When the transfected cells are reintroduced into the patient, there is no fear of immune rejection because the donor cells are an exact genetic match. An ex vivo method was the first successful case of gene therapy in humans (**BOX 6.2**).

Modes of Delivery

Safe and effective delivery of therapeutic genetic material is a necessary element in gene therapy. We are still in the relatively early stages of this field of medical biotechnology. Researchers are investigating a number of approaches in order to develop vectors that can deliver the "payload" consistently in all patients.

Viral Integration

Viruses are one of the most intensely studied modes of delivery in the field of gene therapy. Depending on their life cycle, some viruses are able to integrate their genetic information into the genome of its host. Therefore, integration results in the permanent alteration of the host genome. Other viruses are more transitory and only temporarily alter gene expression in the host (**FIGURE 6.10**). Both life cycles can be useful in gene therapy when they are used appropriately. Furthermore, there are classes of viruses that only infect certain types of cells. These viruses offer the potential to target specific locations during gene therapy. For example, a herpes virus known to infect only nervous

Box 6.1 Bioethics in Genetic Counseling and Gene Therapy

As we increase our ability to test individuals for genetic defects, questions arise.

- Who will have access to gene therapy? Will there be limitations based upon age, type of disorder, or its stage of progression?
- Should gene therapy be used for treatment of disease only, or should it also be used to prevent disease?

The Human Genome Project recognized the need early on to establish guidelines pertaining to societal issues posed by genetics. The Ethical, Legal, and Social Implications (ELSI) joint working group of the Human Genome Project (HGP) addresses concerns and seeks to preemptively set policies that allow for advancement in genetic research while minimizing risks to society.

In 1996, ELSI published recommendations designed to prevent genetic discrimination. While several bills based upon these recommendations have been introduced since that time, none have passed. The only existing policy regarding genetic discrimination is an executive order issued by President Clinton in 2000, though it only applies to current and potential federal employees. The general consensus in the legislative community purports existing discrimination laws could be modified to include genetic disorders. They argue this would be easier to achieve than enacting an entirely new genetic discrimination law.

ELSI has identified the following issues regarding genetic discrimination:

- *Confidentiality of genetic test results:* Who should have access? How will the information be used?
- *Privacy of individual genetic dispositions:* What must be disclosed to an insurer or employer?
- *Neonatal testing:* What tests should be performed? What actions should be taken in the event of a negative test result?
- *Genetic susceptibility and predisposition:* Should there be different policies for confirmed genetic disorders versus genetic predisposition? Should a person susceptible to a genetic condition that is otherwise healthy be subject to special considerations?
- *Stigmatization:* If genetic information is publicly available or otherwise discovered, what psychological effects might result? Will we experience widespread discrimination of individuals with genetic defects, similar to historical discrimination of other subclasses of society?
- *Philosophical:* Will our understanding of human genetics change ideas of free will versus predetermination? Is there any validity to blaming one's genetic make-up on his or her actions and behaviors?
- *Reproductive rights:* How will negative genetic test results affect the choice to carry a pregnancy to term? Should parents receiving IVF procedures be allowed to pick and choose the implanted embryos based on their genetic profiles? Will it be acceptable to one day select desired traits in offspring, such as eye and hair color, musical ability, or athletic prowess? Where is the line drawn between acceptable reproductive genetics and the potential for eugenics?

As with all bioethical issues, the best possible outcome is achieved when all members of society have the opportunity to weigh in. As medical biotechnology innovations progress into the future, the impacts upon society will be fascinating to witness.

Box 6.2 The First Gene Therapy Patient

Ashanti DaSilva was four years old when she was treated with gene therapy for severe combined immunodeficiency (SCID). Representing the first gene therapy procedure in humans, she was treated by a team at the National Institutes of Health (NIH) through ex vivo methods. SCID is a disorder where the immune system is greatly compromised due to a defective adenosine deaminase (ADA) gene. The ADA enzyme normally metabolizes deoxyadenosine triphosphate (dATP), a nucleotide which is toxic to T cells of the immune system. Patients with SCID are unable to metabolize dATP so it accumulates in their bodies, resulting in T cell death and its associated depression of the immune system.

SCID is a prime candidate for gene therapy for several reasons. The disease has been intensely studied and is well understood. It is caused by the disruption of a single gene, so the genetic repair is relatively straightforward. The gene is always turned on, so there was no need for an additional step of manipulating gene regulation. Lastly, there was no concern regarding the quantity of ADA expression. The human body can metabolize ADA with small amounts of its corresponding enzyme, but it also won't respond negatively if large quantities were to be expressed as a final outcome of therapy.

The ex vivo procedure performed on Ashanti began with extraction and culture of viable T cells. The culture was transfected with a retrovirus carrying the normal ADA gene and allowed to grow further. The retrovirus integrated the ADA gene permanently. The transfected cells were returned to Ashanti's body. The NIH team repeated this procedure on Ashanti several times to ensure the successful uptake of transfected cells. Subsequent tests of her T cell levels were promising and she continues to lead a normal life. Without treatment, SCID patients typically die of the disorder in their teen years. The same procedure has saved many other children affected by SCID.

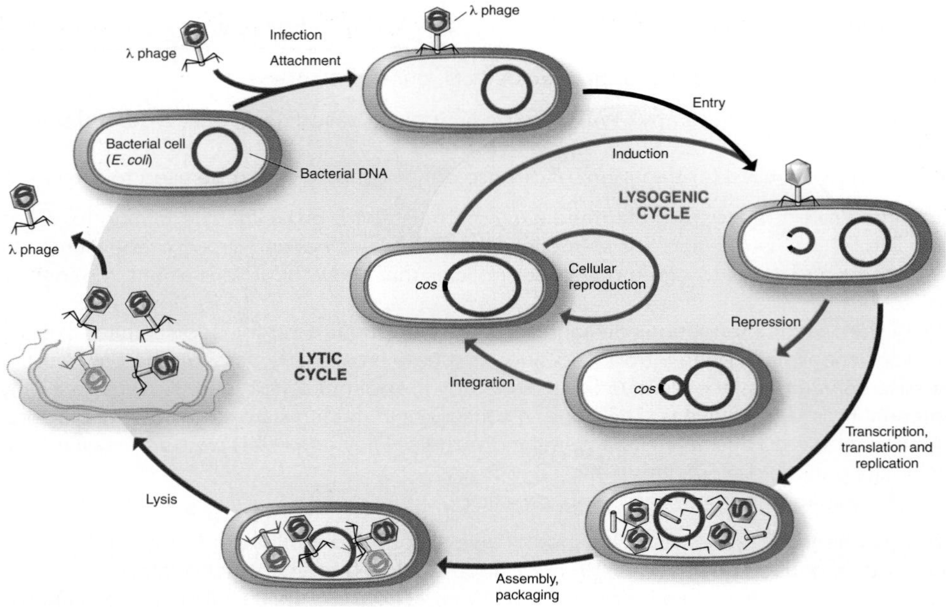

FIGURE 6.10 The bacteriophage lambda exhibits both viral life cycles. The lysogenic cycle involves integration of viral DNA into the host genome. The lytic cycle results in host cell death (apoptosis). Adapted from A. Campbell, *Nat. Rev. Genet.* 4 (2003): 471–477

tissue cells would be desirable for treatment of neurological disorders such as Parkinson's disease. Viruses can also be engineered to display recognition proteins on their surfaces for increased binding to target cells.

Regardless of the virus chosen for payload delivery, researchers undergoing their development must be certain to inactivate the virus so it cannot produce disease nor spread to unintended regions of the body.

The following virus families may serve as vectors in gene therapy:

- Adenoviruses
- Adeno-associated viruses
- Retroviruses

TABLE 6.3 summarizes the advantages and disadvantages of each virus family as applied to gene

TABLE 6.3

Advantages and Disadvantages of Viral Families in Gene Therapy Delivery

Virus Family	Advantages	Disadvantages
Adenoviruses	• relatively mild pathogenicity • nononcogenic • well understood • genome sequenced • easy to culture • high growth rates • targets both dividing and non-dividing cells	• does not integrate into host genome • short-term therapeutic effects • multiple treatments required • must remove pathogenic genes • must be engineered to prevent host cell lysis
Adeno-associated Viruses	• minimal inflammation • does not stimulate immune response • integrates into host genome • specific integration sites • reduced host gene disruption • small size yields higher penetration rates • targets both dividing and non-dividing cells	• requires a second helper virus for replication • may become latent in host cell • small genome size • limited carrying ability
Retroviruses	• does not stimulate immune response • minimal inflammation • integrates into host genome • permanent genetic alteration	• random integration • potential host gene disruption • small genome size • limited carrying ability • cannot infect non-dividing cells

therapy. **FIGURE 6.11** presents the basic steps of viral gene therapy.

Adenoviruses are the family of viruses that cause the common cold (**FIGURE 6.12**). Therefore, previously obtained knowledge of adenoviruses, which is great, facilitates the research process. They will target both dividing and non-dividing cells, making adenoviruses a more well-rounded delivery vehicle compared to some other virus families. Adenoviruses do not integrate into the host genome so the treatment effects are transient. Multiple treatments are usually necessary. Normally adenoviruses lyse the cells they infect as part of their life cycle (**FIGURE 6.13**), but bioengineers have discovered ways to alter the life cycle so that host cells are not killed during therapy. Medical biotechnologists must also engineer adenoviruses to decrease the likelihood they will infect unintended cells and spread disease (**FIGURE 6.14**). To avoid this, genes encoding for viral pathogenicity are removed before the virus is delivered. Thankfully, adenoviruses are nononcogenic, meaning they do not stimulate tumor development.

Adeno-associated viruses, like adenoviruses, will infect dividing and nondividing cells. One drawback to using this class of viruses in gene therapy is they require a second helper virus in order to replicate within the host cell. Much like retroviruses, adeno-associated viruses integrate into the host genome, thus altering genetic material within the host permanently. Adeno-associated viruses do not appear to integrate as randomly as retroviruses. Studies indicate they prefer to integrate into a specific region on chromosome 19. This reduces the risk of insertion into vital host genes. Another advantage of adeno-associated viruses in gene therapy is that they do not typically invoke an immune response in the patient.

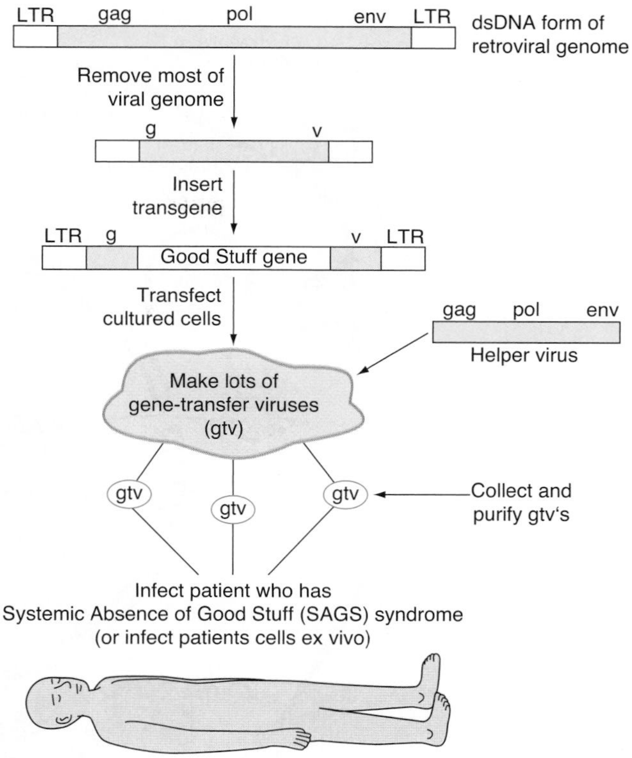

FIGURE 6.11 The general steps of viral gene therapy.

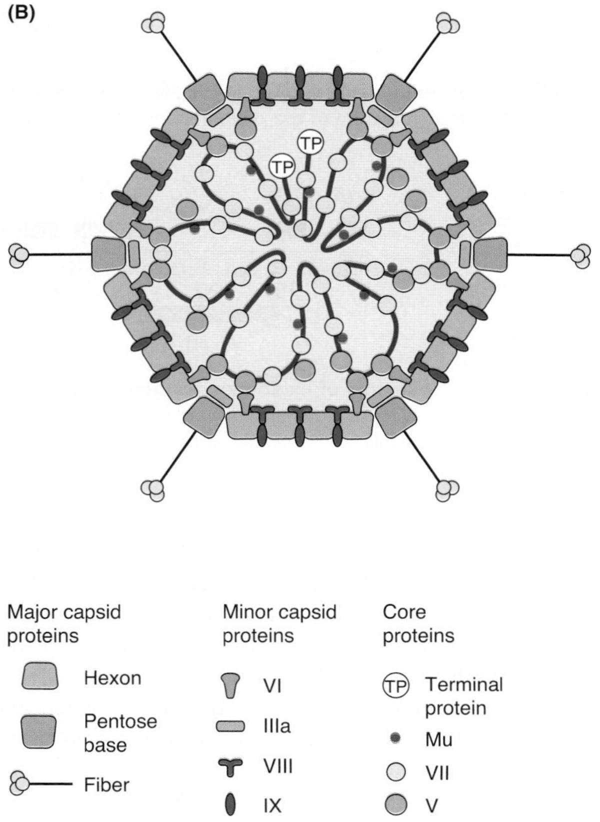

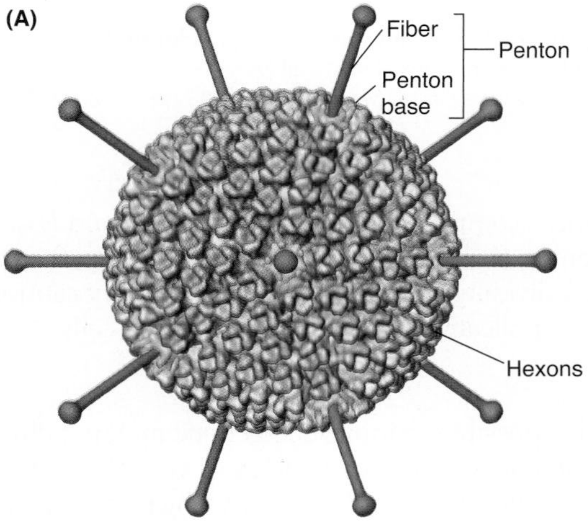

FIGURE 6.12 Structure of an adenovirus. (A) Surface view. (B) Internal view showing structural protein locations. Sources: (A) Photo courtesy of Carmen San Martín, Centro Nacional de Biotecnología (CNB-CSIC). (B) Reprinted from *Virology*, vol. 384, G. R. Nemerow, et al., Insights into adenovirus host cell..., pp. 380–388, © 2009, with permission from Elsevier (http://www.sciencedirect.com/science/journal/00426822)

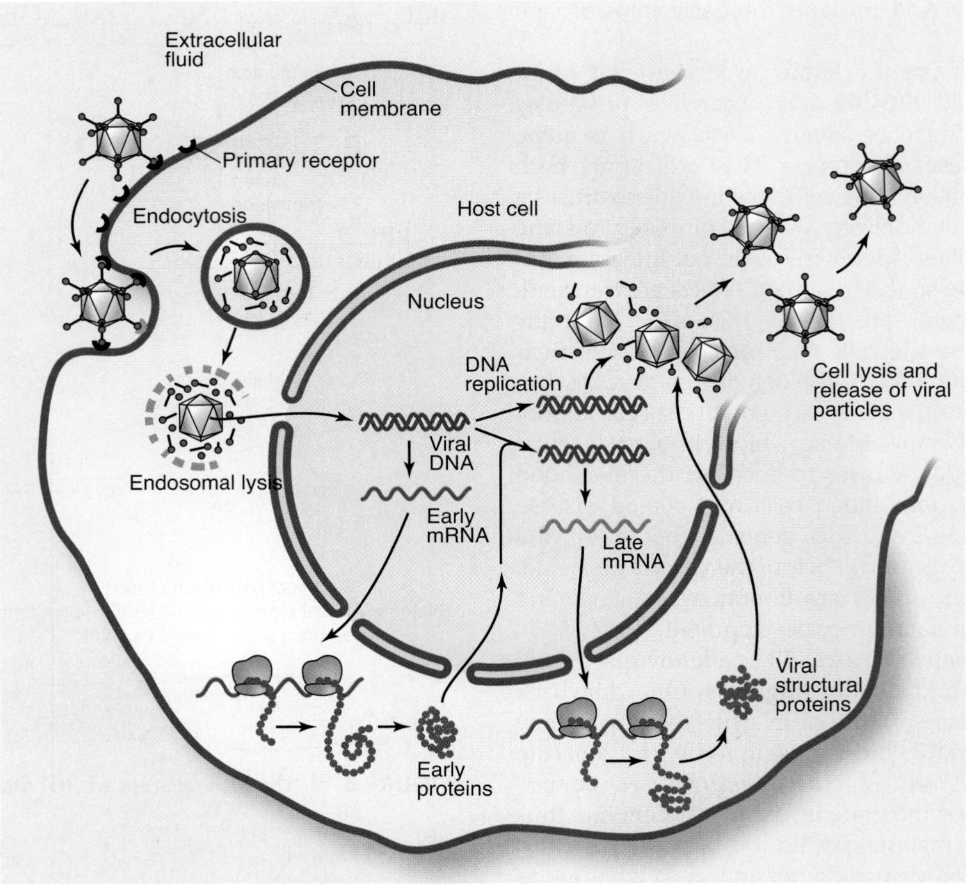

FIGURE 6.13 Adenoviruses exhibit a lytic life cycle that normally lyses host cells. Integration does not occur. Adapted from M.E. Mysiak, Molecular architecture of the preinitiation complex in adenovirus DNA replication (master's thesis, Universiteit Utrecht, 2004)

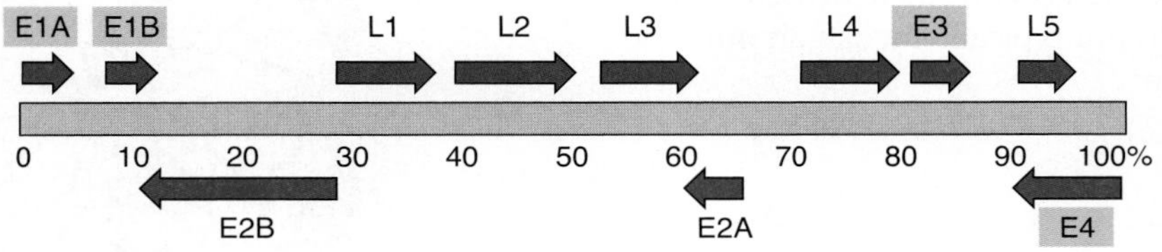

FIGURE 6.14 An example of an adenovirus gene map. Early genes (green) are labeled "E" and late genes (purple) are labeled "L." Arrows show the direction of transcription. Genes in boxes can be removed to prevent normal host cell lysis. Reproduced from K.J. Wood and J. Fry, Gene therapy: potential applications in clinical transplantation, *Expert Rev. Mol. Med.*, vol 1, no. 11, pp. 1–20, 1999. © Cambridge Journals, reproduced with permission

Retroviruses possess RNA as their genetic material instead of DNA (**FIGURE 6.15**). These viruses synthesize complementary DNA copies of their RNA genome upon infection of a host cell. This DNA is then randomly integrated into the host genome (**FIGURE 6.16**). Retroviruses are useful in gene therapy applications where permanent expression of an introduced gene is desired. Because integration is random, there is some concern the transfection gene might insert itself into a host gene and disrupt its function. Another possibility is the inserted gene might activate an oncogene, stimulating the cell cycle and increasing the likelihood of tumor formation. Because such a small fraction of the human genome is composed of genes, however, the odds of this happening are not great. Retroviruses specifically attack actively dividing cells, so this form of delivery cannot be used in applications targeting nondividing cells.

Naked DNA

It is also possible to introduce genetic material directly, without a biological vector such as a virus. Naked DNA is yet another mode of delivery, in which the therapeutic gene is inserted into a plasmid (**FIGURE 6.17**). Cells are then treated with the plasmids in both ex vivo and in vivo procedures. When performed in vivo, the success rate of plasmid uptake is quite low compared to

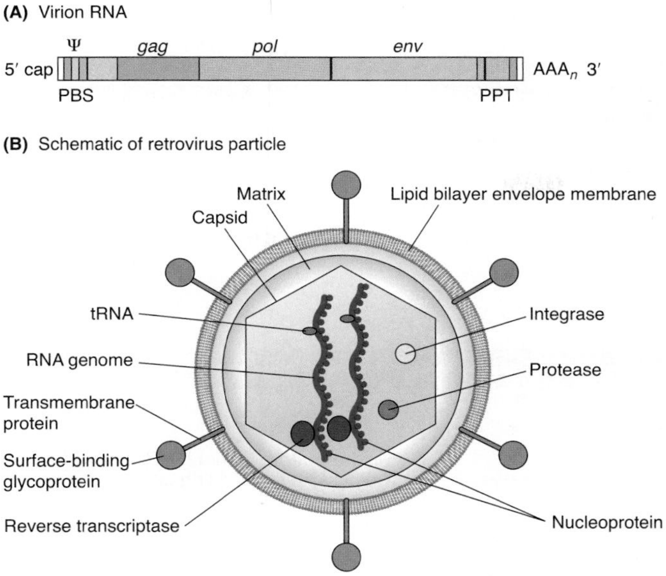

FIGURE 6.15 Generalized retrovirus. (A) RNA genome map. (B) Internal structure. Adapted from F.S. Pedersen and M. Duch, *Encyclopedia of Life Sciences*, [DOI: 10.1038/npg.els.0003843]. Posted May 3, 2005

other methods. Electroporation and microinjection are two techniques that can increase plasmid uptake.

Receptor-Mediated Uptake

DNA may be attached to a cell recognition protein. Cell surface receptor proteins have the natural ability to bind to cell recognition proteins. Indeed, understanding the exact biochemistry of receptor proteins on the intended target cells is a required element in selecting the appropriate recognition protein. Once DNA is attached, the complex is delivered and enters the target cells. Therapeutic DNA essentially hitches a ride into the cell.

Polymer-Complexed DNA

If only temporary expression of gene therapy is desired, polymer-complexed DNA is an advantageous delivery vehicle. Therapeutic genetic material is treated with a positive-charged polymer (remember, DNA carries a negative charge) in order to increase its stability. The complex is then taken up by cells in a manner similar to bacterial transformation. As with naked DNA, electroporation and microinjection are methods to increase polymer-complexed DNA uptake.

Encapsulated Cells

In another delivery vehicle, whole cells expressing a desired protein can serve as a vector in gene therapy. To prevent immune rejection, the cells are encapsulated with selectively permeable polymer coats. Pores in the polymer are large enough for the desired protein to pass through and reach target cells, but small enough to prevent immune system components such as antibodies or white blood cells from entering. In this way, there is no direct contact between host and encapsulated cells (**FIGURE 6.18**).

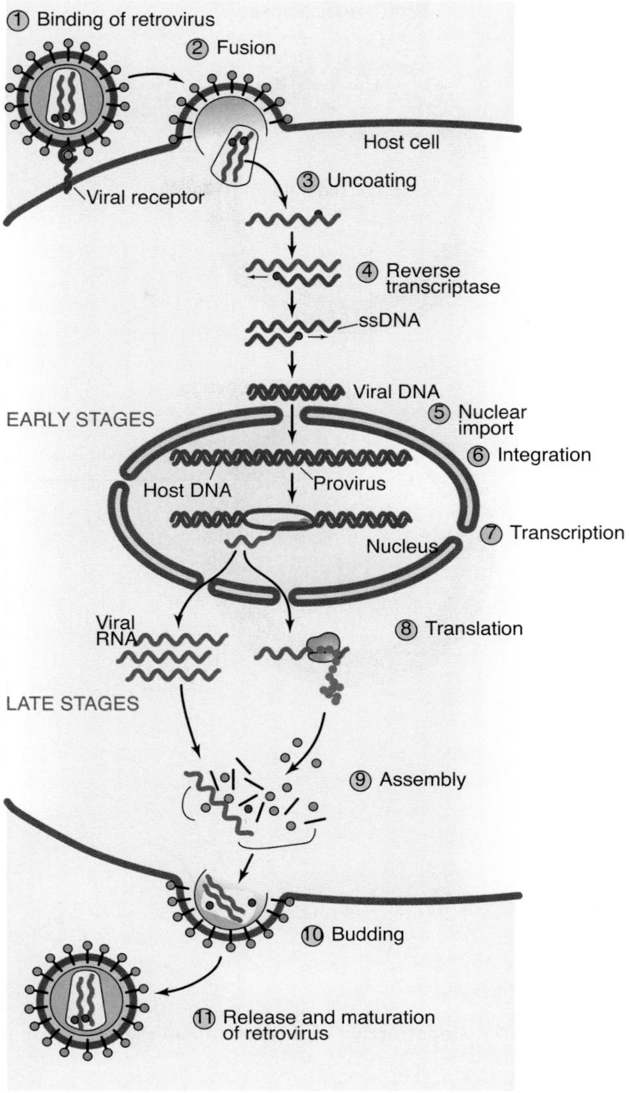

FIGURE 6.16 Retroviruses exhibit a lysogenic life cycle that includes integration into the host genome. Adapted from S. Nisole and A. Saïb, *Retrovirology* 1 (2004): 1–9

Liposomes

Liposomes (from the Greek *lipos*, "oil or fat," and *soma*, "body") are similar to microspheres (described below). Some even classify liposomes as a type of microsphere. While "microsphere" is a general term used in nanotechnology, "liposome" refers specifically to a system that delivers therapeutic genetic material. Liposomes are small hollow spheres made of lipids (**FIGURE 6.19**). Because they resemble the composition of plasma membranes of cells, liposomes readily fuse with target cells in a manner equivalent to endocytosis normally performed by the cell. This mode of action allows the nearly spontaneous introduction of therapeutic genetic material.

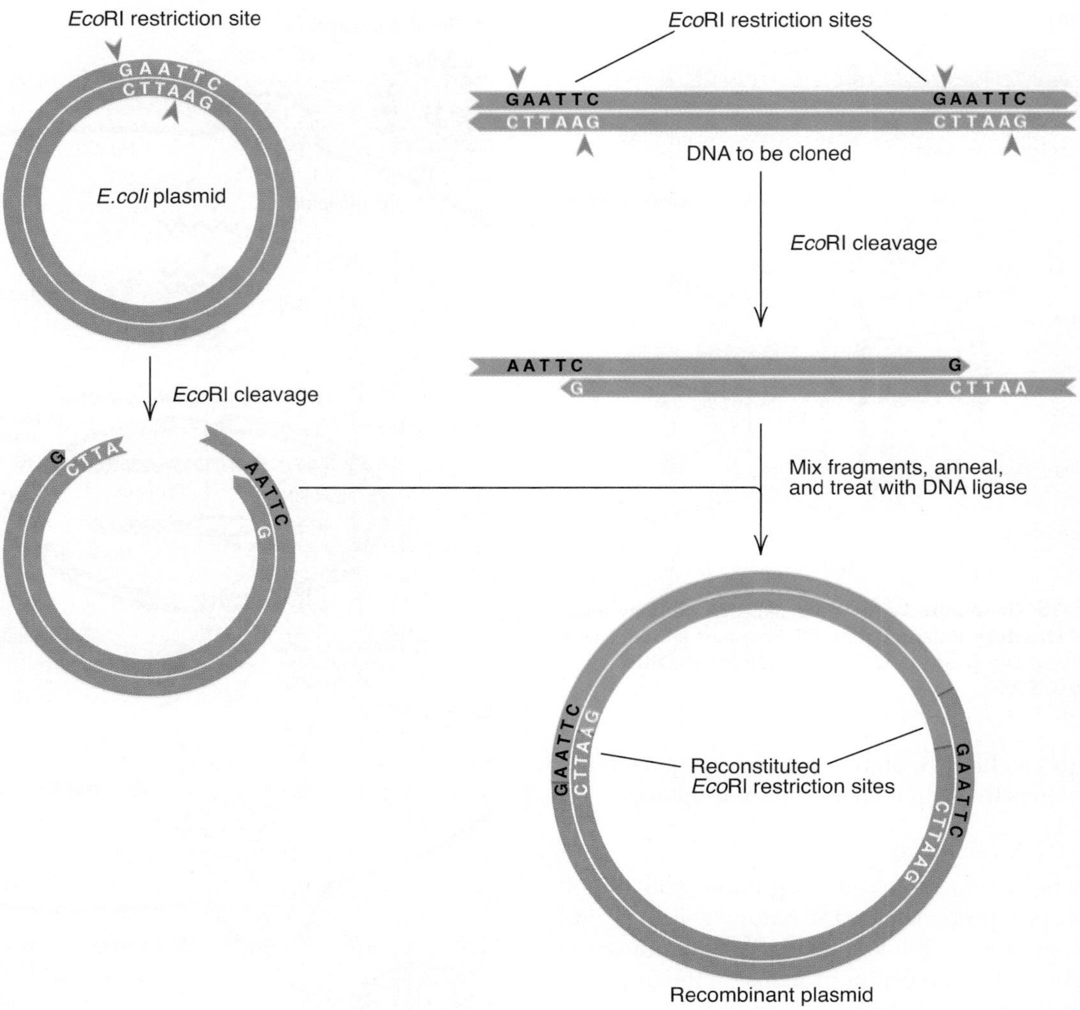

FIGURE 6.17 Construction of a recombinant plasmid such as those used to deliver naked DNA in gene therapy.

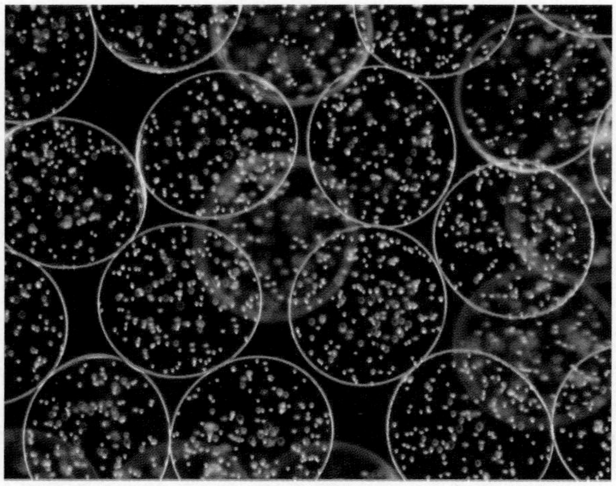

FIGURE 6.18 Encapsulated cells constructed by the laboratory of Suwan Jayasinghe at the University College London, UK. © Suwan Jayasinghe/Science Source

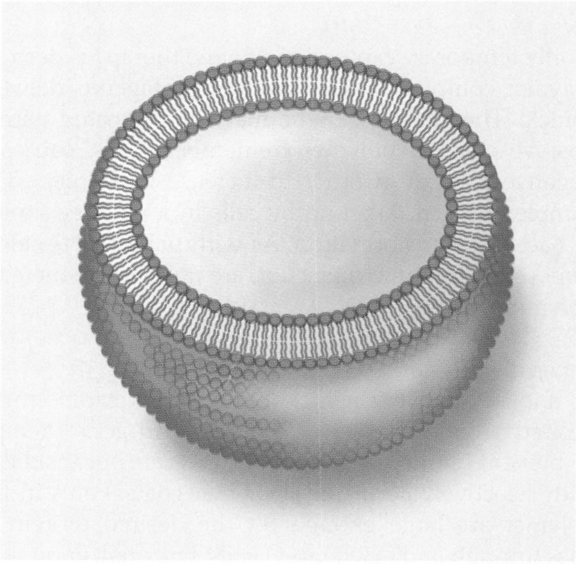

FIGURE 6.19 Liposome structure. The hollow region within the liposome is used as a delivery vector in gene therapy.

Particle Bombardment
Medical biotechnology has developed a procedure known as particle bombardment. Nanoparticles of gold or biodegradable gelatin are coated with naked DNA or other therapeutic genetic material and then shot into cells with a gene gun (**FIGURE 6.20**). The gene gun operates under high pressure so that the delivery vector is injected into target cells without lysing them. A gene gun can deliver liposomes into cells, too. Particle bombardment is also known as biolistics.

Bacteria
Bacteria may be engineered to express therapeutic genetic material. The mode of action is similar to introduction of naked DNA via plasmid vectors, only bacteria act as the delivery vehicle of the transgenic plasmid. To facilitate entry, the bacteria can also be engineered to express recognition proteins specific to receptor proteins on the target cells. Clearly, nonpathogenic strains of bacteria must be used to avoid negative side effects.

Therapeutic Approaches

In addition to selecting an appropriate mode of delivery, gene therapists must also choose the best therapeutic approach for a patient. It may be advantageous to repair or replace a defective gene or it may be preferable to silence the gene or its product instead. The various modes of action utilized in gene therapy are outlined next.

Gene Replacement
One approach to gene therapy is to replace the defective gene. In many instances, only small regions within the gene are damaged. In an extreme example, the gene may contain a point mutation where only a single nucleotide or base pair is the culprit. Gene patching is a mode of action where the defective portion of the gene is replaced in lieu of replacing the entire gene. One method of gene patching is known as **oligonucleotide crossover**. Oligonucleotides containing a normal version of the nucleotide sequence are used to construct RNA-DNA hybrids. Chimeric duplexes (containing both forms of nucleic acid) are known to participate in crossing over more frequently than double-stranded DNA, so researchers hypothesize the hybrids will exhibit high incidences of crossing over. During this normal cell function event, the chimeric duplex can replace mutated DNA with its normal payload sequence.

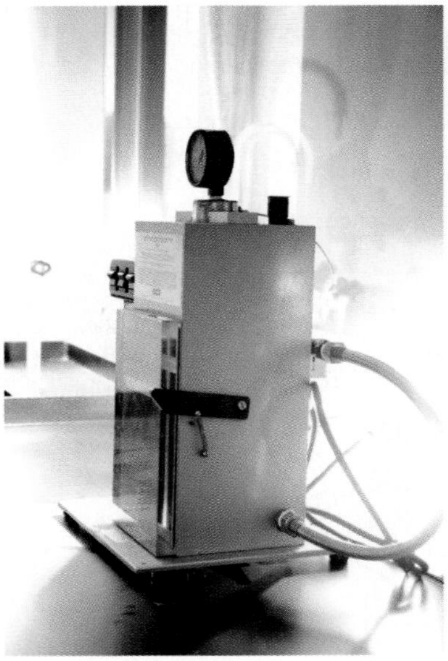

FIGURE 6.20 A gene gun used in biolistics, the particle bombardment technique. © Pan Xunbin/Science Photo Library

Gene Repair
Spliceosome-mediated RNA trans-splicing, abbreviated SMaRT, is an example of gene repair. SMaRT acts by targeting and repairing messenger RNA transcribed from a defective gene. The spliceosome is a naturally occurring enzymatic complex utilized in the normal process of alternative splicing during gene expression. Researchers have successfully engineered a spliceosome with therapeutic properties. First, the location and sequence of introns and exons present in the defective transcript must be known. An RNA strand engineered to be complementary to an intron neighboring the mutated exon to be repaired is allowed to bind. The RNA strand also contains a repaired version of the mutated exon. When binding occurs, this triggers the spliceosome to splice the intron and defective exon. The repaired version of the exon now joins with the remaining exons of the transcript as the process of alternative splicing unfolds. The result is a mature mRNA transcript lacking any genetic defect. The mRNA can now be translated to synthesize healthy functional proteins.

Gene Silencing
Multiple human diseases are caused by a mutated gene that encodes a dysfunctional or otherwise harmful protein. Antisense therapy is a method that blocks translation of an undesired protein. The gene encoding the undesired protein is transcribed into "sense" messenger RNA. The sequence of RNA complementary to sense mRNA is known as **antisense RNA**. When introduced into target cells, antisense RNA binds to sense mRNA and blocks translation (**FIGURE 6.21**). Because antisense therapy prevents protein synthesis, it is also known as gene silencing. Antisense technology is based upon naturally occurring antisense gene regulation in microbes, plants, and animals.

Another form of gene silencing is **RNA interference (RNAi)**. This method of gene therapy uses RNA as the silencing molecules (**FIGURE 6.22**). After transfection, a native enzyme called Dicer cleaves the RNA into small interfering RNAs (siRNAs). Small interfering RNAs bind to a quaternary protein complex called

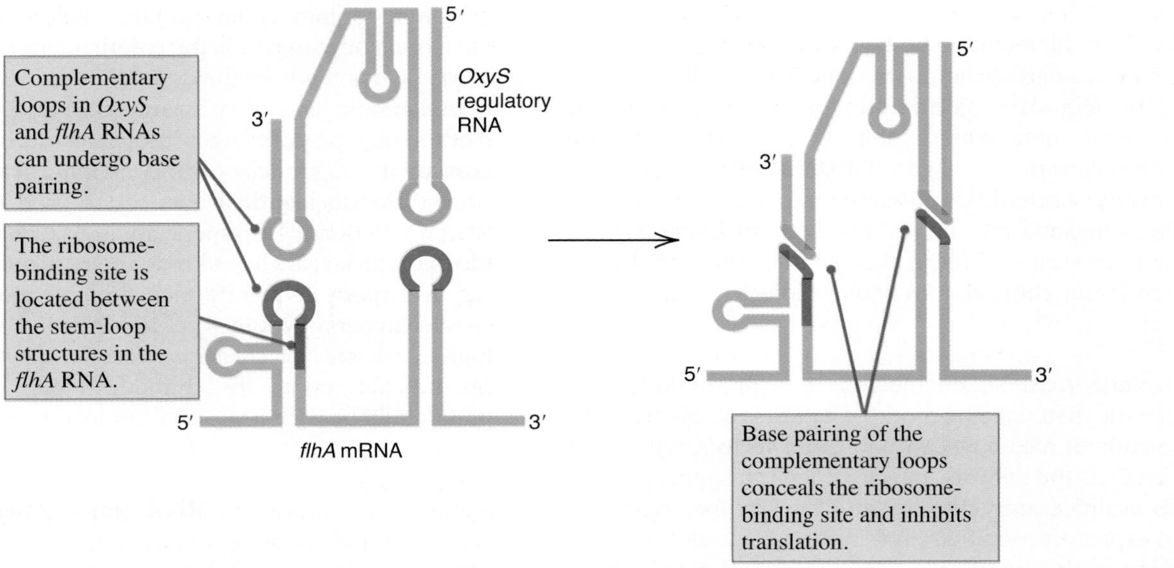

FIGURE 6.21 Antisense RNA is used to block translation in order to silence a defective gene. Adapted form S. Altuvia and E.G.H. Wagner. 2000. *Proc. Natl. Acad. Sci.* USA 97: 9824

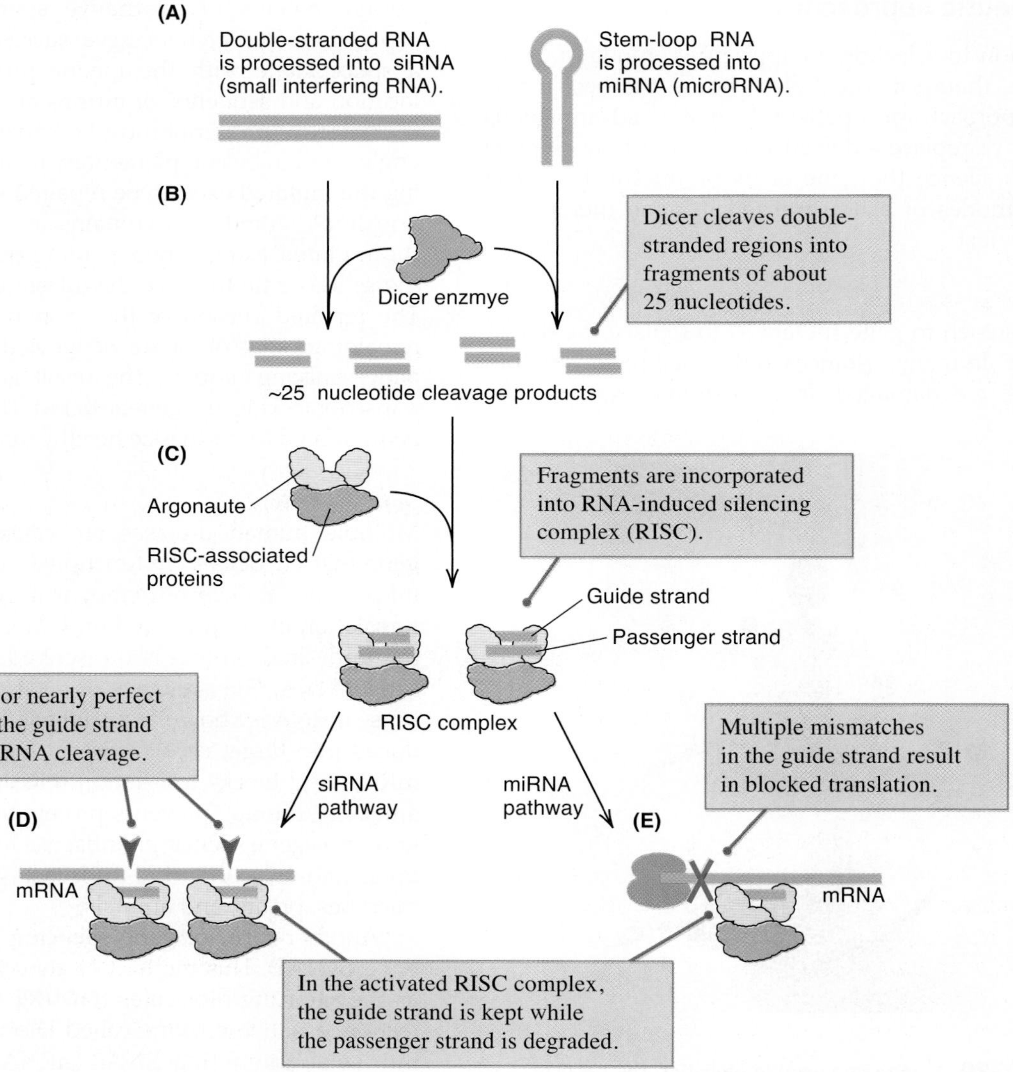

FIGURE 6.22 RNA interference using siRNA (left) and miRNA (right).

the RNA-induced silencing complex (RISC). The RISC delivers siRNAs to their target mRNA and both types of RNA bind through complementary base pairing. RISC then degrades the triplex of RNA, preventing translation and the synthesis of an unwanted protein. RNA interference may also be achieved with native microRNAs (miRNAs) instead of introduced siRNAs. The mechanism of gene silencing inside the cell is similar. But because miRNAs already exist in a cell, it is only necessary to stimulate their activity. RNA interference has proven successful in research studies for the suppression of over a dozen viruses, so it will likely also be used in the future for treatment and/or prevention of viral infections. Examples of viruses suppressed in culture via RNAi include hepatitis C, human papillomavirus, and HIV.

DNA can be used in lieu of RNA in order to silence a gene. In triple-helix-forming oligonucleotide therapy, DNA oligonucleotides are used to block transcription of a defective gene. The therapeutic oligonucleotide is engineered to readily bind within the grooves formed between the two strands of DNA present at the mutated gene locus. Upon binding a triple helix (also called triplex) of DNA is formed. The triplex prevents transcription and thus silences the defective gene.

Ribozyme therapy is another technique used to prevent translation of a mutated messenger RNA transcript. Ribozymes are modeled after enzymatic RNA normally found in cells. For example, transfer RNA is a native ribozyme that catalyzes synthesis of a polypeptide chain during translation. Ribozymes therefore have catalytic properties. RNA is engineered to function as a ribozyme which targets the defective mRNA. In other words, the engineered ribozyme is complementary to the mutated transcript. The ribozyme is introduced into affected cells where it binds and then cuts the transcript. Once cleaved, the messenger RNA cannot be translated and the defective protein will not be synthesized. While naturally occurring catalytic nucleotides are made of RNA, medical biotechnologists have also engineered DNA with enzymatic properties known as deoxyribozymes.

Aptamer therapy silences a gene by inactivating its transcription factor. Transcription factors are enzymes that promote gene expression. Without an active transcription factor, gene expression does not proceed. Aptamers are oligonucleotides with the ability to bind to a target protein, such as a transcription factor protein, thereby disabling its functionality. Researchers are able to engineer aptamers that can preferentially bind to a specific transcription factor associated with the gene to be silenced. Pegaptanib is a recently approved aptamer used in gene therapy to treat macular degeneration.

A notable roadblock of gene silencing is the rapid degradation of RNA inside the cell. RNA, whether introduced or naturally occurring, doesn't persist long enough to have an extended or permanent effect on the patient. Thus, gene silencing with RNA is better suited to cases of gene therapy where short-lived treatment is desired. However, deoxyribozymes promise to overcome this hurdle because they are more persistent within the cell. Because gene silencing targets a single gene, the above methods are most appropriate for single gene disorders and are not suitable for polygenic conditions or chromosomal abnormalities. Also, silencing a gene may not be enough to make a patient healthy. A second step, where a repaired gene is introduced, is often desirable. Several gene silencing treatments are in clinical trials, but as of yet none are approved for market.

Virotherapy

In virotherapy, viruses are engineered to recognize diseased or cancerous cells. The engineered virus binds to target cells and infection occurs. As the virus carries out its life cycle, multiple copies of the virus are replicated within the host cell. The host cell is destroyed as part of the process. Replicated viruses go on to infect and destroy neighboring cells which they are engineered to target. Virotherapy is useful in destroying unwanted cells while keeping healthy cells intact.

Pharmaceuticals and Therapeutics

The fields of genomics and proteomics have contributed volumes of data to facilitate new drug discovery and methods of disease treatment. Identification of useful proteins in various organisms allows pharmaceutical researchers to locate the encoding gene, which then can be inserted into a host organism for production. In many cases, recombinant DNA technology is a viable alternative to extracting proteins from the original source organism. Perhaps it is no surprise that the first patented commercial product produced by recombinant DNA was a drug (human insulin).

Another example is the anticancer drug Taxol, derived from the bark of the Pacific yew tree. To produce one kilogram of the drug, 30,000 pounds of bark are needed; that much bark destroys from 2,000 to 4,000 trees. Furthermore, the tree is so slow growing that it limits the production of Taxol even more. Producing taxol through recombinant DNA technology circumvents these obstacles. In the near future medical biotechnology may be used to replace traditional chemical synthesis of various drug compounds.

Biopharma

Biopharma is the development of pharmaceuticals via biotechnology. It has existed in progressing forms for centuries. The first vaccine was introduced in 1796 by Edward Jenner for prevention of smallpox infection. Doctors used to perform the procedure of bloodletting by applying leeches to ill patients. We have used microbes to synthesize antibiotics for the past hundred years. Thus, the commandeering of biological systems for medicinal purposes is not novel, but it has experienced dramatic leaps of innovation since the advent of molecular biotechnology. Today, numerous prescribed medications are produced through fermentation of transgenic microbes and by genetically-modified plant and animal tissue culture.

Recombinant DNA Proteins

Microbes, plants, and animals may all be genetically modified through recombinant DNA in order to synthesize a desirable protein product. The sector of biotechnology engaged in production of recombinant DNA proteins represents a $32 billion industry. Erythropoietin, which stimulates red blood cell synthesis, and insulin, which controls blood glucose levels, are by far the top selling drugs produced with this type of molecular biotechnology (**TABLE 6.4**).

Production of recombinant DNA proteins as well as other molecules by genetically modified microbes has been detailed. The methods employed in industrial biotechnology to synthesize molecular products from bioengineered bacteria and fungi are equivalent to those performed in medical biotechnology. Microbes provide a well understood and quick method of protein product. However, in the case of many bacterial host organisms, their prokaryotic nature prevents them from properly folding and modifying proteins. Without these events, the protein product is not congruent with human forms. Additional processing must take place to account for these shortcomings.

TABLE 6.4

Some Recombinant DNA Proteins Used in Medical Biotechnology

Protein Product	Application
α and β interferons	• Hepatitis C • Multiple sclerosis • Hairy cell leukemia
Calcitonin	Osteoporosis
Coagulation factor IX	Christmas disease
DNase	Cystic fibrosis
Epidermal growth factor	Skin wounds and burns
Erythropoietin	Anemia
Granulocyte colony-stimulating factor	• Bone marrow transplants • Immune disorders
Insulin	Diabetes
Relaxin	Childbirth
Tissue plasminogen activator	• Heart attack • Stroke
Tumor necrosis factor	• Tumors • Rheumatoid arthritis

Pharmaceutical production can also occur within animal host organisms. For example, animals ranging from insects to goats to chickens have served this purpose in research studies. Insects are valued for their relatively high protein synthesis rates but, as with bacteria, the proteins are modified quite differently than mammalian proteins. Solutions to this hurdle are currently under investigation. Genetically modifying mammals may at first seem like an easy fix to the problem of protein folding and modification. Mammals are costly research subjects because they are larger and take longer to develop. Also, mammals may be inadvertent contamination sources of viruses and/or prions to which humans may be susceptible. Nevertheless, milk and egg producing animals are of particular interest to researchers, as they would make taking medicine as simple and routine as grabbing lunch. Eggs and milk can be produced in large volumes, so the amount of protein synthesized may offset the high costs of growing and caring for transgenic animals. Lastly, studies must carefully monitor transgenic animals for any adverse effects on their health. There is always the possibility that insertion of a human protein transgene can disrupt normal functions within the modified organism. Would this constitute animal abuse? Ethical treatment must be considered.

Both crops and specialty plants are genetically modified to produce medications. Plants have the advantage of being free of viruses and prions with the ability to infect humans. Some specialty plants, such as the tobacco plant, are highly prized due to their high seed production; each transgenic seed will grow up to become a pharmaceutical factory. However, the threat of outcrossing is real unless steps are taken to prevent release of altered genetic material. The risk of contaminating wild plant species can be mitigated by selecting a plant host organism without any nearby wild relatives. Another option is to genetically modify the plant so it expresses the recombinant protein within its chloroplasts, which are passed onto offspring by the mother. Therefore, male pollen does not contain any transgenes. When pollen is carried to neighboring fields, there is no possibility that transgenes will drift into other plants. Yet even with this type of safeguard in place, pharmaceutical plant products must be tightly controlled through chain of custody to avoid their accidental release into the food supply. It would be impossible to differentiate a transgenic potato containing a drug with a normal baking potato picked up in the produce department.

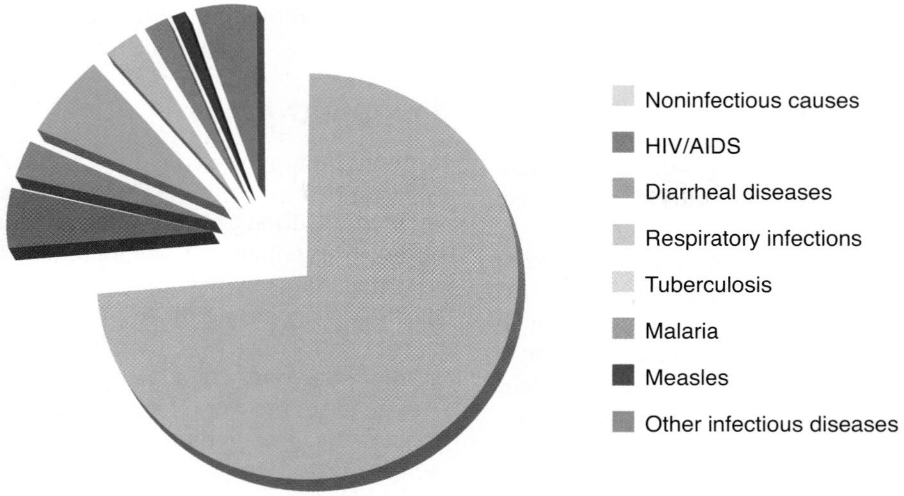

FIGURE 6.23 The majority of deaths worldwide are the result of noninfectious agents.

Pharmacogenomics

Biopharma employs pharmacogenomics, the customization of disease treatment based upon individual genetic profiles. Currently, drugs are prescribed in a one-size-fits-all manner. As a result, subsets of people taking a specific drug will experience side effects that may even be fatal. Over 100,000 deaths occur each year due to drug incompatibility. With pharmacogenomics, researchers aim to catalog genetic profiles with known reactions to different chemical compounds. Then a patient may be individually tested for known genetic mismatches to drug candidates in his or her treatment. This field of medical biotechnology puts us on the path to designer pharmaceuticals and creates the real possibility of reducing or even eliminating drug side effects. While gene therapy is promising in treating genetic disorders, pharmacogenomics is a viable alternative for treatment of noncommunicable diseases, which comprise the highest death rates worldwide (**FIGURE 6.23**).

The genetic subsets of our population are known as **haplotypes**. Each haplotype group possesses a different version of a known DNA polymorphism. Haplotypes occur at a frequency of at least 1% of the total population. Pharmacogenomics will improve as our knowledge of haplotype groups increases. The HapMap Project is a joint effort, funded by a number of pharmaceutical companies, academic institutions, and research organizations, which seeks to create a genetic map of existing haplotypes for use in pharmacogenomics.

Gleevec is a prime example of success in the field of pharmacogenomics. The drug is used in treatment of CML. The translocation event between chromosomes 9 and 22 creates a fusion protein expressed by the affected genes BCR and ABL. Gleevec targets the fusion protein and inactivates it, resulting in an increased survival rate of CML patients (from 30% to almost 90%). While pharmacogenomics is providing innovative treatment methods for both inherited and communicable disease, research is also underway to develop treatment for multifactorial disease with genetic and lifestyle contributing factors. For example, the protein leptin is being investigated as a treatment for obesity (**BOX 6.3**).

Nanomedicine

Nanotechnology is technological innovation at nanoscale. One nanometer is one billionth of a meter. Nanotechnology is commonly used in electronics and

Box 6.3 Leptin

Industrialized nations are in the midst of an obesity epidemic. Approximately 30% of North Americans and 20% of Europeans are overweight. Diet fads and "magic bullet" diet pills and supplements thus far have been ineffective in curbing this trend. Indeed, the trend continues to progress in intensity. The human protein leptin has proven successful in obesity reduction in laboratory mice. The protein is encoded by the *ob* gene (ob stands for "obese"). Leptin reduces appetite and food consumption, leading to weight loss. Pharmacogenomic studies have revealed some humans have a leptin deficiency that could be treated through gene therapy. Recombinant strains of the nonpathogenic bacterium *Lactococcus lactis* are used to infect human patients. Because *L. lactis* bacteria are used in the dairy industry, it already has a GRAS (generally recognized as safe) status with the FDA. The bacterium carries and secretes an engineered leptin gene into patient tissue. A nasal spray containing genetically modified *L. lactis* is in clinical trials.

engineering and shows potential for medical biotechnology as well. The subdiscipline of nanotechnology relating to health care is known as **nanomedicine**. Researchers are investigating the development of nanodevices that could be introduced to human systems in order to repair tissue damage, attack infections and/or cancerous cells, or to monitor vital signs such as blood pressure or blood glucose levels.

Nanomedicine has already introduced the technology of microspheres. Microspheres are nanoparticles used for drug delivery. More commonly, pills and capsules are ingested and must be digested and circulated through the bloodstream for a drug to reach its intended target. This method essentially dilutes the medication and only a fraction of the active ingredients reach the actual ailment.

Microspheres offer a solution to this problem. They are tiny spheres of lipid molecules that resemble the plasma membranes of cells. Microspheres are introduced by inhalation spray through either the nasal passageways or the mouth. Once the inhaled microspheres reach the lungs, there is direct entry into the bloodstream via pulmonary alveoli. Microspheres can be applied to dermal patches or injected directly into tissue layers. It is also possible to coat microspheres with antibodies and other recognition proteins for specific targeting of pathogens or cancer. This increases the effectiveness of the drug while greatly diminishing side effects, because the treated microspheres only deliver the drug to affected areas of the body.

Another area of nanomedicine combines the use of stem cells and therapeutic cloning with tissue engineering. Stem cells produced by somatic cell nuclear transfer (see below) are used to grow cloned tissue. The patient's stem cells are applied to nanoscaffolds and treated with growth factors. Collagen and synthetic polymers are used to create the network of tiny filaments of the nanoscaffold, resulting in a repetitive matrix (**FIGURE 6.24**). The matrix acts as a substrate upon which the applied stem cells may grow. The cells grow between and around the matrix in a predictable pattern, creating an engineered tissue similar in structure to natural tissue.

FIGURE 6.24 SEM image of a polymeric nanoscaffold used in tissue engineering. © Suwan Jayasinghe/Science Photo Library

Immunotherapy

Immunotherapy is a field of medical biotechnology seeking to treat immune disorders. Through a host of different methods, **immunotherapy** treats disease by stimulating, enhancing, or suppressing the immune system, depending upon the needs of the patient. One area of research in immunotherapy is discovery and utilization of antibodies, which are produced by the immune system in response to a foreign antigen (**FIGURE 6.25**). Antigens are present on many cell surfaces and act as a means of self-recognition for an organism. If the antigen is not recognized as "self," an immune response is activated. The production of antibodies is a long-term immune response in that production increases with every exposure to the antigen.

If for whatever reason an individual is unable to produce a needed antibody or cannot produce the antibody in sufficient quantities, antibodies offer a form of medical treatment. Immunotherapy may also introduce the antigen into a patient as a means of stimulating an immune response. This has occurred for hundreds of years in the form of vaccines, although more recently this approach is preferred not only for routine prevention of disease but also in its treatment.

Detecting Antigens

In the past 200 years or so since the smallpox vaccine was introduced, myriad antigens have been discovered and characterized. The work of identifying new, previously unknown antigens continues as technology improves our ability to recognize them. Two innovative techniques, in vivo induced antigen technology (IVIAT) and differential fluorescence induction (DFI), aid in antigen detection. Once a new antigen is discovered, researchers can evaluate the utility of constructing a monoclonal antibody or vaccine designed to inactive the antigen.

In Vivo Induced Antigen Technology

In vivo induced antigen technology (IVIAT) identifies antigens that are produced by a pathogen during times of infection. Previous antigen detection methods would study a disease-causing pathogen in culture. Thus, only antigens produced while the pathogen is free-living were found. IVIAT takes serum from patients infected with the disease of interest. The serum contains antibodies produced by the patient's immune system in response to the pathogen. Sera are mixed with a culture of the pathogen, causing some antibodies in the sera to bind with antigens presented by the pathogen. The remaining antibodies correspond to antigens only produced by the pathogen during times of infection. These antibodies can be isolated and sequenced. The complementary sequence represents

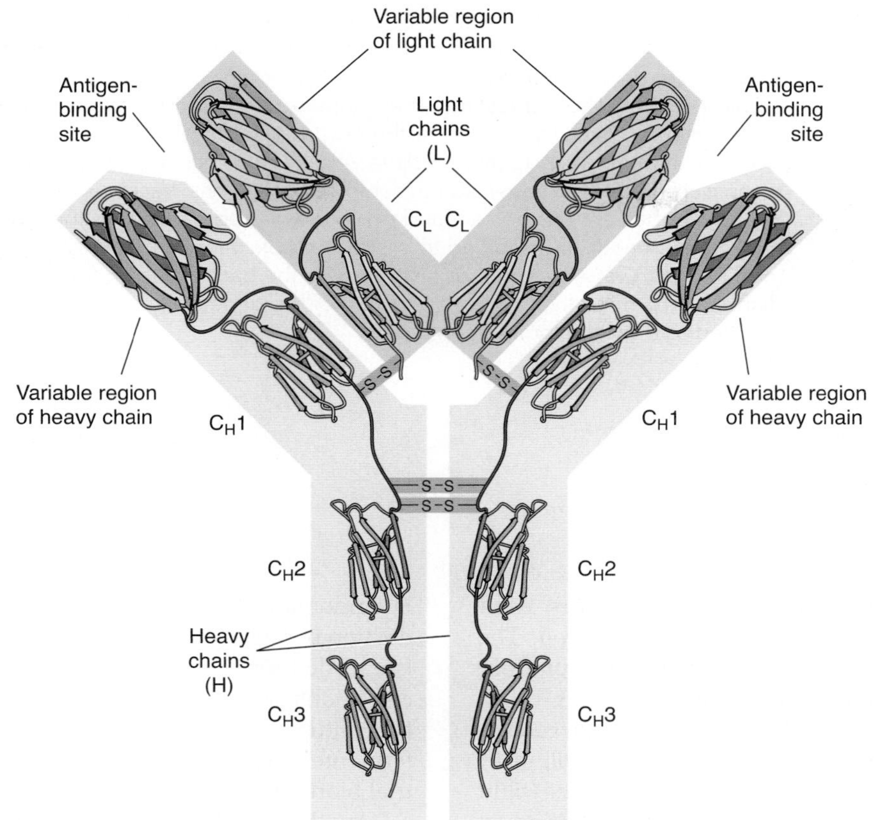

FIGURE 6.25 A diagram of the human antibody IgG, also known as immunoglobulin G displays antigen-binding sites.

that of the previously unidentified antigen. These antigens become candidates for vaccine or monoclonal antibody development.

Differential Fluorescence Induction

Another method for detecting antigens is **differential fluorescence induction (DFI)**. In DFI, a gene library of the pathogen under investigation is constructed and treated with green fluorescent proteins (GFPs). The library is used to transform bacterial cells which are grown in culture. The culture is subjected to a range of environmental conditions, particularly fluctuations in pH. It is known that bacteria seek out acidic environments when they are attacked by white blood cells as part of the immune response. Thus, the first treatment of the GFP-linked bacterial culture is acidic conditions. Cells with high GFP presence indicate high gene expression. These cells are sorted out and treated with neutral pH as a test. As long as GFP levels drop, this indicates it truly was the drop in pH that increased gene expression. At this point, only the cells with low GFP levels are sorted out because they represent populations with active pathogenic genes. They are once again treated with acidic conditions and cells with high GFP are sorted out once more. Each step in the DFI process acts to concentrate the culture so the only cells remaining are those expressing antigens present in infectious conditions (i.e., low pH). As with IVIAT, these antigens become candidates for biopharma applications.

Vaccines

Vaccines introduce foreign infectious agents into the human body in order to stimulate an immune response. Vaccines are so named because the first vaccine contained cowpox derivatives (*vacca* is the Latin word for cow). Vaccination for certain diseases is a routine medical procedure (**BOX 6.4**). There are four major categories of vaccines. They are grouped based upon their mode of action (**TABLE 6.5**).

Monoclonal Antibodies

Monoclonal antibodies (MAbs) comprise a single type of antibody for treatment (or identification) of a particular genetic disorder. The first procedure to synthesize monoclonal antibodies for clinical use was developed in 1975. Murine rodents such as mice or rats are injected with a purified antigen that is complementary to the antibody under production. Time is given for the murine's immune system to respond, which

Box 6.4 Edible Vaccines

Bananas are one of the most well studied host organisms for the production of edible vaccines. The tropical fruit is appropriate for growth in tropical and subtropical environments, which happen to be the most fertile breeding grounds for many of the world's communicable diseases. Edible vaccines also eliminate the need for skilled technicians trained to administer injectable vaccinations. Vaccines for cholera, hepatitis B, and diarrhea are examples found in bioengineered bananas. Other host organisms have been genetically modified to produce edible vaccines. For instance, one genetically modified corn variety expresses a vaccine for the bacterial toxin that causes traveler's diarrhea. The natural reproductive system of the genetically modified plant may be hijacked as a vaccine production facility.

results in biosynthesis of the desired antibodies. After several weeks, the animal's spleen is removed. The spleen is the location of B lymphocytes that produce antibodies, so it is the richest source available. The B lymphocytes are cultured in the presence of cancerous myeloma cells, which are immortal and will grow in culture indefinitely. Scientists can create the right environmental conditions in culture to induce fusion of B lymphocytes to the myeloma cells, resulting in hybrid cells known as **hybridomas** (**FIGURE 6.26**). The fusion can be encouraged by electric shock, introduction of a virus, or through treatment with polyethylene glycol. Hybridoma cultures are living factories for MAb production because the B lymphocyte portion of the hybrid is encoded for the antibody. As long as environmental conditions are maintained, the myeloma portion keeps the culture dividing ad infinitum. Hybridomas can be fermented in batch or continuous culture within industrial bioreactors to increase production.

Another cell type was engineered in order to synthesize monoclonal antibodies. Chinese hamster ovary (CHO) cells contain recombinant DNA that enables the production of MAbs with human antigens (**FIGURE 6.27**). These antigens prevent rejection of CHO-produced

TABLE 6.5

Types of Vaccines

Type of Vaccine	Active Component	Examples
Attenuated	live but weakened bacteria or viruses	• Polio (Sabin vaccine) • MMR • Chickenpox • Cholera • Tuberculosis
Inactivated (also known as killed)	dead bacteria or viruses	• Polio (Salk vaccine) • Rabies • DPT • Influenza
Subunit	portions of bacterial or viral structures, such as proteins or lipids	• Hepatitis B • Anthrax • Tetanus • Meningitis
DNA	bacterial or viral DNA	• West Nile virus • Canine melanoma

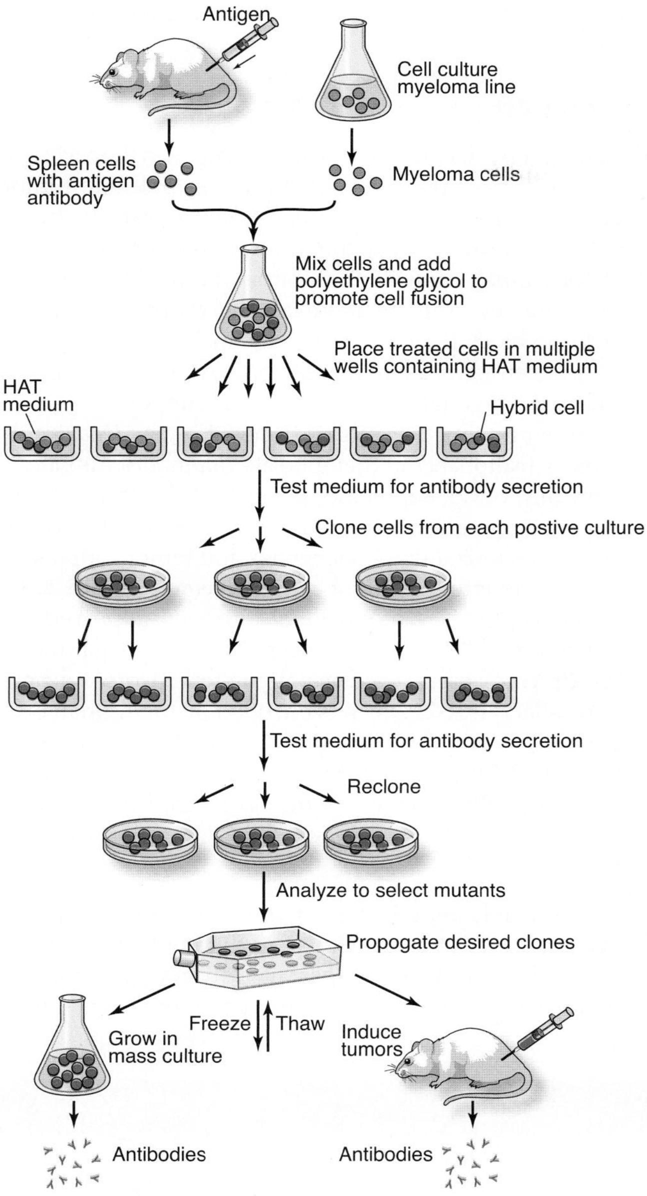

FIGURE 6.26 The establishment of a hybridoma cell line. Adapted from C. Milstein, *Sci. Am.* 241 (1980): 66–74

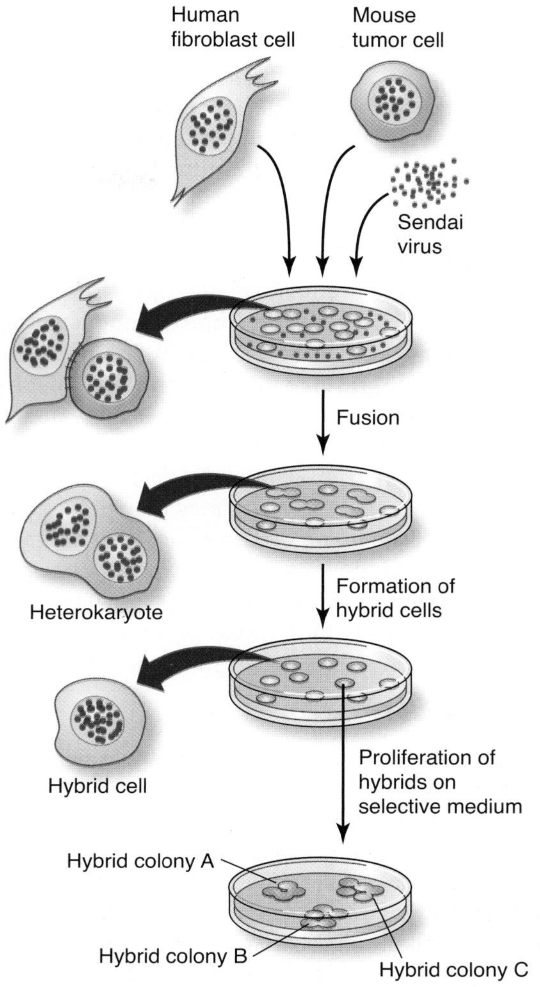

FIGURE 6.27 Formation of a chimeric CHO cell. Adapted from J.F. Griffiths, et al. *Modern Genetic Analysis: Integrating Genes and Genomes*, Second Edition. W. H. Freeman & Company, 1999

antibodies in human patients. Because CHO monoclonal antibodies are part human and part hamster, they are termed **chimeric antibodies**.

The first monoclonal antibody approval by the Food and Drug Administration (FDA) was in 1986. OKT3 was used to treat organ transplant rejection, a significant medical concern with immunotherapy implications. Another approved MAb, Herceptin (highlighted below), treats breast cancer while Rituxan treats lymphoma. Current monoclonal antibody research focuses on increasing their effectiveness by attaching drugs or radioactive isotopes. Another recent application of monoclonal antibodies is in treating chemical dependencies. A person may be treated with an antibody specific to the substance to which he or she is addicted, including both legal substances such as nicotine and illegal drugs like heroin or cocaine.

Monoclonal antibodies are also used to detect the presence of biomarkers. For example, home pregnancy tests contain MAbs to detect human chorionic gonadotropin (hCG), a hormone synthesized in pregnant women. Monoclonal antibodies can also identify the presence of cancerous cells. MAbs treated with fluorescent probes bind to the cancerous cells and may be visualized with magnetic resonance imaging techniques.

Xenotransplantation

Patients receiving donor tissues or organs are subject to severe immune responses leading to rejection. Around 150,000 donor recipients are hospitalized each year and approximately 10,000 individuals die per annum due to tissue and organ rejection. Xenotransplantation is a biopharma technique utilizing animals as donors. The donor organisms are genetically modified to express human antigens and/or shield proteins on their cell surfaces as a means to reduce incidences of rejection. Shield proteins inactivate complement proteins that would otherwise initiate an immune response.

Because pigs have organs similar in size to their human counterparts, these animals are the subject of xenotransplantation research. Another advantage of using pigs is there are relatively few diseases transmitted between us and them. One line of genetically modified (GM) pigs, known as the Astrid line after its first member, was engineered to express four shield proteins. Additional GM pig lines were engineered since Astrid was born in 1992, although xenotransplantation is still in clinical trial phases. Ethical concerns over breeding animals as organ factories may prevent adoption of xenotransplantation technology. Therapeutic cloning (discussed below) uses a patient's own genetic material and serves as a promising alternative for the generation of transplant tissues and organs.

Treating Cancer

Genetic counseling overlaps with the treatment of cancer. Certain defective genes are known to cause susceptibility to cancer. Research is underway to characterize genetic changes present in different forms of cancer, as well as to identify subsets of genetic types for each form (**TABLE 6.6**). One major research project is The Cancer Genome Atlas Project (TCGA). Recognizing that a single type of cancer can result from different genetic changes, TCGA promises to enhance applications of pharmacogenomics specifically related to cancer treatments.

Direct Attack

Cancerous cells may have active oncogenes that actively express transcription factors known to promote tumor growth. Researchers are investigating cancer treatments that would cease the action of these transcription factors. Small molecule inhibitors are a class of drugs that bind to proteins produced by cancerous cells in order to block their ability to function.

Cancerous cells may also possess mutated tumor suppressor genes. Normal tumor suppressor genes express proteins that inhibit tumor growth. Small molecule activators act in the opposite manner of inhibitors. They turn tumor suppressor genes back on, thus activating the expression of suppression proteins. Discovery and characterization of oncogenes and tumor suppressor genes is an intense area of cancer research (**TABLE 6.7**).

Herceptin is a monoclonal antibody approved for the treatment of breast cancer. About one-quarter to one-third of breast cancer patients overexpress a protein called HER-2, which is encoded by the human epidermal growth factor receptor 2 gene. Herceptin binds to HER-2 and inhibits tumor growth and metastasis (spread to other tissues) of the breast cancer.

Suicide

Genasense is an antisense therapy in development for the treatment of cancer. Over 20 different clinical trials are under way to investigate its efficacy in a number of forms of cancer, including prostate, liver, colon,

TABLE 6.6

Inherited Cancer Syndromes

Syndrome	Primary tumors	Associated tumors	Chromosome	Gene	Proposed function
Li-Fraumeni syndrome	Sarcomas, breast cancer	Brain tumors, leukemia	*17p13*	*p53*	Transcription factor
Familial retinoblastoma	Retinoblastoma	Osteosarcoma	*13q14*	*RB1*	Cell-cycle regulator
Familial melanoma	Melanoma	Pancreatic cancer	*9p21*	*p16*	Inhibitor of Cdk4 and Cdk6
Hereditary nonpolyposis colorectal cancer (HNPCC)	Colorectal cancer	Ovarian, glioblastoma	*2p22* *3p21* *2q32* *7p22*	*MSH2* *MLH1* *PMS1* *PMS2*	DNA mismatch repair
Familial breast cancer	Breast cancer	Ovarian cancer	*17q21*	*BRCA1*	Repair of DNA double-strand breaks
Familial adenomatous polyposis of the colon	Colorectal cancer	Other gastrointestinal tumors	*5q21*	*APC*	Regulation of β-catenin
Xeroderma pigmentosum	Skin cancer		Several complementation groups	*XPB* *XPD* *XPA*	DNA-repair helicases, nucleotide excision repair

TABLE 6.7

Cell Cycle Regulatory Genes Affected in Tumors

Protein oncogenes	Alteration	Consequence
Cyclin D	Amplification or overexpression	Promotes entry into S phase.
Cdk4	Amplification	Promotes entry into S phase.
Cdk4	Mutation	Cdk4 resistant to inhibition by p16; promotes entry into S phase.
EGFR (epidermal growth-factor receptor, a tyrosine kinase receptor)	Amplification	Promotes proliferation by constitutive activation of pathway from growth-factor receptor.
FGFR (fibroblast growth-factor receptor)	Amplification	Promotes proliferation by constitutive activation of pathway from growth-factor receptor.
Ras	Amplification	Promotes proliferation by constitutively transmitting growth signal.
Ras	Mutation	Inactivates GTPase activity; constitutive activation of pathway from growth-factor receptor.
Bcl2	Overexpression by translocation next to strong enhancer	Blocks apoptosis.
Mdm2	Amplification	Mimics loss of p53 with loss of G_1/S, S, and G_2/M checkpoint functions.
Telomerase	Overexpression	Cells no longer undergo senescence.
Tumor-suppressor genes	**Alteration**	**Consequence**
p53	Mutation	Loss of G_1/S, S, and G_2/M checkpoint functions.
p21	Mutation	Loss of G_1/S and S checkpoint functions.
RB	Mutation	Promotes proliferation; E2F uninhibited.
Bax	Mutation	Failure to promote apoptosis of damaged cells.
Bub1p	Mutation	Loss of spindle assembly checkpoint function.
E-cadherin	Mutation	Loss of contact inhibition; tissue invasion and metastasis.

breast, and kidney. Antisense RNA, which targets genes overexpressed in a variety of cancers, would have the most widespread application. Genasense targets the gene encoding the protein bcl-2, which inhibits apoptosis (cell death). Without normal cell death, cancerous cells continue to divide. This is one of their most troublesome characteristics. By silencing the *bcl-2* gene, medical biotechnologists hope to stimulate apoptosis of cancerous cells in order to kill tumors.

Virotherapy is also being investigated as a means to destroy cancerous cells. As discussed above, engineered viruses are designed for specificity to the cancerous cells. The virus infects and destroys only the unhealthy cells. This method would alleviate the disastrous symptoms associated with traditional cancer treatments such as chemotherapy and radiation, which blindly destroy all cells in their path.

Regenerative Medicine

Regenerative medicine involves biosynthesis of tissues and organs. The field offers an alternative to tissue and organ donation and xenotransplantation. A major benefit of regenerative medicine is the ability to create replacement tissues and organs that are an exact genetic match to the patient, eliminating the concern of life-threatening immune rejection. Stem cells and cloning are important elements of regenerative medicine.

Stem Cells

Stem cells are the subject of much controversy but also the harbingers of biomedical innovation with seemingly endless potential to cure disease. Stem cells are undifferentiated cells, meaning they have not developed into their specialized form. **Pluripotent** stem cells have the ability to differentiate into the 200 or so cell types present in the human body. Some stem cells are **totipotent** and can form adult cell types as well as cells necessary for embryonic development, such as those obtained in chorionic villus sampling. Others are **multipotent**, with more limited abilities to develop into a narrow range of cell types. Lastly, a stem cell may be **unipotent** with only a single cell type as its ultimate fate. Another major characteristic of stem cells is their immortality. Similar to cancer cells, stem cells do not appear to have a shelf life and can divide endlessly in culture.

The study of stem cells is leading to a greater understanding of human development. In characterizing human development, focusing on abnormal patterns can illuminate mechanisms behind presentation of disease. Stem cells have the potential to introduce normal, healthy genetic variants to a patient, making them prime candidates for gene therapy applications. Stem cells are also the raw material for cloning and regenerative medicine. They are widely investigated and used as tools to replace injured or diseased tissues and organs. This method is preferable to receiving donor tissues and organs as it eliminates the possibility of immune rejection. Stem cells are divided into classes based upon their tissue source method of harvest.

Embryonic Stem Cells

Embryonic stem cells (ESCs) are collected from eight-celled embryos prior to in vitro fertilization (IVF) implantation. As discussed earlier, parents undergoing the IVF procedure can opt to perform preimplantation genetic testing. Any embryo with negative test results is either destroyed or donated to stem cell research. Embryonic stem cells are also harvested from the blastocyst stage of embryonic development. Embryonic stem cells are an origin of controversy regarding stem cells, because aborted embryos are a primary source. The first ESC cell line was established at the University of Wisconsin-Madison in 1998. ESCs are pluripotent.

Embryonic stem cells may be used to repair diseased tissue. For example, ESCs can be transplanted into an affected region of the body where they are allowed to divide and differentiate. The differentiated cells would replace unhealthy ones, which would naturally die off due to their diseased state. There are still hurdles to overcome with ESCs. Issues include tissue rejection by the patient's immune system, the potential for uncontrolled tissue growth, and misdifferentiation of ESCs into unintended cell types.

Fetal

Eight weeks following fertilization, the developing embryo is now considered a fetus. Fetal stem cells are controversial as well, because they are harvested from terminated pregnancies. For a segment of our population, the belief exists that human life begins at fertilization. Thus, some members of our society oppose the use of embryonic and fetal stem cells on the grounds that abortion is equivalent to homicide. As they oppose any form of abortion, they also object to collection of stem cells during the procedure. Like ESCs, fetal stem cells are pluripotent.

Umbilical Cord Blood Stem Cells

There is an increasing trend to store stem cells retrieved from umbilical cord blood. It is possible to collect cord blood stem cells from a newborn and store them in designated facilities. Indeed, cord blood banks are becoming more common. Though the cost is prohibitively high for many individuals, those with sufficient means are opting to store their newborn's stem cells for future use in the event the child requires gene therapy or regenerative medicine. The hope is that by the time the child needs any such treatment—if ever—research will have progressed far enough to provide safe, effective, and approved methods.

Umbilical cord blood stem cells can also be donated for research purposes. An advantage to using cord blood stem cells is they have yet to express antigens that might elicit an immune response in the receiving patient. One drawback to cord blood is their multipotent nature. Umbilical cord blood stem cells can differentiate into many types of cells, but not all. This limits their potential applications.

Adult Stem Cells and Induced Pluripotent Stem Cells

Stem cells exist in the adult human as well. There are stem cells in bone marrow known as mesenchymal stem cells used for the production of blood cells, bone, cartilage, and fat. Blood disorders such as leukemia and myeloma can be treated with adult stem cells obtained from bone marrow. A patient's own bone marrow is preferred over using a donor because there is virtually no chance of rejection by the immune system. This is particularly advantageous as chemotherapy is used to kill diseased bone marrow prior to the procedure. The patient's immune system is basically wiped out in the process. Adult stem cells are present in the blood and skin, too. They are of far fewer number than the adult stem cells found in bone marrow, so they are more difficult to harvest. Nevertheless, they are subjects of investigation as well. However, adult stem cells are not as useful in medical biotechnology since they are multipotent.

Recent advances are changing the utility of adult stem cells. A Japanese research team successfully reprogrammed adult fibroblast cells to enter an undifferentiated state. Retroviruses acted as vectors to deliver four transgenes encoding developmental transcription factors (**FIGURE 6.28**). These proteins are not expressed

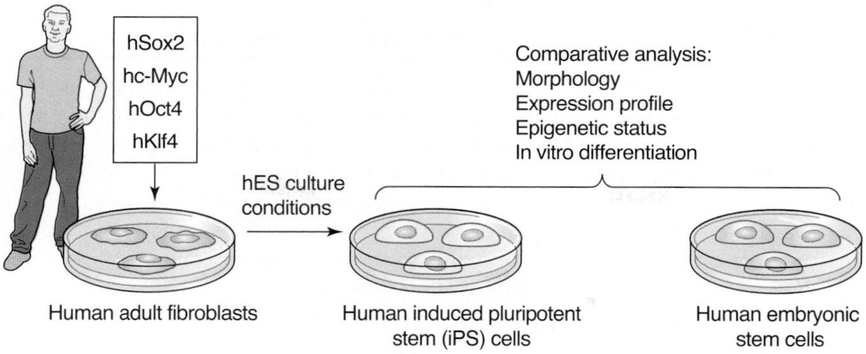

FIGURE 6.28 Transcription factors inserted into human fibroblasts result in induced pluripotent stem cells. Reproduced from *Cell*, vol. 131, H. Zaehres and H. R. Schöler, "Induction of Pluripotency: From Mouse to Human," pp. 834–835, © 2007, with permission from Elsevier [http://www.sciencedirect.com/science/journal/00928674]

in adults, just in embryonic development. The resulting cells exhibit indefinite lifespans and pluripotency much like human ESCs. They are known as induced pluripotent stem cells (IPSCs). Subsequent experiments have demonstrated instances where only one or two of the transgenes may be inserted and still result in induced pluripotency. Medical biotechnologists have differentiated IPSCs into specialized cell types such as neural cells, cardiac muscle cells, and hepatocytes (liver cells). Use of IPSCs in stem cell research assuages any concerns over deriving cell lines from terminated pregnancies. Another significant advantage is their potential in boosting pharmacogenomics research. Creation of induced pluripotent stem cells from individual patients would allow for customized genetic testing and identification of the most effective pharmaceutical treatments. Or, a patient's cells may be grown in culture and allowed to differentiate into a desired (genetically compatible) cell type that is then transplanted to the patient. As with ESCs, IPSCs do have potential drawbacks, such as the unintended proliferation of cells or problems with differentiation to the desired cell type. Furthermore, genetically modified IPSCs may unintentionally transfer information to normal cells within the patient.

Cloning

Say the word "cloning" and science fiction scenarios are bound to come up in the flow of conversation. Cloning is often misrepresented as referring only to the cloning of humans. Nothing could be further from the truth. Cloning is nothing more than the replication of an identical copy, whether that copy is a nucleotide sequence, a gene, a cell, a tissue, organ, or whole individual. In fact, asexual organisms are essentially cloning themselves when they undergo binary fission or budding. An exact genetic replica is produced; offspring of asexual organisms are clones of the parent. Therapeutic cloning and reproductive cloning are the two main types in use today.

Therapeutic Cloning

Therapeutic cloning synthesizes cloned stem cells and tissues for replacement of damaged or diseased areas of the body (**FIGURE 6.29**). Examples of therapeutic cloning include repairing damaged cardiac tissue resulting from heart attack and disease and regenerating healthy neurons in patients suffering neurological conditions such as neuropathy and Alzheimer's disease. Advances in therapeutic cloning are limited by the availability of stem cell lines and funding opportunities (**BOX 6.5**).

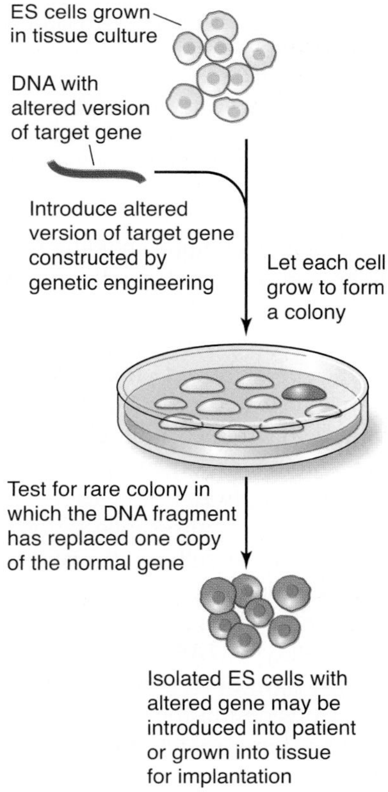

FIGURE 6.29 Therapeutic cloning genetically modifies stem cells. The altered cells may be grown into tissues for implantation. Adapted from B. Alberts, et al. *Molecular Biology of the Cell*, Fourth Edition. Garland Science, 2002

Box 6.5 Regulation and Funding of Stem Cells for Research

The first embryonic stem cell line established in 1998 prompted the introduction of federal oversight for stem cell research. In 2000, the National Institutes of Health announced the availability of federal funding for research along with a set of strict guidelines tied to funding opportunities. Frozen eight-celled embryos from IVF donors were an accepted source of ESCs at the time. Just a year later, President George W. Bush announced the reversal of the NIH policies. Federal funds could no longer be applied to ESC research with one exception. Any existing ESC cell lines established prior to August 9, 2001, were exempt. From that date onward, research teams using federal funds could either use a reported 78 existing cell lines or opt to study non-embryonic stem cells, such as adult stem cells. The number of cell lines available dwindled down to 21 by 2007 for a variety of reasons:

- Some cell lines were actually duplicates (genetic clones).
- Some cell lines failed to grow.
- Certain laboratories would not cooperate in allowing access to some cell lines.
- The patent status of some cell lines has been challenged. Prohibitive licensing fees limit who can afford to use them.

Further dampening the availability of the remaining cell lines is the issue of mutations accumulating over time. The cell lines are also not genetically diverse enough to provide a large enough sample in research studies. A particular genetic disorder of interest may not be present in existing cell lines, which prevents its study.

Two bills repealing the funding ban were approved by the House and Senate, once in 2005–2006 and once again in 2007, although both were vetoed by President Bush. In March 2009, newly elected President Obama issued an executive order to lift the ban and tasked the NIH with developing federal funding guidelines, which they accomplished within a few months. This did not settle the matter for good. A federal judge blocked funding of stem cell research in 2010 based upon an existing executive order that bans funding of research that requires the destruction of embryos. This ruling and the executive order it was based upon were lifted in 2011 by the U.S. Court of Appeals. NIH director Francis Collins stated, "We are pleased with today's ruling. Responsible stem cell research has the potential to develop new treatments and ultimately save lives. This ruling will help ensure this groundbreaking research can continue to move forward."

The ten year moratorium on federal funding in the United States left a void filled by individual states as well as private foundations and international research teams. For example, the state of California earmarked $3 billion for stem cell research in 2004. The state of New Jersey has appropriated $50 million in support of stem cell and cloning research. Private foundations such as the Howard Hughes Medical Institute and the Michael J. Fox Foundation for Parkinson's Research are significant contributors to research funding. Establishment of new stem cell lines and therapeutic cloning are legal in the United Kingdom, India, South Korea, China, Singapore, and Israel. Countries including Belgium, Switzerland, Turkey, and Sweden have stem cell research centers. Yet there are countries that have banned therapeutic cloning, such as Brazil, Australia, and the European Union (EU). Member countries like Belgium and the United Kingdom get around the EU ban by establishing new cell lines from unused IVF embryos donated by the parent(s).

The breakthrough technique enabling both methods of cloning is known as **somatic cell nuclear transfer (SCNT)**. Somatic cell nuclear transfer transfers the nucleus from a donor cell into a recipient egg cell which has had its nucleus removed. The recipient egg cell is therefore an enucleated cell. The engineered cell is grown in culture to produce multiple copies. Genetic material from the donor organism is present in all daughter cells resulting from cell division. Cloned stem cells produced by SCNT can be collected within five to six days from start of the culture for use in therapeutic treatments. Introduction of cloned stem cells in concert with growth factor infusions can promote differentiation into the desired cell type (**TABLE 6.8**).

Whole organs have been constructed by using this biotechnology. Considering the large numbers of individuals awaiting a donor match, therapeutic cloning of organs would alleviate their wait times and increase the availability of organ replacement. Over 60,000 patients are on organ donation wait lists and an estimated 100,000 more are in need of organ transplantation but are not currently on a wait list. On average, 10 people die each day while waiting for organ transplantation. Cloned organs may be preferable to xenotransplants. Immune rejection would be virtually zero and there is no concern over using live animals as organ donors.

Reproductive Cloning

Reproductive cloning results in a whole organism, a completely new individual that is genetically identical to the original. It is this type of reproductive cloning that can invoke controversy, because human cloning is often the first example to come to mind. As mentioned, asexual reproduction is a natural form of reproductive cloning. Bioreactor fermentation also involves reproductive cloning because identical copies of cells in culture are synthesized. Cultivating multiple plants from a plant cutting is another type of reproductive cloning that has been used for generations. Reproductive cloning of microbes and plants is relatively simple compared to animal cloning. In 1997, a sheep named Dolly was the first mammal created through reproductive cloning (**FIGURE 6.30**). Though biotechnologists have

TABLE 6.8

Growth Factors Useful in Stem Cell Differentiation

Growth Factor	Action	Treatment Candidates
Nerve Growth Factor	Promotes growth of nerve cells	• Brain injury • Spinal cord injury • Neuropathy • Alzheimer's disease • Parkinson's disease • Huntington disease
Fibroblast Growth Factor	Promotes cell cycle (increases cell division and growth)	• Broken bones • Ulcers • Blood vessels
Epidermal Growth Factor	Promotes skin cell division	• Persistent wounds • Burns • Scars
Transforming Growth Factor β	Aids in cell differentiation	Any

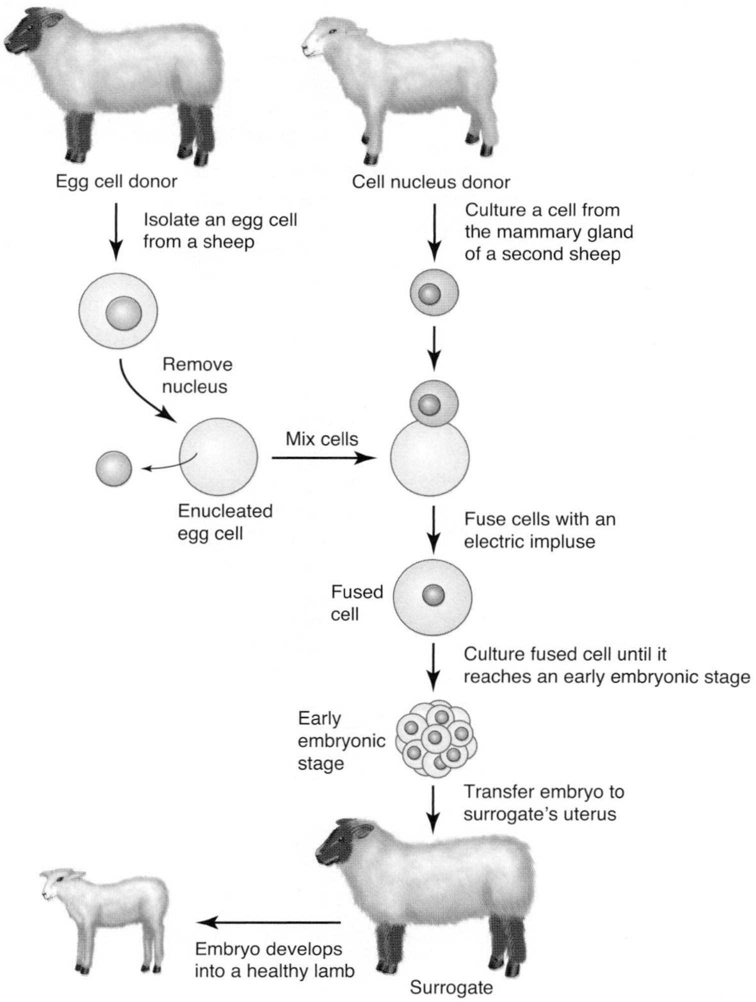

FIGURE 6.30 Reproductive cloning process. Somatic cell nuclear transfer is utilized to create the fused cell, which is cultured through the early embryonic stage. The embryo is then implanted and allowed to develop into a cloned progeny. Adapted from K.R. Miller and J.S. Levine. *Biology*. Prentice Hall, 2003

successfully cloned multicellular animals, this form of cloning is still rudimentary. The costs are high and success rates are low. Cloned animals typically do not live out their full life expectancies and they often have lung defects. Dolly was the single successfully cloned sheep out of nearly 300 attempts. She lived only half of her expected 12 years. Refer to our discussion of conservation biotechnology for more details on reproductive cloning of animals.

SUMMARY

- Genetic counseling seeks to detect genetic disorders and advise individuals on medical treatment options and reproductive planning. Genetic counseling involves the construction of a genetic profile for a patient and may include identification of genes linked to disease, biomarkers, and chromosomal abnormalities. Genetic counselors may test a developing fetus using amniocentesis, karyotyping, and/or chorionic villus sampling. Fluorescence in situ hybridization (FISH) is useful for genetic testing in adults, along with microarrays of DNA and/or proteins, restriction fragment length polymorphism (RFLP) analysis, allele-specific oligonucleotide analysis (ASO), and epigenomics.
- Gene therapy may be performed within the body or outside the body. In vivo gene therapy delivers genetic information directly into affected cells or tissues. Ex vivo gene therapy removes diseased cells, treats them in culture, and returns the transfected cells back to the patient.
- Gene therapy involves techniques that correct defective genes, or alter their expression, in order to treat or cure disease. It can include replacement of faulty genetic sequences or the interference of their expression. Theoretically, gene therapy can treat somatic or germ line cells, but due to ethical concerns only somatic cells are routinely treated. Modes of delivery include viral integration, introduction of naked or polymer-complexed DNA, receptor-mediated uptake, microcapsules, particle bombardment, and bacterial infection. Gene therapy utilizes techniques that either replace, repair, or silence defective genes.
- Biopharma utilizes molecular biotechnology such as genetic engineering via recombinant DNA technology in order to produce medications. It has revolutionized modern medicine as a growing number of prescribed medications are produced through fermentation of transgenic microbes and by genetically modified plant and animal tissue culture.
- Pharmacogenomics amounts to personalized medicine and entails customized treatments based upon individual genetic profiles. It promises to minimize or eliminate harmful side effects which are common with existing prescription medications. Nanomedicine harnesses the small scale of nanotechnology for applications in the medical field. For example, nanodevices could be implanted into human systems in order to repair tissue damage, attack infections and/or cancerous cells, or to monitor vital signs such as blood pressure or blood glucose levels. Nanomedicine also utilizes microspheres, which are nanoparticles used for drug delivery. Microspheres may also be coated with antibodies and other recognition proteins for specific targeting of pathogens or cancer.
- Immunotherapy treats disease by stimulating, enhancing, or suppressing the immune system. One area of immunotherapy under investigation is antigen detection, which utilizes in vivo induced antigen technology (IVIAT) and differential fluorescence induction (DFI). Vaccines and monoclonal antibodies are additional areas of immunotherapy.
- Xenotransplantation uses animals as tissue and organ donors. In order to reduce incidences of immune rejection in the patient, the donor organisms are genetically modified to express human antigens and/or shield proteins on their cell surfaces.
- Advanced cancer treatments based upon medical biotechnology include direct attack and suicide. With direct attack, small molecule inhibitors bind to proteins produced by cancerous cells in order to block their ability to function. Small molecule activators act in the opposite manner by turning on tumor suppressor genes, thus activating the expression of suppression proteins. Treatments such as antisense therapy and virotherapy are suicide therapies that can cause cancer cells to self-destruct.

- Regenerative medicine involves biosynthesis of tissues and organs, a major benefit of which is the ability to offer donations with an exact genetic match to the patient. This eliminates the concern of life-threatening immune rejection. Stem cells and cloning are important elements of regenerative medicine. Embryonic stem cells (ESCs) are collected from eight-celled embryos prior to in vitro fertilization (IVF) implantation and are pluripotent, meaning they are able to differentiate into any of the cell types present in the human body. Fetal stem cells are harvested after the eight week embryonic period of development and are also pluripotent. Umbilical cord blood stem cells and adult mesenchymal stem cells can differentiate into many types of cells, making them multipotent. Induced pluripotent stem cells are taken from multipotent adult stem cells and genetically engineered to behave as pluripotent stem cells.
- Therapeutic cloning synthesizes cloned stem cells and tissues for replacement of damaged or diseased areas of the body, whereas reproductive cloning results in a completely new individual, a whole organism that is genetically identical to the original.

KEY TERMS

Allele-specific oligonucleotide analysis (ASO)
Amniocentesis
Antisense RNA
Aptamer therapy
Biomarkers
Chimeric antibodies
Chorionic villus sampling
Differential fluorescence induction
Epigenomics
Eugenics
Fluorescence in situ hybridization (FISH)
Gene therapy
Genetic counseling
Haplotype
Hybridoma
Immunotherapy
In vivo induced antigen technology
Karyotype
Monoclonal antibody (MAb)
Multipotent
Nanomedicine
Oligonucleotide crossover
Pluripotent
Protein microarray
Reproductive cloning
Ribozyme therapy
RNA interference (RNAi)
Somatic cell nuclear transfer (SCNT)
Spliceosome-mediated RNA trans-splicing (SMaRT)
Therapeutic cloning
Totipotent
Transfection
Unipotent

DISCUSSION QUESTIONS

1. Visit the NCBI Map Finder at http://www.ncbi.nlm.nih.gov/mapview/map_search.cgi?chr=hum_chr.inf&query. Select a chromosome. List and describe known genes located on your chosen chromosome.
2. What are the different types of stem cells? How are they used in gene therapy and cloning?
3. What bioethical issues surrounding medical biotechnology concern you most and why? What issues do not concern you and why?
4. Should medical biotechnology be used solely for treatment of existing conditions or also in order to prevent disease? Should medical biotechnology be used in eugenics?

REFERENCES

Al-Suhaimi, E. A. and A. Shehzad. "Leptin, Resistin and Visfatin: The Missing Link Between Endocrine Metabolic Disorders and Immunity." *European Journal of Medical Research* 18, no. 1 (2013): 12.

Anderson, W.F. "Gene Therapy." *Scientific American* (September 1995): 124–128.

Bartels, D.M., et al., eds. *Genetic Counseling: Ethical Challenges and Consequences*. Piscataway, New Jersey: Transaction Publishers, 2010.

Biotechnology Industry Organization. "2007–2008 BIO Milestones." Industry Intelligence and Analysis. July 21, 2008. Accessed February 21, 2011, from http://www.bio.org/articles/2007-2008-bio-milestones

Cutter, Stephanie. "A Victory for Stem Cell Research and Patients." *The White House Blog*. July 27, 2011. Accessed February 21, 2014, from http://www.whitehouse.gov/blog/2011/07/27/victory-stem-cell-research-and-patients

DeRouen, M.C., et al. "The Race Is On: Human Embryonic Stem Cell Research Goes Global." *Stem Cell Reviews and Reports* 8, no. 4 (2012): 1043–1047.

Droogendijk, H.J., et al. "Imatinib mesylate in the Treatment of Systemic Mastocytosis: A Phase II Trial." *Cancer* 107, no. 2 (2006): 345–351.

Gey, Annerose, et al. "Identification of Pathogens in Mastitis Milk Samples with Fluorescence In Situ Hybridization." *Journal of Veterinary Diagnostic Investigation* 25, no. 3 (2013): 386–394.

Guan, Zeng-jun, et al. "Recent Advances and Safety Issues of Transgenic Plant-Derived Vaccines." *Applied Microbiology and Biotechnology* (2013, March 1): 1–24. Accessed February 21, 2014, from http://nextgenzorg.files.wordpress.com/2013/03/recent-advances-and-safety-issues-of-transgenic-plant-derived-vaccines.pdf

Huang, Leaf, et al. *Nonviral Vectors for Gene Therapy*. Philadelphia, Pennsylvania: Academic Press, 2005.

Jacob, Siju S., et al. "Edible Vaccines Against Veterinary Parasitic Diseases—Current Status and Future Prospects." *Vaccine* 31 (2013): 1879–1831.

Khokhar, Aroob. "Barack Obama Executive Order 13505, November 2008." *Embryo Project Encyclopedia* (2012).

Lattime, Edmund C. and Stanton L. Gerson. *Gene Therapy of Cancer: Translational Approaches from Preclinical Studies to Clinical Implementation*. Philadelphia, Pennsylvania: Academic Press, 2013.

Levine, A.D. *Cloning*. New York: The Rosen Publishing Group, 2009.

Ma, Li-Bing, et al. "The Development and Expression of Pluripotency Genes in Embryos Derived from Nuclear Transfer and In Vitro Fertilization." *Zygote* April (2013): 1–9.

Mahlknecht, Georg, et al. "Aptamer to ErbB-2/HER2 Enhances Degradation of the Target and Inhibits Tumorigenic Growth." *Proceedings of the National Academy of Sciences* 110, no. 20 (2013): 8170–8175.

Martin, Aaron, et al. "Manipulating T Cell-Mediated Pathology: Targets and Functions of Monoclonal Antibody Immunotherapy." *Clinical Immunology* 148, no. 1 (2013): 136–147.

McKinlay Gardner, R.J., et al., ed. *Chromosome Abnormalities and Genetic Counseling*. London: Oxford University Press, 2011.

Metzger-Filho, Otto, et al. "Magnitude of Trastuzumab Benefit in Patients with HER2-Positive, Invasive Lobular Breast Carcinoma: Results from the HERA Trial." *Journal of Clinical Oncology* (abstract) Accessed February 21, 2014, from http://jco.ascopubs.org/content/early/2013/04/15/JCO.2012.46.2440.abstract

Mohamed, Sharmarke and Basharut A. Syed. "Commercial Prospects for Genomic Sequencing Technologies." *Nature Reviews Drug Discovery* 12, no. 5 (2013): 341–342.

National Center for Biotechnology Information. *Genes and Disease*. Bethesda, Maryland: National Center for Biotechnology Information; 1998.

National Institutes of Health. "Federal Policy." *Stem Cell Information*. Last modified November 14, 2011. Accessed February 21, 2014, from http://stemcells.nih.gov/policy/Pages/Default.aspx

Takahashi, Kazutoshi and Shinya Yamanaka. "Induction of Pluripotent Stem Cells from Mouse Embryonic and Adult Fibroblast Cultures by Defined Factors." *Cell* 126, no. 4 (2006): 663–676.

Tantravahi, Ramana V., et al., "Mechanisms of Resistance to Targeted B-Raf Therapies." *Molecular Mechanisms of Tumor Cell Resistance to Chemotherapy*, pp. 69–88. New York: Springer, 2013.

Tiwari, Ashutosh and Atul Tiwari, eds. *Nanomaterials in Drug Delivery, Imaging, and Tissue Engineering*. Wiley-Scrivener, 2013.

Wirth, Thomas and Seppo Ylä-Herttuala. "History of Gene Therapy." *Gene* (2013).

Yu, Junying, et al. "Induced Pluripotent Stem Cell Lines Derived from Human Somatic Cells." *Science* 318, no. 5858 (2007): 1917–1920.

7 Environmental Biotechnology and Conservation

LEARNING OBJECTIVES

Upon reading this chapter, you should be able to:

- Identify the problems environmental biotechnology seeks to address.
- List examples of environmental pollutants.
- Differentiate among bioremediation, biostimulation, and bioventing.
- Compare and contrast in situ bioremediation with ex situ bioremediation.
- Explain the role of microcosms in environmental biotechnology.
- Characterize the unique requirements of contaminated soil and water bioremediation.
- Distinguish among bioaugmentation, phytoremediation, and mycoremediation.
- Discuss the role of metagenomics in environmental biotechnology and its methods.
- Describe examples of conservation biotechnology.
- List alternative sources of energy produced through conservation biotechnology.

Since the Industrial Revolution, our environment has experienced unprecedented destruction at alarming rates. Human activities have led to undesirable alterations of air, land, and water. Industrial processes resulting in waste as well as individual resource consumption contribute to our mounting pollution problems. Government intervention has taken several forms, including the creation of the Environmental Protection Agency in 1970 and the Clean Air and Water Acts, enacted in 1970 and 1972, respectively. Yet a variety of chemical contaminants persist in the environment. Contamination of our environment with toxic chemicals is a serious threat to biodiversity. As we have seen, the abundance of various organisms is essential in maintaining an ecological balance as well as serving as the source of bioprospecting.

Compounding environmental pollution is the concern over resource availability. Individuals and industry contribute to an ongoing energy crisis through consumption of fossil fuels such as natural gas, coal, and oil. Due to dwindling reserves and the unsustainable nature of fossil fuels, it is imperative to find alternative energy sources.

The combined effects of industrial and individual activities have also led to significant destruction of natural habitats. Loss of habitat contributes to depletion of biodiversity, resulting in what is considered to be the sixth mass extinction faced by our planet. Approximately 6,000 plant and animal species are listed as endangered worldwide. Nearly 10,000 more species are considered threatened, with the likelihood of becoming endangered if conditions do not improve.

While not a panacea, molecular biotechnology offers some solutions to these pressing issues. Environmental science and conservation are beginning to explore options that involve three main applications of biotechnology. Direct studies of the environment, such as bioprospecting, reveal previously unknown organisms, genes, and gene products. Discovery of natural novelties may then be investigated and applied in environmental cleanup in a process known as bioremediation. Lastly, biotechnology is useful in areas of environmental conservation. Genomics and cloning techniques are valuable applications for the purpose of maintaining biodiversity and protecting threatened and endangered species. Utilization of genetically modified organisms (GMOs) promises to yield new sources of alternative energy, including biofuels, which will deter overconsumption of limited natural resources.

Environmental Biotechnology

A majority of environmental pollution is due to releases from industrial processes in the form of waste. Accidental oil and chemical spills, intentional dumping, radioactive substances, and pesticide use are also contributing factors. Pollutants are often released into waterways with the rationale that water will dilute and disperse toxic chemicals. However, even minute quantities are taken up by aquatic organisms and accumulate in ever higher levels as they move up the food chain in a process known as **biomagnification**. Through consumption of seafood and via drinking water, chemical pollutants result in numerous negative consequences to human health. Other waste products are burned off, releasing chemicals into the air. Air pollutants include ozone, particulate matter, carbon monoxide, nitrogen oxides, sulfur dioxide, and lead. Chemicals in the air have the potential to precipitate with rainfall, leading to further pollution of land and water.

There is no denying that individuals contribute to the problem as well. It is estimated that the average American disposes of five pounds of refuse on a daily basis, which adds up to approximately 0.85 tons per year. Though there are restrictions on the disposal of household chemicals such as paints and batteries, due to poor oversight much of it is thrown away in regular trash receptacles and ultimately ends up in landfills. Landfill contents seep into deeper and deeper strata, making clean up an increasingly daunting task.

Environmental Pollutants

Environmental pollutants come from multiple sources and include heavy metals, radiation, and organic compounds. In the United States alone, an estimated 200 million tons of hazardous substances make their way into the natural landscape each year. Some of this pollution is the result of neglected or abandoned industrial sites, which are classified as Superfund sites by the Environmental Protection Agency (EPA). Since 1980, the EPA has actively engaged in cleanup of Superfund sites in an effort to reduce pollution and reclaim undesirable lands. If contaminants are successfully reduced to acceptable levels, Superfund sites have the potential for redevelopment. This will theoretically reduce the need for virgin lands that might otherwise result in further habitat destruction. Because 20% of Americans live within four miles of a Superfund site, remediation of these areas is also a public health concern. Soil contamination can seep down into groundwater, tainting water supplies. Contaminants may also evaporate into the air and be carried by wind or end up precipitating with rain, further compounding the issue as it affects even greater areas. While the EPA has successfully remediated over 1100 Superfund sites, more than 1300 still await treatment. Dozens more are in proposal stages for consideration of Superfund status (**TABLE 7.1**). If the past is any indicator, the 30+ years of activity thus far is just the beginning.

Superfund sites are not the only environments in need of bioremediation. Oil spills, whether on land or in marine environments, require swift action to minimize their devastating impacts. Often oil spills are associated

TABLE 7.1

EPA Superfund Sites by Status

Status	Non-Federal	Federal	Total
Proposed Sites	49	5	54
Active Sites	1155	156	1311
Completed Sites	1074	71	1145

Data are from the Environmental Protection Agency. "National Priorities List." Last updated February 22, 2014. Available at http://www.epa.gov/superfund/sites/npl/index.htm.

with release into aquatic ecosystems, but pipeline spills onto land are common (**FIGURE 7.1**). Even still, it is unknown how long it takes to completely restore an area contaminated with crude oil. Biotechnology used in oil spills is discussed below.

Landfills are another significant application for bioremediation. Household products thrown out with the garbage leach into the soil and run off into streams and rivers. Water quality tests frequently contain chemicals found in consumer products like lotions, hairsprays, cosmetics, paints, detergents, and other cleaning supplies. Other sites serving as suitable candidates for bioremediation include:

- Chemical spills
- Wastewater treatment facilities
- Industrial and military sites (including their downstream effluents)
- Groundwater contamination
- Landfills
- Pesticides and other runoff
- Farming operations
- Mining sites
- Radioactive release

Some of the most commonly treated chemical pollutants are listed in **TABLE 7.2**. Molecular biotechnology has introduced biosensor techniques to detect pollutants in the environment (**BOX 7.1**).

Bioremediation

Many chemicals released into the environment are large, organic molecules that are not immediately biodegradable. These are known as **persistent organic pollutants (POPs**; **TABLE 7.3**). Others are heavy metals and radioactive materials. Though normally found in minute quantities, these troublesome chemicals become concentrated as they are routinely released as waste. These persistent substances accumulate over time since they are not quickly broken down through natural processes. **Bioremediation** harnesses the metabolic pathways of living organisms in order to break down persistent chemicals that pollute the environment. It may involve the use of naturally occurring organisms that possess the ability to metabolize certain stubborn pollutants, or it may require the use of

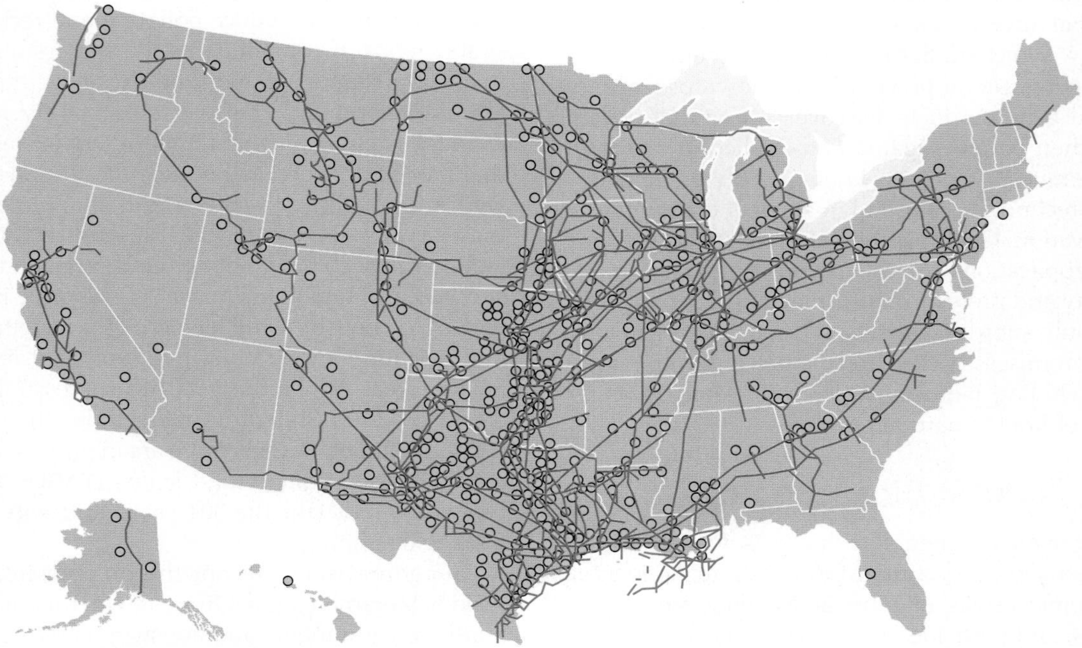

FIGURE 7.1 Toxic liquid spills from pipelines, by state, 1990–2011.

TABLE 7.2

The Most Prevalent Environmental Pollutants

Chemical	Source
Acetone	Plastics, fibers, solvents, drugs, and other chemicals
Benzene	Plastics, resins, detergents, nylon, and other chemicals
Carbon Tetrachloride	Refrigerants, propellants, pesticides, cleaning products, degreasers, fire extinguishers
Chromium	Anticorrosive, electroplating, tanning
Cyanide	Plastics and mining
DDT, DDE	Pesticides
Dichloroethane (1,1- and 1,2-)	Food wrap, plastic packaging, flame retardants, PVC, coatings, adhesives, upholstery, and other chemicals
Detergents	Paper, textiles, paints, plastics, and other chemicals
Mercury	Thermometers, dental fillings, batteries, creams, and other chemicals
Napthalene	Crude oil and petroleum refineries, other chemicals
Pentachlorophenol	Pesticides, wood preservatives
Polychlorinated biphenyls (PCBs)	Cooling and insulating systems, transistors
Polycyclic aromatic hydrocarbons (PAHs)	Car oil and exhaust, crude oil and petroleum refineries
Radioactive compounds	Nuclear power plants, medical facilities, research institutions
Toluene	Paints, cleaners, glues, inks, adhesives, coatings, and other chemicals
Vinyl chloride	Plastics (PVC)

Data from the Environmental Protection Agency. "Common Chemicals Found at Superfund Sites." Last updated August 9, 2011. Available at http://www.epa.gov/oerrpage/superfund/health/contaminants/radiation/chemicals.htm.

Box 7.1 Biosensors

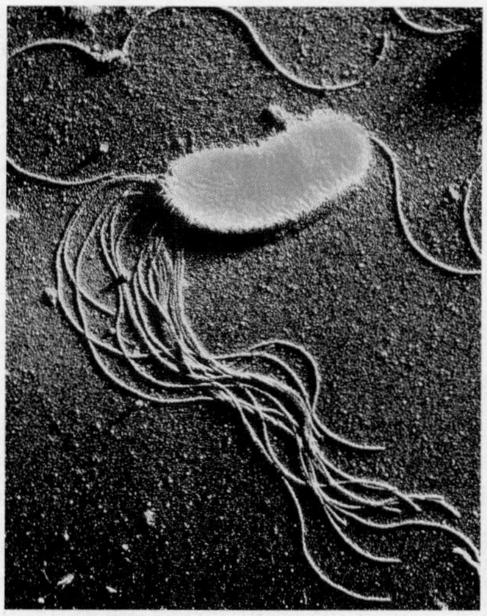

The bacterium *Pseudomonas fluorescens* expresses the *lux* gene. The fluorescence is used as a reporter gene in biosensors and other applications. © Dr. Tony Brain/Science Photo Library

Biosensors are living indicators of chemical pollution. They are used as warning signs that an environment has become contaminated. For example, genetically modified bacteria containing a phosphorescent reporter gene is an effective monitor of pollution levels. The *lux* gene isolated from bioluminescent marine bacteria encodes for the enzyme luciferase, which emits light when expressed. When the *lux* gene is spliced into a gene under investigation, it acts as a "reporter": the modified organism will glow when the investigatory gene is expressed. *Pseudomonas fluorescens* containing the *lux* reporter gene acts as a biosensor for polycyclic aromatic hydrocarbons (PAH) in the environment. *Lux* is spliced to the PAH metabolic pathway, resulting in glowing *P. fluorescens* when PAH is present. The more intensely they glow, the more PAH. The cut and paste nature of recombinant DNA technology gives environmental biotechnologists the tools to create biosensors specifically designed to detect various pollutants in the environment.

TABLE 7.3

The 12 Key POPs: The Dirty Dozen

POP	Use
Aldrin	crop insecticide (corn, cotton)
Chlordane	crop insecticide (vegetables, citrus, cotton, potatoes)
DDT	crop insecticide (cotton)
Dieldrin	crop insecticide (cotton, corn)
Endrin	crop insecticide (cotton, grains)
Heptachlor	insecticide (termites and soil insects)
Hexachlorobenzene	fungicide for seed treatment
Mirex	insecticide (termites, fire ants)
Toxaphene	insecticide (livestock and crops)
PCBs	industrial chemical (heat exchange fluid for electrical transformers, paint and plastic additive)
Dioxins	unintentionally produced during combustion
Furans	unintentionally produced during combustion

Data from the Environmental Protection Agency. "Persistent Organic Pollutants: A Global Issue, a Global Response." Available at http://www.epa.gov/international/toxics/pop.html

genetically engineered organisms specifically designed to break down chemicals, often at a faster rate. Indeed, it is becoming more common to enhance the properties of so-called indigenous organisms, which exist in nature, through genetic modification, biostimulation, or both. **Biostimulation** accelerates bioremediation by introducing nutrients that promote the growth of organisms used for cleanup. **Bioventing** is a similar process that increases available oxygen by adding either air or hydrogen peroxide. Hydrogen peroxide quickly degrades into water and oxygen that is taken up by aerobic organisms. Recombinant DNA technology allows biotechnologists to combine metabolic pathways in order to engineer organisms with the ability to degrade multiple pollutants at once. The *Focus on Careers* box discusses careers related to bioremediation.

Bioremediation can occur through two primary methods. In situ bioremediation takes place on site, where the pollution is actually located. For example, in situ bioremediation can occur at the site of a chemical spill in order to clean up the affected soil and/or water. Ex situ bioremediation is the other option. In ex situ methods, the polluted soil or water is removed from its contamination site and treated elsewhere. Because of the extra steps involved, ex situ bioremediation is often more expensive and time consuming. Nevertheless, the

FOCUS ON CAREERS
Bioremediation Project Scientists

In 1990, the Environmental Protection Agency approved bioremediation as an environmental cleanup method and the agency continues to play an instrumental role in research and development of bioremediation techniques. Bioremediation projects entail the cleanup of environments via metabolism of pollutants by biological systems. Bioremediation Project Scientists serve to plan and manage cleanup projects by integrating their scientific expertise. Additionally, project scientists require extensive environmental engineering knowledge in order to characterize the physical environment present at a cleanup site. Bioremediation Project Scientists may be employed directly by government agencies, such as the Environmental Protection Agency, U.S. Geological Survey, or state environmental agencies, but more often they are employed as contractors and work directly for environmental consulting and engineering firms. Bioremediation projects are typically contracted out by government agencies such as the Department of Energy and the Department of Defense, but are becoming increasingly common among commercial entities, particularly energy, chemical, and transportation companies. The educational requirement for this type of occupation is at least a masters degree if not a doctorate degree. Due to the multidiscipline nature of bioremediation, backgrounds in ecology, microbiology, geology, or environmental or genetic engineering may be appropriate. Indeed, there is not a single clear-cut track in becoming a Bioremediation Project Scientist. The ideal candidate for a Bioremediation Project Scientist position also has experience with molecular biology techniques, global positioning systems and project management. Similar positions have alternative titles, such as Remediation Engineer, Environmental Scientist, or Environmental Engineer. While this is a senior-level leadership role, opportunities are also available as project team members. These positions are more appropriate for student interns, recent graduates and/or individuals holding bachelors degrees, and even postdoc applicants.

preferred method will depend on the exact nature of the contamination.

Microcosms are often used to characterize a contaminated site in order to determine the most appropriate bioremediation techniques for a particular project. A **microcosm** is a simulated environment created under controlled laboratory conditions, designed to mimic the actual site of pollution. Microcosms may be small enough to fit in a test tube or large enough to fill a bioreactor. Or instead of employing one of these sterile environments, small ponds or soil plots may serve as the microcosm environment. Researchers design environmental conditions and carefully control experimental variables such as moisture, pH, temperature, and nutrient availability. Target pollutants are added along with test species under investigation as bioremediation agents. In addition to determining the efficacy of potential organisms, microcosms are also useful in estimating decomposition rates and calculating the amount of time necessary for cleanup.

In addition to environmental cleanup, bioremediation can also serve as a recycling mechanism. Certain industrial processes involve the use of precious metals including silver and gold. Where these valuable materials would otherwise become a part of industrial waste, bioremediation has the power to extract metals so they may be used again and again. This not only reduces the amounts of metals released into the environment, it lowers operating costs and creates a mutually beneficial arrangement between industry and the environment.

Nevertheless, while bioremediation offers promise it does have its drawbacks. There are always risks involved when foreign organisms are introduced into an ecosystem. Unintended consequences may prevail. It is important to always be mindful of the fragile balance in nature and exercise ample humility and caution as we attempt to reverse the negative modifications we have made to the environment.

Soil Environments

Different strategies are employed in bioremediation depending upon the type of environment involved. As mentioned, in situ methods are usually a less expensive option. In the case of soil environments, in situ bioremediation is preferred. It not only reduces costs, but it allows larger areas to be decontaminated at one time. Indigenous microbes present in the environment will be harnessed as decomposers. These are often encouraged through biostimulation as well as bioventing. However, in situ works best in loose and/or sandy soils because they are less compact and allow these methods to proceed most effectively. Biostimulation and bioventing is not as successful in rocky soils or densely packed soils like clay. When these conditions are present, ex situ bioremediation may be more cost efficient in the long run.

In ex situ bioremediation, the soil is removed from the cleanup site and treated before it is returned to its original location. Slurry-phase bioremediation mixes contaminated soil with water, fertilizers, and possibly oxygen into a bioreactor. Microbial decomposition is monitored in this controlled environment. Solid-phase bioremediation includes landfarming, composting, and biopiles. In landfarming, contaminated soil is spread into a thin layer on top of absorbent pads. Pollutants are allowed to leach out of the soil and collect into the pads. Spreading the soil increases surface area and improves ventilation, in effect creating bioventing conditions that accelerate microbial metabolism and chemical break down. Ex situ composting is a scaled up version of home composting, which degrades organic household waste such as food scraps. The same technique is employed, whereby large heaps of contaminated soil are mixed with organic matter such as hay or straw. The organic materials act as a substrate for microbial growth that facilitates bioremediation of the soil. Lastly, biopiles are an ex situ technique similar to composting, only no organic material is added (**FIGURE 7.2**). Instead the piles undergo bioventing with piping systems and/or fans in order to promote bioremediation.

Aquatic Environments

Aquatic environments include natural bodies of fresh and marine waters, groundwater sources such as aquifers, and human-made locales like reservoirs and wastewater treatment facilities. Wastewater treatment is a well-established application of bioremediation that has been in use for several decades. Septic systems and municipal treatment facilities can use bioremediation techniques. Household sewage collected in underground septic tanks is often treated with consumer products containing preserved microbes. The microbes are activated when added to the septic tank, where they go to work using natural enzymes to break down the sewage. In municipal systems, solid waste is separated from liquid sewage. Solids are known as sludge, whereas the liquid portion is called the effluent. Sludge

FIGURE 7.2 An in situ soil cleanup project utilizing microbial bioremediation. Photo taken at Fawley Refinery, an oil refinery and chemical plant located in Fawley, Hampshire, UK. © Paul Rapson/Science Photo Library.

is moved into bioreactors where it is treated with anaerobic microbes that release methane and carbon dioxide gases as waste products. These gases can then be harnessed as sources of energy for the treatment facility. Effluent is passed through a separate filtration system. Filters, membranes, or even rocks can be coated with active biofilms capable of degrading chemical pollutants. The effluent is slowly filtered across one of these treated substrates in order to remove contaminants. It is then disinfected, often with chlorine, before release back into the environment (**FIGURE 7.3**).

Groundwater can also be cleaned up through bioremediation (see **BOX 7.2**). Groundwater can become polluted when chemicals seep through the ground into the water table. Chemical spills also affect surface water, including freshwater streams and rivers as well as lakes and reservoirs. One common source of this type of pollution comes from seepage of underground gasoline storage tanks. Gasoline will rise to the water's surface, so most of the pollution can be pumped off the top. Any remaining contamination is treated ex situ in batches, where water is pumped into bioreactors containing biofilm filters similar to those described in wastewater effluent treatment. The treated water is then pumped back to its original source, where bioremediation continues in situ.

Microbial Bioremediation

Microbial bioremediation employs bacteria as the living decomposers that they are. One of the primary roles carried out by bacteria is the breakdown of detritus and other matter into its chemical building blocks. Humans have harnessed this natural activity in the process of

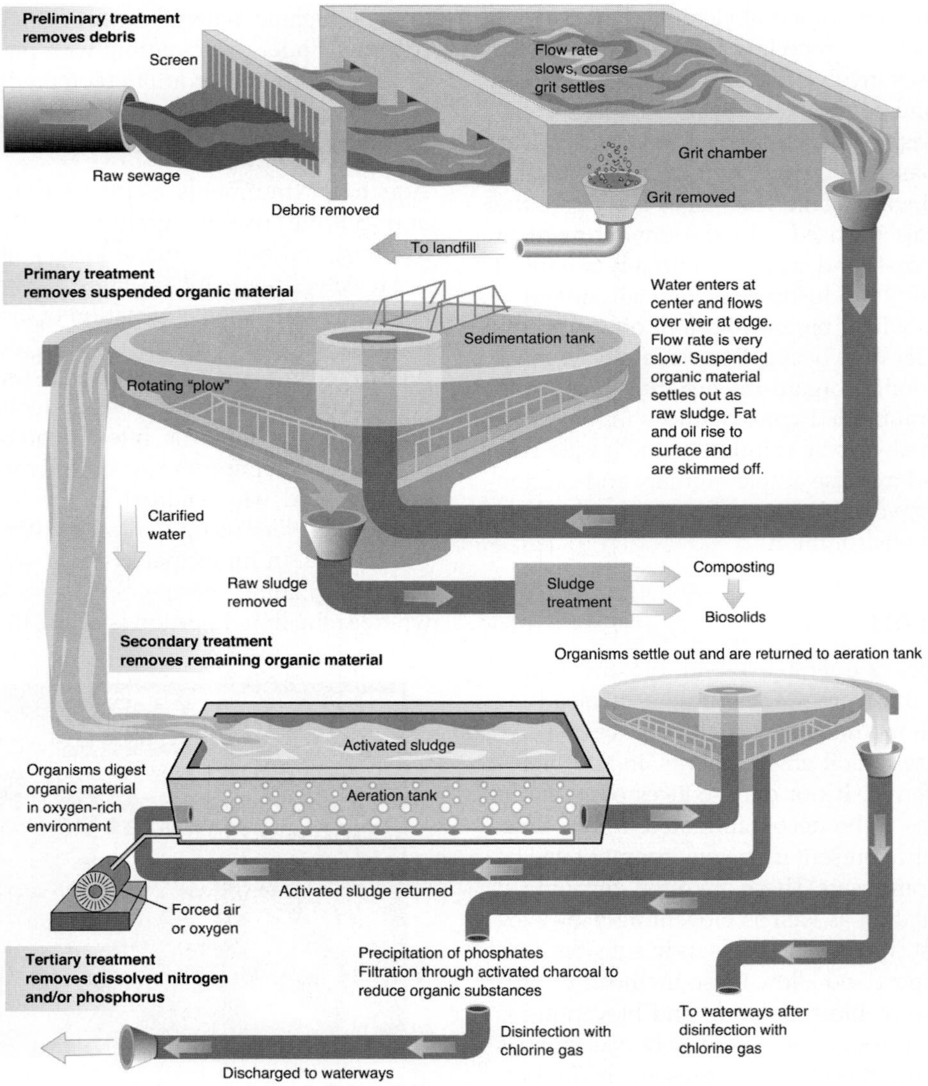

FIGURE 7.3 The process of sewage treatment, seen here, is one method of bioremediation commonly employed by municipalities for waste disposal. Adapted from B.J. Nebel. *Environmental Science: The Way the World Works*, Third Edition. Englewood Cliffs, NJ: Prentice Hall, 1990.

Box 7.2 Bioremediation in Oil Spills

Extraction and refinement of crude oil provides not only fuel to power our cars, boats, and planes, but also serves as the raw material for hundreds of commercial products, including purified chemicals used in cosmetics, plastics, paints, and cleaning products. Crude oil operations pose a risk of leakage and spills at several steps in the overall process. For example, oil seepage occurs naturally along the ocean floor. Oil being pumped out of the ground, whether in terrestrial locations or submerged offshore drills, can result in uncontrolled oil flow from its source when control measures fail. Another avenue for oil spills is during its transport in underground pipelines as well as in large sea vessels known as tankers. There have been hundreds of oil spills recorded since the 1970s. Between 1990 and 2013, 448 spills were recorded. Three ecological disasters in recent decades are discussed below. Each was well publicized and all are the result of massive oil spills. Each spill occurred under unique circumstances which required different remediation approaches.

In 1989, the oil tanker *Exxon Valdez* ran aground off the coast of Alaska in Prince William Sound. Around 260,000 barrels of oil, equal to about 11 million gallons, were released into the pristine arctic environment. Though a tragedy, the Valdez spill gave environmental biotechnologists the unique opportunity to test bioremediation techniques in a real world setting. Physical cleanup, such as collecting oil into containment booms or behind skimmers, was the first step. Then large remnants of the spill were vacuum-pumped into disposal tanks. Lastly, the physical cleanup involved spraying the shoreline with high pressure freshwater in order to disperse the oil into finer droplets. This final step prepared the environment for bioremediation, as the smaller droplets increased the surface area of the remaining oil contamination, which improves microbial access and decomposition rates. Environmental technologists employed biostimulation in the form of nitrogen and phosphorus fertilizers to increase indigenous microbial populations, in particular *Pseudomonas* species known to degrade oil compounds. While over time the bioremediation approach was successful in breaking down persistent chemicals along the shoreline, unknown quantities of oil still seeped into deeper layers of soil and rock. It is unclear how long it will take for complete decomposition to occur. Some environmental scientists suggest the environment may never return to its original condition.

A second oil release took place in war-torn Kuwait during the early 1990s when it was occupied by Iraq. This ecological disaster was unique because the oil releases were intentional destruction on the part of Iraqi forces, which routinely burned Kuwaiti oil fields. Camera crews broadcasted images of burning oil towers. The exact quantity of oil released is estimated to be at least 20 times greater than the Exxon Valdez spill (a low estimate is approximately 250 million gallons). The spills in Kuwait are particularly difficult to remediate because the terrain is sandy and dry. While similar physical methods to the Valdez spill were initially employed, there are no bodies of water to help disperse the oil. Instead, it tends to stick to sand grains and rocks. Also, arid conditions lack many indigenous microbes to aid in bioremediation. Genetically modified strains are being introduced as part of a $1 billion program developed by the Kuwait government.

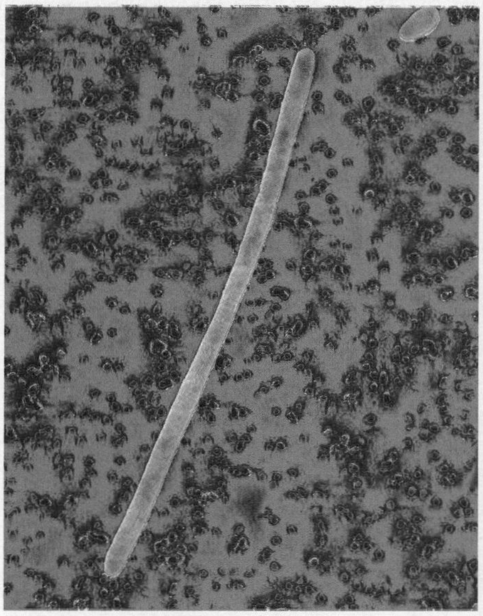

Oil degrading bacterium (yellow) collected from plumes of the Deepwater Horizon spill. The smaller spheres surrounding the bacterium are oil droplets. © Hazen group, Lawrence Berkeley National Laboratory/Science Photo Library

Perhaps the oil spill most frequently cited in recent times is the Deepwater Horizon oil spill, which occurred beneath the Gulf of Mexico on April 20, 2010. Ironically, this was two days before Earth Day. The British Petroleum oil rig exploded and released massive amounts of oil into the water. There are conflicting estimates of the exact quantity released, though it is generally agreed to be at least 4.9 million barrels (205.8 million gallons). The oil spill was dramatic and highly publicized due to the fact that it went on for nearly three months. The relentless outpour was not capped until mid-July. Physical cleanup as described above was employed, followed by a combination of wave action and chemical dispersants to disperse the remaining oil. Some surface oil simply evaporated, which was aided by controlled burns. Bioremediation was a final measure and once again a tragic event provided a living laboratory. Biostimulation was part of the approach. The massive plumes of oil flowing from the ocean floor became hot beds of microbial activity. Researchers identified over 1,500 genes encoding for oil degradation in samples taken near the plumes. Over 900 bacterial families were detected and several new species were discovered. Scientists studying Deepwater Horizon estimate microbes effectively metabolized 200,000 tons of methane in addition to ethane and propane released during the spill. However, regardless of efficacy, public concerns over adding chemicals and fertilizers to open oceans are still prevalent. Levels of these substances as well as monitoring of the oil itself, will be studied well into the future as researchers determine the long-term effects of oil spills.

bioremediation. Microbial bioremediation can involve both aerobic and anaerobic bacteria. Aerobic bacteria thrive in oxygen-rich environments while anaerobic bacteria necessitate conditions devoid of oxygen. When approaching a contaminated site, full treatment may require both forms of bacteria. This will depend on the specific conditions found on site (or ex situ). For example, surface water and soils typically have enough oxygen to maintain aerobic bacteria populations. In contrast, severely contaminated areas, deep soil layers, and groundwater sources such as aquifers are usually poor in oxygen content and thus require use of anaerobic microbes. Also, anaerobic microbes are naturally attracted to and readily uptake contaminants such as methane, sulfates, nitrates, and iron, making them especially well-suited for bioremediation purposes.

Another component of bioremediation is a process known as bioaugmentation, also known as seeding. **Bioaugmentation** introduces additional microbial strains into the cleanup site in order to enhance the activity of indigenous populations. The added strains may be GMOs designed to complement naturally occurring microbes. For example, the indigenous microbes may metabolize one pollutant, such as vinyl chloride, while the seed microbe metabolizes something else, like dichloroethane. Characterization of the contaminated site is critical when bioaugmentation is implemented, as the target pollutants must be identified. Then the appropriate seed microbes can either be sourced for introduction or custom engineered for a specific project.

Indigenous Microbes

As discussed, biostimulation is a method designed to accelerate the bioremediation process. Also known as enrichment, biostimulation is the addition of nutrients (and possibly oxygen) intended to spur growth of microbial decomposers. The supposition here is that indigenous microbes must be adapted to existing pollution in the environment if they are presently surviving. If adapted, the indigenous microbes must be able to metabolize chemical contaminants and, therefore, all that may be required is to provide them with a boost. Microbial population size is positively correlated to decomposition rates. As the indigenous microbes increase in number, the quantity of chemicals metabolized also goes up. If indigenous microbes are aerobic, bioventing can also be employed to encourage growth.

Another advantage of indigenous microbes is their applications for research. Bacteria strains that naturally possess the ability to breakdown environmental pollutants may be isolated and cultured in the laboratory. Genomic studies and metabolic pathway investigations can reveal the raw materials for creating engineered varieties with enhanced bioremediation properties.

Genetically Modified Microbes

Molecular biotechnology is tapping into indigenous microbes as sources of bioremediation tools. Strains of *Escherichia coli* and species of *Pseudomonas* have provided environmental biotechnologists with genes encoding enzymes able to degrade persistent substances. For example, *Pseudomonas putida* metabolizes styrene, a component of many plastics, into a biodegradable form of plastic. The same species is also able to break down toluene, xylenes, benzoates, and methyl benzoates. The first patent awarded to a bioengineered organism was granted to an oil degrading bacterium (**BOX 7.3**).

Box 7.3 The First GMO Patent

Although today it is not unusual to patent a genetically modified organism, the first patent application for such a biotechnology product was the subject of debate and scrutiny. In the 1970s, a General Electric scientist named Ananda Chakrabarty developed a recombinant strain of *Pseudomonas* capable of degrading several organic molecules present in petroleum. This strain became the first living organism to be patented in the United States. Chakrabarty did not utilize recombinant DNA to produce his novel strain. Instead, he went through the laborious process of isolating various indigenous strains with bioremediation properties. He hypothesized that each strain possessed a unique plasmid encoding specific metabolic pathways for different chemicals. Then, he mated different types to facilitate conjugation events. Through multiple rounds of conjugation, Chakrabarty's experiments led to a *Pseudomonas* strain with multiple plasmids, making it capable of metabolizing several chemical constituents present in crude oil. In fact, it was utilized in the cleanup efforts following the *Exxon Valdez* oil spill in 1989 (see Box 7.2). Upon submitting a patent application, the decision whether or not to award a patent was held up in courts until 1980. The primary issue stalling a final decision revolved around the patentability of a living organism. The final ruling awarded by the U.S. Supreme Court stated, "anything under the sun that is made by man" is patentable and argued that "a live, human-made microorganism is patentable subject matter…as a manufacture or composition of matter." Even so, Chakrabarty's strain has seen limited applications in subsequent oil spill cleanups. While the strain metabolizes some chemicals in oil, it does not degrade them all. Crude oil is a highly complex blend of chemicals that is difficult to characterize in its entirety. Therefore, a better approach may be to utilize the patented *Pseudomonas* with additional microbial agents in a form of bioaugmentation. Additionally, public concerns have limited bioremediation of oil spills. The fear of releasing GMOs into the environment has been a significant roadblock in the adoption of bioremediation technology.

Box 7.4 Paul Stamets Uses Fungi to "Save the World"

> Fungi are the grand recyclers of the planet and the vanguard species in habitat restoration...If we just stay at the crest of the mycelial wave, it will take us into heretofore unknown territories that will be just magnificent in their implications.
>
> —Paul Stamets

He has been called the "Johnny Appleseed of mushrooms." Paul Stamets is an American mycologist on a mission. Through his books, speaking engagements, and appearances in documentaries such as *The Eleventh Hour*, he seeks to promote the use of fungal species for a number of useful purposes. His work has received patents for using fungi as pesticides and he has performed research on using fungi as antiviral agents as both an advisor and consultant to the Program for Integrative Medicine at the University of Arizona Medical School. Stamets personally maintains over 250 species of fungi in culture and has constructed fungal gene libraries through his company, Fungi Perfecti.

With this background, it is not surprising he is a pioneer in the field of mycoremediation. Fungal mycelia are vast networks a single cell layer thick residing beneath the Earth's surface. These natural decomposers are effective at breaking down organic matter such as cellulose and lignin found in plant cell walls. Another of his patents uses oyster mushrooms to degrade diesel oil. One initial study of this application compared fungal decomposition with microbial bioremediation as well as enzymatic action. After eight weeks, 95% of the hydrocarbons had broken down and showed statistically significant data indicating mycoremediation in this case was more effective than the other two treatments (bacteria and enzymes). The study area was deemed nontoxic by the Washington State Department of Transportation. By partnering with both governments and private entities, Paul Stamets is at the forefront of establishing fungi as a viable bioremediation tool.

Through recombinant DNA technology, environmental biotechnologists can engineer GMOs designed to metabolize specific chemicals. Those chemicals are more persistent in the environment and are not easily broken down by indigenous microbes. In particular, large organic molecules with multiple bonds and radioactive compounds can last hundreds of years. Genetically modified strains with customized metabolic properties are useful for persistent substances. It is also possible to introduce **stacked traits**, where more than one transgene is inserted in order to confer multiple abilities in the GMO. In this way, molecular biotechnology can produce one strain capable of decomposing multiple contaminants during a single bioremediation project.

Phytoremediation and Mycoremediation

Environmental biotechnology has successfully carried out bioremediation using plants and fungi in addition to bacterial species. Known as phytoremediation and mycoremediation, respectively, these cleanup methods have advantages in particular applications. Fungi are another of nature's decomposers (**BOX 7.4**). Therefore, they are excellent candidates for bioremediation as indigenous organisms as well as research material. Because plants and fungi are eukaryotic organisms, their cellular capabilities vary from those of bacteria. These organisms thus encode metabolic pathways for substances that bacteria may be unable to break down. Examples of such environmental pollutants are highlighted in **TABLE 7.4**.

TABLE 7.4

Phytoremediation and Mycoremediation Applications

Method	Chemicals	Example Organisms
Phytoremediation	• Arsenic • Atrazine • Radioactive compounds (Cesium, Strontium) • Phenanthrine • PAHs • PCBs	• Juniper trees • Alfalfa • Prairie grass • Sunflowers • Water hyacinths • Indian mallow • Barnyard grass
Mycoremediation	• Creosote • Pentachlorophenols • Asbestos • Heavy metals • Sewage • PCBs	• *Fusarium oxysporum* • *Phanerochaete spp.* • *Mortierella spp.*

Plants have bioremediation potential not seen in bacteria or fungi. Chemicals are absorbed via plant roots as they absorb water from the ground. The plant may metabolize pollutants if able; otherwise the plants themselves become hazardous material needing disposal. Phytoremediation is advantageous because it can also alleviate air pollution. Plants are well known as "carbon sinks" that sequester carbon dioxide from the atmosphere. Awarding carbon credits in exchange for phytoremediation projects is one attempt to reduce levels of this greenhouse gas in the environment. Another benefit of phytoremediation is it typically results in a more aesthetically pleasing landscape. Planting grasses, shrubs, and trees at polluted sites can improve appearances while effectively degrading environmental pollutants (**FIGURE 7.4**).

Even plants tolerant of hazardous chemicals have limits, which is one drawback of phytoremediation. The use of plants is most appropriate when toxicity levels are relatively low. Also, because plant roots only grow to certain depths, phytoremediation is most efficient when contamination is near the surface. When used as a stand-alone method, phytoremediation is a relatively slow process compared to microbial remediation or mycoremediation.

Metagenomics

The myriad advances in molecular biotechnology may give the impression that our understanding of the microbial world is well underway. In fact, only about 0.1 to 1% of microorganisms have been isolated from the environment and cultured for laboratory investigation. Due to their innate abilities to adapt to extreme environments, microbiologists continue to discover new species in the harshest locations on Earth, once previously thought too inhospitable for life to exist. Furthermore, many microorganisms are resistant to culture and are therefore nearly impossible to study under controlled conditions. It would be advantageous to construct a method of identifying novel microbial strains within their natural environments, thus eliminating the need to isolate individual species or grow them in culture. The field of metagenomics serves this very purpose.

FIGURE 7.4 Trin Warren Tam-Boore in Royal Park, Melbourne, Victoria, Australia, is an example of a phytoremediation project. The artificial wetland serves as an entry point into Port Phillip Bay, where the planted species filter and clean pollution run-off.

Metagenomics is the study of genomes of whole microbial communities in situ, with the aim of identifying genetic sequences instead of particular organisms. Also known as gene mining, metagenomics utilizes a number of molecular biotechnology techniques. Whereas the field of genomics compares genetic sequences of individual organisms, bacteria, viruses, viroids, prions, and free DNA may be present in one metagenomic sample. Biotechnologists specializing in this field are more concerned with identifying genes, proteins, enzymes, and metabolic pathways than characterizing a single species. Discoveries made through metagenomics can be applied to bioremediation when constructing GMOs for this purpose.

Techniques

Both DNA and mRNA may be detected in metagenomic studies. DNA provides the raw material for sequence-dependent techniques, while mRNA may be used in function-dependent techniques.

Sequence-Dependent

Some sequence-dependent techniques, such as DNA microarrays and fluorescence in situ hybridization (FISH) have already been presented. These techniques have a multitude of applications including metagenomics. Additional sequence-dependent techniques include integron analysis, phylogenetic anchors, and polymerase chain reaction (PCR) primers.

Integrons are mobile elements similar to transposons. They carry **gene cassettes**, which are mobile genetic elements containing one or more genes. Integrons readily relocate within genomes and are often transferred via plasmids in bacterial transformation. This is possible because in addition to the gene cassette, integrons also encode the enzyme integrase, which allows integrons to insert themselves into new locations within the genome. They are also flanked by restriction sites, allowing for cut-and-paste methods as desired. Lastly, the integron contains a promoter region to allow expression of the gene cassette (**FIGURE 7.5**). In **integron analysis**, open reading frames in the metagenomic sample are identified through complementation with a known integron. These open reading frames are an indicator of a potentially novel gene discovery.

Often, PCR primers are used in combination with integron analysis. A primer is added in order to bind

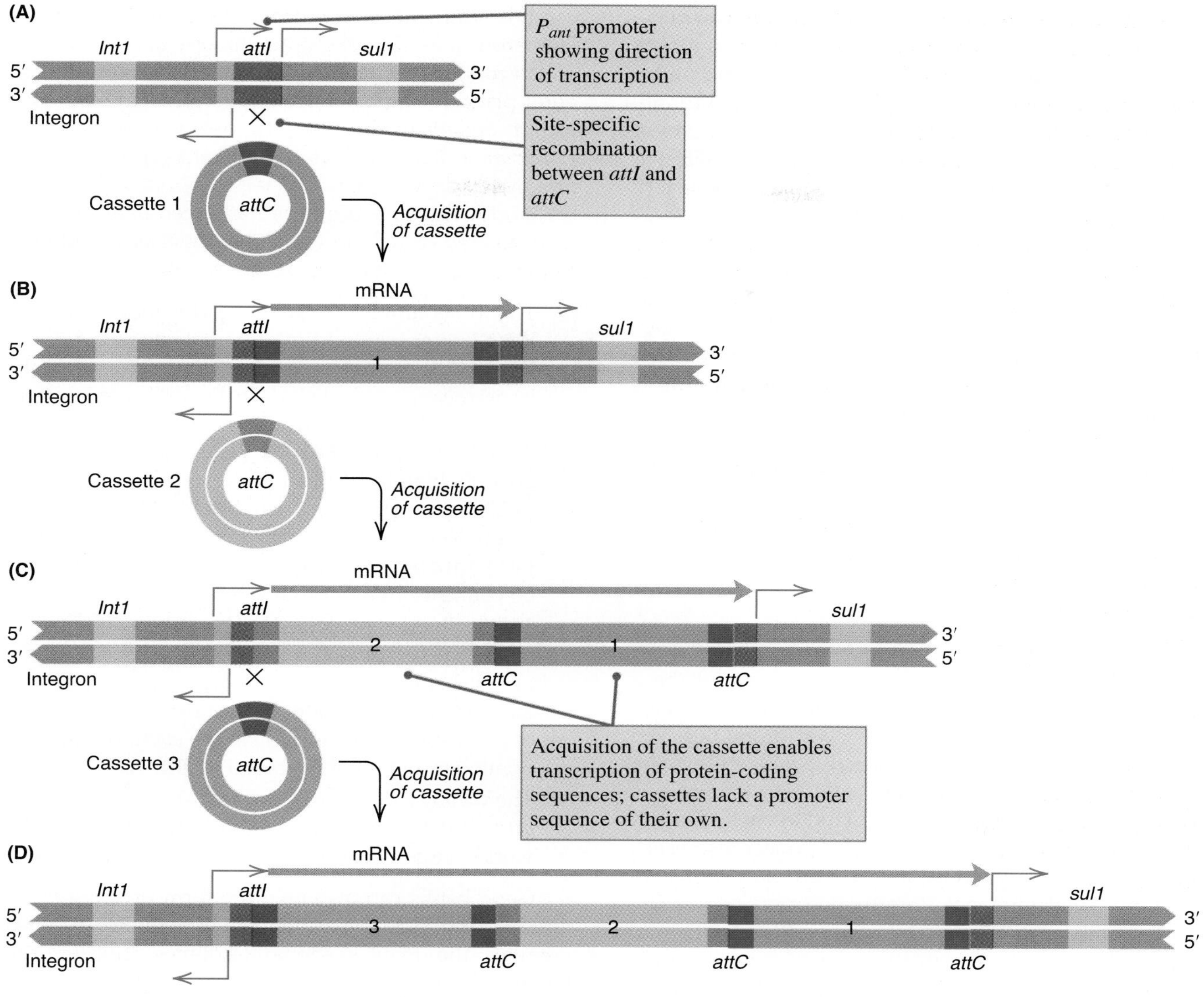

FIGURE 7.5 The mechanism for gene cassette recombination at site-specific integrons. Integrase inserts the cassette plasmids between the restriction sites *attI* and *attC*.

the integron to the identified open reading frame. Then the polymerase chain reaction amplifies the genetic sequence. The amplified sample may be used to screen existing libraries in order to find a genetic match. If no match is found, the novel gene is studied further to identify its capabilities, often through function-dependent techniques as described below.

Sequencing of metagenomic DNA can be accomplished with phylogenetic anchors. First, known gene sequences in a sample are identified. For example, a genetic marker is located. The known region serves as the **phylogenetic anchor** because it represents a conserved sequence present in several organisms. Once identified, the upstream and downstream regions of the phylogenetic anchor are sequenced. These regions may be rich in gene sequences, which can be identified and, if necessary, characterized further via function-dependent analysis.

Function-Dependent

Function-dependent techniques rely on identification of mRNA. The logic behind function-dependent screening is that any detected mRNA represents expressed genes active in the sample environment. Function-depending techniques are especially useful when novel genes are identified because the sequence alone does not directly indicate the activity it encodes. Expression screening and phage biopanning are methods used in function-dependent metagenomic analysis.

Expression screening involves cloning of novel genes identified in sequence-dependent techniques. The cloned genes are inserted into vectors such as plasmids that contain stop and start codons, allowing for the expression of the clones. Screening of the expressed clones requires various environmental treatments to determine the function of the gene. For example, clones may be grown in the presence of a toxic chemical to determine resistance properties. In other screening techniques, the gene encoding green fluorescent protein is inserted along with the cloned gene under investigation. The transgenic cells are subjected to different environmental conditions. When they are grown in a medium containing a target substrate, the cells will glow. This indicates the cloned gene encodes the ability to metabolize the target substrate.

Another function-dependent technique is phage biopanning. Phages, viruses that use bacteria as their hosts, are engineered to express fusion proteins on their surfaces. The fusion proteins contain the naturally-occurring phage coat protein as well as an inserted cloned gene under investigation. **Phage biopanning** is similar to chromatography methods as coated beads are used as binding substrates. Beads may be coated with specific chemical pollutants and then treated with the engineered phages. If the phages bind to the beads, the cloned gene is identified as a match suitable for metabolism of that particular pollutant.

Genome Projects

Numerous independent genome projects are under way to identify novel microbial genes. Some of these will be useful in bioremediation, particularly for bioengineering of GMOs with enhanced properties. The Environmental Genome Project, organized by the National Institute of Environmental Health Sciences (a division of the NIH), aims to study genes involved in contaminant decomposition. The project also seeks to characterize human genes that are susceptible to environmental influence. Information obtained in the project will aid in our understanding of environmental factors that contribute to disease as well as how specific chemicals may alter gene expression in different haplotype groups of the human population.

Conservation Biotechnology

The field of conservation biotechnology seeks to maintain biodiversity of the planet while encouraging protection of our existing environments. During the world's sixth mass extinction (underway now), conservationists are pushing back the tide of depletion with the help of molecular biotechnology. Direct measures include the creation of genomic libraries of threatened, endangered, and even extinct species. Cloning, seed banks and living laboratories are other direct methods. Indirectly, conservationists are attempting to promote biodiversity by protecting ecosystems and natural resources. Creation of biofuels is an indirect method designed to offer alternative energy sources.

Gene Banks

Preserving biodiversity is a major goal of conservation efforts. Gene banks are designed to store genetic information from as many species as possible. Gene banks can take different forms, such as seed banks and living laboratories. Seed banks store preserved seeds, each of which contains the entire genome of the corresponding adult plant. Living laboratories are facilities that house extant organisms in order to conserve biodiversity. Other gene banks include tissue banks, pollen banks, and frozen depositories.

Worldwide there are over 1,400 seed banks storing an estimated six million samples, known as **accessions**. Some of the largest banks storing seeds include:

- The Millenium Seed Bank Project is located in West Sussex, United Kingdom outside of London, England. It houses a multistory underground vault impermeable to nuclear destruction. There is space to store billions of seeds and is the largest seed bank in the world (**FIGURE 7.6**). It achieved its goal of obtaining 10% of the world's plant species in 2009 and aims to collect 25% by 2020. Operated by the Kew Royal Botanical Gardens, the project has partnered with research organizations in over 80 countries in order to reach target storage levels.
- The Svalbard Global Seed Vault resides within a sandstone mountain on the arctic Norwegian island of Spitsbergen. Cold temperatures and an isolated location make it ideal for protecting the biodiversity of plant species, particularly in the event of nuclear bombings and other warfare (**FIGURE 7.7**).
- The Vavilov Institute of Plant Industry in St. Petersburg, Russia was established in 1921 and houses the world's largest collection of European plant species.

Gene banks can also take the form of living laboratories. The U.S. Department of Agriculture (USDA) operates the Plant Genetic Resources Unit on the Geneva, New York campus of Cornell University. Over 300 species are growing on site in the form of approximately 20,000 accessions. Part of the U.S. National Plant Germplasm System, the unit maintains genetic diversity as resources for crop improvements. It conducts research in genomics and bioinformatics as well

FIGURE 7.6 Seeds being stored in the Millenium Seed Bank demonstrate the diversity of plant life to conserve. © Mint Frans Lanting/Mint Images/age fotostock

FIGURE 7.7 The inconspicuous entrance into the Svalbard Global Seed Vault is protected with two security doors and three air locks before admittance into the cave. © Rolf Adlercreutz/Alamy

FIGURE 7.8 One of two surviving African wildcats cloned at the Audubon Nature Institute, the male "Jazz" still lives at the research center in New Orleans. © Karen Ross/Audubon Nature Institute

as offering public outreach including science education. Nations and private institutions across the world maintain their own living laboratories.

Frozen tissue, sperm, and egg samples are another gene bank option typically used for preserving animal biodiversity. For example, the United Kingdom has established the Frozen Ark Project to collect endangered specimens and a separate semen bank of ram species. It currently holds over 28,000 frozen DNA samples, over 7,000 of which were obtained from endangered species. The San Diego Zoo has maintained a "frozen zoo" of 800 species since the early 1970s. Another means of preserving animal biodiversity we have discussed is the Genome 10K Project. The i5K project at the Human Genome Sequencing Center of Baylor College of Medicine is a similar project designed to collect and preserve arthropod genomic data. It has the goal of cataloging at least 5,000 insect genomes.

Cloning Endangered and Extinct Species

Although reproductive cloning techniques are expensive, time-consuming, and relatively unsuccessful, research seeks to improve its current status. Several endangered species are under consideration for breeding programs involving reproductive cloning, and some have already succeeded. Some notable endangered species are highlighted below.

- The Audubon Center for Research of Endangered Species successfully cloned African wildcats in 2004 using a domestic cat as the surrogate. Two out of seventeen cloned wildcats survived (**FIGURE 7.8**).
- Seoul University (Korea) scientists cloned two gray wolves using domestic canine surrogates in 2005.

Pandas are under current investigation for cloning programs, but to date no successful live births have been announced. Rabbits and black bears are being considered as potential surrogates. Other endangered species that may be cloned in the future include gorillas and ocelots. The Royal Zoological Society of Scotland established the Institute for Breeding Rare and Endangered African Mammals with the hopes of cloning the Ethiopian wolf, the white rhinoceros, and pygmy hippopotamus.

There have also been successful clones of extinct species. In 2001, an extinct species of wild ox native to Southeast Asia known as the gaur was cloned. The calf, named Noah, lived for just two days. Born to a domestic cow surrogate, Noah passed away from dysentery believed to be unrelated to the cloning process (**FIGURE 7.9**). The extinct Iberian mountain goat called the ibex has also been cloned. Declared extinct in 2000, researchers had the forethought to preserve the last remaining specimens. Using preserved DNA, members of the cloning project implanted 57 ibex embryos into domestic goat surrogates in 2009. Seven pregnancies resulted with only one live birth. Unfortunately the revived organism died shortly after due to respiratory complications, much like the first cloned mammal (Dolly the sheep). Another extinct species with the

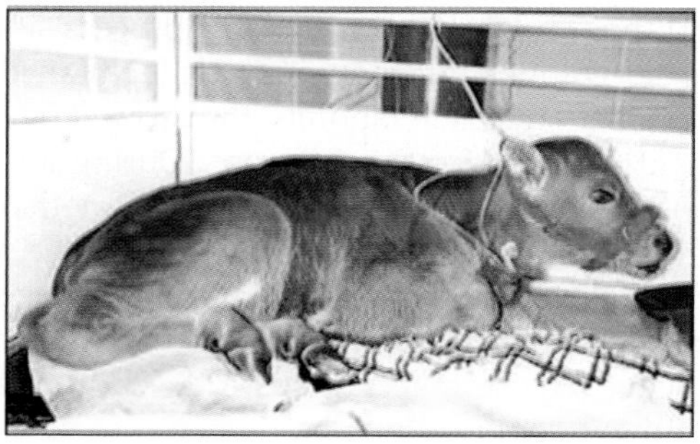

FIGURE 7.9 The cloned gaur calf, Noah, receives oxygen one day after his birth. He would die the very next day due to dysentery. © Iowa State

potential for revival is the wooly mammoth. Because preserved specimens have been found in frozen tundra, DNA is available that could be used with extant relatives, that is, elephants, as donor egg and surrogate sources. As discussed elsewhere in the text, another project involved in extinct species revival is the Long Now Foundation's Revive and Restore Project. It is currently researching reproductive cloning of the extinct passenger pigeon.

Long-term health effects of cloned animals are unknown because reproductive cloning technology is relatively new. Interspecies clones pose additional problems as they possess chimeric DNA; their nuclear DNA is of the cloned organism while their mitochondrial DNA was inherited from the surrogate mother. Another issue is the need for foster mothers of infant clones. If we are to clone endangered or extinct animals for conservation purposes, we will need to provide them with nurturing and socialization. Ideally this would involve members of their own species, but due to low numbers of endangered organisms it may not be feasible. In the case of extinct species, close extant relatives would be most desirable.

Biofuels

As we have seen, the use of oil as the world's primary energy source has negative consequences on the environment, including the release of carbon dioxide and other greenhouse gases and the potential for toxic oil spills. Indeed, these detrimental impacts are commonly viewed as necessary evils—sunk costs incurred in exchange for the outpouring of available energy treated as though it were in endless supply. As the world's less developed countries experience improvements in infrastructure, demand for fossil fuels is expected to spike. In addition to oil, natural gas and coal are fossil fuels used to generate electricity. **Biofuels** are under investigation as sustainable, alternative energy sources designed to fulfill our insatiable needs. A number of source organisms are the subjects of research studies.

Household and Human Waste: Methane

Bioreactors containing anaerobic bacteria may be used to decompose household wastes in a process similar to household composting methods. Rather than disposing of organic wastes in landfills, they would be sent to treatment facilities for this purpose. During anaerobic metabolism, natural byproducts such as methane gas can be collected as a source of energy. Methane may be used to produce electricity. It is a routine energy supply in many industrial facilities, including wastewater treatment plants. The sewage sludge produced in wastewater treatment could instead be tapped into as a public supply of energy. Excess electricity produced by these facilities could serve as an additional revenue stream for municipalities.

Polluted Sediments: Microbial Fuel Cells

Sediment found at the bottoms of open water bodies contains concentrated amounts of detritus. As dead organisms settle to the bottom, their natural decomposition results in a mosaic of chemical components. Anaerobic organisms indigenous to this environment are able to metabolize organic molecules. Through oxidation, organic molecules are converted to carbon dioxide and electrons. Under normal conditions, the electrons would be taken up by an electron acceptor such as iron. But when these electrons are transferred to electrodes, it results in the generation of electricity. Organisms with this ability are now referred to as **electrigens** because they serve as potential power plants for our species. Researchers are actively culturing electrigens for use in microbial fuel cells. Microbial fuel cells harness the oxidative power of electrigens such as *Geobacter metallireducens* and *Rhodoferax ferrireducens* (**FIGURE 7.10**).

Biomass: Ethanol and Biodiesel

Due to economic conditions in the United States, namely corn subsidies, surplus corn was an early candidate for production of biofuels. Fermentation of corn was already a well understood process that could be commandeered for the purpose of ethanol production. Ethanol is an alternative fuel with a proven track record. Although not as efficient as gasoline and known to corrode engines, ethanol is nevertheless under consideration as a mass-produced biofuel and is currently added to gasoline blends (**FIGURE 7.11**). Another major concern is the impact on food prices. Diverting portions of the nation's corn supply, which would otherwise be used as livestock feed, into ethanol production is attractive to corn farmers because biofuel projects fetch higher prices than farming operations. The cascading effect is higher prices for feed and therefore meat, dairy, and eggs. In an effort to alleviate spikes in food prices, biofuel researchers have shifted their focuses to other sources of biomass, such as stalks

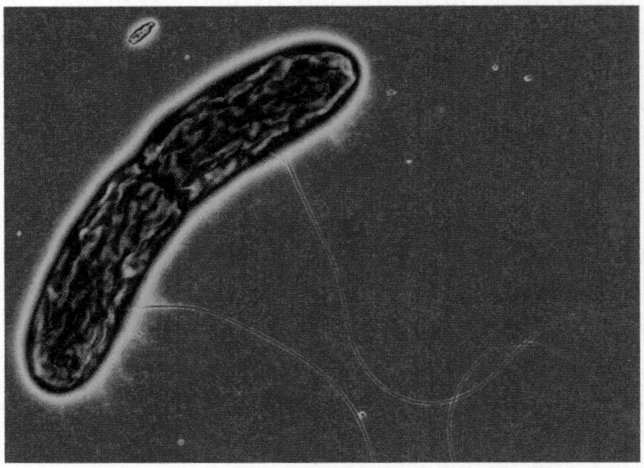

FIGURE 7.10 The bacterium *Geobacter metallireducens* is an electrigen discovered by Dr. Derek Lovley in 1987. The species is used in microbial fuel cells. © Derek Lovley/Science Source

FIGURE 7.11 E-85 at the pump: a blend of 85% ethanol and 15% gasoline is readily available at U.S. filling stations, although it is corrosive to engines.© Carolina K. Smith/ ShutterStock, Inc.

from corn (instead of cobs) and other crop plants including sugarcane, rice hulls, and fast-growing plants like switchgrass and bamboo that can be grown specifically for ethanol production. In an example of biotechnology aiding itself, enzymes such as cellulase and lipase produced via microbial fermentation are added to biomass to aid in ethanol production. Fungal species are also under investigation to help break down cellulose and lignin found in biomass.

Biodiesel is another potential alternative energy source (see **BOX 7.5**). Naturally-derived oils harvested expressly for this purpose or purified spent oils are able to power diesel fuel engines. Biodiesel is produced in soy, sunflower, canola, and palm species and can also be sourced from restaurants in the form of used fats and oils. Studies suggest algal biodiesel has the greatest potential to compete with gasoline, based on cost estimates. Algal biodiesel prices could be as low as $50 per barrel, whereas gasoline prices hit $100 per barrel back in 2008. They also have the potential to outperform other sources of biodiesel. Algae may yield as much as 20,000 gallons of oil per acre, whereas soy fields produce only 117 gallons of oil per acre. If the United States were to adopt algal biodiesel on the existing millions of acres of marginal agricultural lands, it would be feasible to eliminate the need for imported sources of oil. A further advantage of growing such a large quantity of algae would be its ability to sequester carbon dioxide from the atmosphere, potentially reducing greenhouse gases.

Nevertheless, ethanol and, to a lesser extent, biodiesel have drawbacks as fossil fuel alternatives. One consideration is the net energy gain of these fuel sources. Corn ethanol yields 25% more energy than it takes to produce, whereas biodiesel produced from soybeans offers 93% more. Biodiesel releases a fraction of particulate air pollutants as compared to ethanol. Biodiesel also outperforms ethanol in terms of reduced greenhouse gas emissions as compared to traditional fossil fuels, with 41% reduction compared to 12% with ethanol. This data suggest biodiesel as a superior alternative compared to ethanol. Despite these numbers, if the United States were to devote all of its corn and soybean crops to the production of ethanol and biodiesel, these fuels would meet 12% of the nation's gasoline demand and 6% of the diesel fuel demand. While these are improvements over fossil fuels, alone they are not enough to fulfill our national energy requirements.

Box 7.5 How Economics Can Spur or Stall Biotechnology

Algal biodiesel research and development is a historic example of how the nation's economy can affect the pace of biotechnology innovation. The Department of Energy researched algae as a biodiesel source from 1978 to 1996. The Aquatic Species Program was initiated by President Jimmy Carter in response to the 1970s oil crisis. High oil prices created a sense of urgency in producing cheap, home-grown alternative fuel sources. The project identified and cloned several algal genes encoding metabolic pathways that synthesize oil. Over three thousand algal species were maintained by the project, with 51 of those strains exhibiting the greatest promise. During the Clinton administration, oil prices were low and alternative fuels went on the backburner. When funding was dropped in 1996, maintaining the algal cultures was too impractical continue. Only 23 strains are still intact today.

Though discontinued in its original inception, beginning in 2007 the DOE began funding private industry studies that continue today. The American Recovery and Reinvestment Act under President Obama has awarded approximately $82 million to organizations involved in algal biodiesel research, including the National Alliance for Advanced Biofuels and Bioproducts (NAABB) and the National Advanced Biofuels Consortium (NABC). As the earlier work of the Aquatic Species Project experiences a renaissance, current research seeks not only to produce high-yield algal strains, but also to scale up production to a commercially viable level. Our renewed interest in alternative fuels comes when oil prices are at all-time highs and international conflicts burden our national defense. Economic and political conditions have come full circle to once again stimulate research and development of alternative energy.

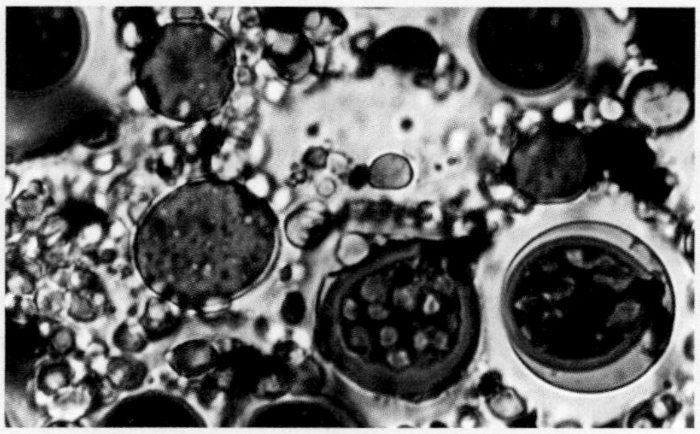

A light micrograph of algal biodiesel. The spheres are lipid droplets synthesized by algae in culture. © NREL/US Department of Energy/Science Source

SUMMARY

- Environmental biotechnology involves three main applications. Bioprospecting reveals previously unknown organisms, genes, and gene products. Discovered natural novelties may then be studied and utilized in bioremediation. Biotechnology is also useful in areas of environmental conservation. Genomics and cloning may be used to maintain biodiversity and protect threatened and endangered species. Genetically modified organisms promise to provide new sources of alternative energy, including biofuels.
- Environmental pollutants include releases from industrial processes in the form of waste, such as accidental oil and chemical spills, intentional dumping, radioactive substances, and pesticides. Air pollutants include ozone, particulate matter, carbon monoxide, nitrogen oxides, sulfur dioxide, and lead. Disposal of household chemicals such as paints and batteries ultimately ends up in landfills, contributing to soil and groundwater pollution.
- Bioremediation takes advantage of living organisms in order to break down persistent chemicals polluting the environment. Biostimulation accelerates bioremediation by introducing nutrients, which promote the growth of organisms used for cleanup. Bioventing is a similar process that increases available oxygen by adding either air or hydrogen peroxide.
- Bioremediation can occur through two primary methods. In situ bioremediation takes place on site, where the pollution is actually located. Ex situ bioremediation is the other option, where polluted soil or water is removed off site, treated, and returned to its original location.
- Microcosms are often used to determine the most appropriate bioremediation techniques for a given location. A microcosm is a simulated environment created under controlled laboratory conditions, designed to mimic the site of pollution. Microcosms are helpful in determining the efficacy of bioremediation candidates and in estimating decomposition rates and cleanup time needed.
- Different strategies of bioremediation are applied to different types of environment. In the case of loose or sandy soil environments, in situ bioremediation is preferred. In situ reduces costs and allows larger areas to be decontaminated at one time. Biostimulation and bioventing is not as successful in rocky soils or densely packed soils like clay. When these conditions are present, ex situ bioremediation may be more cost efficient in the long run. In aquatic bioremediation such as within municipal systems, solid waste is separated from liquid sewage. Solid waste is moved into bioreactors where it is treated with anaerobic microbes that release methane and carbon dioxide gases as waste products. Liquid sewage is passed through a separate filtration system and then disinfected before release back into the environment.
- Bioaugmentation, also known as seeding, introduces nonnative microbial strains into the cleanup site in order to enhance the activity of indigenous populations. In addition to microbes, environmental biotechnology has successfully used plants and fungi. Known as phytoremediation and mycoremediation, respectively, these cleanup methods have advantages in particular applications.
- Metagenomics is the genomics of entire microbial communities, with the aim of identifying genetic sequences instead of particular organisms. Metagenomics is also known as gene mining and utilizes a number of molecular biotechnology techniques. Both DNA and mRNA may be detected in metagenomic studies. DNA provides the raw material for sequence-dependent techniques, while mRNA may be used in function-dependent techniques.
- Conservation biotechnology has the dual goal of maintaining the biodiversity of the planet and protecting our existing environments. Direct measures include cloning, seed banks, living laboratories, and the creation of genomic libraries of threatened, endangered, and even extinct species. Some indirect methods are the protection of ecosystems and natural resources and the creation of biofuels.
- Biofuels are a growing source of alternative energy, such as methane, microbial fuel cells, ethanol, and biodiesel.

KEY TERMS

Accessions
Bioaugmentation
Biodiesel
Biofuels
Biomagnification
Bioremediation
Biosensors
Biostimulation
Bioventing
Electrigens
Gene cassettes
Integron analysis
Integrons
Metagenomics
Microcosm
Persistent organic pollutants (POPs)
Phage biopanning
Phylogenetic anchors
Stacked traits

DISCUSSION QUESTIONS

1. Visit the EPA Superfund "Where You Live" website (http://www.epa.gov/superfund/sites/npl/where.htm). What Superfund sites, if any, are near your city or town? In your state? What types of industry are associated with the sites, and what pollutants are present?
2. Compare and contrast microbial bioremediation, phytoremediation, and mycoremediation. What benefits do they offer? What are their drawbacks?
3. What initiatives is your state involved with as a means to stimulate biofuel research and production? Characterize specific projects, including the organisms employed and predicted applications.
4. Do you think it is appropriate to clone endangered animals? Extinct animals? Why or why not?

REFERENCES

Alvarado, S., et al. "Arsenic Removal from Waters by Bioremediation with the Aquatic Plants Water Hyacinth (*Eichhornia crassipes*) and Lesser Duckweed (*Lemna minor*)." *Bioresource Technology* 99, no. 17 (2008): 8436–8440.

Baker, Linda. "How Mushrooms Will Save the World." *Salon*, November 25, 2002. Accessed February 22, 2014, from http://www.salon.com/2002/11/25/mushrooms

BumHan, B., et al. "Site Investigation and Remedial Study at Darakdae Artillery Shooting Range for Full-Scale Phyto- and Bioremediation." In *Seventh International In Situ and On-Site Bioremediation Symposium, Orlando, Florida, USA, 2-5 June 2003. Part F. Phytoremediation.* Battelle Press, 2004.

Corley-Smith, Graham E., and Bruce P. Brandhorst. "Preservation of Endangered Species and Populations: A Role for Genome Banking, Somatic Cell Cloning, and Androgenesis?" *Molecular Reproduction and Development* 53, no. 3 (1999): 363–367.

Cowan, Don, et al. "Metagenomic Gene Discovery: Past, Present and Future." *TRENDS in Biotechnology* 23, no. 6 (2005): 321–329.

Environmental Protection Agency. "Common Chemicals Found at Superfund Sites." Last updated August 9, 2011. Available at http://www.epa.gov/oerrpage/superfund/health/contaminants/radiation/chemicals.htm

Environmental Protection Agency. "National Priorities List." Last updated April 22, 2013. http://www.epa.gov/superfund/sites/npl/index.htm

Fang, Chengwei, Mark Radosevich, and Jeffry J. Fuhrmann. "Atrazine and Phenanthrene Degradation in Grass Rhizosphere Soil." *Soil Biology and Biochemistry* 33, no. 4 (2001): 671–678.

Gillis, J. and L. Kaufman. "After Oil Spills, Hidden Damage Can Last for Years." *New York Times* (New York, NY), Jul. 17, 2010.

Glaser, J. A. and R. T. Lamar. *Lignin-Degrading Fungi as Degraders of Pentachlorophenol and Creosote in Soil*. Madison, Wisconsin: Soil Science Society of America, American Society of Agronomy, Crop Science of America, 1995.

Gómez, Martha C., et al. "Birth of African Wildcat Cloned Kittens Born from Domestic Cats." *Cloning and Stem Cells* 6, no. 3 (2004): 247–258.

Greer, Diane. "Harnessing the Power of Microbes." *BioCycle* 48, no. 5 (2007): 49.

Harvey, S., et al. "Enhanced Removal of Exxon Valdez Spilled Oil from Alaskan Gravel by a Microbial Surfactant." *Nature Biotechnology* 8, no. 3 (1990): 228–230.

Hill, Jason, et al. "Environmental, Economic, and Energetic Costs and Benefits of Biodiesel and Ethanol Biofuels." *Proceedings of the National Academy of Sciences* 103, no. 30 (2006): 11206–11210.

Hughes, Kevin A., Paul Bridge, and Melody S. Clark. "Tolerance of Antarctic Soil Fungi to Hydrocarbons." *Science of the Total Environment* 372, no. 2 (2007): 539–548.

Iqbal, M. and R. G. J. Edyvean. "Biosorption of Lead, Copper and Zinc Ions on Loofa Sponge Immobilized Biomass of *Phanerochaete chrysosporium*." *Minerals Engineering* 17, no. 2 (2004): 217–223.

Jordahl, James L., et al. "Effect of Hybrid Poplar Trees on Microbial Populations Important to Hazardous Waste Bioremediation." *Environmental Toxicology and Chemistry* 16, no. 6 (1997): 1318–1321.

Kamath, Roopa, Jerald L. Schnoor, and Pedro Alvarez. "Effect of Root-Derived Substrates on the Expression of Nah-Lux Genes in Pseudomonas fluorescens HK44: Implications for PAH Biodegradation in the Rhizosphere." *Environmental Science & Technology* 38, no. 6 (2004): 1740–1745.

Kim, Min Kyu, et al. "Endangered Wolves Cloned from Adult Somatic Cells." *Cloning and Stem Cells* 9, no. 1 (2007): 130–137.

Loi, Pasqualino, Cesare Galli, and Grazyna Ptak. "Cloning of Endangered Mammalian Species: Any Progress?" *Trends in Biotechnology* 25, no. 5 (2007): 195–200.

Madrigal, Alexis. "How Algal Biofuels Lost a Decade in the Race to Replace Oil." *Wired*, December 29, 2009. Accessed February 22, 2014, from http://www.wired.com/wiredscience/2009/12/the-lost-decade-of-algal-biofuel/

Mileski, Gerald J., et al. "Biodegradation of Pentachlorophenol by the White Rot Fungus *Phanerochaete chrysosporium*." *Applied and Environmental Microbiology* 54, no. 12 (1988): 2885–2889.

Qiu, X., et al. "Grass-Enhanced Bioremediation for Clay Soils Contaminated with Polynuclear Aromatic Hydrocarbons." In: *ACS Symposium Series*, vol. 563, pp. 142–142. American Chemical Society, 1993.

Stamets, Paul. *Mycelium Running: How Mushrooms Can Help Save the World*. New York: Random House, 2005.

United States Department of Agriculture. "Plant Genetic Resources Unit." *Agricultural Research Service*. Accessed February 22, 2014, from http://www.ars.usda.gov/main/site_main.htm?modecode=19-10-05-00.

Vogel, Gretchen. "Cloned Gaur a Short-Lived Success." *Science* 291, no. 5503 (2001): 409.

Williams, Peter A. and Keith Murray. "Metabolism of Benzoate and the Methylbenzoates by *Pseudomonas putida* (*arvilla*) mt-2: Evidence for the Existence of a TOL Plasmid." *Journal of Bacteriology* 120, no. 1 (1974): 416–423.

Worsey, Michael J. and Peter A.Williams. "Metabolism of Toluene and Xylenes by *Pseudomonas putida* (*arvilla*) mt-2: Evidence for a New Function of the TOL Plasmid." *Journal of Bacteriology* 124, no. 1 (1975): 7–13.

Yanase, H., et al. "Degradation of the Metal-Cyano Complex Tetracyanonickelate (II) by *Fusarium oxysporum* N-10." *Applied Microbiology and Biotechnology* 53, no. 3 (2000): 328–334.

8 Agriculture and Food Production

LEARNING OBJECTIVES

Upon reading this chapter, you should be able to:

- List agricultural products obtained through molecular biotechnology.
- Discuss examples of microbial biotechnology employed in agriculture.
- Describe methods of introducing transgenes into plant species.
- Illustrate the utility of the Cre/*lox* P system.
- Summarize the process of plant tissue culture.
- Identify genetic modifications used in crop plants.
- Explain the role of molecular biotechnology in aquaculture and livestock production.
- Contextualize food biotechnology throughout human history.

The adoption of agriculture approximately 10,000 years ago began our intimate mutualistic relationship with plants. To facilitate the massive undertaking of controlling our food supply, the domestication of certain animal breeds was also achieved. In order to confer desirable traits, humans have engaged in selective breeding of plants and animals for millennia. Proponents of agriculture biotechnology view advances in food production as a continuation of the previously established practice of plant and animal manipulation for our own benefit. In the scope of molecular biotechnology, agricultural applications are seen as an intensified, accelerated version of an already existing practice. Employing recombinant DNA technology and other molecular biotechnology methods in the allied fields of agriculture and food production introduces specific, desirable traits just as selective breeding has in the past. The significant difference is that we are no longer limited to working with existing variability in one species' gene pool. Plants and animals can be modified with genes of virtually any species. Interestingly, although traditional breeding methods such as cross pollination are no longer necessary, they are still employed. The process is faster now because introduced genes result in a unique plant or animal variety from the start, whereas traditional selective breeding may take generations to achieve. The risks and benefits of traditional and molecular techniques must be weighed in each individual project before determining the most appropriate course of action.

Agriculture biotechnology methods can also involve the use of microbial organisms to improve growth conditions of crop plants. In addition to agriculture, bacteria are used downstream in food processing. Methods of production and preservation both utilize biotechnology, and often the ingredients added to processed food are the products of biotechnology. This chapter discusses some of the major applications of molecular biotechnology to agriculture and food production.

Agriculture Biotechnology

The term *agriculture* is derived from the Latin *ager*, meaning "field," and *cultura*, meaning "to culture." Thus, the word specifically refers to growing plants, most often crop plants for consumption by humans and their livestock. But agriculture also involves the production of plants for commercial uses, such as ornamentals, paper, medicines, the floral industry, fibers including hemp and cotton for use in textiles, and other uses. In a broader context, agriculture is typically thought to include raising of livestock for meat, dairy, and eggs. Today even seafood is farmed as part of an industry known as aquaculture. Aquaculture is also used in the production of cultured pearls and algal species such as kelp and seaweed. Molecular biotechnology methods have come to dominate modern agricultural practices. Indeed, along with medical biotechnology, agricultural biotechnology represents one of the most intense areas of implementation to date.

Using Microbes

One of the first applications received by the U.S. Department of Agriculture–Animal and Plant Health Inspection Service (USDA-APHIS) for the genetically modified organism's (GMO's) field testing was not for a plant, but for a genetically modified bacterium. Known as Frostban, the bacteria were a strain of *Pseudomonas syringe* that contained a transgene for an "ice-minus" protein. One natural enemy of agriculture is freezing temperatures. Frost damage is estimated to cause $1 billion in agricultural crop losses each year in the United States. Ice crystal formation can lyse the cells of crop plants and drastically lower yields. Researcher Steve Lindow created the Frostban strain by inserting the ice-minus gene, hypothesizing the recombinant strain would prevent frost damage when applied to crop fields. *Pseudomonas syringe* is a naturally occurring microbe that grows on crop species including strawberries. Wild-type strains of the bacterium secrete the *Ina* "ice nucleation-active" protein, termed "ice," which causes ice crystals to form on crops. The ice-minus mutant variant is naturally occurring as well, and was used to transform *Pseudomonas syringe* into Frostban. Advanced Genetic Sciences, Incorporated designed a field test using strawberries in order to determine whether Lindow's genetically modified (GM) ice-minus strain would be able to outcompete indigenous strains. If so, the ice-minus strain would eliminate the wild-type strain and effectively reduce frost damage. After successfully enduring public outcries during the required open comment periods, the field test took place. Nearby residents were especially vocal in their opposition, citing concerns over releasing genetically modified bacteria into the environment which might potentially disrupt the natural ecological balance (**FIGURE 8.1**). While the field test was successful in proving the Frostban bacteria were effective in preventing frost damage, the product was not marketed. Indeed, the Frostban variety lowered temperatures at which ice crystals formed an average of two degrees Farenheit and as low as 23°F. Lindow also performed experiments with ice-minus strains using potatoes. These plants were uprooted in an act of vandalism. The public reaction was construed to indicate Frostban would not be commercially viable and the ice-minus strain of *Pseudomonas syringe* was abandoned.

Another bacterial species, though not genetically modified, has served as an example of agricultural biotechnology for over 30 years. *Bacillus thuringiensis* (Bt) may be applied to crops as a natural insecticide and is frequently used by organic gardeners. Bt expresses the Cry protein, which is a crystalline toxin that disrupts

FIGURE 8.1 Field testing of Frostban was so controversial, technicians wore HazMat suits to alleviate perceived fears (foreground). Note the number of onlookers (background).

the digestive tracts of insects. More specifically, different strains of Bt encode different forms of Cry, which affect different species of insects (**TABLE 8.1**). Once Cry takes effect, the insect ceases to feed and eventually dies. Because the optimal environment for Cry proteins is alkaline, Bt is highly effective in the high pH environments found in insects. As birds and mammals (including humans) have acidic digestive environments, the Bt toxin may be safely applied to food crops, because it will deteriorate in such conditions. Genes encoding Cry proteins are also used as transgenes in the genetic modification of crop plants, as discussed below.

Plant Biotechnology

Genetically modified plants are a major component of worldwide agricultural fields. The United States, as the home of recombinant DNA technology, has been especially aggressive in the adoption of agricultural biotechnology. As we have seen, 88% of corn, 93% of cotton, and 94% of soybean crops in the United States are GM. Worldwide, agricultural fields composed of GMOs have increased 100-fold in less than two decades, from 1.7 million hectares to 170 million hectares. Twenty-eight countries across the world plant GM varieties.

TABLE 8.1

Cry Protein Specificity

Bt Subspecies	Cry Protein	Susceptible Organisms
kurstaki	CryI	Lepidoptera (butterflies and moths)
kurstaki	CryII	Lepidoptera and Diptera (flies)
tenebrionis	CryIII	Coleoptera (beetles)
israelensis	Cry IV	Diptera

Transfection Methods

Inserting transgenes into plant species is known as **transfection**. The process is not as straightforward as genetically modifying unicellular organisms such as bacteria. One significant obstacle is that insertion into the plant genome is random; therefore, trial and error is a necessary step in achieving the desired result. Plant biotechnologists seek to introduce a transgene so that it may be highly and consistently expressed in the GM variety, while avoiding insertion into functional genes that may be disrupted.

In addition to inserting a selected transgene, a promoter region must be included to allow expression of the transgene once introduced into the plant. Two types of promoters are commonly used—constitutive promoters and inducible promoters. **Constitutive promoters** turn on a transgene in every plant cell. The gene is expressed in all cell and tissue types throughout the entire life of the plant. **Inducible promoters** may be turned on and off within the transfected plant depending on environmental conditions. One inducible promoter is *cab*, which is a switch for the gene encoding the chlorophyll a/b binding protein. Because it is associated with the chlorophyll pigment, the *cab* promoter is only turned on when a plant is exposed to light. Thus, the chlorophyll a/b binding protein will only be expressed in the presence of light, and light-deprived portions of the plant, such as roots and tubers (e.g., potatoes), will not synthesize the protein. The appropriate promoter for a given transfection event is determined based upon where and when the transgene should be functional.

A final part of the construct delivered into the plant is a selectable marker. This is necessary so the bioengineer may recognize when transfection is successful. One approach includes an antibiotic resistance gene as the selectable marker. Plant cells subject to transfection can then be cultured in the presence of the antibiotic. Only the cells that have taken up the transgene construct will grow, as they contain the resistance gene (**FIGURE 8.2**). These cells can then be isolated and used in plant tissue culture in order to grow a whole plant. However, adding antibiotic resistance genes into our food supply is a valid safety concern. To alleviate the risks associated with antibiotic resistance genes, molecular biotechnology offers a technique to remove selectable marker genes, known as Cre/*lox* P (see below).

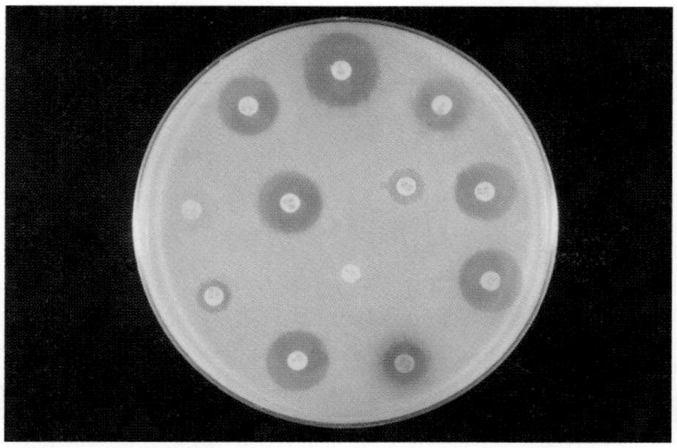

FIGURE 8.2 Selectable marker detection is demonstrated in an *E. coli* culture. The white disks are saturated with an antibiotic. Rings of no growth (surrounding the disks) indicates transfection did not occur. Courtesy of CDC

The Ti Plasmid

While most DNA transfer occurs between closely related species of the same kingdom, there are exceptions to this principle in nature. The genus *Agrobacterium* common to soil environments is the cause of plant tumors. For example, *Agrobacterium tumefaciens* confers crown gall disease. *Agrobacterium* species harbor a plasmid known as Ti (for "tumor-inducing"), which is able to infect plant tissue. This makes the Ti plasmid an effective delivery vehicle for transgenes. Plants with wounded tissue secrete phenolic compounds that attract *Agrobactium spp.*, thus initiating infection. A segment of the Ti plasmid, known as T-DNA, enters the plant cells in a manner similar to bacterial conjugation. By constructing a T-DNA region with a selected transgene, bioengineers are able to insert the desired genetic information into a plant species, thus creating a genetically modified variety (**FIGURE 8.3**).

The Ti plasmid is rather large and difficult to work with as is, so bioengineers have removed a number of genes from the plasmid that are unnecessary in the transfection process. Also, the tumor-inducing genes were spliced out so that the genetically modified plant will not exhibit abnormal growth. These steps resulted in a smaller plasmid that is easier to manipulate. Plant biotechnologists have also simplified the production of recombinant Ti plasmids by cloning them into *E. coli* bacteria. This allows for scaled up production of the Ti plasmids because the parameters of *E. coli* culture are so well understood. The plasmids are then isolated and introduced into plant tissue culture in order for transfection to proceed.

Floral Dip

Another way to transfect plants is the floral dip method. Plants are allowed to grow until they begin to produce floral buds, which are removed. Once the floral buds begin to regenerate, they are dipped in a solution containing recombinant *Agrobacterium spp.* and surfactant molecules. The recombinant bacteria harbor the Ti plasmid as described above, including the modified T-DNA region carrying the promoter, transgene, and selectable marker gene. The surfactant (soap) facilitates adhesion of the recombinant bacterium to the regenerating floral buds. Though the exact mechanism is unknown, transfection occurs in the ovarian tissue present in the immature flowers. This introduces the transgene into germline tissues, allowing the conferred trait to be passed on to successive plant generations. The transfected plant continues to grow and sets seed. Seeds are then harvested and germinated. This generation of plants is screened for selectable markers and those plants expressing incorporated T-DNA are isolated as members of the genetically modified variety.

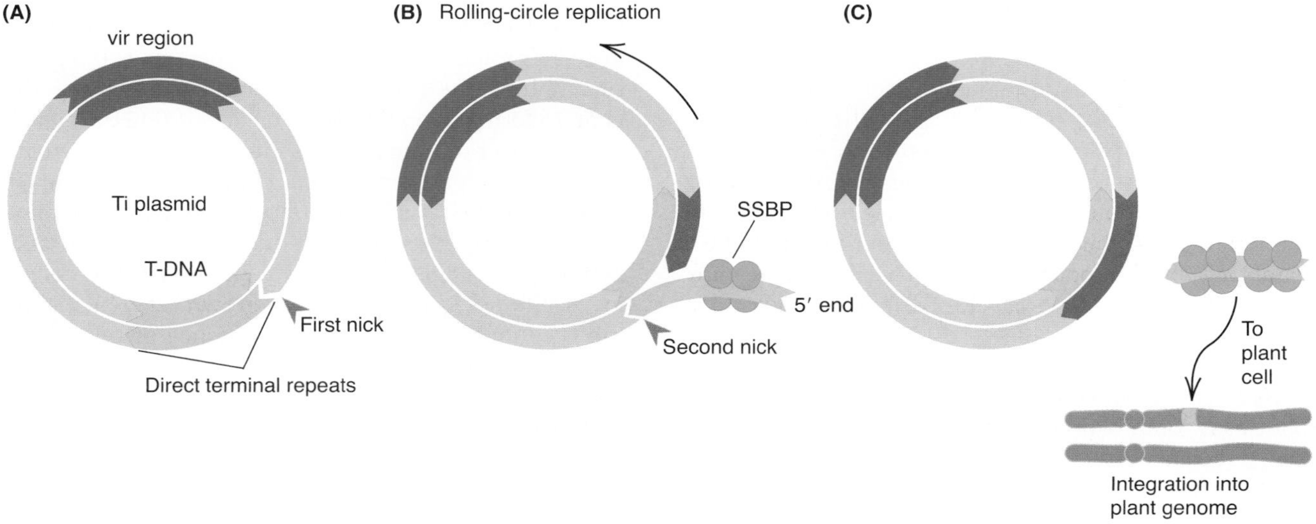

FIGURE 8.3 The Ti plasmid is used to transfect plants. (A) A nick forms at the edge of the T-DNA region. (B) Replication of the T-DNA is carried out. Replication concludes with a second nick. (C) Single-stranded binding proteins (SSBP) stabilize the replicated T-DNA, which inserts into the plant genome.

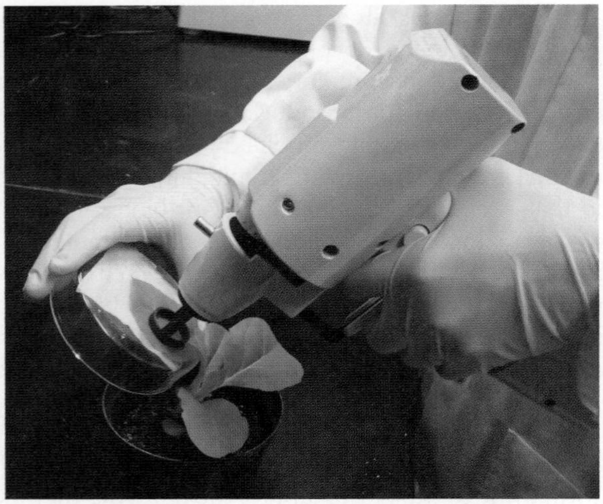

FIGURE 8.4 A gene gun is used to insert genetic information into plant tissue. Courtesy of Dr. Marcia Harrison, Marshall University

Biolistics

The technique of particle bombardment employed in gene therapy is also useful in plant transfection. As discussed previously, particle bombardment is also known as "biolistics." Gene guns are used to blast transgene constructs past cellulosic cell walls into the plant, where it may enter the cytoplasm or directly incorporate into the plant DNA (**FIGURE 8.4**). The high pressure of the gene gun shoots metal particles (made of gold or tungsten) coated with the transgene construct with either a blast of air or helium or a strong electrical pulse.

*Cre/*lox *P*

As mentioned earlier, concerns exist over the use of antibiotic resistance genes as selectable markers. There is a real possibility that outcrossing might spread antibiotic resistance genes to non-target organisms and lead to reduced effectiveness of our existing antibiotic arsenal. Furthermore, despite evidence to the contrary, the fear of ingesting transgenes in GM food persists. Studies indicate transgenes are destroyed early during digestion by gastric juices, removing the possibility of outcrossing into our intestinal flora. Yet even now, it is routine to excise these genes from genetically modified plants as a precautionary measure. This is possible with the Cre/*lox* P technique developed through molecular biotechnology.

The Cre/*lox* P system originates from the *E. coli* bacteriophage P1. This virus contains the gene (*cre*) that "**c**auses **re**combination." The gene encodes for Cre protein, a recombinase enzyme that specifically recognizes a DNA region known as *lox* P. To facilitate the eventual removal of selectable marker genes, plant biotechnologists will attach *lox* P regions to either side of the marker gene when engineering a transgene construct such as T-DNA. In a mechanism similar to intron splicing, Cre proteins bind to the two separate *lox* P regions surrounding the selectable marker gene, causing the gene to form a loop that is then excised (**FIGURE 8.5**).

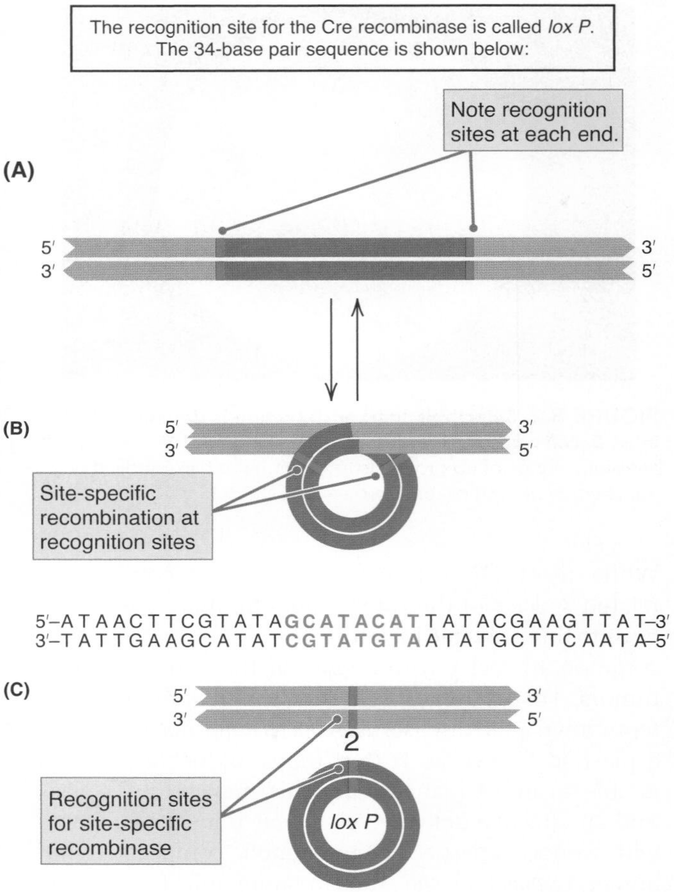

FIGURE 8.5 The Cre/*lox* P system removes selectable marker genes from GMOs, such as antibiotic resistance genes. (A) Regions in blue represent the selectable marker gene to be excised. (B) Cre recombinase recognizes the *lox* P sites and excises the selectable marker gene. (C) The selectable marker gene is now absent from the engineered organism.

To deliver the Cre proteins into the GM plant, cross-pollination is performed. One plant containing the *cre* gene is crossed with the transfected plant containing the transgene construct (including the selectable marker gene flanked by two *lox* P regions). The resulting seeds are germinated and grown, at which time the seedlings are screened for presence of the selectable marker. For example, if the selectable marker is a neomycin-resistance gene, the seedlings will be treated with neomycin to determine sensitivity. If sensitive, this indicates the *cre* gene was inherited by the offspring and therefore the marker gene was excised. These transgenic offspring are isolated and crossed with wild-type plants, resulting in some offspring that contain only the transgene (but not the *cre* gene). Through the

Cre/*lox* P system, plant biotechnologists can produce genetically modified plant varieties that harbor only the desired transgene.

Plant Tissue Culture

Due to the random insertion of a transgene into the plant genome, multiple attempts must be undertaken to achieve a desired level of transgene expression. Each insertion, occurring at different loci within the plant genome, is called an event. Plant tissue culture is a crucial step in characterizing the nature of each event. It allows plant biotechnologists the opportunity to observe the phenotypic outcome of each transgenic event under controlled laboratory conditions, with the ultimate goal of selecting the most ideal phenotype(s) for large scale field testing and (hopefully) commercialization.

In plant tissue culture, an entire plant can be regenerated from a single totipotent cell. Culture can occur in liquid media, known as **suspension culture**, or on a solid medium, referred to as **callus culture**. First, a portion of the plant to be cultured is removed. This piece of tissue is called the **explant**. In callus culture, the explant is in the form of a mass of cells obtained from the immature embryo, apical meristem (where new shoots emerge), or root tip. For suspension culture, the explant must be comprised of individual cells, such as protoplasts (plant cells lacking cell walls) or immature gametes (egg or sperm cells).

Next, the explant is treated with a cocktail of plant hormones customized for a predetermined growth pattern. The specific hormone treatment depends upon the plant species. For example, tomato plants require the hormone cytokinin to dedifferentiate the explant cells into a totipotent state and stimulate cell division. On solid media, a mass of undifferentiated cells known as a callus will form. The callus is treated with additional hormones to initiate shoot formation for a period of time (**FIGURE 8.6**). This is immediately followed with hormone treatment to initiate root growth, after which time the cultured plants can be transferred to soil for its final growth period. With suspension culture, shoot and root formation occurs simultaneously. Plant tissue culture is also a popular method of propagating, or cloning, plants for commercial production (**BOX 8.1**).

FIGURE 8.6 An example of plant tissue culture: Shoots of a tobacco plant emerging from an undifferentiated callus.

Genetically Modified Plants

As our knowledge of genomics and proteomics increases, the applications to plant biotechnology appear limited only by imagination. In reality, economic forces are a primary consideration when proposing a transgenic

Box 8.1 Micropropagation

Plant tissue culture is useful for selecting genetically modified plants with the most desirable transgenic events. It is also used as a method of plant cloning, which serves as an alternative to traditional breeding methods involving seeds. Plant cloning is utilized in the mass production of commercially important plants such as ornamentals, as well as crops that cannot reproduce well on their own, including bananas and strawberries. Plant cloning for these types of commercial ventures is known as **micropropagation**. Micropropagation may take place in laboratory settings, in plant nurseries, or even larger operations including plantations. For example, palm oil plantations in Malaysia were originally established in the 1960s using plant tissue culture methods to create genetically identical palm trees via asexual cloning.

plant variety. The industrial nature of today's agricultural operations has led to nearly universal adoption of monoculture practices. Growing a single plant variety on vast quantities of land creates conditions suitable for particular traits, such as herbicide and insect resistance. The dramatic increase of GM agriculture throughout the world, and especially in the United States, is thanks to these types of traits.

Plant biotechnology has demonstrated success in improving crop yields and by extension is often touted as a solution to world hunger issues. Introducing traits such as stress tolerance, pathogen resistance, and enhanced nutrition are heralded as ways to address the growing demand for food as the human population continues to skyrocket. While agricultural biotechnology holds much promise in these regards, there are still complex issues interfering with the widespread adoption of certain GM plants.

Herbicide Resistance

Herbicides are chemicals designed to kill weeds. Weeds are a constant annoyance in farming operations; they reduce crop yields up to 67% by competing with them for sunlight, water, and nutrients. Weeds are a direct result of the no-till practices common to monoculture agriculture. No-till keeps soil undisturbed and is tremendously beneficial in the prevention of soil erosion. It is estimated that no-till practices can reduce soil erosion by 90% or more. In the United States, soil erosion decreased from 3.5 billion tons in 1938 to about 1 billion tons in 1997. The benefit of reduced soil erosion comes with the cost of increased herbicide applications as a measure to combat weed infestations. Around 85% of U.S. fields are sprayed with herbicides each growing season. Furthermore, if modern agriculture returned to tillage methods, labor costs would increase dramatically, resulting in higher food prices. The National Center for Agriculture and Food Policy estimates that 288 billion pounds of food and fiber would be lost at a cost of $21 billion if herbicides were not used as they are today.

Engineering crop plants with herbicide resistance has facilitated the widespread use of these chemicals. From the years 1966 to 1991, herbicide applications more than quadrupled in the United States. Over 500 million pounds are applied each year. The top herbicide, accounting for over 80 million acres per year, is glyphosate. Monsanto markets glyphosate by its more commonly known brand name, RoundUp. Monsanto is also responsible for RoundUp Ready GM crops including:

- Corn
- Canola
- Cotton
- Alfalfa
- Sugar beets
- Soybeans

Wheat and sorghum are additional crops currently in research and development stages for genetic modification.

Opponents of GM agriculture are quick to point out that Monsanto profits from the sale of the herbicide as well as the herbicide resistant crops. They were indeed designed as complementary products. Farmers are able to plant monoculture stands of RoundUp Ready crop and blindly apply glyphosate to the fields. It is not unusual today for herbicides to be sprayed with crop dusters, whereas before herbicide resistant varieties this would have been unheard of (**FIGURE 8.7**). Traditionally bred plants are just as susceptible to herbicides as are weeds. Insertion of herbicide resistant transgenes allows farmers to apply the herbicide directly to monoculture fields, effectively killing weeds while still maintaining the integrity of their crops.

Glyphosate-resistant plants are engineered to degrade RoundUp into nontoxic byproducts. The herbicide is effective in killing weeds as it inhibits a vital metabolic pathway known as the shikimate pathway (**FIGURE 8.8**). The enzyme EPSPS (5-enolpyruvylshikimate-3-phosphate synthase) normally catalyzes synthesis of aromatic amino acids, but the herbicide inactivates this enzyme. The aromatic amino acids include tryptophan, tyrosine, and phenylalanine. Herbicide resistant varieties degrade glyphosate so that the enzyme remains effective and the pathway is able to synthesize the amino acids. Additional GM plant varieties are resistant to other herbicides including glufosinate (marketed as Liberty), bromoxynil, and sulfonylurea.

Insect Resistance

As discussed above, the bacterium *Bacillus thuringiensis* naturally synthesizes the Cry protein that destroys insect pests (**FIGURE 8.9**). Plant biotechnologists have created GM crop varieties containing the *cry* genes, eliminating the need to apply the bacterium directly. Transfection techniques were employed to insert the

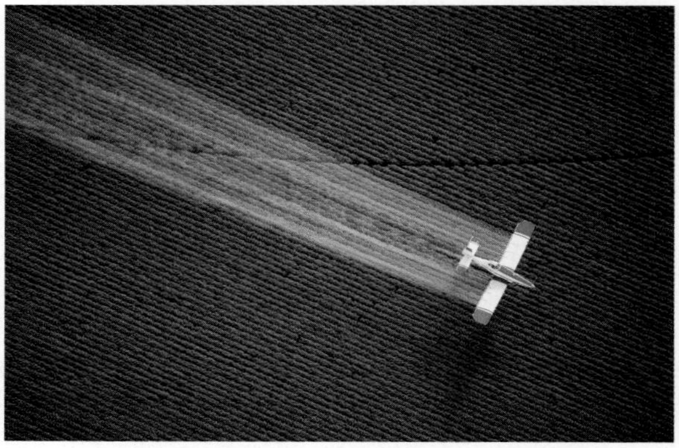

FIGURE 8.7 Crop dusting allows herbicides to be sprayed over entire monoculture stands. © B. Brown/ShutterStock, Inc.

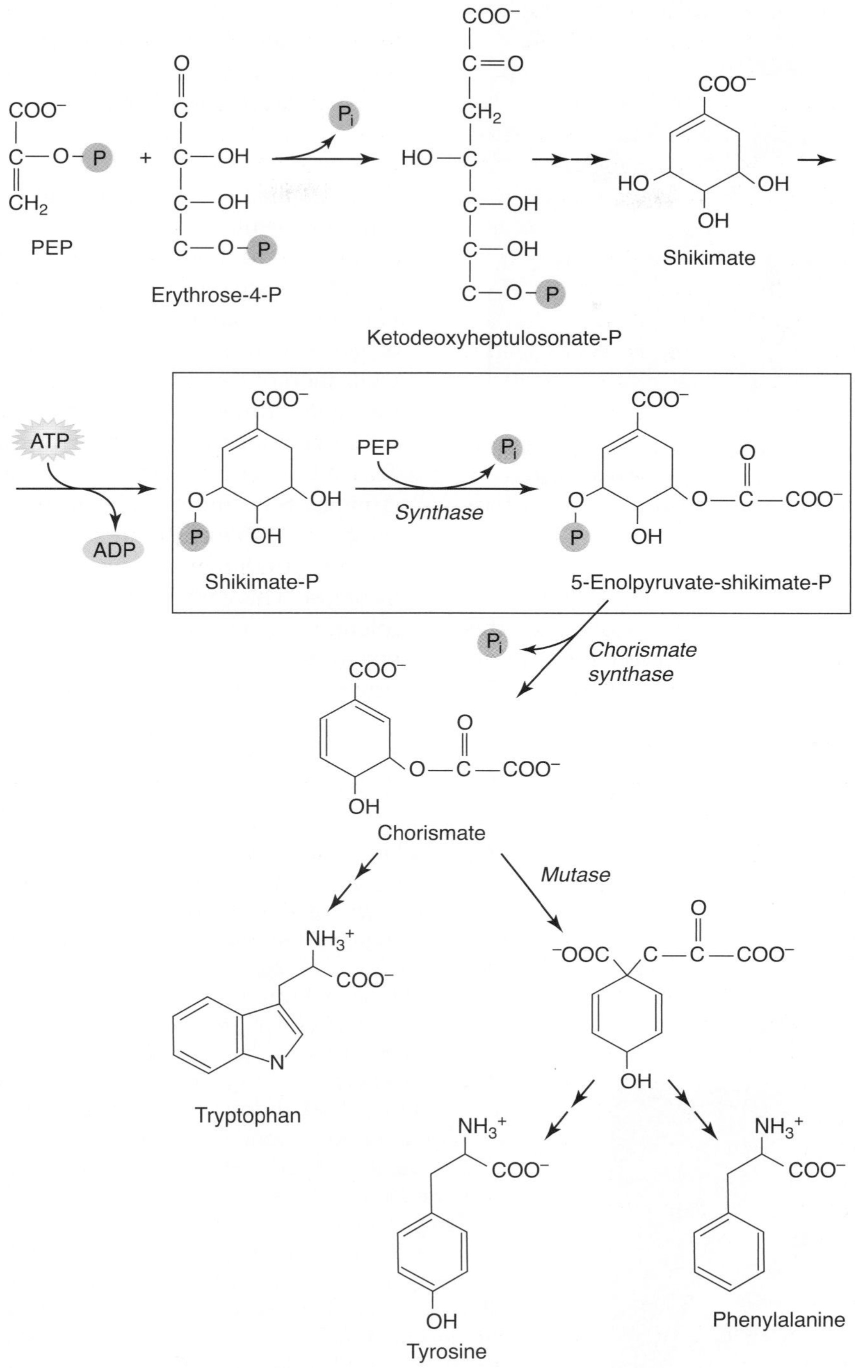

FIGURE 8.8 Glyphosate inactivates the "synthase" enzyme, preventing biosynthesis of aromatic amino acids including tryptophan, tyrosine, and phenylalanine.

cry transgenes so that crops express the Bt toxin, which essentially results in insect resistant plants. In particular, corn and cotton are crops transfected with insect resistance genes. Because several *cry* genes affect different insect species, research is now focusing on delivering multiple *cry* genes into a single GM crop variety. This will improve insect resistance to include more than one species. The herbicide resistant plants engineered by companies such as Monsanto often contain *cry* transgenes as well, serving as an example of stacked traits.

A purported benefit of pesticide resistant crops is the potential for reduced pesticide applications. Although this has yet to be realized, companies marketing these GM varieties stress the lessened impact

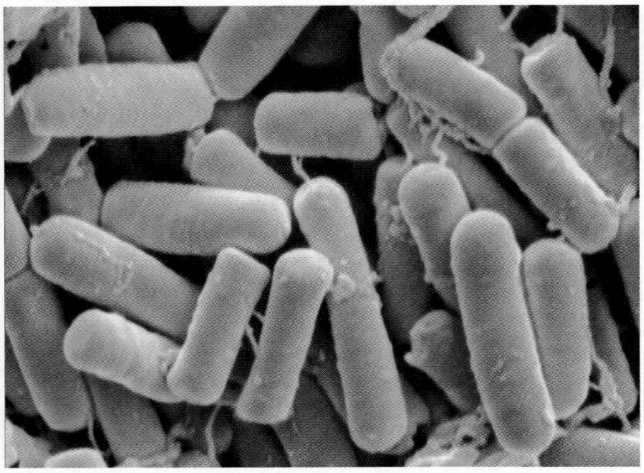

FIGURE 8.9 The bacterium *Bacillus thuringiensis* exists in multiple strains that express various forms of the *cry* gene, producing Bt toxin as an effective pesticide. © SciMat/Science Source

upon the environment as a benefit to pesticide resistance. Opponents of GM agriculture downplay this potential benefit, because it has yet to come to fruition. Another issue is the increased use of herbicides might very well negate any reduction of pesticide applications in the future. Finally, mixed results of environmental studies have raised concerns over Bt toxins harming non-target organisms (**BOX 8.2**).

Stress Tolerance

Existing plant species found in certain environments can be useful as bioprospecting candidates. Natural adaptations to drought, high salinity, high temperatures, and other stressful conditions have a genetic basis. Functional genomic studies have revealed a number of genes that confer resistance to stressful conditions. Utilization of stress tolerant genes for bioengineering of transgenic varieties has the potential to spread agricultural operations into currently inhospitable landscapes. This in turn would increase our capabilities to grow more crop plants on more land, resulting in higher yields and the promise to realistically address world hunger.

One stress tolerance mechanism involves the sugar trehalose. This carbohydrate is present in many drought tolerant plants, fungi, and bacteria. Encoded by two enzymes in its metabolic pathway, trehalose has the property of absorbing and releasing water molecules as

Box 8.2 Bt and Butterflies

Genetically modified food crops are a source of controversy for environmental and health activists as well as individual consumers concerned over what they eat. Plants transfected with *cry* genes expressing the Bt toxin have contributed to this controversy via a well-publicized article in the journal *Nature*. A Cornell University study released in 1999 indicated Bt toxin expressed by *cry* genes had an adverse effect on the growth of Monarch butterfly larvae when consumed. The experiment was designed to mimic natural conditions, where Monarch butterfly larvae are known to feed on milkweed leaves near corn fields. It tested the hypothesis that corn pollen could be carried by wind onto milkweed plants, thus leading to the poisoning of non-target organisms such as the Monarch butterfly. Three subject groups of caterpillars were given milkweed with no pollen, with Bt pollen, and with non-GM pollen, respectively. In particular, researchers noted the larvae were harmed after eating milkweed leaves covered with pollen from Bt corn. They had stunted growth, ate less, and exhibited higher death rates. These results were alarming to GM watchdog groups, especially because Monarchs are migratory pollinators that serve an integral role in the ecology of Canada, the United States, and Mexico. Subsequent studies published in the *Proceedings of the National Academy of Sciences* challenged the validity of the Cornell experiment. Scientists from the U.S. Department of Agriculture–USDA-ARS, North American universities, and industry and environmental groups all participated to achieve the most balanced approach. Flaws in the design of the Cornell experiment were addressed, such as the quantities of pollen applied to milkweed leaves. Researchers concluded the original study included unnaturally high levels of pollen that would not occur under real-life conditions. They also found that caterpillars, when given a choice, will not eat pollen-coated leaves and adult females prefer to lay their eggs on pollen-free leaves. Lastly, researchers tested multiple forms of the Bt toxin, derived from different *cry* genes, and determined that only one variety was significantly toxic to Monarch butterflies. The Cry1Ab protein (event #176), accounting for about 1% of the entire corn crop in 2000, was a culprit strategically removed from fields. Furthermore, this transgenic event is no longer approved by regulatory agencies and is therefore off the market, eliminating any concerns of Monarch butterfly poisoning.

A monarch butterfly caterpillar feeds on a milkweed leaf. © Sleddogtwo/iStockphoto.com

needed. A third gene encoding for trehalase, an enzyme that catalyzes trehalose, has also been isolated through genomic studies. Utilizing these transgenes enables the creation of drought tolerant crop varieties which can grow in areas with little rainfall. It is estimated that 40% of land on Earth is subject to desertification, the conversion of semi-arid land into desert-like conditions. If this trend continues, drought tolerant plants will be crucial elements in feeding the growing world population. A drought tolerant version of rice has been developed using trehalose genes, which has the added benefit of resistance to high salinity conditions.

Pathogen Resistance

Plants are subject to infection by bacterial, fungal, and viral pathogens. Destruction of crops by pathogenic organisms is a serious threat and can result in significant economic losses, on the order of millions of dollars each year. Genomic studies can reveal resistance genes specific to different pathogens; these genes may then be introduced into crop species through transfection methods. Some of these genes are described in **TABLE 8.2**.

TABLE 8.2

Pathogen Resistance Transgenes in Commercial Crops

Resistance Gene Origin	Type of Pathogen	Species
Mouse-ear cress	Bacterial	*Pseudomonas syringae*
Tomato	Fungal	*Cladosporium fulvum*
Flax	Fungal	Rust diseases (multiple species)
Tobacco	Viral	Tobacco mosaic virus

One technique to confer viral resistance involves plant transfection with viral coat protein genes. Viruses normally reproduce by activating the DNA replication mechanisms of their host cells. The viral genome and its protective capsid are synthesized in this way. The capsid is comprised of viral coat proteins. The viral coat protein gene can be switched off through RNA interference. By switching off the gene, the virus is unable to replicate and its life cycle is stopped short, thus preventing further viral infection of neighboring cells within the plant tissue. Squash varieties employing this biotechnology, including yellow squash, zucchini, pumpkins, watermelon, and cucumbers, are on the market that confer resistance to watermelon mottle virus, zucchini yellow mosaic virus, and cucumber mosaic virus, respectively. The USDA Agriculture Research Service (ARS) was granted non-regulatory status for a GM plum resistant to the plum pox virus. After an outbreak of papaya ring spot virus in Hawaii during the 1990s, resistant varieties known as Rainbow and SunUp were developed by plant biotechnologists at Cornell University and offered to papaya farmers at no cost. Virus-resistant potatoes have also been developed, but are no longer on the market (**BOX 8.3**).

Box 8.3 Do You Want Fries with That? The Case of NewLeaf Potatoes

We have all been asked that question at chain restaurants such as McDonald's and Burger King. For U.S. consumers, the answer is apparently no, when the potatoes are genetically modified varieties. Two genetically modified potatoes, known as NewLeafPlus and NewLeafY, were successfully developed and field tested by Monsanto Corporation, and in 1995 they were approved for market. These potatoes exhibited stacked traits, including viral-resistance and Bt toxin production. NewLeafPlus was engineered for resistance to the potato leaf roll virus and NewLeafY garnered resistance to potato virus Y. A *cry* gene was also inserted into the potato varieties to combat infestation by the Colorado potato beetle. At the time NewLeaf potatoes were introduced, an estimated 40% of insecticides applied in the US were to control this particular pest.

When it was publicized that fast food companies were using GM potatoes in their French fries, the negative consumer response was so overwhelming that these companies collectively refused to keep buying NewLeaf potatoes. Proctor and Gamble and Frito Lay, two top food manufacturers, followed suit. These companies pledged not to use GMOs in their menu items in order to satisfy their customers. The main concern appeared to be the presence of the Cry protein; because it is also known as the Bt toxin, this raised red flags. Even though it is less than one-tenth of one percent of the entire protein content of the NewLeaf potatoes, and even though Cry protein is not toxic to mammals, the lack of public understanding, and thus acceptance, was simply too overwhelming to food marketers. As explained by Margaret Mellon, Director of the Food and Environment Program of the Union of Concerned Scientists:

> The Bt potato offers farmers reduced costs; it doesn't offer consumers anything. And so a lot of consumers, if they were given a choice, might say, 'well, it doesn't provide an advantage for *me*, and therefore, not knowing a whole lot about it, I might...simply say no.'

Due to negative associations with the product, Monsanto ceased selling NewLeaf potatoes in 2001. It had captured around 5% of the market before its demise. Monsanto continues to sell GM varieties of corn, soybean, and cotton with unparalleled rivalry.

Enhanced Nutrition

Genetic modification of crops in order to enhance their nutritional profiles is a technique known as **biofortification**. Biofortification is viewed as a potential remediation for malnutrition seen in developing countries. Some biofortified transgenic crops in development include beta-carotene enriched tomatoes and iron-fortified rice. Iron malnutrition results in anemia. Beta-carotene is a precursor to vitamin A. This nutrient is essential in the human diet, though unfortunately vitamin A deficiency is a significant component of malnutrition that can lead to blindness and may eventually be fatal. Iron and vitamin A deficiencies are common in countries where rice is a staple in traditional diets, because existing rice crops lack these nutrients. Other nutrients used for biofortification include protein, zinc, and vitamin E.

Golden Rice, which is genetically modified to express beta-carotene, has become the poster-child of biofortification, due to its intense development spanning several decades at the behest of both humanitarian organizations and agribusiness (**BOX 8.4**). BioCassava Plus is a GM crop with increased expression of beta-carotene, iron, zinc, vitamin E, and protein. It is currently in field trials in regions of sub-Saharan Africa where cassava is a staple food. Biofortified transgenic plants currently in development are exhibited in **TABLE 8.3**.

Aquaculture

Harvesting fish and other seafood from the world's oceans is one of the last vestiges of the hunter-gatherer lifestyle carried out by our ancestors for millennia. The commercial fishing industry represents an enormous

Box 8.4 Golden Rice Offers Hope for Malnutrition

In development for nearly 30 years, Golden Rice is a transgenic crop containing beta-carotene, which after consumption enables the synthesis of vitamin A. Beta-carotene genes derived from the daffodil plant (*Narcissus pseudonarcissus*) and the bacterium *Erwinia uredovora* were used to transfect rice plants, which were then crossed to produce offspring with the full complement of beta-carotene genes (see figure). Development of Golden Rice was funded in part by the Rockefeller Foundation and the European Union in order to explore opportunities for international aid. The prolonged development of Golden Rice is due to historically low expression levels of beta-carotene. Rights to Golden Rice are now owned by Syngenta, a large agribusiness firm. Syngenta is concurrently looking for ways to commercially market Golden Rice in developed countries such as Japan and the United States while also exploring no-cost distribution of the transgenic crop to impoverished nations. Another major goal for Syngenta is to increase the expression of beta-carotene so that serving-size portions of rice will enable biosynthesis of vitamin A according to recommended daily intake values. Researchers are also investigating the introduction of stacked traits into Golden Rice, including transgenes encoding vitamin E, iron, protein, and zinc.

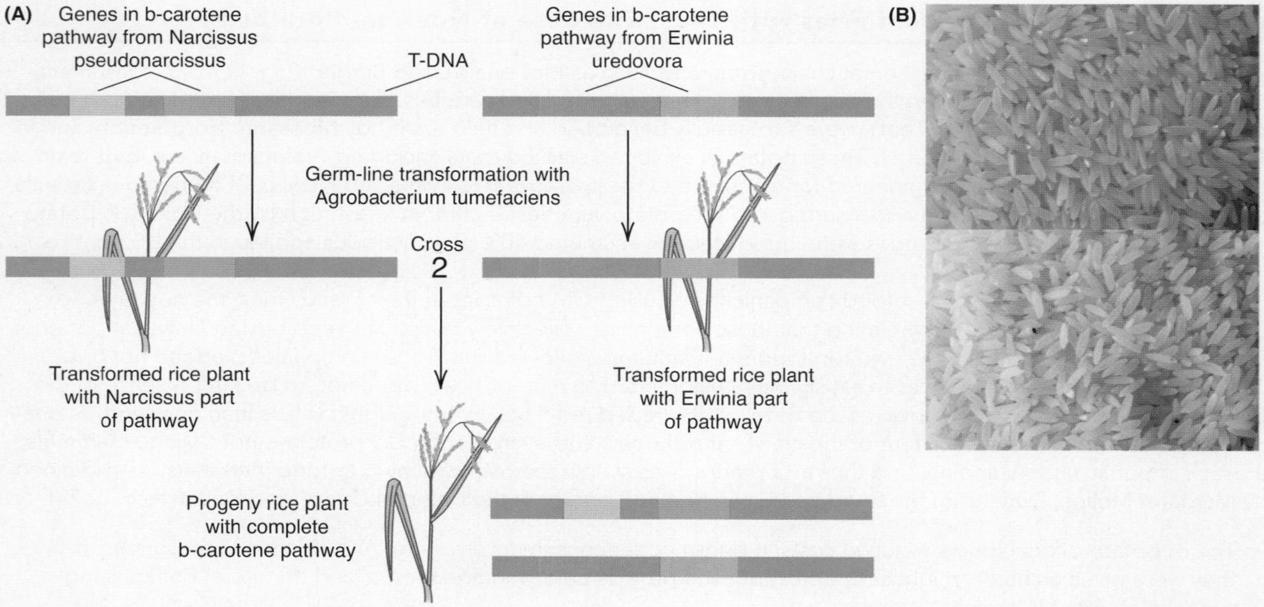

Genetic engineering of Golden Rice. (A) Production of the rice plant involved transfection of two varieties followed by their hybridization. (B) Golden Rice (above) exhibits a trademark yellow color due to the presence of beta-carotene, a vitamin A precursor, which is not seen in non-GM varieties (below). Courtesy of Ingo Potrykus, Institute für Pflanzenwissenschaften, ETH Zurich

TABLE 8.3

Biofortified Transgenic Crops

Crop	Nutritional Enhancement(s)
• BioCassava Plus • Golden Rice	• Beta-carotene • Vitamin E • Protein • Iron • Zinc
• Lettuce • Carrots	Calcium
• Tomatoes	Beta-carotene
• Rice	Iron
• Corn	• Beta-carotene • Folate • Vitamin C
• Potatoes	Protein
• Soybeans	Omega-3 fatty acids

pressure on wild fish stocks. An estimated 100 million tons are harvested each year. As the human population increases, so does our demand for fresh seafood. This is compounded by seafood's attraction as a healthier source of animal protein as incidences of heart disease and obesity likewise go up. These factors in combination require new outlets for seafood production, as demand is expected to increase 70% in the next 35 years. Enter **aquaculture**, the production of fish or shellfish in a controlled environment for human consumption (**FIGURE 8.10**). Aquaculture is also intensely utilized in the culture of pearls and for extraction of bioactive compounds used in pharmaceuticals and nutritional supplements.

FIGURE 8.10 An enclosed aquaculture operation in Italy grows trout for human consumption.

Forms of aquaculture have been carried out by humans for at least 3,000 years. Ancient Chinese cultures farmed carp species in managed ponds. Some evidence suggests that date may be even earlier, with the proposition that Aboriginal peoples of Australia may have raised eels since 6,000 BC. Today, approximately 36 million people are employed in the aquaculture industry. Worldwide aquaculture operations produced over 125 million tons in 2010, compared to just under 10 million tons in 1985. Species which are cultured include:

- Bivalve mollusks: clams, oysters, mussels, scallops
- Gastropods: abalone
- Crustaceans: shrimp, crabs, prawns, sea bream, crawfish
- Finfish: tilapia, salmon, halibut, yellowtail jack, carp, catfish, trout, bass
- Algae: nori, wakame, kombu
- Other: bullfrogs, alligators, water chestnuts, eels, watercress

Genetically modified species are utilized in aquaculture. Salmon containing growth hormone genes enable increased growth at faster rates. At the time of this writing, these GM fish are currently in the final stages of the regulatory approval process and are not commercially available. Fish eggs are transfected to produce GM fish with inserted growth hormone genes. The process is relatively simple because fish eggs are transparent and quite large, making them easy to work with. Microinjection is employed to deliver the transgene into the egg nucleus. Inducible promoters are being tested that would turn on transgenes in the presence of light, specific temperature ranges, or components of their diet.

Another genetic modification of fish is the insertion of "antifreeze" protein, which enables arctic fish such as the winter flounder to survive colder temperatures. Isolation and insertion of the antifreeze gene would increase the available areas for aquaculture operations as well as allowing culture of fish that otherwise do not survive well in controlled environments. Also, existing aquaculture operations are suspended in winter months when subzero temperatures are common. Antifreeze proteins would allow operations to continue year round. As with genetically modified plants, GM seafood is the subject of debate over concerns that they may outcross with wild populations, leading to unintended and potentially devastating consequences. Aquaculturists are investigating ways to prevent this by engineering sterile species.

Genetically Engineered Livestock

To date, no GM livestock has reached approval and marketing stages, although several examples are currently in research and development phases. These projects often involve the introduction of biochemical pathways that are not naturally present. Engineering livestock to synthesize vital macromolecules such as amino acids is currently an intense area of investigation for agricultural biotechnologists.

The pathway for the amino acid cysteine is not present in mammals so it must be added to their diet. Inserting the cysteine pathway is especially desirable for sheep, as cysteine is a limiting factor in wool production. Experiments adding extra cysteine to sheep diets have proven unsuccessful in improving wool output, as the amino acid is degraded by intestinal microbes. As several bacterial species are able to independently synthesize cysteine, genes isolated from *Salmonella* bacteria may be used to construct a genetically engineered cysteine pathway. The construct includes a promoter and the genes *cysE* and *cysK*. One research project has produced transgenic mice containing the cysteine pathway construct; initial results show the mice are able to synthesize their own cysteine when it is excluded from their diet. However, when the construct is used to transfect sheep, results were not as successful. Cysteine synthesis occurred but expression levels were too low to achieve the desired result of increased wool production. Fine-tuning of the transfection event is still necessary before these transgenic sheep will be commercially viable. Other metabolic pathways are being considered as potential genetic modifications to livestock, including those of the amino acids threonine and lysine.

Food Biotechnology

For over 8,000 years, humans have used microbial fermentation for food production. Numerous food items traditional to the world's oldest cultures are derived from fermentation methods (**FIGURE 8.11**). Food biotechnology is decidedly historic in this sense. Fermentation products include:

- Beverages (alcoholic): Beer, wine, cider, and distilled spirits

FIGURE 8.11 Fermentation technology has been used by humans for millennia. An ancient Egyptian hieroglyphic at the Temple of Abydos depicts the pharaoh Seti I pouring wine.

- Beverages (non-alcoholic): Coffee and tea
- Dairy: Cheese, yogurt, kefir, buttermilk, and sour cream
- Soybean products: Tofu, miso, soy sauce, tamari, and tempeh
- Preserved vegetables: Sauerkraut and pickles
- Other foods: Bread, vinegar, olives, salami

Additional methods obtained from molecular biotechnology are employed in food production, both as ways to facilitate production and as a means to obtain food ingredients. Biotechnology is so widespread in the food industry that the average consumer would find it nearly impossible to avoid the purchase and ingestion of GM foods or their derivatives, especially in the United States where labeling is not required.

Processing

In order to improve the efficiency of food production methods, biotechnologists are investigating GM bacterial strains that are resistant to phage infection. Bacteria used in fermentation are subject to viral bacteriophages, which can decimate active cultures utilized in bioreactors. Virus-resistant strains will eliminate this concern. Also, bioreactors can become contaminated with unwanted bacterial species. GM bacteria that produce toxins able to kill these unwanted contaminants can be added to the bioreactor environment as a method of prevention.

Food biotechnology research is investigating several avenues for improving the efficiency of production methods. The benefits include:

- More environmentally friendly processes
- Reduced energy consumption
- Reduced waste and improved waste treatment processes
- Better food safety via improved evaluation methods
- Bioengineering of natural flavorings and food colorings
- Enhanced enzymes, thickeners, emulsifying agents, and preservatives
- More available and affordable ingredients and additives

Food ingredients and additives are discussed next.

Ingredients and Additives

Bioreactors may be employed to culture bacteria that synthesize food ingredients and additives. One example we have seen is the industrial fermentation of chymosin used in cheese production. This enzyme coagulates milk into curds and whey. Lactase is a microbial enzyme used in products marketed to lactose-intolerant individuals. Amylase is employed in the production of high fructose corn syrup.

Other additives produced by cultured bacteria include flavor enhancers, vitamins, and preservatives. Amino acids can also be added as nutritional supplements. GM strains of *Xanthomonas campestris* are fermented for the synthesis of xanthan gum, a food additive used as a thickening agent or extender. Over 50 enzymes produced by microbial fermentation, recombinant DNA technology, or both are now commonly used in food production.

SUMMARY

- Agricultural products obtained through molecular biotechnology include genetically modified plants used for food and commercial products, farmed seafood, and, potentially, livestock. Agriculture biotechnology may use microbial organisms to improve growth conditions of crop plants.
- Microbial biotechnology is used in agriculture to improve aspects of productivity. Frostban bacteria were a strain of *Pseudomonas syringe* with a transgene for an "ice-minus" protein. The recombinant strain prevented frost damage when applied to crop fields, but due to its negative public image the product was not marketed.
- Another example of microbes used in agriculture is *Bacillus thuringiensis* (Bt), which is applied to crops as a natural insecticide. Bt expresses the Cry protein, which is a crystalline toxin that disrupts the digestive tracts of insects, causing them to die of starvation.
- Transfection methods of introducing transgenes into plant species include the Ti plasmid, the floral dip, and biolistics. *Agrobacterium* species contain a plasmid called Ti (for "tumor-inducing"). The plasmid is able to infect plant tissue, making the Ti plasmid an effective mode of delivery for transgenes. A segment of the Ti plasmid, known as T-DNA, enters the plant cells in a manner similar to bacterial conjugation. By constructing a T-DNA region with a selected transgene, bioengineers are able to insert the desired genetic information into a plant species, thus creating a GM variety. The floral dip method allows plants to grow until they produce floral buds, which are removed. When the floral buds start to reappear, they are dipped in a solution including surfactant molecules and recombinant *Agrobacterium spp.* containing an engineered Ti plasmid. The surfactant molecules facilitate adhesion of *Agrobacterium spp.* to the regenerating floral buds. The exact mechanism is unknown, but transfection occurs in the ovarian tissue of the immature flowers which introduces the transgene into germline tissues. The conferred trait is then passed on to future offspring. With biolistics, gene guns are used to blast transgenes into plant cells, where they either enter the cytoplasm or become incorporated into the plant DNA.
- The Cre/*lox* P technique removes antibiotic resistance genes from GM plant species. Because of a concern that inserted antibiotic resistance genes could transfer to other species and lead to increased resistance, this technique helps alleviate the fear of reduced effectiveness of our existing antibiotics. Through the Cre/*lox* P system, biotechnologists produce GM plant varieties containing only the desired transgene.
- Plant tissue culture produces an entire plant from a single totipotent plant cell. Suspension culture uses liquid media, while callus culture takes place on a solid medium. The first step is to remove the portion of the plant to be cultured, which is called the explant. Next, the explant is treated with plant hormones that will yield a predetermined pattern of growth. The specific hormone treatment depends upon the plant species. When solid media is used, a mass of undifferentiated cells called a callus will form. The callus is treated with additional hormones, first to initiate shoot formation and then to initiate root growth. Now the cultured plants can be transferred to soil for its final growth period. In the case of suspension culture, shoot and root formation occurs simultaneously.
- Some of the genetic modifications used in crop plants are herbicide, pathogen, and insect resistance. Plants can be modified for increased tolerance to environmental stress, such as introducing tolerance to drought or high salinity, or for enhanced nutritional properties as well.
- The industry known as aquaculture farms seafood under controlled conditions. In addition to food products, aquaculture is also used in the production of cultured pearls and algal species such as kelp and seaweed. Some fish are genetically modified for faster growth rates or increased tolerance to freezing temperatures. Currently no GM livestock are available for sale, but several examples are currently in research and development phases. Engineering livestock to synthesize vital macromolecules such as amino acids is currently an intense area of investigation for agricultural biotechnologists.
- Humans have used organisms as living factories in the production of food for millennia. Cultured and fermented foods are examples. Domestication and selective breeding of crops and livestock continued the progression toward increased control of our food supply. Notably, agricultural fields composed of GMO crops have increased 100-fold in less than two decades, and 28 countries across the world plant GM varieties. In addition to agricultural biotechnology, bacteria are used downstream in food processing. Production and preservation methods utilize biotechnology, and some ingredients added to processed food are the products of biotechnology.

KEY TERMS

Aquaculture
Biofortification
Callus culture
Constitutive promoter
Explant
Inducible promoter
Micropropagation
Suspension culture
Transfection

DISCUSSION QUESTIONS

1. Agricultural biotechnology can be seen as a progression of manipulating plants and animals for human use that has taken place for thousands of years, or it may be perceived as a radically new and different technology. What point of view do you take, and why?
2. What types of GMO products utilized in agriculture do you find most valuable to society, and why?
3. Describe potential unintended consequences of adopting GMO crops in agriculture. What safeguards might we employ to prevent such scenarios?

REFERENCES

Bawden, Christopher Simon, et al. "Expression of Bacterial Cysteine Biosynthesis Genes in Transgenic Mice and Sheep: Toward a New In Vivo Amino Acid Biosynthesis Pathway and Improved Wool Growth." *Transgenic Research* 4, no. 2 (1995): 87–104.

Clough, Steven J. and Andrew F. Bent. "Floral Dip: A Simplified Method for *Agrobacterium*-Mediated Transformation of *Arabidopsis thaliana*." *The Plant Journal* 16, no. 6 (1998): 735–743.

Colorado State University. "Transgenic Crops: An Introduction and Resource Guide." Department of Soil and Crop Sciences. Last updated March 8, 2004. Accessed February 24, 2014, from http://cls.casa.colostate.edu/transgeniccrops/current.html

Fisheries and Aquaculture Department. "FishStat Database." Food and Agriculture Organization of the United Nations. (ND). Retrieved February 24, 2014, from http://www.fao.org/fishery/statistics/en

Franz, John E., Michael K. Mao, and James A. Sikorski. *Glyphosate: A Unique Global Herbicide*. American Chemical Society, 1997.

Fuchs, Marc and Dennis Gonsalves. "Safety of Virus-Resistant Transgenic Plants Two Decades After Their Introduction: Lessons from Realistic Field Risk Assessment Studies." *Annual Review of Phytopathology* 45 (2007): 173–202.

Gianessi, Leonard P. and Sujatha Sankula. "The Value of Herbicides in U.S. Crop Production." National Center for Agricultural Policy. April 2003. Retrieved February 24, 2014, from http://www.ncfap.org/documents/ExecutiveSummary.pdf

Gilkerson, Jonathon G., Joseph A. Kelley, and Marcia A. Harrison. "Evaluation of Ethylene Production in Tobacco and *Arabidopsis* Induced by Particle Bombardment." (Tech Note) *Bio-Rad Bulletin* (2009): 5847.

Gill, Sarjeet S., Elizabeth A. Cowles, and Patricia V. Pietrantonio. "The Mode of Action of *Bacillus thuringiensis* Endotoxins." *Annual Review of Entomology* 37, no. 1 (1992): 615–634.

Green, Beverly R., Eran Pichersky, and Klaus Kloppstech. "Chlorophyll *a/b*-Binding Proteins: An Extended Family." *Trends in Biochemical Sciences* 16 (1991): 181–186.

Harding, Margaret M., Pia I. Anderberg, and A. D. J. Haymet. "'Antifreeze'Glycoproteins from Polar Fish." *European Journal of Biochemistry* 270, no. 7 (2003): 1381–1392.

Head, Lesley. "Using Palaeoecology to Date Aboriginal Fish-traps at Lake Condah, Victoria." *Archaeology in Oceania* 24, no. 3 (1989): 110–115.

Losey, John E., Linda S. Rayor, and Maureen E. Carter. "Transgenic Pollen Harms Monarch Larvae." *Nature* 399, no. 6733 (1999): 214.

Naimon, J. S. "Using Expert Panels to Assess Risks of Environmental Biotechnology Applications: A Case Study of the 1986 Frostban Risk Assessments." *Risk Assessment in Genetic Engineering: Environmental Release of Organisms.* McGraw-Hill Environmental Biotechnology Series. New York: McGraw-Hill (1991).

Nayar, Anjali. "Grants Aim to Fight Malnutrition." *Nature* (14 April 2011). Published online.

Pollan, Michael. *The Botany of Desire: A Plant's-Eye View of the World.* New York: Random House, 2001.

Sathyanarayana, B.N. and Dalia B. Varghese. *Plant Tissue Culture: Practices and New Experimental Protocols.* New Delhi: I.K. International Publishing House, 2007.

Sears, Mark K., et al. "Impact of Bt Corn Pollen on Monarch Butterfly Populations: A Risk Assessment." *Proceedings of the National Academy of Sciences* 98, no. 21 (2001): 11937–11942.

Thornton, M. "The Rise and Fall of NewLeaf Potatoes." *Biotechnology: Science and Society at a Crossroad.* National Agricultural Biotechnology Council, Boyce Thompson Institute, Ithaca, New York (2003): 235–243.

White, Philip J., and Martin R. Broadley. "Biofortification of Crops with Seven Mineral Elements Often Lacking in Human Diets–Iron, Zinc, Copper, Calcium, Magnesium, Selenium and Iodine." *New Phytologist* 182, no. 1 (2009): 49–84.

Zambryski, P., et al. "Ti Plasmid Vector for the Introduction of DNA into Plant Cells Without Alteration of Their Normal Regeneration Capacity." *The EMBO Journal* 2, no. 12 (1983): 2143.

9 Forensics and Biodefense

LEARNING OBJECTIVES

Upon reading this chapter, you should be able to:

- Define the terms forensics and biodefense.
- Characterize the role of molecular biotechnology in forensics and biodefense.
- Explain DNA fingerprinting and its uses.
- Summarize DNA fingerprinting techniques.
- Differentiate between exclusivity and probability as they apply to criminal justice proceedings.
- List examples of biowarfare.
- Describe the uses for and mechanisms of biosensors.

In a time when social unrest and global conflicts seem to be growing in number and intensity, the fields of criminal justice and homeland security are escalating defensive protocols and offensively responding with more sophisticated technologies. Molecular biotechnology has completely altered the modern justice and defense systems. Where fingerprints and blood types were once the primary forms of biological evidence, today DNA is the most accepted means in criminal proceedings. The use of fingerprints became generally accepted in the late 1800s, while blood typing has been used for over fifty years. These methods are still valuable and have their place, but their accuracy and reliability are nowhere near the levels reached by DNA analysis. Blood typing may only provide a 25 to 50% probability of a biological match. Furthermore, it is often used as a mechanism to rule out suspects rather than positively identifying one. Comparison of DNA samples via DNA fingerprinting was originally developed in Britain during the 1980s by Alec Jeffreys. Since genetic variation is much more variable than blood types are within the human population, DNA analysis can yield a probability of nearly 100%. Forensic analysis of DNA is useful in other applications in addition to criminal justice.

Molecular biotechnology is also providing tools to military and intelligence agencies for the purposes of defense. The field of **biodefense** uses biological systems to defend populations, both by preventative measures and potential offensive attack technologies. Biodefense uses biotechnology to improve national security and includes the development and detection of biological weapons. Another application is the identification of remains resulting from military casualties and acts of war, including terrorist attacks. It can also take advantage of forensic databases in order to identify and/or track individuals who pose a threat to national security, linking together the areas of criminal justice and biodefense.

Forensics

The term **forensics** refers to science and technology used in the investigation of crimes. For the purposes of this discussion, it can be thought of as the biotechnology of law. Forensic analysis of DNA samples is routinely applied in crime scene investigation as a means to either rule out or positively identify a suspect. By taking advantage of existing DNA polymorphisms found in individuals, we can analyze the genetic code and match biological evidence to their donors. Genetic markers such as repetitive segments are a means of distinguishing different people and can be used to establish paternity, other familial relationships, and the identity of criminal suspects or human remains. As genetic profiles are collected in forensic investigations across the world, databases are increasingly used to help communicate information amongst genetic analysis professionals. The fields of genomics and bioinformatics have contributed greatly to the forensic applications of molecular biotechnology.

With the power of this technology come profound ethical considerations, which are the subject of debate. The main divide is between the concern for privacy and the desire to more efficiently identify (and prosecute, if applicable) desired persons. The privacy camp fears the collection of DNA evidence and cataloguing of genetic profiles of suspects and convicted criminals will provide a slippery slope towards an eventual DNA databank of well, everyone. The argument for such a scenario is the simplicity it would provide in facilitating prosecutions and other forensic investigations. Consider which side is most appropriate as you learn the molecular techniques behind forensic biotechnology.

DNA Fingerprinting

Within each person's cells typically reside 23 pairs of chromosomes, where just over 3.2 billion DNA base pairs—around 6.4 billion nucleotides—may be located. About 98% of these nucleotides are non-coding regions of DNA; in other words, they do not contain genes that code for proteins. Almost all of the human genome, except a meager 64 million nucleotides, was once termed "junk" DNA with no readily observable purpose. Through genomic analysis we are now aware that non-coding DNA contains several locations of repetitive segments that are highly variable amongst individuals. Known as **genetic markers**, they are solid evidence for the identification of a particular person. Sampling and analyzing a person's DNA sample in order to detect genetic markers creates a **genetic profile**, also known as a **DNA fingerprint**.

Because the DNA in someone's skin or bone is the same as the DNA in that person's saliva or blood, different biological samples will yield the same DNA fingerprint. This makes DNA fingerprinting especially useful in forensic analysis, where it is often necessary to compare DNA samples, such as those found at a crime scene compared to those collected from different suspects and/or victims. The DNA among different people is composed of unique nucleotide base sequences, distinguishing one person from another. This is compounded by the convincing nature of mathematical probability. Forensic analysts can readily assign a numeric value to the likelihood of a genetic match between biological samples, which gives additional validity to this field. Because DNA fingerprinting is such compelling evidence, it can also be used to successfully determine familial lineage and identify victims of crime or disasters, whether natural or human-made.

Genomic Tools

Genetic markers only account for about one-tenth of one per cent of a person's DNA. The technology used to analyze these minute polymorphisms must be incredibly precise. Restriction enzymes are irreplaceable as a means to target polymorphism regions for further analysis. Forensic biotechnology must also be able to generate results with miniscule quantities of DNA if necessary. If only a tiny drop of blood or just one strand of hair is available, methods to amplify the small sample amounts are employed to create a workable quantity for subsequent testing methods.

The discussion on genomics presented a number of genetic markers found in the human genome. Polymorphisms include copy number variants, single nucleotide polymorphisms, and dispersed repetitive sequences. Two genetic markers in particular are of prime interest to forensic analysts: short tandem repeats (STRs) and restriction fragment length polymorphisms (RFLPs). Short tandem repeats are a type of tandem repeat polymorphism belonging to a class of genetic markers known as microsatellites. The two terms are essentially interchangeable; **microsatellites** are one- to six-nucleotide repeats dispersed through the genome, qualifying them as STRs. The most commonly identified STR in DNA fingerprinting is the **tetranucleotide repeat**, which consists of a four base pair sequence that repeats anywhere from 5 to 50 times. RFLPs are a unique form of single nucleotide polymorphism. Each of the two preferred genetic markers are associated with a particular DNA fingerprinting technique.

Testing Methods

Through the course of practicing and refining DNA fingerprinting, two testing methods have emerged as the most frequently used and most generally accepted forms of genetic profiling. The first method utilized for this purpose was the Southern blot, which focuses on restriction fragment length polymorphisms. The more recently introduced method is the polymerase chain reaction, which is best suited to small biological samples and analyzes short tandem repeats. Because Southern blotting compares RFLPs, the method yields data on fragment lengths that vary among individuals. With polymerase chain reaction (PCR), data are not based directly upon DNA fragment length. Instead, this method provides information on the presence of STRs, including the number of repeats in each sample, which do vary by fragment length.

Southern Blot

The Southern blot method of DNA fingerprinting was the first to be developed and practiced. Recall that restriction fragment length polymorphisms are genetic markers that occur at restriction sites, so they can be detected with restriction enzyme digestion. This is the first preparatory step for the Southern blot technique. Each known RFLP in the human genome is cut by a specific restriction endonuclease; thus, there are many restriction enzymes available for Southern blotting (**TABLE 9.1**). In preparation for the DNA fingerprint, the analyst must decide whether to probe for a single RFLP or for multiple sites. Detection of a single RFLP site is known as **single-locus probing (SLP)**, whereas a DNA fingerprint locating more than one RFLP is known as **multiple-locus probing (MLP)**. MLP was used prior to SLP, although SLP is now preferred. The reasons why will be explained shortly. Once the probing method is decided upon, a forensic biotechnologist will select the restriction enzyme(s) that target the RFLP(s) to be identified. The Federal Bureau of Investigation (FBI) focuses on four specific RFLPs for use in case comparisons. An additional preparatory step now occurs. Following enzyme digestion, the resulting DNA fragments are subjected to gel electrophoresis.

TABLE 9.1

Types of Restriction Endonucleases

Type	Structure	Cofactor(s)	Recognition Sequence	Cleavage
I	R_2M_2S	ATP (hydrolysis) AdoMet, Mg^{2+}	Asymmetric interrupted	Cut DNA at sites distant from the recognition sequence
Example	*Eco*B		-TGA(N_8)TGCT-	
II	R_2	Mg^{2+}	Palindrome 4–8 bp	Within the recognition sequence to produce blunt or staggered ends
Example	*Eco*RV		-GATATC-	
III	RM	ATP (no hydrolysis) Mg2+, AdoMet	Asymmetric	Cut DNA close to the recognition sequence
Example	*Eco*PI		-AGACC-	

Abbreviations: R, restriction endonuclease subunit; M, modification subunit; S, specificity subunit; AdoMet, S-adenosylmethionine.
Data from A. Pingoud, and A. Jeltsch, *Eur. J. Biochem.* 246 (1997): 1–22

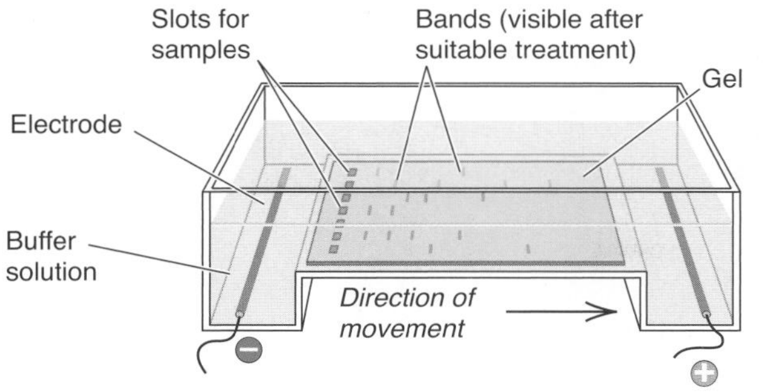

FIGURE 9.1 Gel electrophoresis applies an electrical current to DNA samples. Negatively charged DNA travels toward the positive pole based on size: smaller fragments travel faster. The resulting bands indicate relative sizes.

DNA is a negatively charged molecule and therefore it will travel toward the positively charged pole of an electrophoresis chamber (**FIGURE 9.1**). This process uses an electrical current to separate the fragments according to size: smaller fragments will travel faster through the gel and larger fragments travel more slowly. In other words, the farther the distance traveled, the smaller the fragment (**FIGURE 9.2**). The fragments are now visualized through the actual Southern blotting step.

The fragments are transferred to nylon paper, whereupon radioactively-labeled DNA probes are applied. The DNA probes will bind to any complementary DNA present in the fragments. The nylon paper is then covered with a radioactive-detecting film, creating the final product of Southern blotting (**FIGURE 9.3**). The **autoradiograph** resulting from Southern blotting visualizes all DNA fragments bound to the radioactive probes. The observable locations are referred to as "bands." In SLP, one or two bands will appear. When two bands are present, this indicates a heterozygous genotype (two variants of an RFLP are present). When one band is exhibited, the person must be homozygous for the RFLP. When homozygous, the person possesses two copies of a single variant of the RFLP. In such instances, the single band is thicker since it is actually two bands on top of each other in the same location. The limited forms of results obtained from SLP make statistical analysis and interpretation relatively straightforward. On the other hand, MLP fingerprints result in a plethora of bands. This makes interpretation much more difficult because individual RFLPs cannot be distinguished (**FIGURE 9.4**). For this reason, SLP is most commonly used today. Autoradiographs are typically produced to include DNA markers, a control, and experimental samples. The **DNA marker** or standard contains fragments of known size and serves as a "ruler" to the other samples (**FIGURE 9.5**). A control sample provides DNA from a source that is well known to exhibit high fidelity in Southern blotting. When the control yields predicted results, this ensures a level of certainty that the blot was carried out successfully. The experimental samples are those being probed and may include DNA from crime scene evidence, suspects, victims, or other persons of interest.

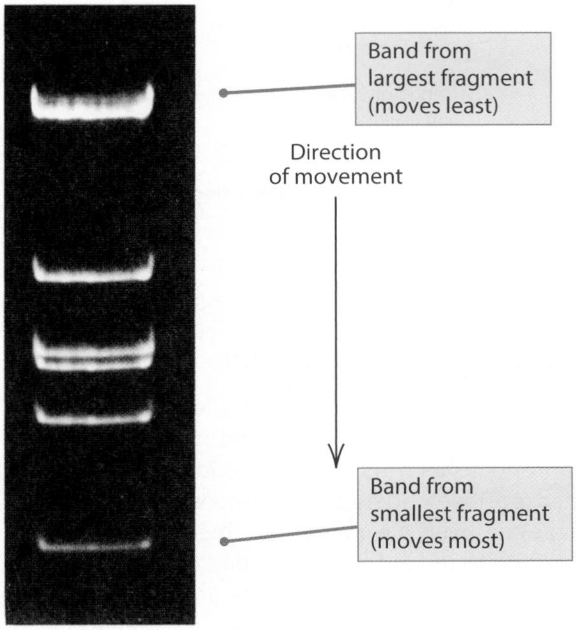

FIGURE 9.2 The bands of DNA resulting from gel electrophoresis. The farther the distance traveled, the smaller the DNA fragment. Courtesy of Daniel Hartl

Polymerase Chain Reaction

PCR is a preferred method employed in DNA fingerprinting when biological samples are sparse. In cases where the DNA may be degraded, PCR is the more effective choice. PCR amplifies DNA samples through exponential rounds of replication. It is possible to analyze the DNA from a single cell using PCR. PCR is typically used to detect short tandem repeats in DNA samples. While STRs are highly variable amongst individuals, the regions of DNA to either side of these polymorphisms are highly conserved in the human genome. These areas are known as **flanking regions** (**FIGURE 9.6**). PCR primers are commercially available for the various known flanking regions, making it easy to detect and amplify STRs in a biological sample. Thus, only STRs and their flanking regions are amplified through PCR methods in order to isolate the variable areas. In fact, it is possible to amplify several STR loci through multiple PCR reactions all at once. Known as **multiplex PCR**, commercial kits are available that contain multiple primers for various flanking regions. Once PCR is complete, the amplified DNA samples are then probed to determine STR presence and quantities thereof. The probing step is similar to Southern blotting,

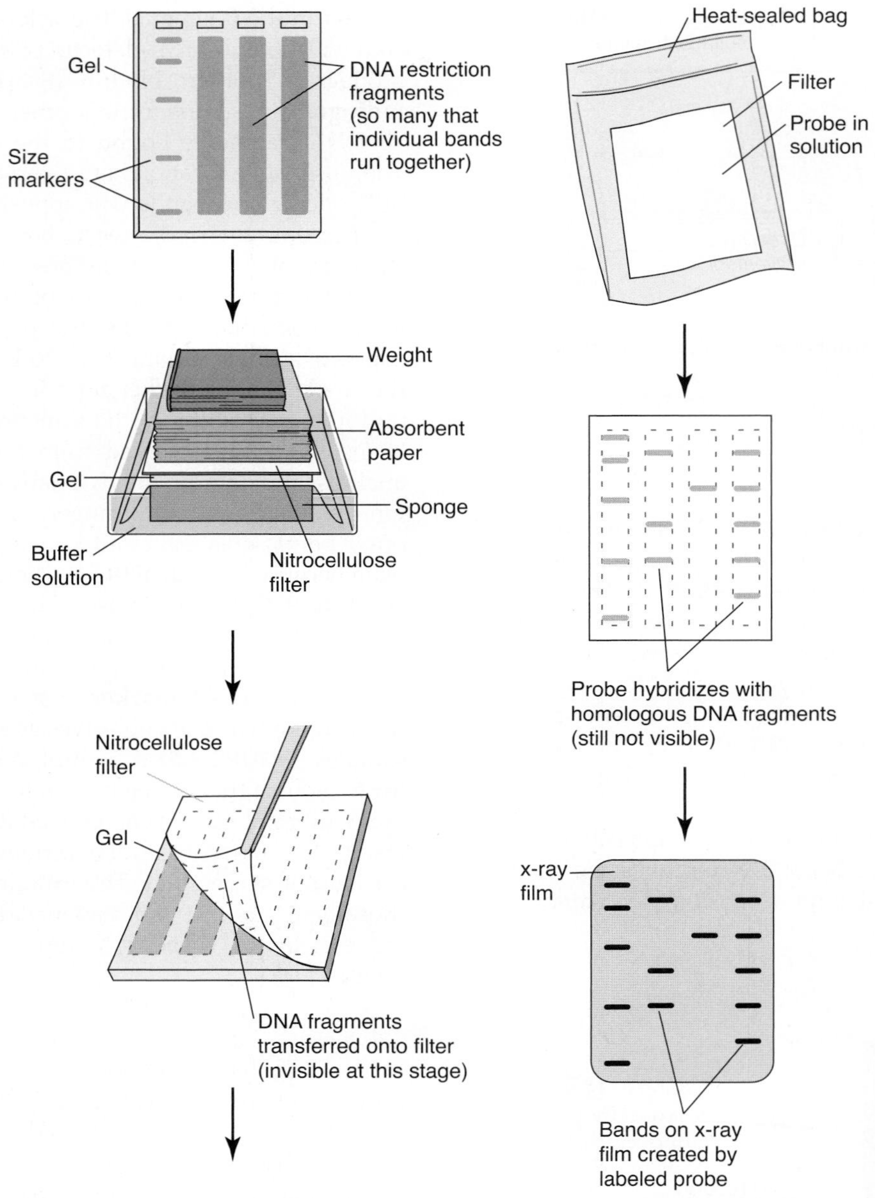

FIGURE 9.3 The steps of Southern blotting results in an autoradiograph ("x-ray film"), a form of DNA fingerprint.

as the sample DNA will bind to complementary probes and result in a visual match. The match indicates the presence of a known STR and the number of matches correlates to the number of STR repeats present in the sample. The amplified DNA regions resulting from PCR may also be sequenced and then analyzed through computational genomics. Both probing and sequencing will allow the comparison of different DNA samples, thus typically only one of these methods is necessary for analytical interpretation. Thirteen STRs were chosen by the FBI in the creation of their DNA profile database **CODIS (Combined DNA Index System)**, described below.

Exclusivity and Probability

As we have seen, traditional identification methods such as fingerprinting and blood typing are often more appropriate as a means to rule out individuals rather than positively identifying someone. For example, we can definitively say person A's fingerprints don't match those collected at the crime scene much easier than we can say person B's fingerprints are a match. This was especially true prior to computer-aided analyses, when specialists known as latent print experts would rely on visual comparisons and metrics obtained with hand tools. The same is true for blood typing. A person with

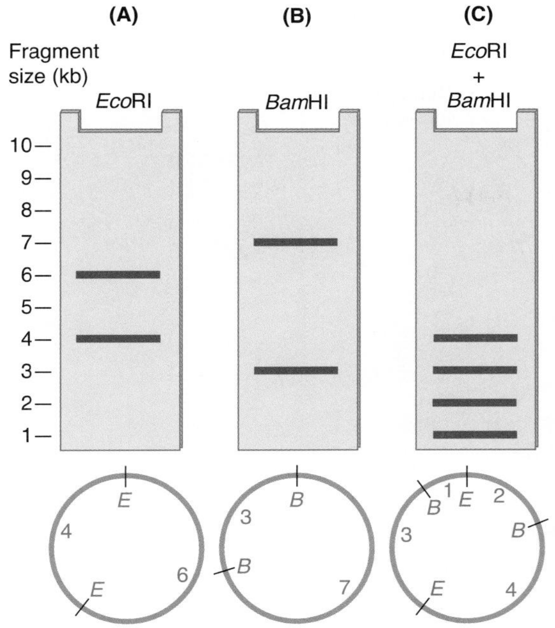

FIGURE 9.4 Results of single-locus probing compared to multi-locus probing. Gel electrophoresis bands are shown in blue and restriction maps produced by each digest are shown directly below. (A) A SLP using the restriction endonuclease EcoRI results in two bands, indicating a heterozygous genotype. (B) A SLP using the restriction endonuclease results in two bands, indicating a heterozygous genotype. (C) A MLP using both enzymes results in four bands, but without results from A and B, it is not possible to tell which bands are the result of which enzyme digest.

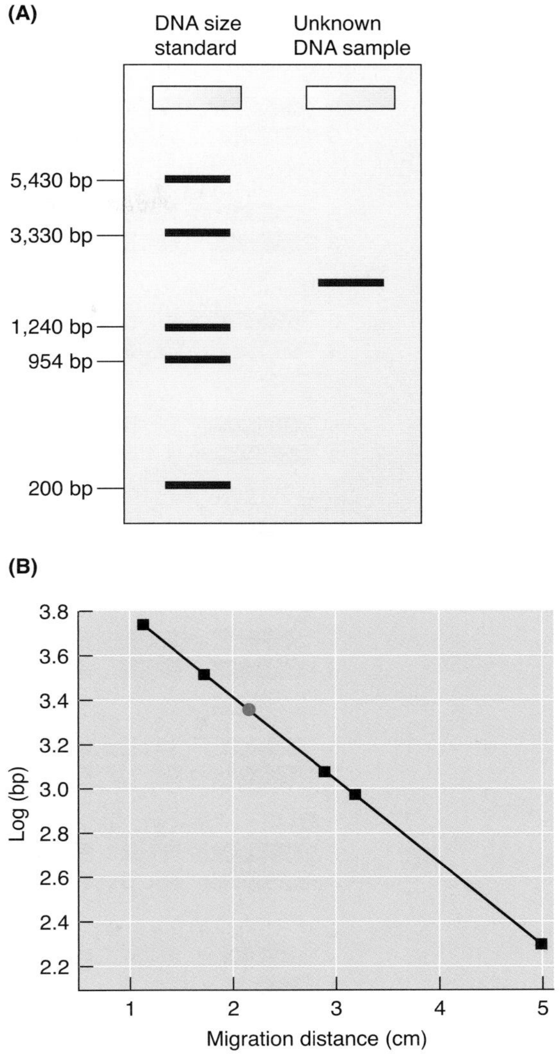

FIGURE 9.5 Inclusion of a DNA marker or standard in the Southern blot technique allows the determination of unknown DNA fragment sizes. (A) The marker on the left acts as a ruler to compare known sizes to the unknown sample on the right. (B) Known sizes are plotted in black and used to estimate the unknown size in green. Modified from illustrations by Michael Blaber, Department of Biomedical Sciences, Florida State University. http://www.mikeblaber.org/oldwine/bch5425/lect20/lect20.htm. Accessed March 1, 2014.

type B blood is likely not the murderer when type A is found at the crime scene. Ruling out individuals is a form of **exclusivity**; they are excluded from the pool of potential suspects.

Forensic analysis of DNA provides much more compelling evidence in the form of a positive match. Instead of simply ruling out a person, a positive match identifies someone with a high degree of certainty. There is always the possibility that by random chance two people have the same genetic profile; this is especially concerning when two relatives are involved because they have a greater likelihood of possessing the same genetic markers. However, in most cases the chances are astronomically low. Some polymorphisms have hundreds of different variants, which increase the likelihood of two people possessing strikingly different genetic profiles. Measures of probability seek to provide the numeric value of this chance.

In order to determine a level of probability, forensic analysts will first compare a person's DNA to a random sample of DNA from members of the same ethnicity. Comparative genomic studies have revealed that each ethnic group of the human population has distinguishable frequencies of the different known genetic markers. Frequency data from various ethnicities are collectively known as **reference databases**. Most of these data were obtained from volunteers as part of ongoing efforts to increase the sample sizes of different ethnic groups. The larger the sample size, the more statistically relevant the data becomes. Comparing DNA to the same ethnic group narrows down the chances of random differences. It creates the appropriate conditions for **random match probability**, defined as the probability that a given DNA sample will match a random sample obtained from the same ethnic population. Analysts then compare the sample DNA to the

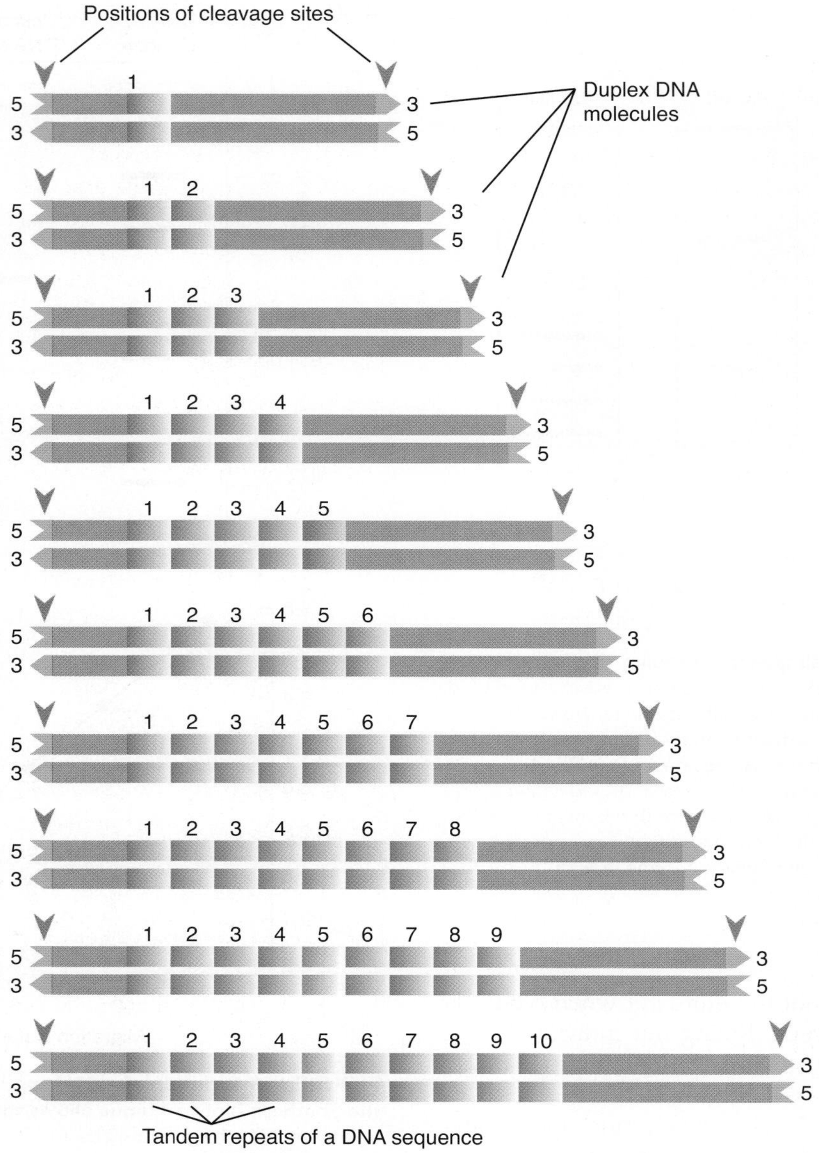

FIGURE 9.6 Variability of polymorphisms seen in a tandem repeat genetic marker. Tandem repeats are seen in red and vary from one to ten repeats in a given DNA sample. Flanking regions are conserved regardless of the repeat number (shown in blue).

frequencies of detected genetic markers in the ethnic group. For example, it may be determined that ethnic group A, to which the DNA sample belongs, has one particular genetic marker at a frequency of 80%. This step is repeated for each genetic marker under investigation. The product rule is applied as a final step in determining overall probability:

Probability of Events 1–3 = (Probability of Event 1) × (Probability of Event 2) × (Probability of Event 3)

Each "event" refers to the presence of one genetic marker variant. Multiplying the probability of each unique polymorphism present in the DNA sample will give the overall probability of these same polymorphisms being present in combination. This gives a numeric indication of how likely this specific combination of polymorphisms, that is, the individual's genetic profile, would be in the given ethnic group. The greater the number of genetic markers included in the product rule calculation, the smaller the random match probability becomes. The probability is usually given as the odds of a genetic profile match between the sample and a randomly selected member of the same ethnic group. The value can be as low as one in one quintillion (1 in 1,000,000,000,000,000,000), representing a nearly 100% certainty of a match.

Applications

Forensic analysis of DNA is useful in criminal justice to exclude or positively identify a suspect. While thousands of cases have relied on DNA evidence for criminal prosecution, paternity disputes are actually a much more frequent application of DNA fingerprinting. Genetic profiles are often stored in databases for easy sharing and retrieval, paving the way for DNA profiling of individuals. For privacy reasons, storage is currently limited to convicted criminals and unidentified DNA samples obtained as evidence. As we saw with pharmacogenomics, neonatal testing of DNA and its subsequent entry into medical records is under consideration. If this were to become routine or even legally mandated, we would have a national DNA fingerprinting system in place. The ethical implications of such a scenario raise many concerns.

Crime

Forensic analysis of crime scene evidence is an overwhelmingly conclusive method of investigation. The integrity of DNA is a function of its age and exposure to a variety of environmental factors. DNA degrades over time, which is amplified in humid or wet conditions, in the presence of sunlight, and/or extreme pH or temperatures. Because DNA has a shelf life, collecting and preserving samples as quickly as possible is desired, as is preventing contamination of the sample by ensuring a documented chain of custody. This was especially the case when Southern blotting was the

FOCUS ON CAREERS

Forensic Scientists

With popular television shows like *CSI* and *Dexter* bringing the field of forensic investigation into our homes, it is not surprising that crime scene investigation and other jobs associated with forensic analysis have become popular career choices in recent years. Along with its popularity, however, comes some degree of misleading and unsuitable training programs. Becoming a forensic scientist requires meeting the specific expectations of hiring authorities, so it is wise to become familiar with the exact requirements of one's desired future employer before deciding on the most appropriate education. Seeking a program accredited by the Forensic Science Education Programs Accreditation Commission (FEPAC) is recommended. A list of accredited programs is available at http://fepac-edu.org/accredited-universities.

Job positions exist beginning at the bachelor degree level of education. As with most careers, the higher the degree level, the higher the earning potential and the opportunities for advancement. Degrees in hard science are preferred, such as biology or chemistry. An ideal forensic scientist will be rational and unbiased, possess extraordinary attention to detail, and be at ease working within the broader context of the criminal justice system. Major components of this line of work are to maintain accurate records, apply stringent quality control of crime scene and laboratory analysis, and provide reliable documentation and testimony. Ethical integrity is equally as important as technical ability and scientific knowledge. Likewise, it is important upon choosing this career path to realize that the forensic investigation portrayed in entertainment outlets do not paint a realistic picture, and that the true nature of being a forensic scientist is not always a glamorous or high-profile proposition.

Multiple professional organizations exist to empower and connect individuals in forensic careers, including the American Academy of Forensic Sciences, the Forensic Sciences Foundation, and the International Association of Forensic Science. There are also a variety of specialized forensic organizations that are more specific to the subdisciplines of the science, such as pathology, toxicology, and genetics. Several of the subdisciplines promote their own boards for certification of professionals in addition to requiring continuing education of their certified members. Forensic scientists may be employed by state or federal law enforcement agencies or work in academic settings as educators and/or researchers. Some work for medical examiners, hospitals, or corner's offices, or it is possible to work as an independent consultant. For budding forensic scientists, the Young Forensic Scientists Forum is a highly useful resource for learning more about specific career paths and educational programs. Visit their website at http://yfsf.aafs.org/ to learn more. The American Academy of Forensic Sciences (http://www.aafs.org/) and the Forensics Sciences Foundation (http://fsf.aafs.org/) also offer career information online.

only DNA fingerprinting method available. With the advent of PCR, it is also now possible to analyze partially degraded DNA. Biological evidence that has sat in storage for years may be damaged, but PCR allows amplification of small and/or incomplete DNA. Cold cases—those unresolved due to exhausted investigation options—can be revisited to include DNA testing that may not have been available when the crime was committed. DNA databases such as CODIS are also an important tool for identifying repeat offenders. In 1985, the United Kingdom established the first DNA database, known as NDNAD (National DNA Database). Other countries which maintain their own DNA databases include France, New Zealand, Canada, Sweden, Germany, and Norway. The reasons for including an individual in a database vary by country. Once a convicted criminal is in the database, that person's genetic profile will remain. With over 13 million entries, CODIS is the largest DNA database in the world. Law enforcement officials can compare an existing genetic profile to new evidence obtained from separate crimes, or even match evidence from previous crimes that occurred in different geographic areas. By linking distant jurisdictions, criminals have a much harder time remaining hidden from the law, even if they relocate. Some states such as Virginia go a step further and store genetic profiles of people convicted of nonviolent crimes and/or individuals who are arrested but not convicted. To learn more about careers in crime scene analysis and forensic investigation, see the *Focus on Careers* box.

Forensic analysis of DNA is also being used to exonerate convicted criminals. The first convicted felons exonerated based upon DNA evidence were released in 1989. In that first year, a total of 20 individuals were exonerated. Since that time, over 1,300 convicts have been exonerated (**FIGURE 9.7**). In effect, molecular biotechnology helped to save these lives. Securing DNA testing is the responsibility of the inmate and costs are prohibitive. Furthermore, the criminal prosecutor(s) involved with the original case must approve an inmate's request for DNA testing. Organizations such as the Innocence Project have the goal of financially and legally assisting inmates, particularly when they were convicted based upon weak or insubstantial evidence. Solidifying the importance of DNA analysis in the criminal justice system is the number of truly guilty individuals identified as a result of exoneration. The Innocence Project notes that close to 50% of exoneration cases were ultimately solved through additional DNA testing.

Paternity

Just as genes are inherited from both parents, so too are polymorphisms. Therefore, every person possesses some genetic markers from his or her mother and some from his or her father. They are inherited as pairs in the same manner genes are passed down from each parent. DNA fingerprinting technology may be used to determine the identity of a biological father when paternity is under dispute. DNA from the mother, child, and suspected father(s) are collected and subjected to DNA fingerprinting. The child's genetic profile is compared to the mother; any matching polymorphisms were inherited maternally. The remaining unmatched genetic markers found in the child's DNA fingerprint are compared to those of the suspected father(s). A polymorphism match that is unique to father and child reveals a positive paternity test. DNA fingerprinting is a more reliable method of paternity testing compared to blood typing, as blood typing has the potential to be inconclusive. For example, two suspected fathers might have the same blood type. Because genetic markers are much more variable than blood type, DNA fingerprinting eliminates any uncertainty in paternity testing. Probability again plays an important role in determining the extreme unlikelihood that a suspected father might coincidentally have the same genetic markers as another man in addition to the child.

In much the same way as paternity testing, familial relationships can also be established through forensic analysis of DNA (**FIGURE 9.8**). For example, several self-described descendants of Thomas Jefferson have identified themselves in recent years. They claimed

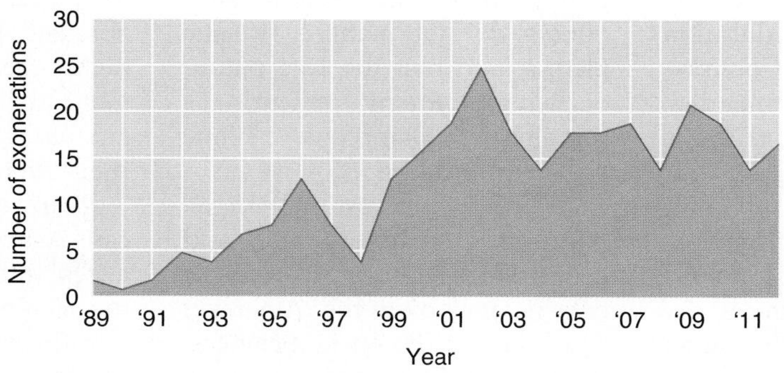

FIGURE 9.7 Graph showing the number of exonerations due to DNA evidence by year from 1989 to 2011 in the United States.
Data from The Innocence Project.

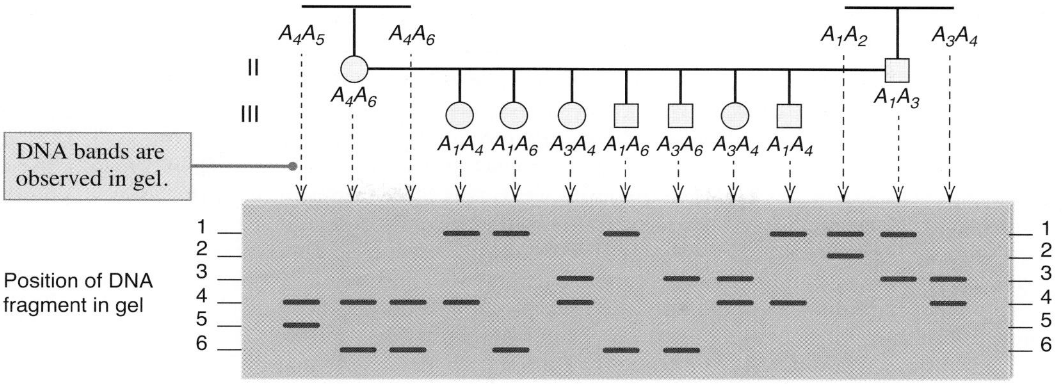

FIGURE 9.8 A human pedigree shows the presence of six variants of a repetitive genetic marker. Only one or at most two markers can be inherited by any one member of this family.

to be descended from Jefferson and his female slave, Sally Hemings. DNA testing revealed only one of two claimant families was truly related to our founding father. Forensic analysis has also been used to solve the mystery of the Romanovs, the last royal family of Russia (**BOX 9.1**).

Profiling

Profiling individuals based upon their genetic make-up bridges the areas of forensics and biodefense. While information might be obtained for forensic investigations, the continued availability of the information in databases such as CODIS opens the door

Box 9.1 Exhuming the Romanovs

In July 1918, the Bolshevik revolution in imperial Russia culminated with the execution of the last royal family. A team of soldiers invaded the palace and executed Tsar Nicolas II, his family, and servants by firing squad. Following the execution, the bodies were buried nearby. The mass grave was rediscovered in 1989 and nine skeletons were excavated two years later. The royal family included the Tsar, his wife Tsarina Alexandra, and five children: Olga, Tatiana, Marie, Anastasia, and Alexei. Simple math would suggest the nine skeletons included the family and two servants. However, upon forensic analysis it was found that four servants were buried in the mass grave. Anastasia and Alexei were not found with the remains.

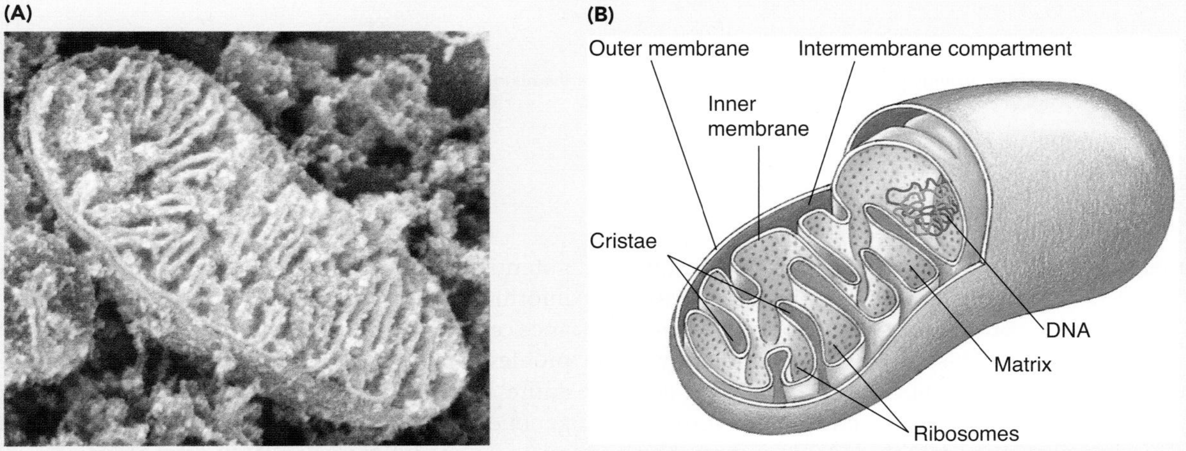

The mitochondrion. (A) An electron micrograph of the cellular organelle, present in eukaryotic organisms. (B) A schematic drawing of the mitochondrion shows the presence of mitochondrial DNA, inherited through the maternal line.

(continues)

Box 9.1 Exhuming the Romanovs (continued)

Several women worldwide came out stating they were the true Anastasia, having escaped the execution and fleeing Russia. Forensic analysis, especially DNA fingerprinting, falsified all of these claims. In this case, however, genetic markers were used in concert with **mitochondrial DNA (mtDNA)** analysis. Mitochondria are organelles found in eukaryotic cells, such as human cells, that possess their own DNA, separate from nuclear DNA. While both parents contribute to the inheritance of nuclear DNA in the form of two sets of 23 chromosomes, mtDNA is inherited through the mother's egg. Sperm lack mtDNA as a trade-off for more energy and speed. Thus, mtDNA is exclusively maternal in origin. Investigators were able to obtain samples of mtDNA from a living descendant of Tsarina Alexandra: Prince Philip, Duke of Edinburgh, husband to Queen Elizabeth II of England, and great nephew of Alexandra. It was necessary to find a relative on the maternal side of the perished royal family in order to trace mtDNA. This DNA is conserved over generations because it does not recombine with DNA from the father, as is the case with nuclear DNA. The investigators hypothesized that mtDNA of Prince Philip should be a match with that of Tsarina Alexandra, as well as her children. Comparisons of Prince Philip's mtDNA with the royal remains confirmed their identities; comparison with the women claiming to be Anastasia proved their allegations were all false. It is now believed that both Anastasia and Alexei were indeed executed, but their bodies were burned beyond recognition. Forensic analysis settled a mystery that had been unsolved for decades and squashed any aspirations to deceptively claim royal lineage and the rights assigned to such a designation. Tsar Nicolas II, Tsarina Alexandra, and their three identified children, Olga, Maria, and Tatiana, were reburied in 1998 at the Peter and Paul Fortress of St. Petersburg in a grand funeral attended by over one million people, finally laying to rest the last royal family of Russia.

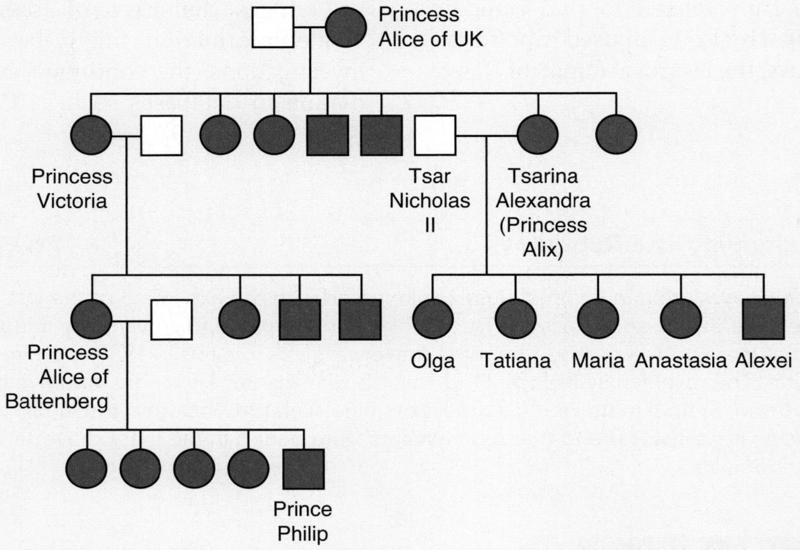

A pedigree of royal families, including Russian and British relations, shows the maternal lineage of Tsarina Alexandra. Members shown in purple share mtDNA with Alexandra.

for homeland security operations such as counterterrorism as well as other intelligence activities. Having access to the genetic profiles of persons of interest is a major advantage. It may also be possible to identify a person through a family member's DNA that has been stored in the database. Because genetic profiles compile genetic marker data, they mainly include information from non-coding DNA. Therefore, a person's genetic make-up, in the form of their coding DNA (actual genes), remains private. This is a distinct characteristic to highlight when faced with opposition from privacy advocates. Regardless of this assertion, opponents of a national database point out the information could potentially be misused. Hiring managers might use the information to screen potential job applicants, insurance companies could discriminate against certain profiles when assigning coverages and premiums, or entire subsets of the population could be subject to genetic oppression. The concept of eugenics once again raises its head. Time will tell the ultimate fate of genetic databases used for profiling.

Biodefense

Biodefense is the use of biotechnology in improving national security. The terrorist attacks on the World

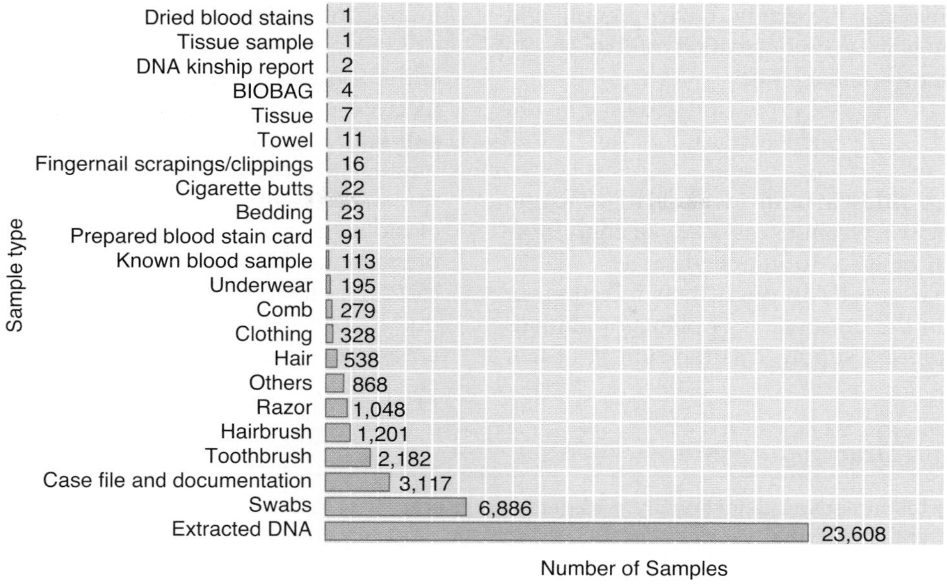

FIGURE 9.9 Types of biological samples collected at the World Trade Center site and their relative numbers as reported by the New York City Chief Medical Examiner. Modified from the President's DNA Initiative: DNA Identification in Mass Fatality Incidents. Data from: New York City Office of the Chief Medical Examiner/Department of Justice

Trade Center on September 11, 2001 (9/11), set into motion rapid innovations in this field. The "war on terror" and the creation of a new Cabinet department, Homeland Security, demanded cutting edge technologies that would position the United States at the forefront of biodefense initiatives. Notably, the identification of the World Trade Center's victims was one major task aided by biotechnology. The limited remains found at Ground Zero amongst 1.5 million tons of debris made visual identification impossible, so forensic methods became invaluable. Family members of victims played a crucial role in providing biological samples of their loved ones that were used in comparative analyses of the thousands of bone fragments, teeth, and tissue samples collected at the site (**FIGURE 9.9**). Even with advanced forensic methods, fewer than 1,700 of the over 2,800 victims were ultimately identified through DNA profiling methods (**FIGURE 9.10**). Nevertheless, the enormous efforts carried out following the 9/11 attacks led to the development of Mass Fatality Identification System software. The same software is being used in the DNA Shoah Project to identify victims of the Nazi Holocaust with the aim of reuniting long lost family members. Similar methods are employed to identify military casualties, perhaps most dramatically exemplified by the identification of remains held in the Tomb of the Unknown Soldier. DNA analysis confirmed the remains belonged to an Air Force pilot shot down over Vietnam in 1972 (**FIGURE 9.11**). In addition to identifying remains of war and other disasters, the field

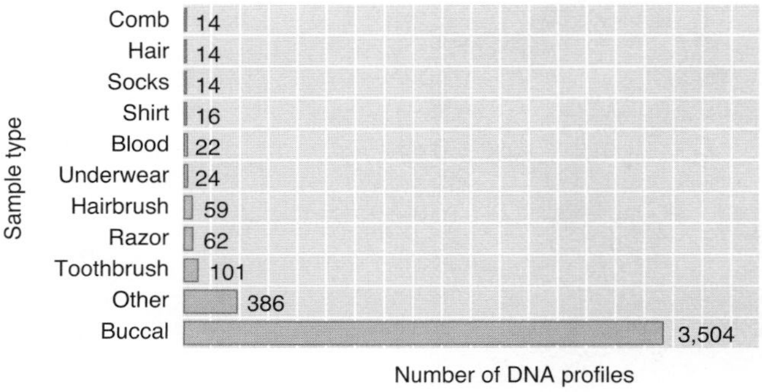

FIGURE 9.10 The samples collected at Ground Zero and the relative quantities of DNA profiles constructed from each type. Modified from the President's DNA Initiative: DNA Identification in Mass Fatality Incidents. Data from: New York City Office of the Chief Medical Examiner/Department of Justice

FIGURE 9.11 While the inhabitant of the Tomb of the Unknown Soldier has been identified, hundreds more have been laid to rest in Arlington National Cemetery. Forensic DNA techniques have the potential to rectify their anonymity. © jcarillet/iStockphoto.com

of biodefense involves the study, development, and prevention of biowarfare and creation of biosensors designed to detect warfare agents.

Biowarfare

Biological warfare (**biowarfare**) involves the creation of biologically active weapons as well as their storage and eventual delivery to intended targets. Agents may target people directly through their infection, or indirectly through the decimation of plants or animals that make up the food supply. It is the subject of fear mongering and psychological threats much more than it is an actual method of engagement. Consider the anthrax letters of 2001: the attack killed five people but received enormous media attention (**BOX 9.2**). Other naturally spreading pathogens kill many times that number of people each year but are not sensationalized in the media. Historically, many alleged incidences of biowarfare are purely speculation or even a form of propaganda. Furthermore, even the well documented cases of biowarfare had little impact. This could be due to a misunderstanding of the biological action of an agent. During the Middle Ages, Tartars catapulted

Box 9.2 Tracking Anthrax

Biosensors allow for immediate detection of biowarfare agents and other toxins. This is well and good for applications operating in present time. When an agent's history is in question, other methods must be employed. In the wake of the World Trade Center attacks of 2001, the U.S. Postal Service became the distribution vehicle for a bioattack. Anthrax spores were sent in envelopes to two U.S. senators—Tom Daschle of South Dakota and Patrick Leahy of Vermont—and two media outlets, NBC News and the New York Post. *Bacillus anthracis* (anthrax) is a deadly bacterial pathogen listed as a Category A bioterrorism agent by the Centers for Disease Control, representing its most stringent level of containment. It became clear that whoever sent the anthrax letters must have had security clearance to obtain the spores. In order to identify the perpetrator, considered a terrorist, the origin of the anthrax spores had to be determined.

The Federal Bureau of Investigation, in cooperation with the CDC, sequenced the genome of the spores sent in the mail. Because all genomes mutate and evolve, investigators hypothesized the anthrax used in the attacks would exhibit genetic evidence of such mutations. Generally, microbes undergo mutations every time they are exposed to a new environment or undergo stress, such as when strains are relocated to different laboratories. This would theoretically produce a genetic "time stamp," allowing investigators to track the movement of all existing anthrax strains. As anthrax is a controlled organism that is only permissible in certain laboratory locations, this allowed the FBI and the CDC to narrow down all possible sources. Existing stocks of anthrax were likewise sequenced and genomic comparisons, along with electron microscopy analysis, aided in determining the true source of the anthrax spores sent in the mail. In total, four distinct mutations became the key evidence in creating a traceable path of movement from the original laboratory to the final handler, a U.S. Army researcher with access to anthrax named Bruce Ivins. His guilt is still in question as he was never tried nor convicted; Ivins committed suicide soon after a previous suspect, Dr. Steven Hatfill, was exonerated. This final act is seen by some as proof of Ivan's guilt, but unfortunately we will never be absolutely certain. The FBI did award Hatfill a damages settlement of $5.8 million for the false accusations against him, yet his career was never the same. Nevertheless, the anthrax letters of 2001 enabled various agencies to improve their response capabilities when faced with a bioattack.

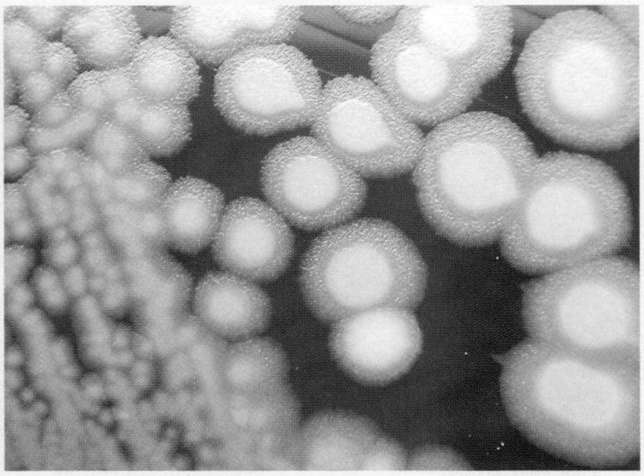

Bacillus anthracis **growing in culture on sheep's blood agar.** Anthrax infects herbivorous mammals such as sheep, in addition to humans. Courtesy of Larry Stauffer, Oregon State Public Health Laboratory/CDC

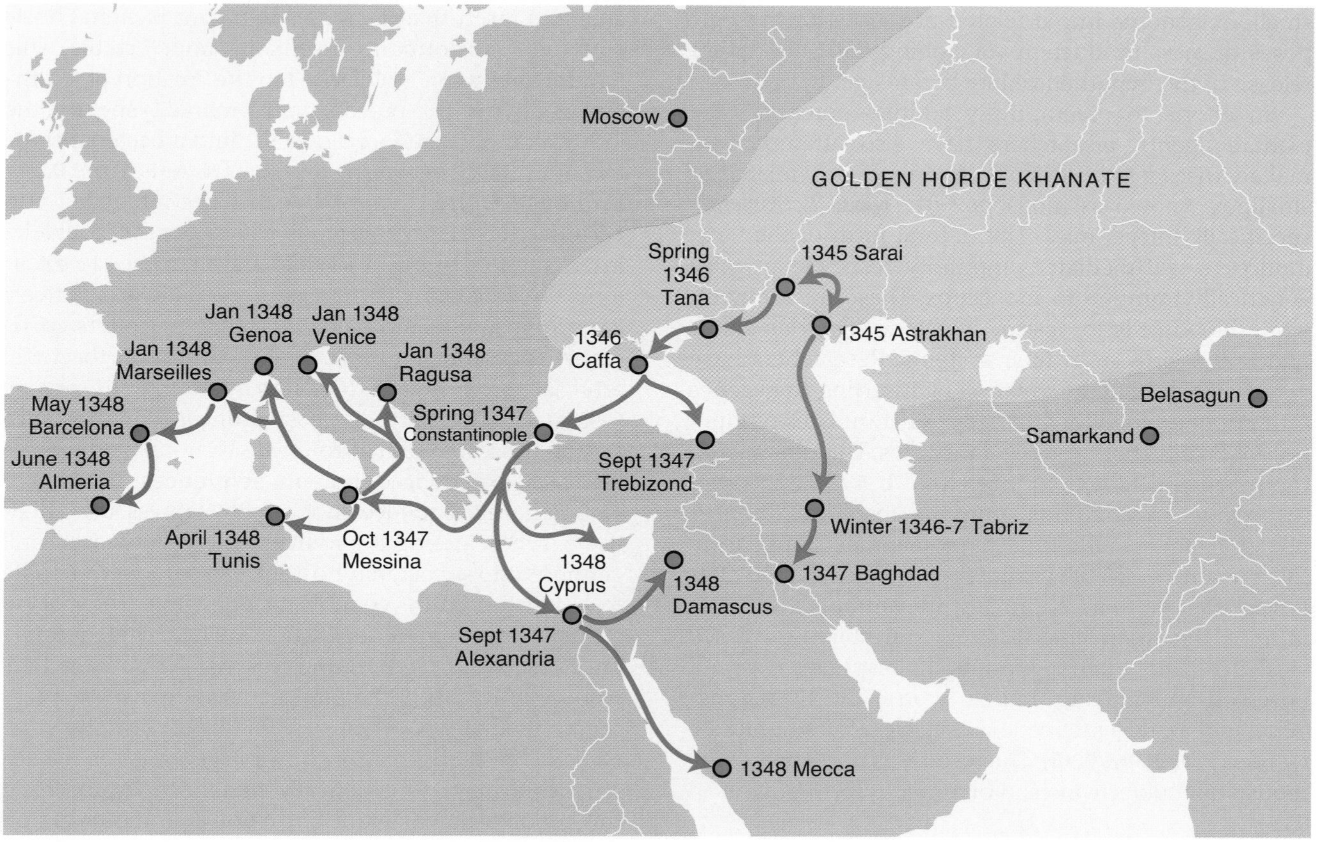

FIGURE 9.12 During the war at Caffa in 1346, Tartars used trebuchets to catapult plague-ridden corpses toward their enemies. This map traces the chronology of the spread of the plague throughout Europe, as hypothesized by biological warfare researchers. Note the origin point at Caffa.

corpses riddled with Bubonic plague into walled cities in an attempt to infect inhabitants (**FIGURE 9.12**). There was no negative effect though, because the plague (the bacterium *Yersinia pestis*) spread through fleas carried by rats, not via exposure to diseased persons, whether dead or alive. Another storied example holds little water, because there are no definitive corroborating accounts. White colonialists in early North America are purported to have given Native Americans blankets riddled with smallpox (the virus *Variola spp.*). Even if absolutely true, smallpox had previously swept through the colonies and would likely already have reached Native American populations. Biowarfare would not have been very successful in a time when people lived in extreme filth by today's standards, when pathogenic diseases already had free reign to spread and infect humans on their own. With modern hygienic practices and widespread availability of vaccines, in the industrialized world we are rarely in danger of contracting deadly pathogens. Today, biowarfare poses more of a realistic threat. Several federal agencies disseminate information on preparing for and responding to a biowarfare attack, including the Centers for Disease Control, the Department of Homeland Security, and the National Institutes of Health. Yet this discussion will focus on the molecular biotechnology contributing to biowarfare rather than attempt to explain the political underpinnings and social implications of the subject.

Genetically Modified Agents

Genetic modification of biowarfare agents could be carried out in several ways. The organism may be modified to increase pathogenicity and/or toxicity, to increase its ability to reproduce and spread, or to decrease its response to medical treatments, such as by making it resistant to current vaccinations, although such genetic modifications are not strictly necessary. Many deadly pathogens already exist in their natural forms; any one would certainly do the trick if it got into the right hands. One particular case of genetic modification was seen in the Cold War era Soviet Union, where government researchers are believed to have modified the smallpox virus in order to confer vaccine resistance. An outbreak of the engineered virus in 1971 is documented through eyewitness accounts but also was contested by government officials. Smallpox was declared eradicated by the World Health Organization in 1980, but some stocks are still maintained in research laboratories. If given the opportunity, it is technologically possible to insert the smallpox genome into a related pox virus for the purposes of mass production via cloning and subsequent release of the hypothetical biowarfare agent.

In efforts to preemptively address potential biowarfare agents, researchers have also attempted to make viruses less destructive. A virus closely related to smallpox, known as mousepox (the virus *Ectromelia*), specifically infects mice. The natural form of the virus induces a cell-mediated immunity response in mice genetically immune to mousepox. These mice rely on white blood cells to destroy infected cells, thus eliminating the virus from their systems. Researchers have genetically modified *Ectromelia* by inserting transgenes for human interleukin 4 (IL-4). IL-4 stimulates production of B cells that produce antibodies. The rationale of the bioengineers was to insert the IL-4 gene in order to provide an additional immune response (mousepox antigen recognition). As sometimes happens in the pursuit of science, unintended consequences followed. Rather than improving the immune response in mice, the engineered mousepox exhibited increased virulence. The intended antibody response was not observed and making matters worse, the IL-4 transgene appeared to intercede any cell-mediated immune response (**FIGURE 9.13**). The result was a weapon to the mouse test subjects instead of a milder form of virus.

Camouflaged Viruses

A **camouflaged virus** is the insertion of a pathogenic viral genome into a harmless bacterium, which may then be used as a delivery vector for the purposes of biowarfare attack. Recall that lysogenic viruses have the ability to insert their genomes into their bacterial hosts and then lie dormant, essentially undetectable. This natural ability can be hijacked in the creation of a camouflaged virus. The process would mimic **lysogeny**, the integration of bacteriophage DNA into a bacterial chromosome, which results in prophage DNA (**FIGURE 9.14**). **Prophage DNA** is viral DNA that has integrated into bacterial DNA. An infectious viral genome could be inserted into a bacterial plasmid and then used to transform the bacterium. If the viral genome is prohibitively large for insertion, molecular biotechnology offers **artificial chromosomes** for bacterial modification (bacterial artificial chromosomes, BACs) in addition to eukaryotic systems in the form of yeast artificial chromosomes (YACs). Research aimed to create camouflaged viruses has also experienced **poison sequences**, which are viral genome sequences not readily expressed by their hosts. When attempting to insert the yellow fever viral genome into a bacterial host, researchers were unable to clone the entire genome at one time. The poison sequence prevented replication of the viral genome once inserted. To get around this problem, researchers were able to insert the viral genome in two separate segments that were then successfully replicated by the bacterial host. Expression of viral genomes in bacterial hosts may be genetically enhanced by inserting the genome downstream of a strong promoter region within the plasmid vector.

Biosensors

Biosensors are any device with the ability to detect the presence of toxins, viruses, and other biowarfare agents.

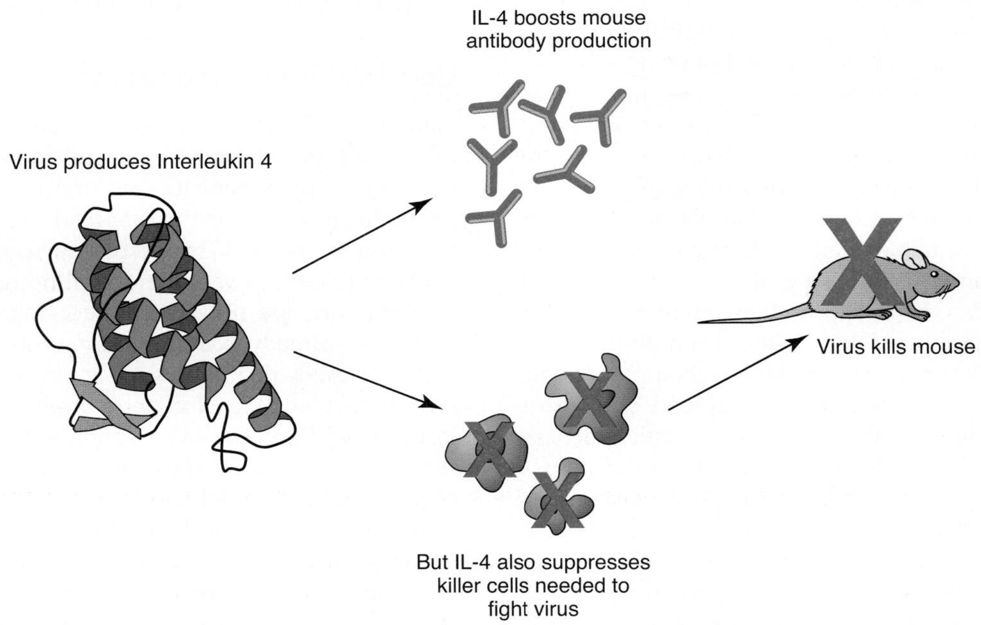

FIGURE 9.13 Genetically modifying the mousepox virus (*Ectromelia*) resulted in unintended consequences. Instead of decreasing virulence, it became greater. The engineered IL-4 proteins interfered with cell-mediated immunity.

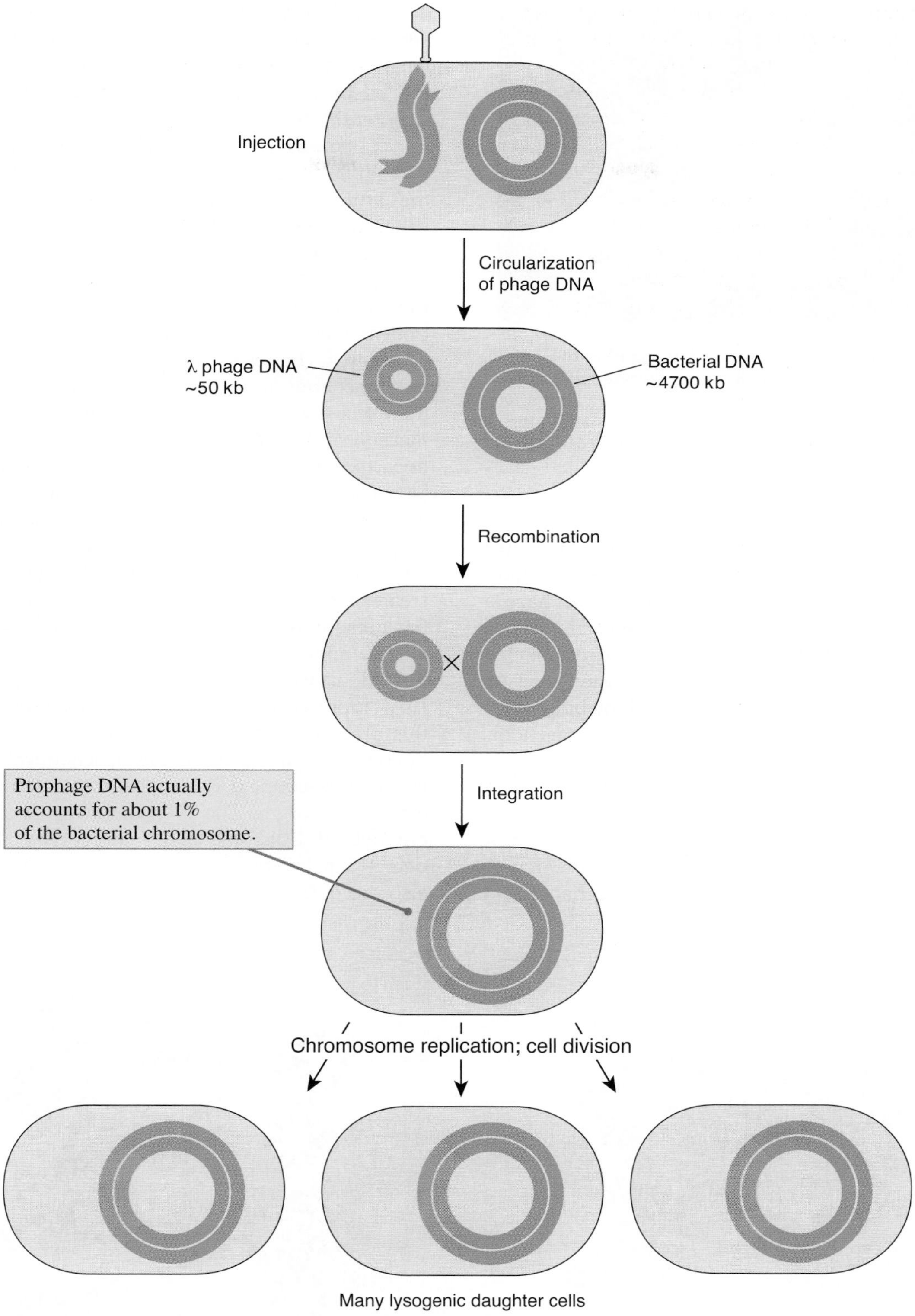

FIGURE 9.14 The process of lysogeny: a bacteriophage inserts its DNA into a bacterium, whereupon integration occurs, resulting in prophage DNA.

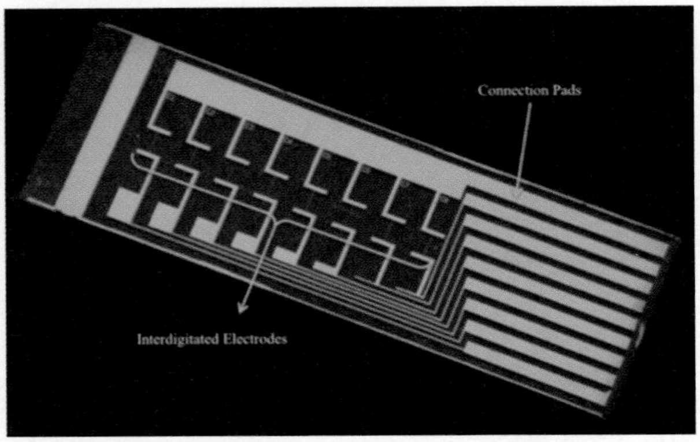

FIGURE 9.15 The biochemiresistor developed at the University of New South Wales has interdigitated electrodes that contain biosensing nanoparticles. Reproduced from B. Dr. Leo M. H. Lai et al., 2012. The Biochemiresistor: An Ultrasensitive Biosensor for Small Organic Molecules. Angewandte Chemie, 51 (26), 6456–6459

In the discussion on environmental biotechnology, we have seen examples of biosensors used in the detection of chemical pollutants as well. All biosensors have a biological component and a physicochemical component. The physicochemical component is responsible for the physical detection of a targeted substance based upon chemical reactions taking place. Many biosensors in use today are designed to detect enzymatic reactions, as these can generate an electrical signal which can be recognized by the physicochemical component.

A recently developed biosensor consists of molecular biotechnology known as a biochemiresistor (**FIGURE 9.15**). Constructed by Australian researchers at the University of New South Wales, the hand-held device uses magnetic nanoparticles covered in gold to attract the very substances it is seeking to detect. The magnets are peppered with surface antibodies that are selective for the target substance. Clearly, there is potential to customize the device with any variety of antibodies in order to probe for any desired material. While in use, the biochemiresistor unit disperses the antibody-displaying magnets into the sample. When target molecules or organisms are detected, the antibodies bind and the complex detaches from the nanoparticles. A separate opposable magnet attracts the complexes onto a film residing between two electrodes, which measures electrical resistance of the complexes. The resistance is negatively correlated with the concentration of the target substance present; the more substance is detected, the lower the resistance measured (**FIGURE 9.16**). The biochemiresistor is a departure from existing biosensor technology. Most biosensors rely on the target substance to "find" its way to the sensing apparatus, but the biochemiresistor employs nanoparticle magnets that can seek out the target substance. This improves detection speeds and creates a much more sensitive measurement of quantities present. A drawback of this technology is that the sample must be in solution, so scanning of surfaces or individuals would not be appropriate.

A more established form of biosensor gets around this problem. While still employing antibodies as the biological component, the physicochemical component is a charge-coupled device that detects movement of electrical charges. The biosensor unit is coated with engineered cells displaying antibodies designed to detect a substance or organism of interest. There is also the potential to design a single biosensor with an

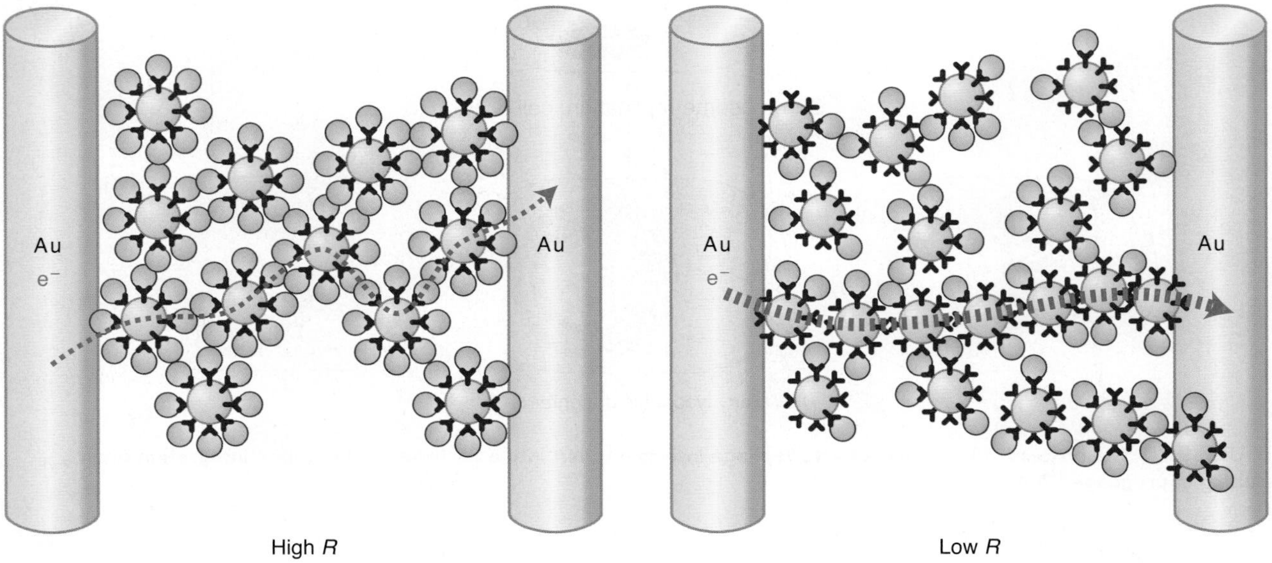

FIGURE 9.16 Gold-plated (Au) electrodes detect concentrations of nanoparticles. The more present, the lower the resistance measured. Reproduced from B. Dr. Leo M. H. Lai et al., 2012. The Biochemiresistor: An Ultrasensitive Biosensor for Small Organic Molecules. Angewandte Chemie, 51 (26), 6456–6459

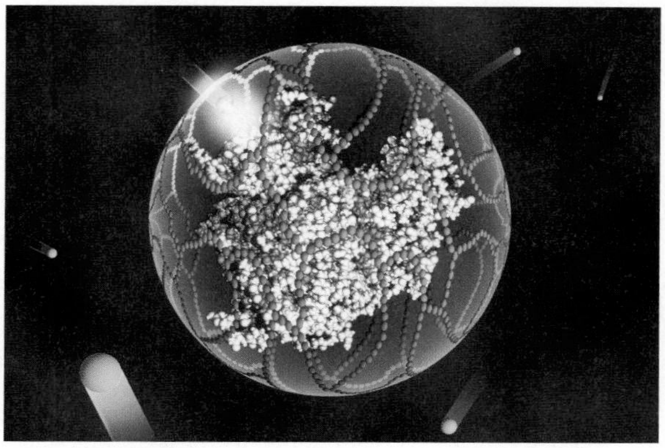

FIGURE 9.17 The engineered cells used in one form of biosensor detect calcium ions (yellow), causing emission of blue light. © Kenneth Eward/Science Source

array of antibodies so that multiple substances can be detected in one swipe. In addition to the antibodies, the engineered cells coating the biosensor unit are also genetically modified to express aequorin. Aequorin is a light-emitting protein naturally occurring in the jellyfish *Aequorea victoria*. When the antibody binds to the target substance, a signal cascade is initiated that releases calcium ions. The release of calcium ions serves as a trigger to the aequorin proteins, causing them to emit blue light (**FIGURE 9.17**). The physicochemical receptor then detects the presence and intensity of the blue light in order to provide a reading.

SUMMARY

- Forensics utilizes science and technology in the investigation of crimes. Biodefense is the practice of using biological systems to defend populations through preventative measures and offensive attack technologies. Both have been greatly influenced by advancing biotechnology in the past few decades.
- Molecular biotechnology has changed criminal justice proceedings by introducing new forms of evidence. Fingerprints and blood types, once the primary forms of biological evidence, have been supplanted by DNA. DNA evidence is statistically more reliable than fingerprints and blood types as it introduces the probability of a match rather than simply excluding an individual as a match.
- Molecular biotechnology provides tools to military and intelligence agencies for the purposes of defense. Biodefense may include biological warfare, biological agents used to attack opponents, and biosensors, which can detect biowarfare substances such as genetically modified pathogens and camouflaged viruses.
- Non-coding DNA contains genetic markers, which are repetitive segments exhibiting high degrees of variability amongst individuals. Genetic markers provided solid evidence for the identification of a particular person. Sampling and analyzing a person's DNA sample in order to detect genetic markers creates a genetic profile, also known as a DNA fingerprint. DNA fingerprints are commonly used as crime scene evidence, in paternity cases, and in the identification of unknown individuals. Profiling of individuals or groups can be accomplished with DNA fingerprinting.
- DNA fingerprinting techniques often include restriction fragment length polymorphism analysis, including both single locus and multi-locus probing, followed by visualization using Southern blotting. The polymerase chain reaction may also be necessary with DNA fingerprinting in order to amplify a small DNA sample.
- In forensics, ruling out a suspected individual is known as exclusivity; the person is excluded from the pool of potential suspects. Certainty goes a step further as it does not merely exclude a suspect. Forensic analysis of DNA yields more compelling evidence as it provides a positive match. Instead of simply ruling out a person, a positive match identifies someone with a high degree of certainty. The level of certainty can also be quantified, whereas exclusivity garners qualitative information.
- Biowarfare has occurred throughout history. For example, the bubonic plague and smallpox have been purposely introduced into populations as a means of attack. The anthrax letters of 2001 are a more recent case of biowarfare. With biotechnology, genetically modified pathogens and camouflaged viruses are now part of the potential biological warfare arsenal.
- Biosensors are any device with the ability to detect the presence of toxins, viruses, and other biowarfare agents. Many biosensors are designed to detect metabolic reactions, as these can generate an electrical signal which may be transmitted as an alert.

KEY TERMS

Artificial chromosomes
Autoradiograph
Biodefense
Biosensors
Biowarfare
Camouflaged virus
CODIS
DNA fingerprint
DNA marker
Exclusivity
Flanking regions
Forensics
Genetic markers
Genetic profile
Lysogeny
Microsatellites
Mitochondrial DNA
Multiple-locus probing
Multiplex PCR
Poison sequences
Prophage DNA
Random match probability
Reference databases
Single-locus probing
Tetranucleotide repeat

DISCUSSION QUESTIONS

1. What is your stance on a national DNA database? Do you believe it is more important to safeguard individual privacy or ensure the safety of all citizens? Provide examples to support your answer and discuss in small groups.
2. In this chapter, forensic analysis projects such as the DNA Shoah Project and the Innocence Project were presented. Identify and characterize a unique project and present the information to your classmates.
3. In 2005, the DNA Fingerprint Act was signed into law. Research the Act using a reputable source. What are the major mandates prescribed in the Act? How will it affect you as an individual? How might it affect our society?

REFERENCES

Ainsworth, Claire. "Disasters Drive DNA Forensics to Reunite Families." *Nature* 441, no. 7094 (2006): 673.

Alonso, Antonio, et al. "Challenges of DNA Profiling in Mass Disaster Investigations." *Croatian Medical Journal* 46, no. 4 (2005): 540.

Butler, John M. *Fundamentals of Forensic DNA Typing*. Waltham, Massachusetts: Academic Press, 2009.

Centers for Disease Control and Prevention. "Bioterrorism Overview." *Emergency Preparedness and Response*. Last updated February 2007. Available at http://www.bt.cdc.gov/bioterrorism/overview.asp

Close, Dan M., Steven Ripp, and Gary S. Sayler. "Reporter Proteins in Whole-Cell Optical Bioreporter Detection Systems, Biosensor Integrations, and Biosensing Applications." *Sensors* 9, no. 11 (2009): 9147–9174.

Daoudi, Y., et al. "Identification of the Vietnam Tomb of the Unknown Soldier: The Many Roles of Mitochondrial DNA." In: Proceedings of the Ninth International Symposium on Human Identification. 1998.

Department of Justice. "DNA Identification in Mass Fatality Incidents." President's DNA Initiative. September 2006. http://massfatality.dna.gov/

Federal Bureau of Investigation. "Combined DNA Index System (CODIS)." Laboratory Services. (ND). http://www.fbi.gov/about-us/lab/biometric-analysis/codis

Foster, Eugene A., et al. "Jefferson Fathered Slave's Last Child." *Nature* 396, no. 6706 (1998): 27–28.

Henderson, D.A., and R. Preston. *Smallpox: The Death of a Disease: The Inside Story of Eradicating a Worldwide Killer*. New York: Prometheus Books, 2009.

The Innocence Project. "About the Innocence Project." Benjamin N. Cardozo School of Law at Yeshiva University. http://www.innocenceproject.org/about/

Jackson, Ronald J., et al. "Expression of Mouse Interleukin-4 by a Recombinant Ectromelia Virus Suppresses Cytolytic Lymphocyte Responses and Overcomes Genetic Resistance to Mousepox." *Journal of Virology* 75, no. 3 (2001): 1205–1210.

Lai, Leo M.H., et al. "The Biochemiresistor: An Ultrasensitive Biosensor for Small Organic Molecules." *Angewandte Chemie International Edition* 51, no. 26 (2012): 6456–6459.

Riedel, Stephan. (January 2005). "Smallpox and Biological Warfare: A Disease Revisited." *Baylor University Medical Center Proceedings* 18, no. 1: 13–20.

Shaler, Robert. 2005. *Who They Were: Inside the World Trade Center DNA Story: The Unprecendented Effort to Identify the Missing.* New York: Free Press, 2007.

Wheelis Mark. "Biological Warfare at the 1346 Siege of Caffa." *Emerging Infectious Disease*. September 2002. http://wwwnc.cdc.gov/eid/article/8/9/01-0536.htm

World Health Organization. "Smallpox." Global Alert and Response. 2013. http://www.who.int/csr/disease/smallpox/en

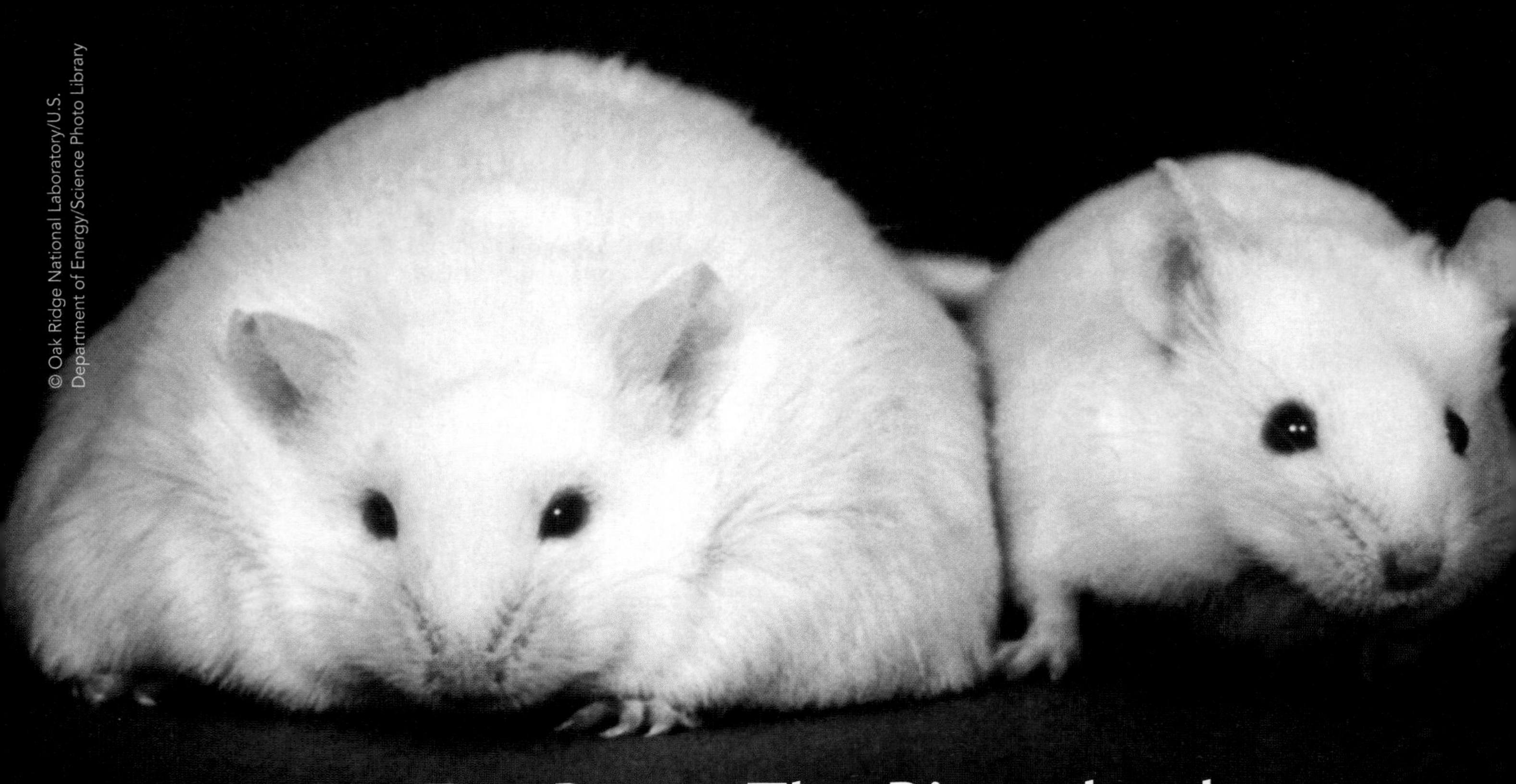

10 Evo Devo: The Biotechnology of Evolution and Development

LEARNING OBJECTIVES

Upon reading this chapter, you should be able to:

- Define evo devo.
- Relate the terms phylogeny, ontogeny, exaptation, and homologous structure.
- Compare and contrast macroevolution with microevolution.
- Identify the evidence for macroevolution. List and describe the mechanisms of microevolution.
- Detail the methods used to research the process of development.
- Discuss major discoveries in the study of development.

"I should infer from analogy that probably all the organic beings which have ever lived on this earth have descended from one primordial form, into which life was first breathed."

—Charles Darwin, *On the Origin of Species*, 1859

Sitting in a dusty cabinet for 165 years, 314 slides of specimens collected by Charles Darwin and botanist Joseph Hooker were found by Dr. Howard Falcon-Lang of the British Geological Survey in April 2011. The slides were prepared by polishing fossils into thin translucent layers which could then be observed under a microscope (**FIGURE 10.1**). Hooker was apparently responsible for cataloguing and registering the slides into the British Geological Survey's official "specimen register," but he embarked upon a research expedition before ever getting around to it. In his absence, the slides remained unregistered and were relocated several times to different science museums and institutions before their rediscovery by Dr. Falcon-Lang. The slides are now on display in an online collection, making them available to a worldwide audience.

Darwin made use of slides including fossils and embryos in order to gather evidence for his burgeoning theory of evolution. Fossils serve as a record to past life forms, which allowed Darwin to elucidate evolutionary relationships among extinct organisms and their extant descendants. Comparative embryology is another form of evidence often cited in support of evolutionary theory. The development of organisms from fertilization through the embryo and fetus stages is considered to support the idea that living things share a common ancestor. This idea was championed by an early contributor to embryology, Ernst Haeckel, in addition to Darwin. Haeckel is probably best known for his assertion that **ontogeny** (development) recapitulates **phylogeny** (evolution). More recently, evolutionary biologist Stephen Jay Gould coined the term **exaptation**, which is a pre-existing character. Exaptations are derived from **homologous structures**, those that are evolved from the same ancestral structure. An example is reptile scales and bird feathers, which are homologous. Feathers are an adapted form of scales, making them an exaptation because scales are pre-existing. Feathers are merely a modified form of an already present structure.

FIGURE 10.1 A slide mounted from a fossilized tree specimen collected by Charles Darwin in South America (1834) was lost for over 150 years. This slide is now part of the British Geological Survey's J.D. Hooker Slide Collection. © British Geological Survey.

Haeckel's comparative embryology drawings were referenced by Darwin in his writings on evolution (**FIGURE 10.2**). Haeckel's hypothesis is oversimplified and not strictly supported by currently available evidence. A more accurate picture highlights the fact that major changes observed during development follow a generalized pattern; this indicates life forms share the essential genes that orchestrate the entire process. We can see from Haeckel's drawings that early ontogeny is similar among related organisms and becomes progressively specialized during later development, leading to the distinct characteristics of each species. It is now known that gene regulation, the timing and intensity of gene expression, plays a vital role in development. Biotechnology has contributed greatly to this understanding. Comparative genomics has revealed approximately 500 genes known as "immortal genes." The **immortal genes** emerged early in the history of life on Earth and have been conserved through at least two billion years of evolution. Thus even with the immense biodiversity witnessed on our planet, there is a degree of genetic unity. The immortal genes are strong evidence for Darwin's precept of common descent. Darwin never knew the genetic link to his theory; while he understood inheritance somehow occurred, he did not know the mechanisms behind it. Even though Gregor Mendel developed his classical genetics during Darwin's lifetime, the concepts went unrecognized until the early twentieth century. The discovery of immortal genes heralded the formation of a new branch of biological inquiry known as evo devo. **Evo devo** is the study of interrelationships between evolution and development. As a modern reincarnation of Haeckel's precept, it seeks to discover the molecular mechanisms that contribute to these processes and acts as the convergence of ontogeny and phylogeny.

Evolution

Evo devo is the direct result of the modern synthesis of evolution that occurred in the mid-twentieth century. The discovery and characterization of chromosomes and then nucleic acids (DNA and RNA) aided to reconcile Darwin's descent with modification observed at the macro level with the newly emerging field of molecular biology that viewed evolution at the micro level. The main tenets of the modern synthesis are summarized in **TABLE 10.1**. Just as with evolution itself, enabling adaptation to changing environments, the modern synthesis has morphed into the field of evo devo as molecular biotechnology approaches were introduced. (See the *Focus on Careers* box at the end of this chapter.)

Macroevolution: Tracing the Past

Without genetic knowledge, Darwin was limited to explaining his theory of evolution. Study of the fossil record provided a storyline of **speciation**, the emergence of new organisms, and **extinction**, the disappearance of species or higher groups. Naturalists of his time, and even earlier beginning with Carolus Linneaus in the eighteenth century, were actively involved in classifying organisms into categories based upon their relationships. Without any other means, **taxonomy**, placing organisms into hierarchical groups, was historically based upon **morphology**, the outward appearance of living things. Observation of anatomy and the identification of homologous structures is one way morphology may be compared (**FIGURE 10.3**). It also involves comparative embryology and cell structure.

Taxonomy delineated organisms into increasingly exclusive categories until the level of species was reached, thus providing a framework for the concept of speciation. The combination of speciation with extinction explains evolution primarily in terms of large changes occurring over long periods of time, a branch of study known as **macroevolution**. Paleontologists have constructed the geological time scale based upon the fossil record resulting from macroevolution (**TABLE 10.2**). Prior to radioactive isotope dating, the ages of unearthed fossils were determined based upon their relative depths in layers of strata. The rationale of early paleontologists was the deeper the fossil was embedded, the older it must be (**FIGURE 10.4**). Darwin

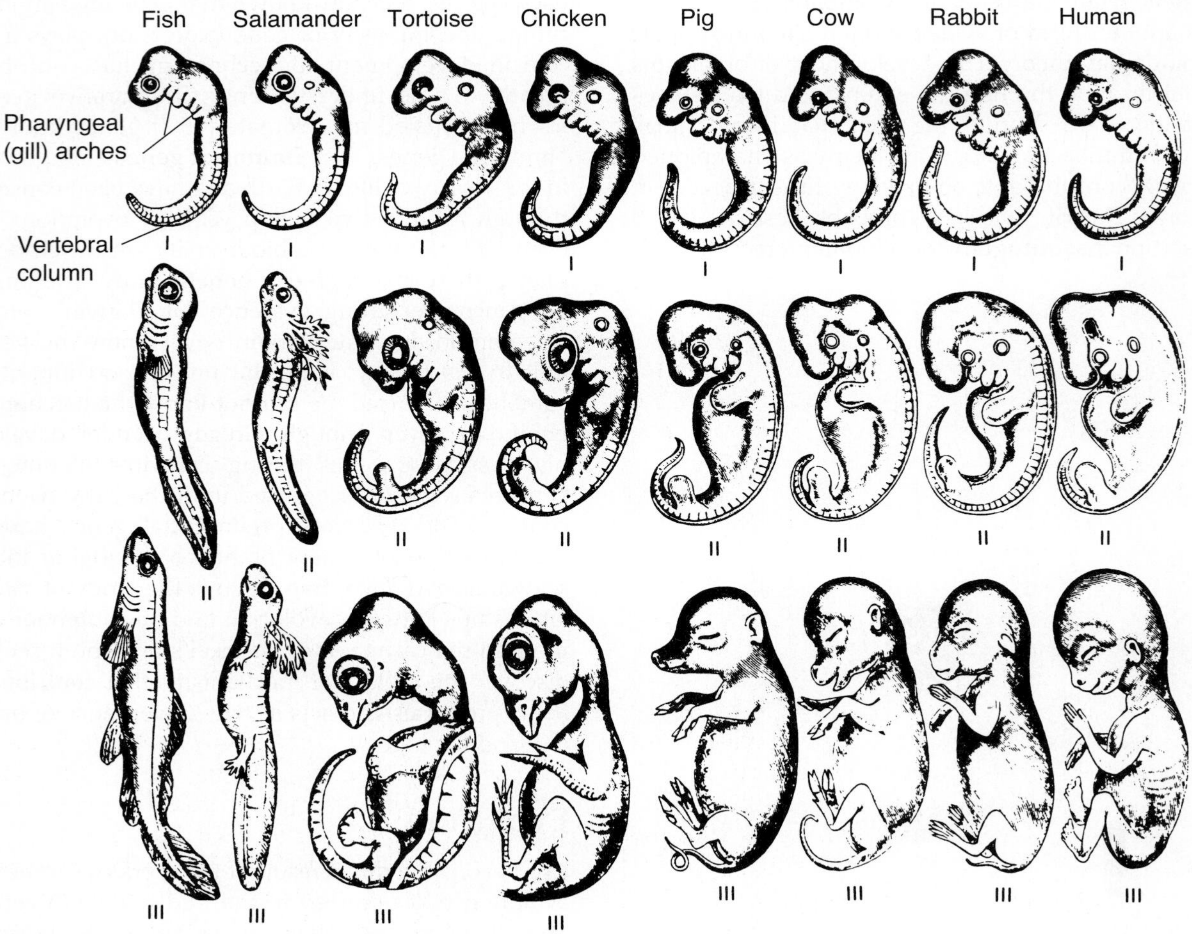

FIGURE 10.2 Comparative embryology drawings by Ernst Haeckel from his 1874 work *Anthropogeny*. The drawing depicts (from left to right): fish, salamander, turtle, chick, pig, cow, rabbit, and human embryos at "very early" (top), "somewhat later" (middle), and "still later" (bottom) stages. Adapted from Romanes G. J. Darwin, and After Darwin. Open Court, 1910.

TABLE 10.1

Major Elements of the Modern Synthesis of Evolution

Organisms exist as individuals. Individual multicellular organisms develop, grow, mature, reproduce, and die.
Natural selection acts on individuals but individuals do not evolve. Individuals pass on their genes to individuals of the next generation.
Individuals exist in populations. Populations usually include different age classes of a single generation and individuals from other generations ("grandparents, grandchildren").
Populations of a sexually reproducing organism consist of individuals that are not identical to one another; they are not clones. Populations of asexually reproducing individuals may be clones.
Populations do not reproduce, individuals reproduce.
Resources are often limited.
Not all individuals in a population will survive to reproduce and contribute offspring to the next generation.
Variation is an essential prerequisite for evolution to act. Natural selection allows some variants to survive and others not.
Differential reproduction results in survival to the next generation of those individuals best suited to the conditions of their existence.
Because the genetic background of individual sexually reproducing organisms differs, those that are selected are more likely to pass their genes to the next generation.
Because of differential reproduction, mutations, and exchange of genes between populations, the genetic composition of a population will change gradually from generation to generation.
Populations may subdivide into smaller groups and so reinforce genetic differences that can provide the basis for speciation.
Populations or subsets of populations may "crash" or become extinct following environmental catastrophes.

Note: Dates derived mostly from Gradstein et al. *A Geological Time Scale.* Cambridge University Press, 2004 and *Geologic Time Scale*, available from http://www.stratigraphy.org, Accessed January 2010

was keenly aware of the large gaps present in the fossil record, which represented swaths of time without the presence of transitional organisms. Darwin predicted that transitional fossils would be found in the future to fill in these gaps.

Three Domains

Traditional classification of organisms relied upon morphology, but with evo devo there has been an explosion of interest in reexamining established taxonomic groupings in light of new evidence. Linnean classification was based on a hierarchical scheme where kingdoms were the most inclusive taxon and the species represented the least inclusive taxon (**FIGURE 10.5**). Five kingdoms were in this scheme, as proposed by plant ecologist Robert Whittaker in 1959: Monera (prokaryotic microbes), Protista (eukaryotic microbes), Plantae (plants), Animalia (animals), and Fungi. Comparative genomics provides the relative similarities and differences among organisms at the genetic level, and has revealed a new layer of information to consider when determining evolutionary relationships among organisms. Inclusion of genomic data in the classification process has necessitated revisions in certain cases.

Perhaps the greatest amendment to the Linnean classification system is the formation of the three domain classification system. First introduced by microbiologist Carl Woese in 1977, the system groups organisms into three domains as the most inclusive level of organization. This makes a domain one level higher than a kingdom. Determination of which domain an organism belongs to is based upon analytical comparisons of ribosomal RNA (rRNA). Ribosomes are composed of two subunits which contain proteins and ribonucleic acid (RNA); the subunits vary between prokaryotes and eukaryotes. Ribosomes are a major player in the synthesis of proteins, demonstrating the interdependence among nucleic acids (rRNA) and polypeptides (proteins). This harkens the central dogma of molecular biology: DNA is transcribed into messenger RNA, which is translated into a polypeptide with the aid of ribosomes (and hence rRNA). Prokaryotic ribosomes have a large 50S subunit and a small 30S subunit, where S is the sedimentation coefficient (**FIGURE 10.6**). Eukaryotic ribosomes have a large 60S subunit and a small 40S subunit (**FIGURE 10.7**).

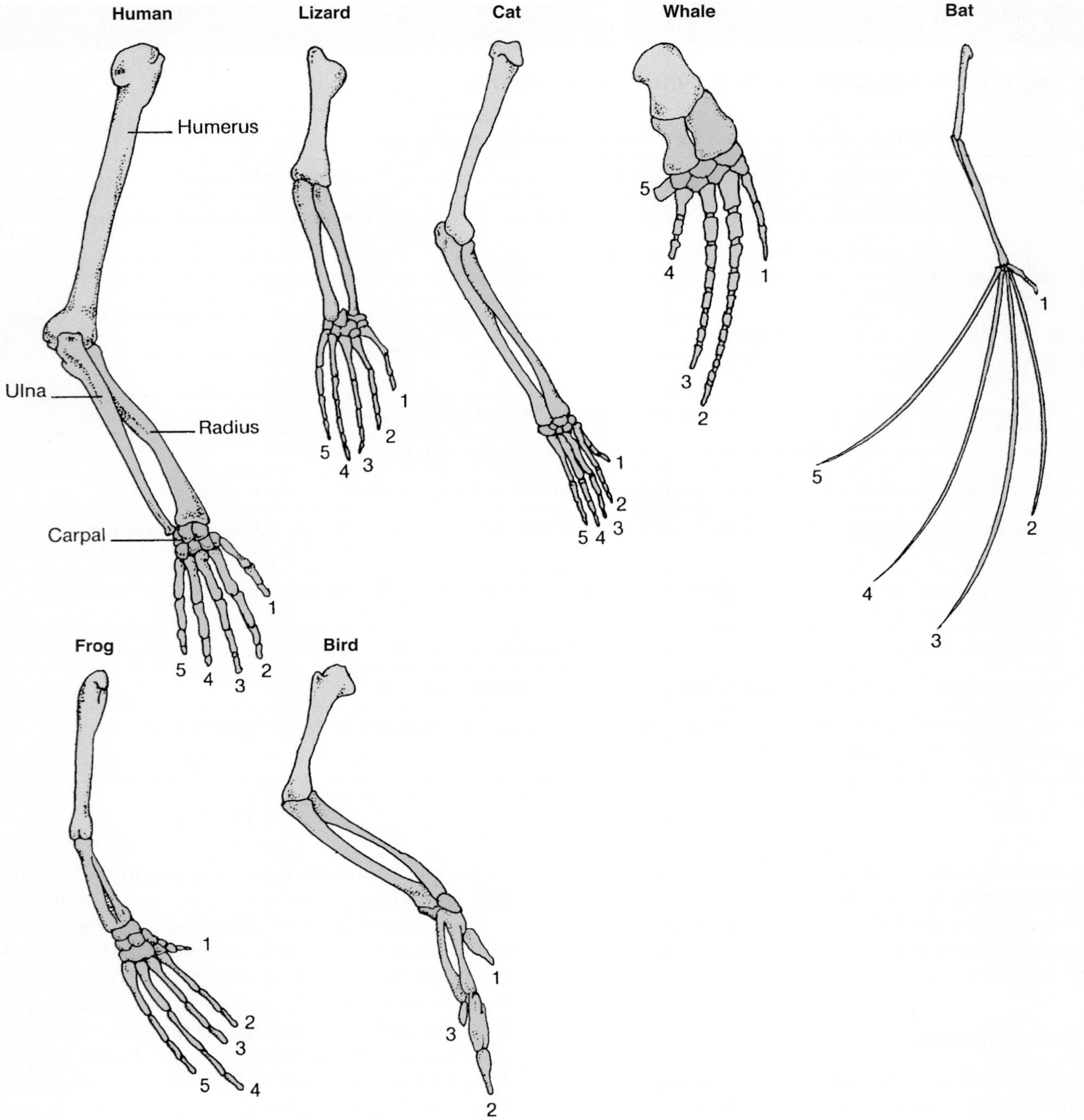

FIGURE 10.3 The composition of forelimbs in different organisms demonstrates homology. The presence of skeletal homologous structures indicates all of these organisms are descended from a common ancestor.

Sequence comparisons seen in rRNA necessitated some reorganization of the existing classification scheme characterized by Whittaker. Genomic studies of a unique group of extremophilic prokaryotes, initially referred to as archaebacteria, led to the discovery they were strikingly different than other prokaryotes. Extremophiles thrive in extreme, harsh environments where other bacteria cannot prosper, such as high salt concentrations, acidic pH levels, or boiling hot temperatures. Woese and colleagues suggested the prokaryotes be divided into two domains: Archaea, representing the extremophiles previously known as the archaebacteria, and Bacteria, the remaining prokaryotic organisms. The third domain was named Eukarya as it encompasses all eukaryotic organisms, whether unicellular or multicellular. Even the three domain system

TABLE 10.2

Major Events in Organismal Evolution in Relation to the Geological Time Scale

Time Scale						
Eon	**Era**	**Period**	**Epoch**	**Millions of Years Before Present (approx.)**	**Duration in Millions of Years (approx.)**	**Some Major Organic Events**
Phanerozoic	Cenozoic	Quaternary	Recent (last 5,000 years)	0.01	1.8	Appearance of humans
			Pleistocene	1.8		
		Tertiary	Pliocene	5.3	3.5	Dominance of mammals and birds
			Miocene	23.8	18.5	Proliferation of bony fishes (teleosts)
			Oligocene	34	10.2	Rise of modern groups of mammals and invertebrates
			Eocene	55	21	Dominance of flowering plants
			Paleocene	65	10	Radiation of primitive mammals
	Mesozoic	Cretaceous		142	77	First flowering plants; extinction of dinosaurs
		Jurassic		206	64	Rise of giant dinosaurs; appearance of first birds
		Triassic		248	42	Development of conifer plants
	Paleozoic	Permian		290	42	Proliferation of reptiles; extinction of many early forms (invertebrates)
		Carboniferous	Pennsylvanian	320	30	Appearance of early reptiles
			Mississippian	354	34	Development of amphibians and insects
		Devonian		417	63	Rise of fishes; first land vertebrates
		Silurian		443	26	First land plants and land invertebrates
		Ordovician		495	52	Dominance of invertebrates; first vertebrates
		Cambrian		545	40	Sharp increase in fossils of invertebrate phyla
Precambrian	Proterozoic	Upper		900	355	Appearance of multicellular organisms
		Middle		1,600	700	Appearance of eukaryotic cells
		Lower		2,500	900	Appearance of planktonic prokaryotes
	Archean			4,000–4,400	1,400	Appearance of sedimentary rocks, stromatolites, and benthic prokaryotes
	Hadean			4,560	160–560	From the formation of Earth until first appearance of sedimentary rocks; no observable fossil organisms

Data from Gradstein et al. *A Geological Time Scale.* Cambridge University Press, 2004 and *Geologic Time Scale.* http://www.stratigraphy.org/index.php/ics-chart-timescale. Accessed January 2010.

has undergone revisions as new data emerges. Initially the phylogenetic tree of life under Woese's scheme had a universal ancestor at its base (**FIGURE 10.8**). It now radiates from a center point in light of new information: Archaea is more closely related to Eukarya than it is to Bacteria (**FIGURE 10.9**). This is the same Archaea that was once called archaebacteria, so the difference of this new scheme cannot be understated. All three domains are at even par in this "universal tree." Another significant update reclassified ciliates and slime molds into domain Eukarya, where they had once been considered prokaryotic organisms.

Further complicating this more recent phylogeny is the consideration of horizontal gene transfer (HGT). Phylogenetic trees are often developed with a linear thought process in mind: one organism begets the next

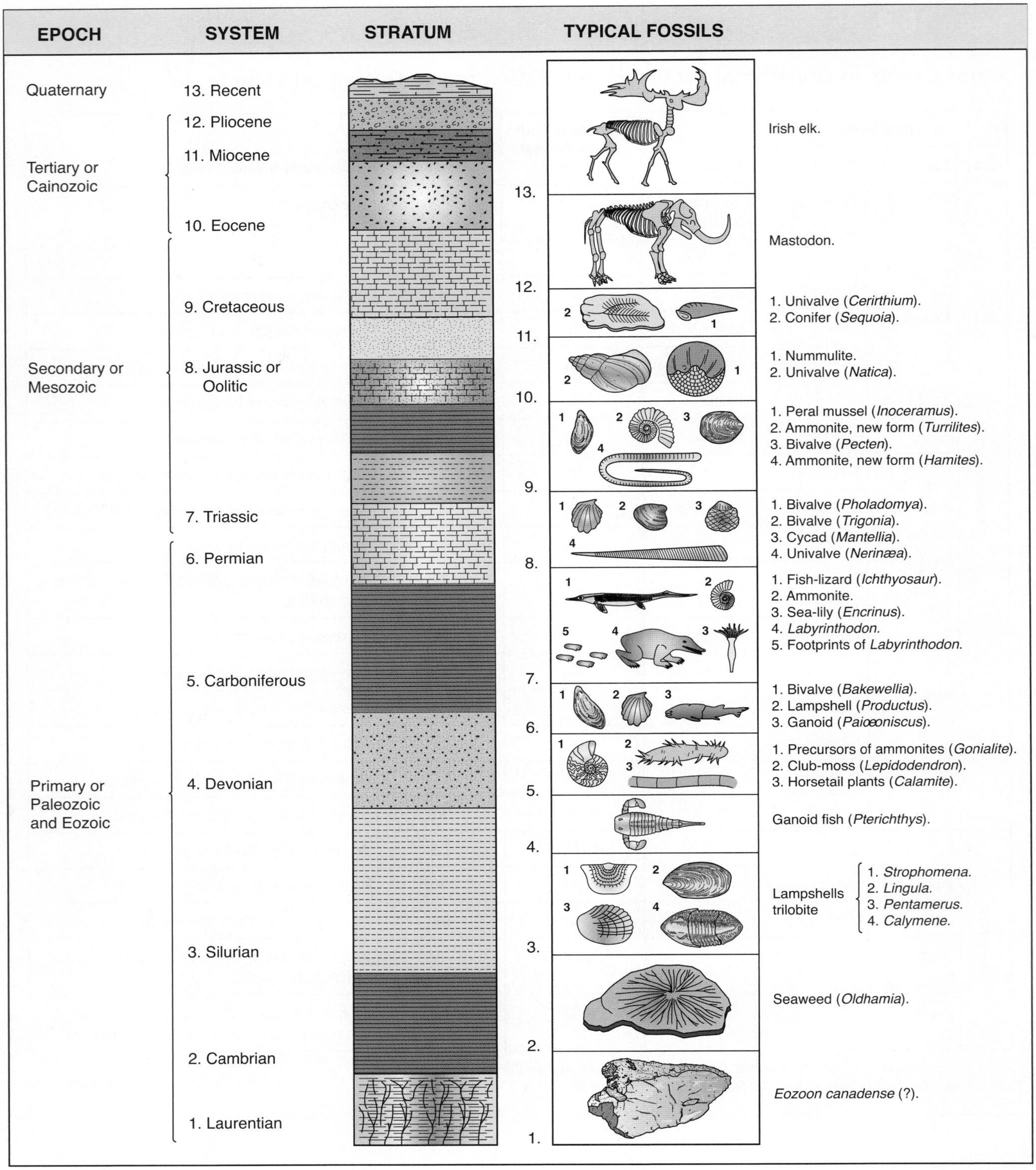

FIGURE 10.4 This chart of the fossil record is adapted from a nineteenth century illustration. It depicts layers of strata and their relative depths used as a correlation in determining fossil age. The deeper the fossil was found, the older it must be. Adapted from Clodd, E., 1888. The Story of Creation. Longmans Green, London.

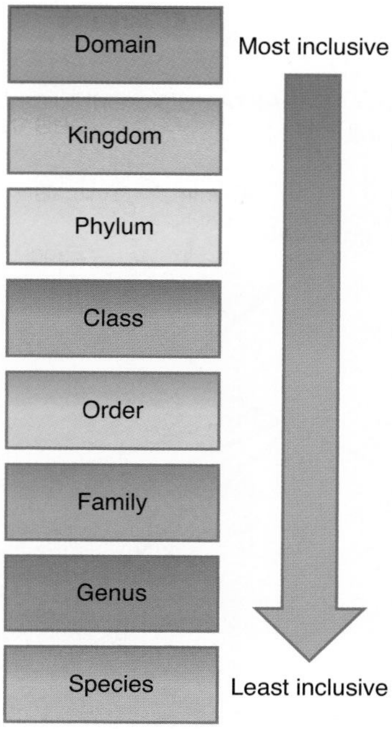

FIGURE 10.5 The Linnean classification system defined seven hierarchical taxa ranging from most inclusive to least inclusive: kingdom, phylum, class, order, family, genus, and species.

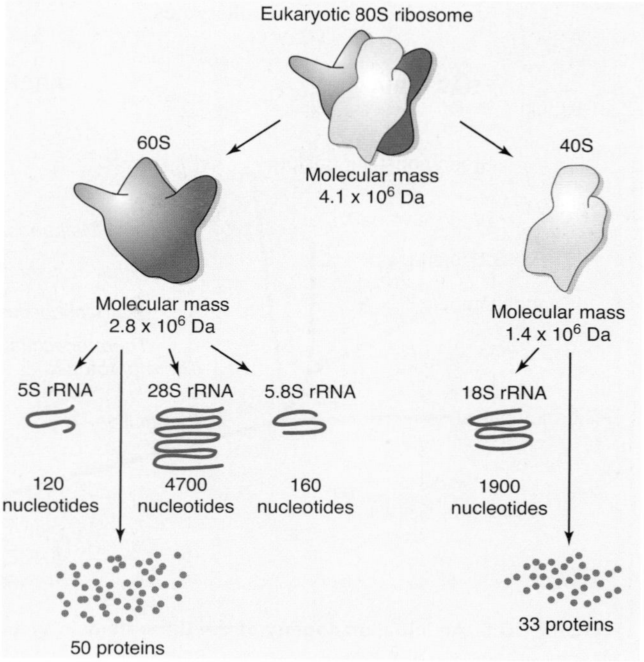

FIGURE 10.7 The structure of a eukaryotic ribosome as seen in a representative taxon, class Mammalia. There is a large 60S subunit, which varies in composition among eukaryotes, and a small 40S subunit. The subunits consist of ribosomal RNA and protein molecules.

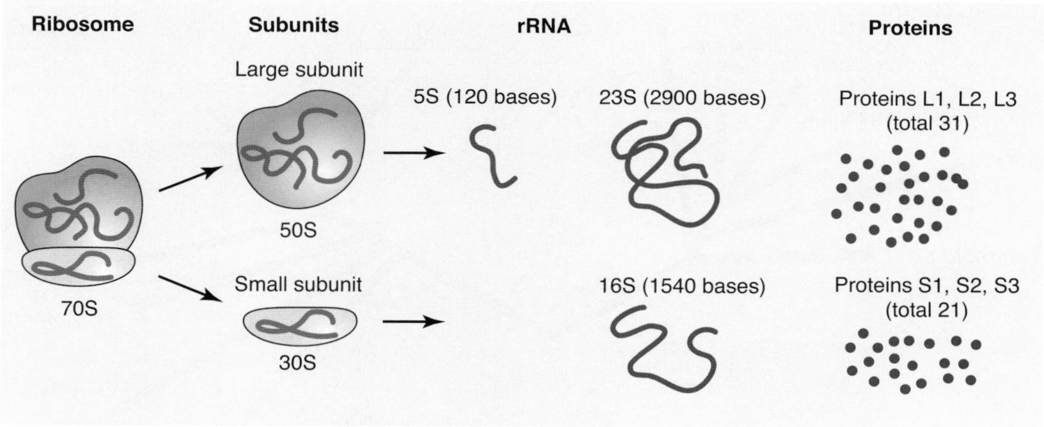

FIGURE 10.6 The structure of a prokaryotic ribosome as seen in a representative bacterium, *E. coli*. There is a large 50S subunit composed of two smaller units (5S and 23S) and a small 30S subunit. The subunits consist of ribosomal RNA and protein molecules. Adapted from an illustration by Kenneth G. Wilson, Department of Botany, Miami University.

in the line of descent. This concept mimics generational inheritance, that is, parents beget offspring and pass down genetic information. Asexual reproduction doesn't follow this pattern, and most of Earth's inhabitants are asexual. The earliest forms of life, unicellular organisms, are asexual. Thus, in the course of evolution, HGT resulting from asexual reproduction took place again and again, continuing right into the present. An interwoven version of the three domain tree of life accounts for HGT (**FIGURE 10.10**).

Transitional Organisms

Another contribution to our understanding of macroevolution provided by molecular biotechnology involves transitional fossils. Gaps in the fossil record

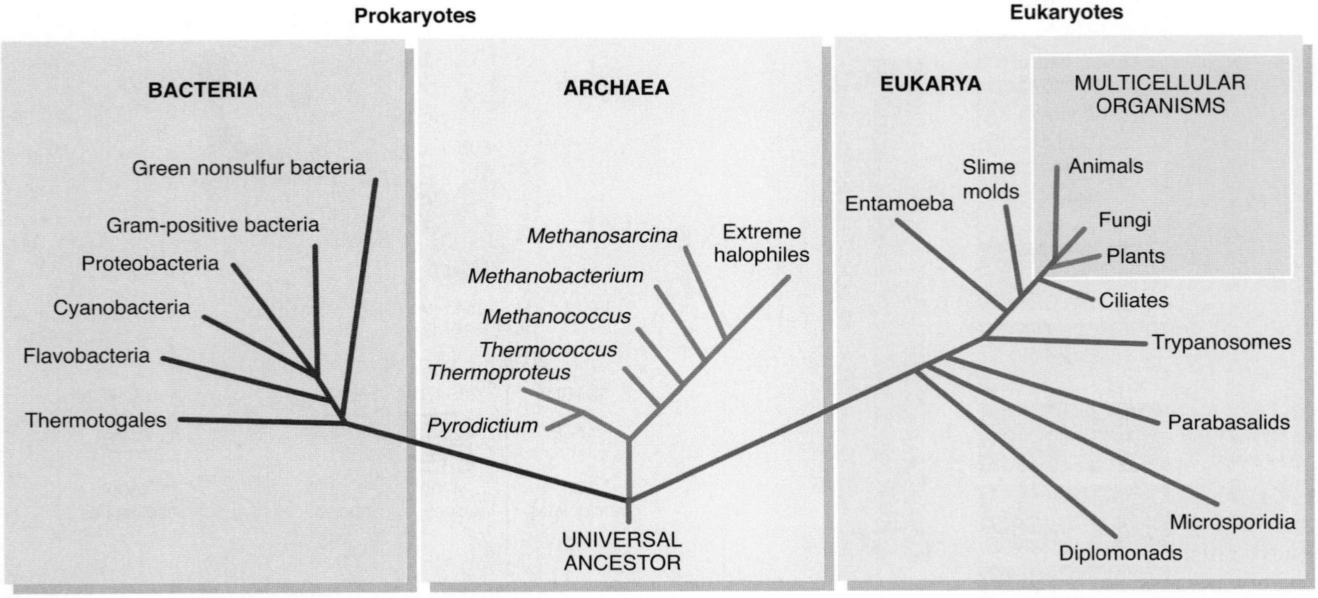

FIGURE 10.8 An initial phylogeny of the three domain system as proposed by Woese and colleagues. There is a universal ancestor at the base of the tree.

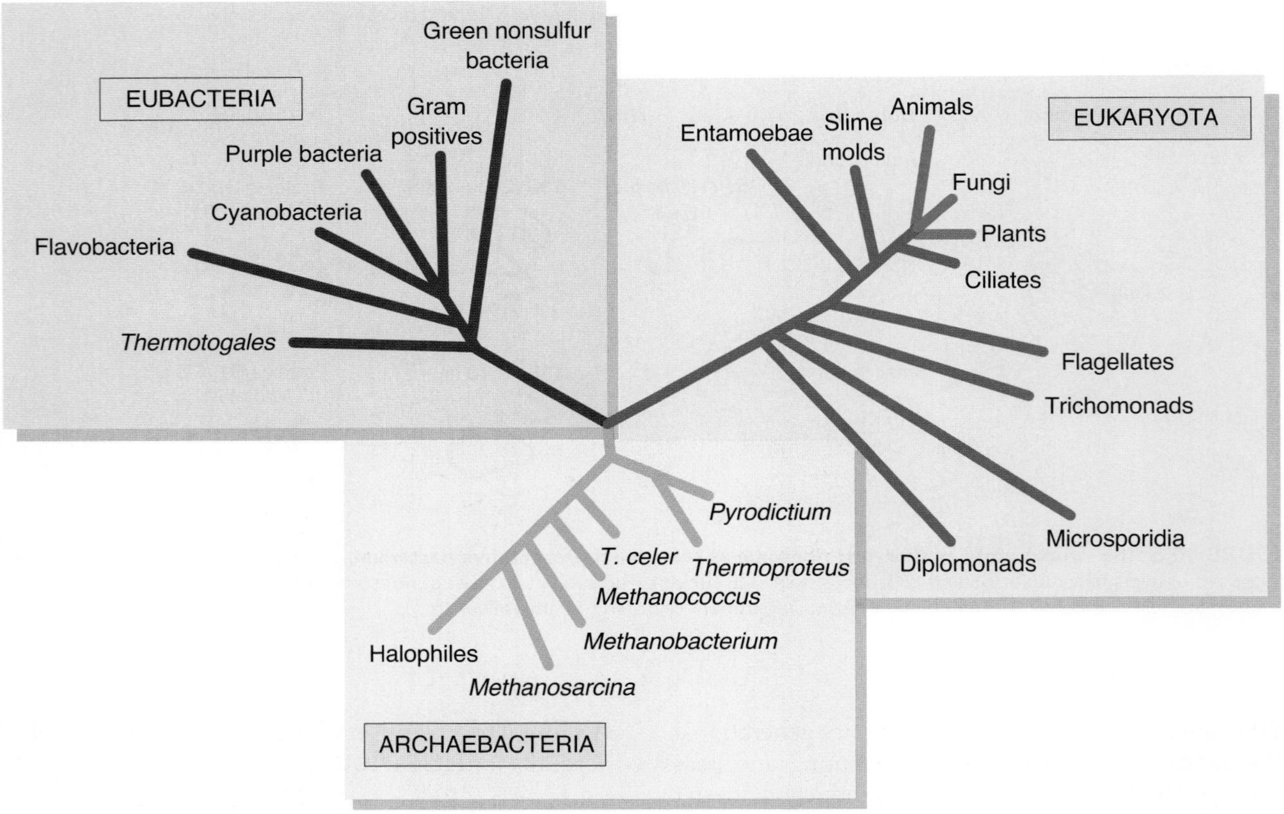

FIGURE 10.9 A revised phylogeny of the three domain system shows all domains deriving from a center point of origin. Ciliates and slime molds, once classified as prokaryotes, are now placed in domain Eukarya. Adapted from Sogin, M. L. 1991. *Curr. Opin. Genet. Dev.* 1, 457–463, and Wheelis, M. L. et al., 1992. *Proc. Natl. Acad. Sci.* USA 89, 2930–2934.

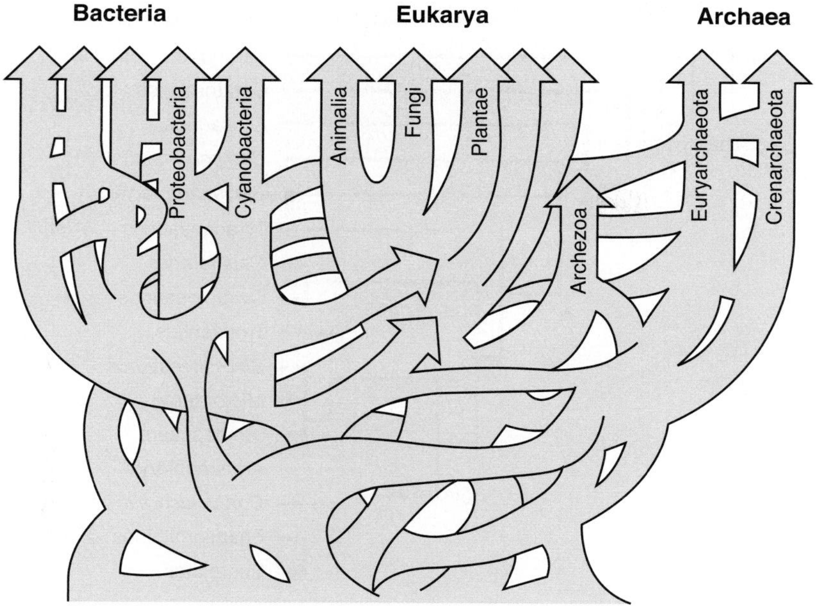

FIGURE 10.10 Horizontal gene transfer taking place over the course of evolution likely resulted in a highly interbranched phylogeny as seen here.

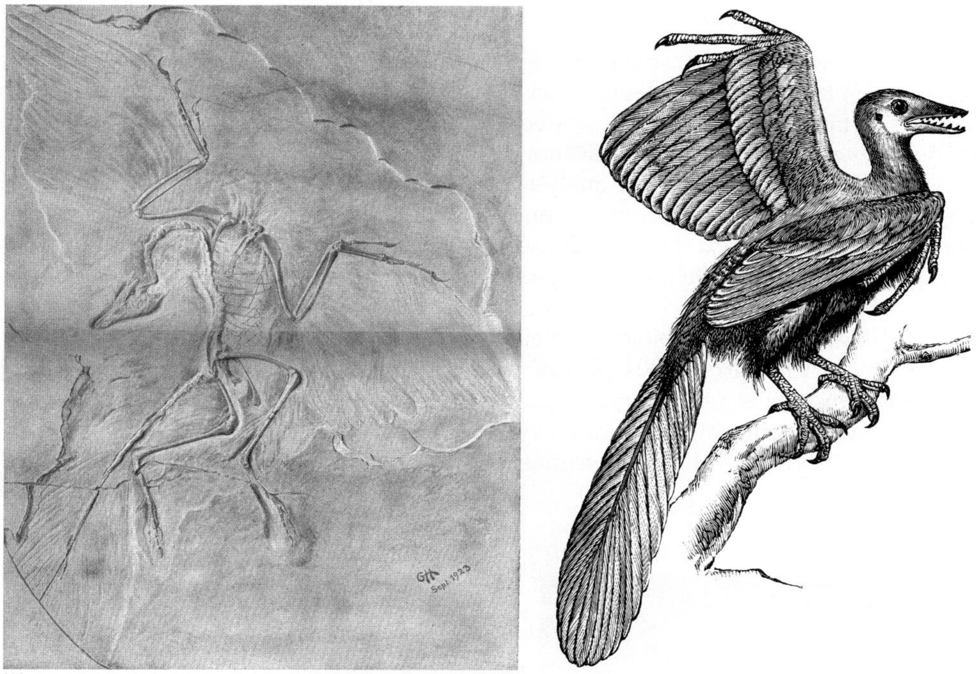

FIGURE 10.11 The transitional organism *Archaeopteryx*. (A) The fossil specimen. (B) An artist's rendering of what *Archaeopteryx* may have looked like. Reproduced from Heilmann, G. *The Origin of Birds*. Appelton, 1927 (Reprinted Dover Publication, 1972).

observed by Darwin have been filled to a higher degree in the past 150 years, though there are still significant "missing links." Fossils of transitional organisms exhibit characteristics of both earlier and later species, suggesting a step-by-step fashion in the process of evolution. One in particular was discovered in Darwin's lifetime and provided clues that helped to shape his common descent hypothesis. The archaeopteryx fossil has some reptilian traits such as teeth and claws in addition to certain avian features including feathers and a wishbone (**FIGURE 10.11**). But this knowledge is obtainable through traditional paleontology

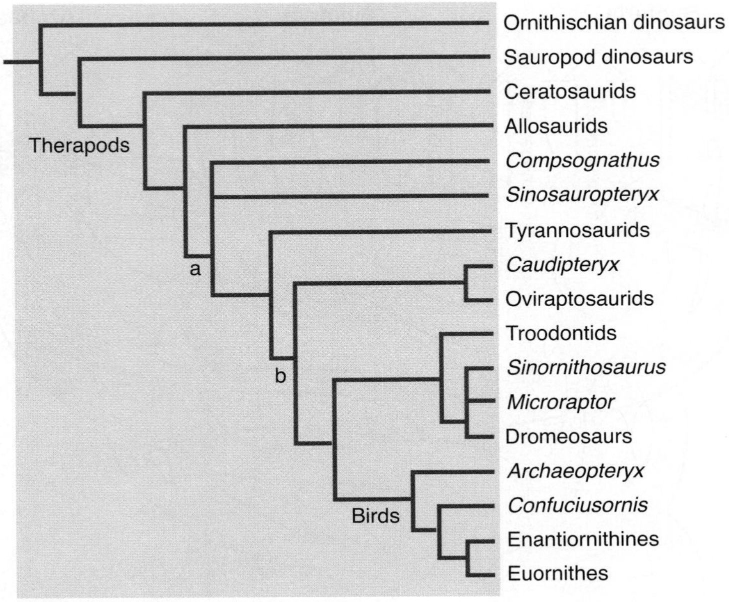

FIGURE 10.12 The phylogeny of dinosaurs and birds demonstrates the transition from an extinct group of reptiles into an extant group of warm-blooded fliers. Adapted from Carroll, Robert, 1997. *Patterns and Processes of Vertebrate Evolution*. Cambridge University Press, Cambridge, UK.

methods. Transitional fossils may also be studied using biotechnology as a means to add genetic detail to scientific investigations. In fact, unearthing of a related specimen (*Xiaotingia zhengi*) revisited the classification status of Archaeopteryx. Mitochondrial DNA analysis places both of these extinct organisms in the same taxonomic branch as dinosaurs and extant birds. These studies confirm that birds really are flying dinosaurs (**FIGURE 10.12**).

The discovery of *Tiktaalik roseae* by biologist Neil Shubin of the University of Chicago in 2004 provides another example. *Tiktaalik roseae* is a transitional fossil connecting fish to terrestrial organisms approximately 375 million years ago (**FIGURE 10.13**). The specimen is considered to be the common ancestor to all land dwellers, even humans. It gave rise to anatomical features such as limbs, including wrists, ankles, and digits, and the neck. It was given an Inuktitut name after the inhabitants of Ellesmere Island, in the Canadian arctic where it was found. Shubin and his team have dubbed *Tiktaalik roseae* a "fishapod" because it was partially like a fish but is the earliest known organism to possess tetrapod limbs.

FIGURE 10.13 An artist's rendering of *Tiktaalik roseae*, moving from its aquatic environment onto land. © Science Source

Microevolution: Genetic Shifts

Small changes occurring over short periods of time are collectively known as the process of **microevolution**. These minute changes are occurring at the genetic level and represent any permanent alteration of the genome. Genetic change occurs at the individual level and within populations (groupings of a single species), although the specific characteristics of these changes will vary depending on the vantage point taken (**TABLE 10.3**). When Darwin formulated the steps of natural selection leading to adaptations, he considered variations among individuals in a population. However, he made clear that changes over time involved all reproductive members of the population and ascertained it is only populations that adapt, not individuals.

While Darwin was concerned with phenotypic adaptations, he knew there was an underlying mode for inheritance of these adaptations. It was not until the modern synthesis that researchers focused in on the genotype as an agent of change. Genotypes consist of alleles, or alternate forms of a gene. It is the frequency

TABLE 10.3

Comparison of Characteristics of Individuals and Populations

Characteristic	Individual	Population
Life span	One generation	Many generations
Spatial continuity	Limited	Extensive
Genetic characteristics	Genotype	Gene frequencies
Genetic variation	Expressed in one lifespan	Expressed in evolutionary change
Evolutionary characteristics	No changes, because an individual has only one genotype and is limited to a single generation	Can evolve (change in gene frequency), because evolution occurs between generations
Selection	Operates on phenotype in one generation	With mutation, leads to change in genotype from generation to generation
Mutation	Somatic mutation transferred through cell reproduction	Gametic inheritance/gametic
Variation	Phenotype transferred via cell reproduction	Genotype inherited/genotypic

of alleles in a population that can change over time, thus constituting the mechanism for microevolution. The field of population genetics investigates changes in allele frequencies over time, which is described by the Hardy-Weinberg equilibrium principle (**FIGURE 10.14**). Named for the two men who independently conceptualized the principle, Englishman Godfrey Hardy and Wilhelm Weinberg of Germany, it makes a series of assumptions about a population. As long as all assumptions are met, the population is in equilibrium, meaning there are no changes to allele frequency. When considering the assumptions made, it is quite obvious that they will never all be met in a living, dynamic population. Therefore, any condition leading to an assumption not being met is in essence a mechanism for microevolution to occur. These mechanisms are now described.

Natural Selection

Natural selection was first conceived independently by Charles Darwin and Alfred Russel Wallace in the late nineteenth century. The process results in adaptation of a population to its environment, thus benefitting the survival of the species rather than any one member. The most reproductively fit individuals are more successful at obtaining resources from the environment and passing on their genes, causing an increase in their form of variation. Variation is the raw material for natural selection; it must exist for it to be selected upon. The most successful variants survive to reproduce, and over the generations become more common in the population. What Darwin and Wallace did not know is that variation is partially or wholly a result of genotype (environment is another factor that can play a role). As genotypes consist of allelic combinations, allele frequencies directly affect genotype frequencies. In other words, natural selection is operating on the scale of microevolution just as much as it influences macroevolution. The genetic changes taking place within populations can accumulate until a tipping point is reached: divergence and speciation. Since the time of Darwin and Wallace, study after study has confirmed the process of natural selection is a present force. It has occurred time and again before our very eyes. As more data mounted, specific forms of natural selection became apparent (**FIGURE 10.15**). These include disruptive selection, directional selection, and stabilizing selection.

Disruptive Selection

Disruptive selection results in one average phenotype splitting into two equally common forms. Rock pocket mice, *Chaetodipus intermedius*, are a recent species to have undergone disruptive selection. Populations in Arizona deserts exist in two phenotypic variations. Researchers at the University of Arizona observed a dark variety that primarily forages on black lava rocks along with a lighter version of the mouse that feeds in sandy areas (**FIGURE 10.16**). They hypothesized the species was being selectively disrupted upon. Each phenotype was adapted to its feeding area by being of a similar color. They could blend into the background and avoid being eaten by predators. To support this notion, they collected DNA samples and analyzed both nuclear and mtDNA sequences. Mitochondrial differences, associated with slow change, did not correlate to the distinct color variations, so the researchers concluded the colors were recent adaptations. This

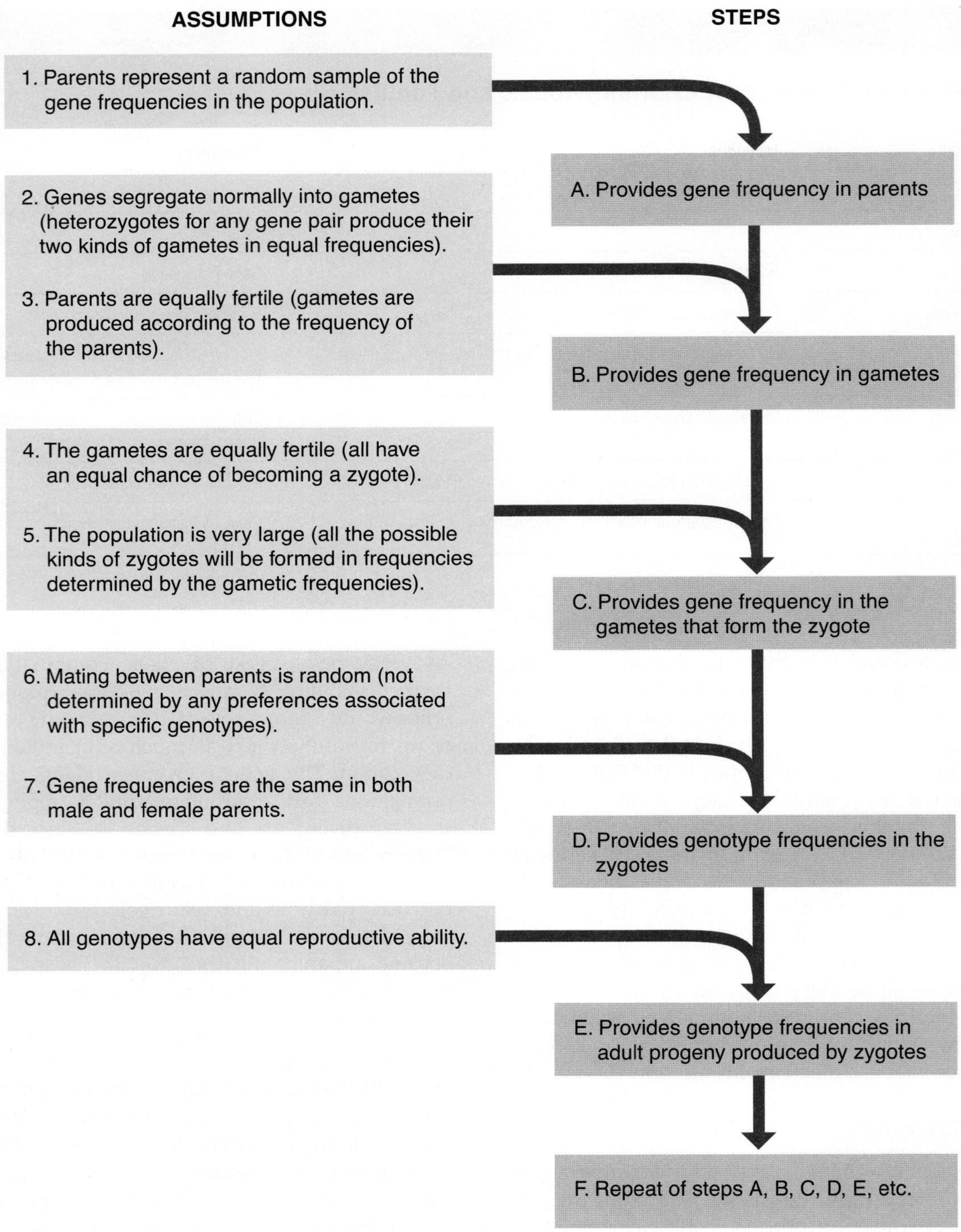

FIGURE 10.14 Assumptions of the Hardy-Weinberg equilibrium principle. Based on Faconer, D. S., and T. F. C. Mackay, 1996. Introduction to Quantitative Genetics, 4th ed. Longman, Harlow, Essex, UK.

suggested disruptive selection. Analysis of nuclear DNA revealed a gene that codes for melanocortin 1 receptor. *MC1R* assists in the regulation of melanin deposition in the mice's fur. DNA analysis aided by biotechnology allowed the researchers to prove the genetic causes contributing to disruptive selection of rock pocket mice.

Directional and Stabilizing Selection

Directional selection may be the most well-known form of natural selection thanks to media exposure in recent years about antibiotic resistant strains of bacteria. Bacteria that can survive an attack by antibiotic medications are resistant. As we use antibiotics more frequently, bacteria adapt. Eventually, antibiotic resistant

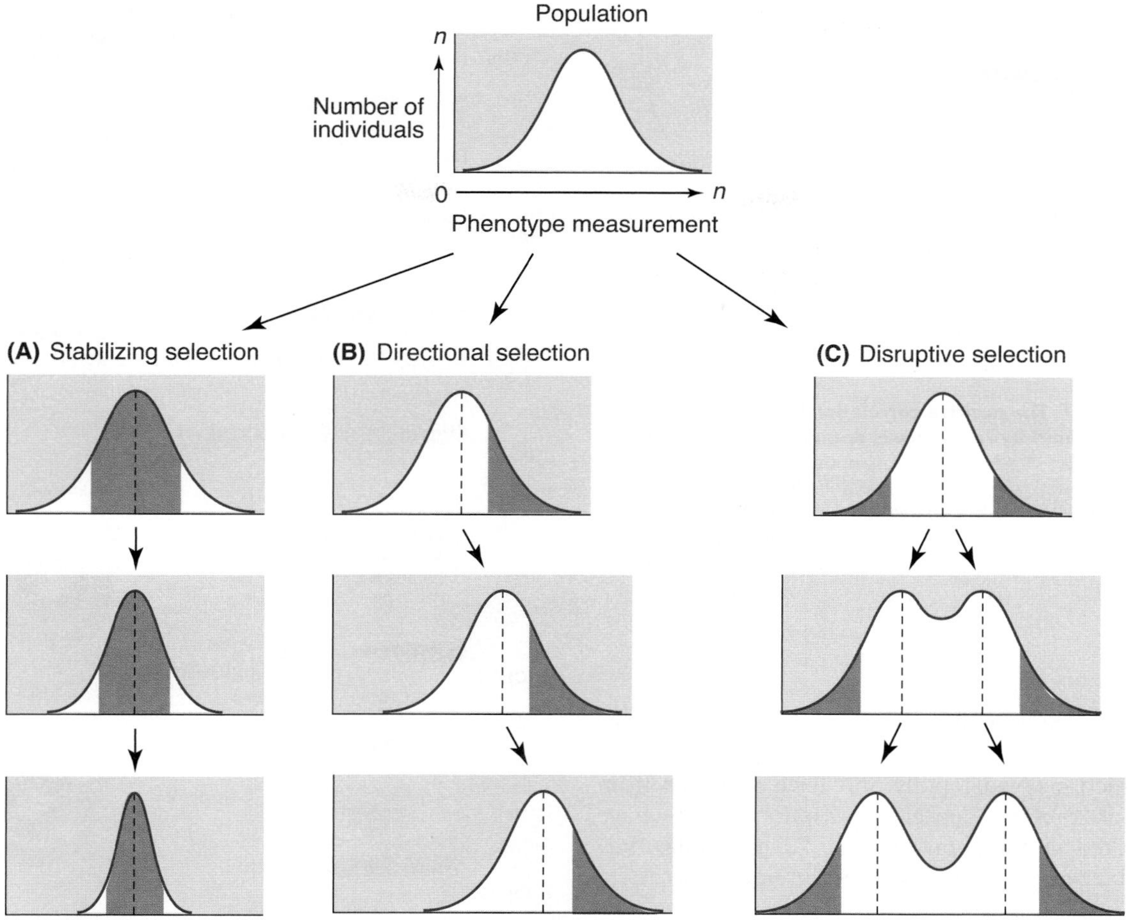

FIGURE 10.15 The types of natural selection demonstrate their different effects on a normally distributed population (at top). (A) Stabilizing selection causes the average phenotype to become more and more common. (B) Directional selection causes the average phenotype to shift in one direction. Although this image shows an increase in phenotype measurement over time, a decrease is also possible. (C) Disruptive selection causes the average phenotype to split into two equally common forms. Two phenotypes are characteristic of disruptive selection.

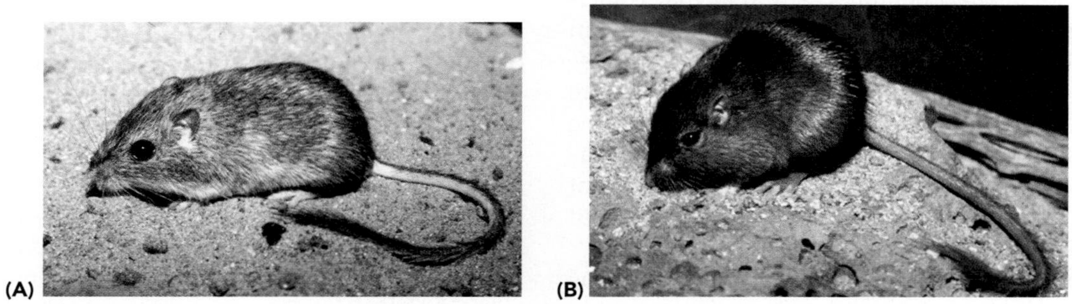

FIGURE 10.16 The two phenotypic variants of rock pocket mice. (A) The lighter form feeds on sandy substrates. (B) The dark variety feeds on black lava rock. © R B Forbes and the Mammal Images Library of the American Society of Mammalogists.

strains become more prevalent in their populations. A similar example was documented in the 1940s. The use of dichlorodiphenyltrichloroethane (DDT) as an insecticide had become more common in agricultural operations. Houseflies adapted to the presence of DDT in the environment and resistant strains increased (**FIGURE 10.17**). Resistance was measured by LD50, which equates to the dosage level needed to kill 50% of a population.

Stabilizing selection causes one average phenotype to become increasingly common. In contrast, other varieties become less common or even obsolete. Genomic

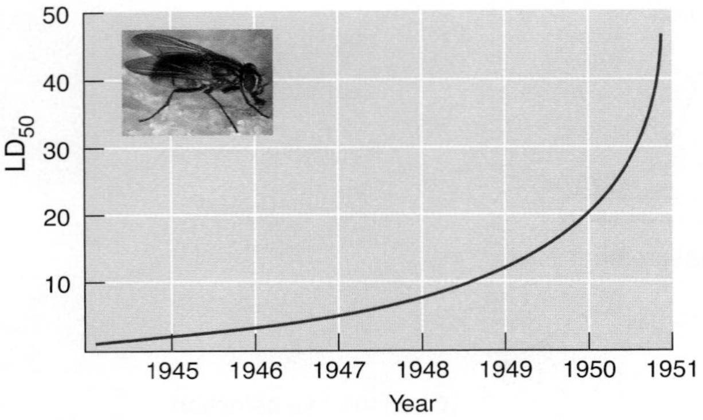

FIGURE 10.17 The positive correlation between DDT use and household fly resistance is depicted. As more DDT insecticide is used, it takes higher quantities to kill their targets. Adapted from Strickberger, M. W. *Genetics*, Third Edition. Macmillian, New York, 1985. (Based on data from Drecker, G. E., and W. N. Bruce, 1952. Housefly resistance to chemicals. *Am J. Trop. Med. Hyg.*, 1, 395–403.) (Insert) © Frank B. Yuwono /ShutterStock, Inc.

comparisons of humans with their primate relatives identified several transcription factors shared in common, which evolved rapidly and have persisted from that point forward. These transcription factors have been locked in for approximately 70 millennia; they have stabilized because of their evolutionary success.

Mutation

Mutations are permanent alterations of an individual genome. They may be detrimental to the organism if an essential gene is disrupted. They can also exist silently, causing no observable changes. Thus, mutations can occur that are in effect neutral. Lastly, they may coincidentally provide a benefit to the organism, resulting in a happy accident. However, the mutation must be present in gametes (sex cells) for it to be passed on to future generations. When they provide a benefit, they may be continuously selected for because they are advantageous, causing them to persist in the population. Silent mutations may also persist, as they provide no benefit nor cause any harm; they simply lie dormant.

Types of mutations are classified by the effect to the genome. Some of the most significant mutations in terms of genomic alteration are those that cause changes to chromosome structure (**FIGURE 10.18**). Some are so dramatic in their effects they can cause speciation. Two species of muntjac deer serve as an example. Karyotypes of the two species reveal a severe reduction of chromosome number as a result of translocation (**FIGURE 10.19**). The Chinese muntjac deer (*Muntiacus reevesi*) has 23 pairs of chromosomes. After translocation events, the Indian muntjac deer (*Muntiacus muntjac*)

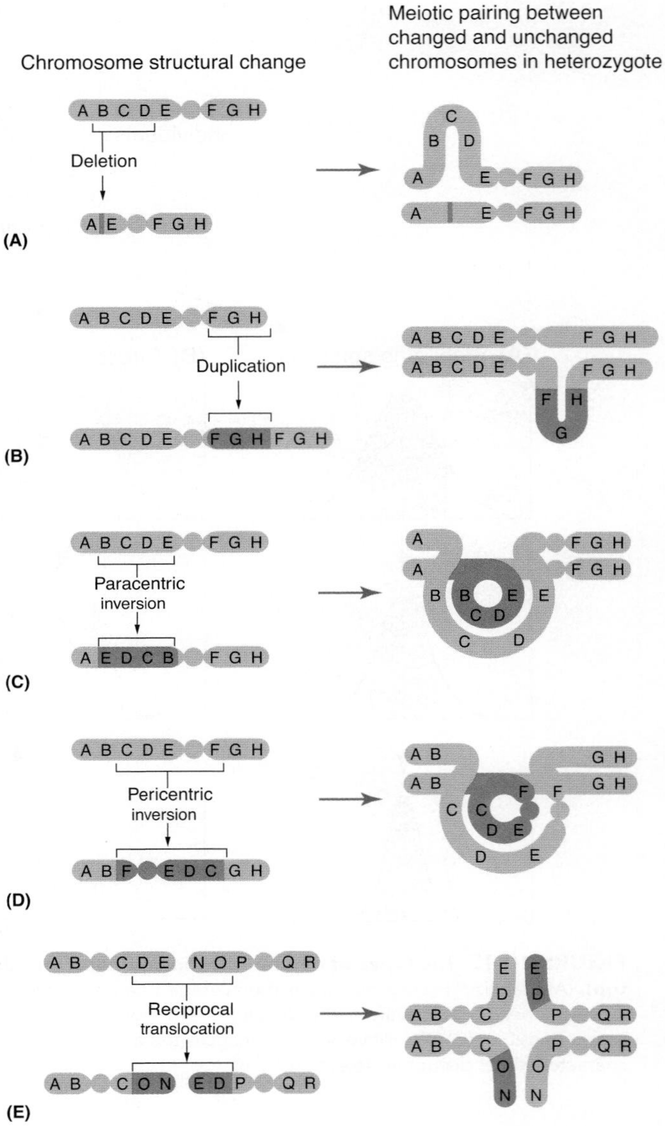

FIGURE 10.18 Mutations that alter chromosome structure (left) and the challenges they bring to pairing in diploid organisms (right).

has significantly fewer chromosomes but similar genes, indicating their common descent.

As we sequence genomes of more and more organisms, we can make genomic comparisons that readily identify mutations occurring over time. Genetic sequences are unparalleled as a tool in determining evolutionary changes, because the genetic code is universal. Proteomics can likewise provide information on descent with modification; instead of nucleotide sequences, we compare amino acid sequences. The enzyme triosephosphate isomerase is utilized in glycolysis, the breakdown of sugars to produce adenosine 5′-triphosphate (ATP energy). This is a necessary metabolic process in a wide variety of organisms, so the enzyme is highly

FIGURE 10.19 Translocation of chromosomes in muntjac deer differentiates two species. (A) The Chinese muntjac deer (*Muntiacus reevesi*) has 23 pairs of chromosomes. (B) The Indian muntjac deer (*Muntiacus muntjac*) has a severely reduced number of chromosomes but similar genes, indicating their common descent. (A) Courtesy of Chuck Dresner/Saint Louis Zoo; (B) Courtesy of Saint Louis Zoo.

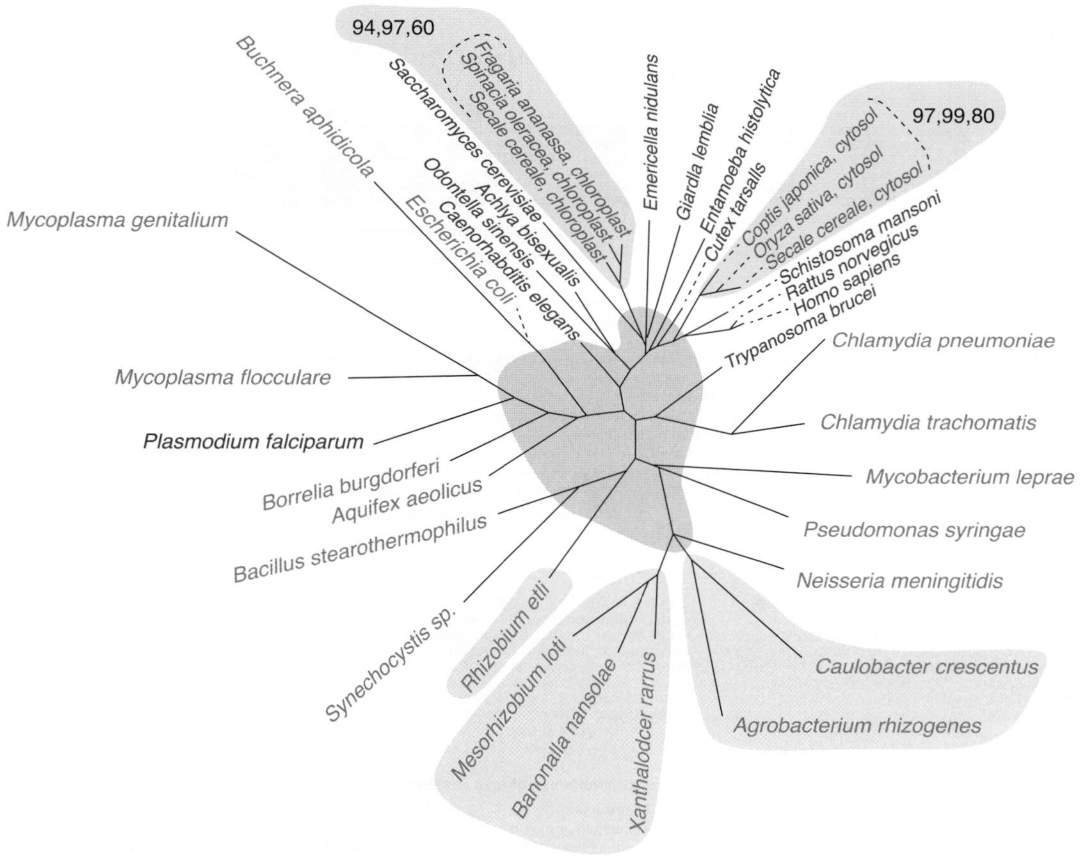

FIGURE 10.20 A phylogeny of distant organisms based on analysis of the triosephosphate isomerase enzyme. Eukaryotes are in brown; bacteria are blue; proteobacteria in blue background; green plants are in green background. The gray area is unknown. Reproduced from B. Canback, S. G. Anderson, and C. G. Kurland, 2002. The global phylogeny of glycolytic enzymes. *Proc. Natl. Aacd. Sci. USA* 99, 6097–6102

conserved among taxonomic groups. Comparisons of the enzyme's structural sequence allow geneticists to reassess previously valid relationships and estimate the distance between each organism in evolutionary time (**FIGURE 10.20**).

Analysis of genetic changes over time is often referred to as the study of molecular clocks. Any genetic sequence, whether nucleotides in RNA and DNA or amino acids found in proteins, can act as a **molecular clock**. Various species may be compared by using molecular clocks, which are conserved through evolutionary time. The genes of organisms under investigation are aligned to detect mutations. Sequence changes are positively correlated to the amount of evolutionary time. Therefore, greater amounts of observed mutations indicate more distant relations, and vice versa. Molecular clocks are used to determine the relative lengths of time that would be necessary for genetic changes to accumulate as species evolve. Carl Woese called them "evolutionary chronometers" because they measure evolutionary time between speciation events.

According to this concept, a common ancestor would possess the original sequence of genetic material, which then mutates over time as new species emerge. In fact, these genetic changes are what ultimately lead to divergence of organisms and the speciation that follows. Sequences that change slowly are preferred, such as mitochondrial DNA and ribosomal RNA. The slower the changes arise, the easier they are to distinguish as separate events occurring at different times. Analysis of molecular clocks is facilitated by aligning genetic sequences so mismatches are easily recognizable. Of course, bioinformatics makes it truly possible due to the enormous quantity of data that must be combed through. Changes as small as a single nucleotide or base pair can be detected, allowing us to narrow down organisms all the way through the phylogenetic tree. In one study, twelve species of *Drosophila* (fruit flies) were identified based upon comparative genomics (**FIGURE 10.21**).

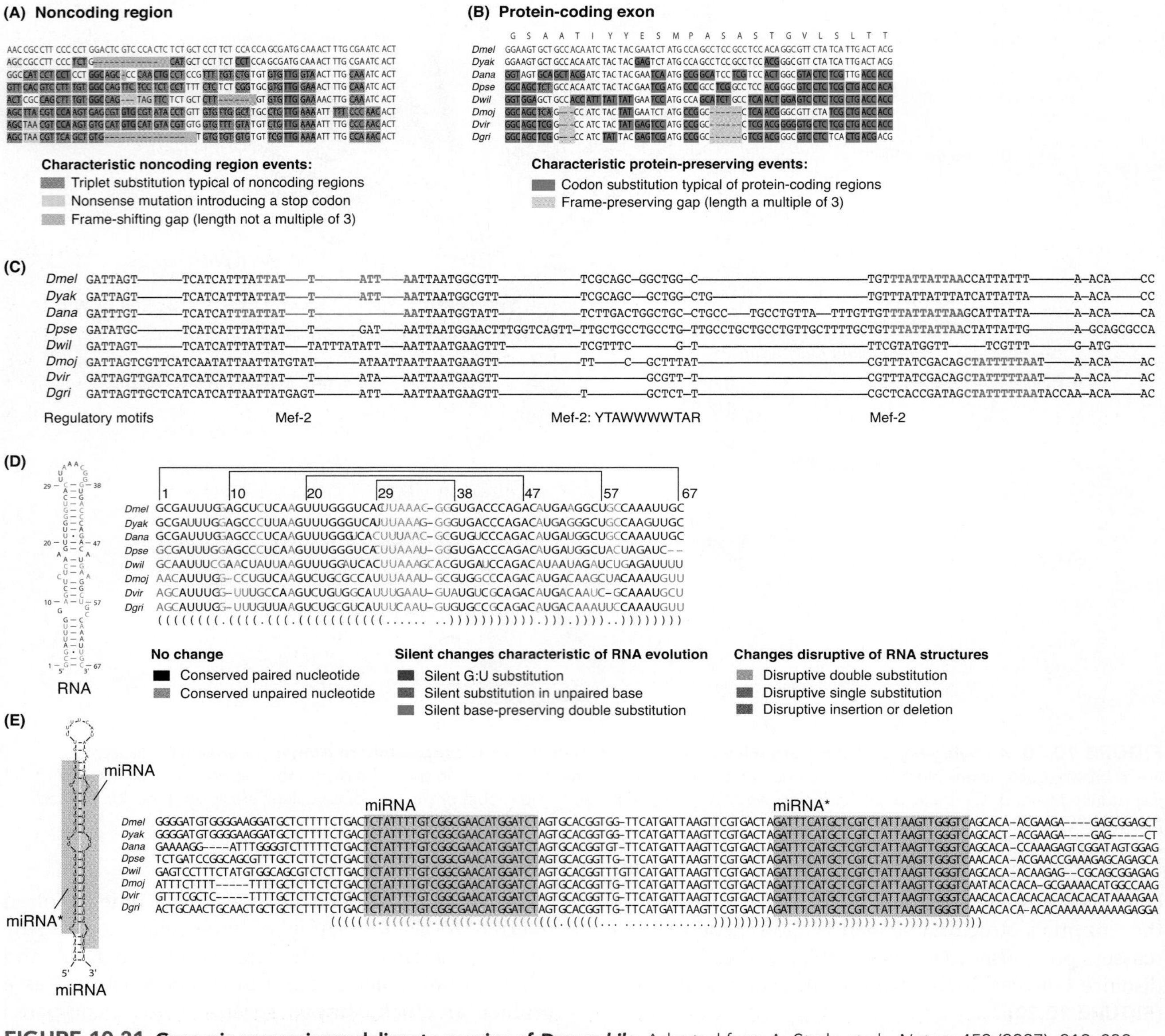

FIGURE 10.21 Genomic comparisons delineate species of *Drosophila*. Adapted from A. Stark, et al., *Nature* 450 (2007): 219–232

Nonrandom Mating and Sexual Selection

Another assumption in the Hardy-Weinberg equilibrium principle is random mating. For allele frequencies to remain in stasis, members of a population cannot preferentially select a mate for reproduction. This is alarmingly implausible. Examples of mate preference abound in nature. The nonrandom nature of mate selection can even go so far as to constitute sexual selection. **Nonrandom mating** is the choosing of a particular phenotype (or phenotypes) in a sexual partner over any other variety. When a particular variety is favored, this specific phenotype is selected upon and becomes stabilized. The fixing of a trait due to mate preference is known as sexual selection. We are all familiar with the brilliant plumage of peacocks. It is only exhibited in the male because they are the gender being chosen. The female birds that nonrandomly select the extreme male phenotypes are not in the crosshairs of sexual selection, so peahens remain rather drab in their appearance (**FIGURE 10.22**)

Gene Migration and Phylogeography

Gene migration is the movement of alleles across gene pools. A gene pool represents all alleles in a population. When members of a population relocate to another gene pool, the allele frequencies of both populations are changed. Microevolution occurs. Genomic databases and computational analysis allows gene migration to be traced across geographic regions. Combining gene migration studies with phylogenetics is a field of study called **phylogeography**. It tracks the movement of alleles and places these patterns into a geographic history (**FIGURE 10.23**). Interpretations of gene migration across distances can then be used to infer phylogeny (**FIGURE 10.24**).

Genetic Drift

The word "drift" implies randomness, going with the flow. **Genetic drift** is change in allele frequency happening without any clear pattern or selective agent. It is often cited as one of the most influential forms of microevolution. It can be thought of as non-adaptation, or change that happens just because. It is not typically seen as having a clear direction; it is aimless. Allele frequencies fluctuate but can return to their previous position, just as easily as they might fluctuate again. Because the concept of genetic drift is somewhat nebulous, many definitions exist (**TABLE 10.4**). Despite their differences, one underlying commonality in explanations of genetic drift is the increased likelihood when populations are small and/or isolated. Isolation limits gene migration and small population sizes can cause shifts in less amounts of time than is seen in large populations. One reason for this is inbreeding. The smaller a population, the greater the likelihood of closely related members reproducing becomes. Inbreeding has a tendency to concentrate recessive alleles.

Development

The ways in which organisms develop has been the subject of intense curiosity and mystifying complexity. The adage that ontogeny recapitulates phylogeny is not strictly true, but it did serve as a guidepost for pioneering developmental biologists looking for answers. Observing the similarities among organisms in their early development confirmed the notion of a

FIGURE 10.22 Sexual selection caused by the nonrandom mating preferences of peahens (A) results in an extreme phenotype in the peacock (B). (A) © InavanHateren/ShutterStock, Inc.; (B) © Igor Grochev/ShutterStock, Inc.

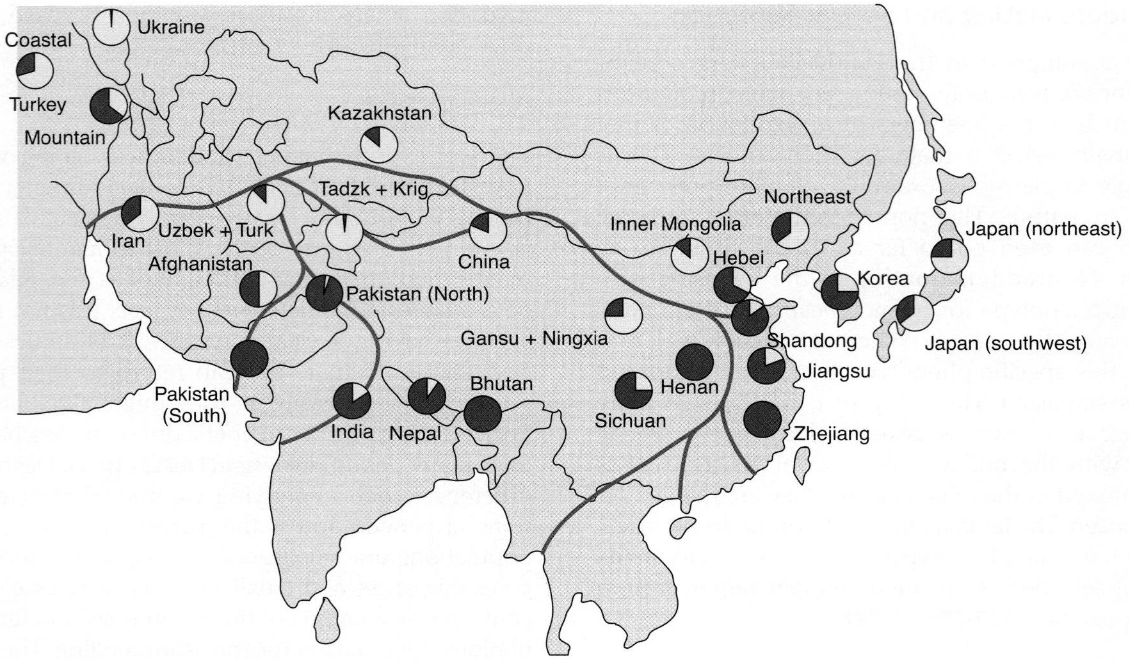

FIGURE 10.23 Comparisons of enzymes in Asian common wheat (*Triticum aestivum*) reveals a phylogeographic history linked to the Silk Road trade route (shown in brown). Adapted from Ghmire, S. K., Y. Akashi, C. Maitani, M. Nakanishi, and K. Kato, 2005. Genetic diversity and geographical differentiation in Asian common wheat (*Triticum aestivum L.*), revealed by the analysis of peroxidase and esterase isozymes. *Journal of Plant Breeding and Crop Science*, 55, 75–185.

TABLE 10.4

Definitions of Genetic Drift and Comments on Them

Definition	Comment
1. Random change in gene frequencies in populations	Far too broad and nonspecific
2. Variation in the genetic makeup of a species over time, often as a consequence of environmental change or isolation	How does environmental change affect gene frequency? Isolation of the species, a population, a subset of individuals?
3. Stochastic fluctuations produced by sampling in finite populations	Assumes that genetic drift is an anomaly of sampling
4. Changes in gene frequencies in a small population resulting from chance processes and not from natural selection	Sets genetic drift apart from selection, which is good, but confines the process to small populations
5. Gene frequency changes in populations resulting from random events in the gene pool of small populations rather than natural selection, especially the effects of sampling error	Confines drift to random events but then sees drift as a sampling problem not a natural process
6. Random fluctuations in gene frequencies, most evident in small populations	Does not confine the process to small populations
7. Change in frequency of a gene in a population through mutation, regardless of the adaptive value of the mutation	Does not distinguish genetic drift from mutation

common ancestor. Later stages of development display the inevitable specialization into unique organisms. The biggest question continues to be asked: how is the body accomplishing these intricate activities? How does a single cell, the zygote, ultimately divide and morph into a complex multicellular being? With the inception of biotechnology, researchers have had a previously locked door opened to them in the form of genomics and proteomics. Clearly, cells don't "know" how to orchestrate their development; there must be triggers manipulating ontogeny. The ability to isolate genes and determine their functionality, including when they are turned on and to what degree, enables scientists to put together the puzzle of ontological processes. Research

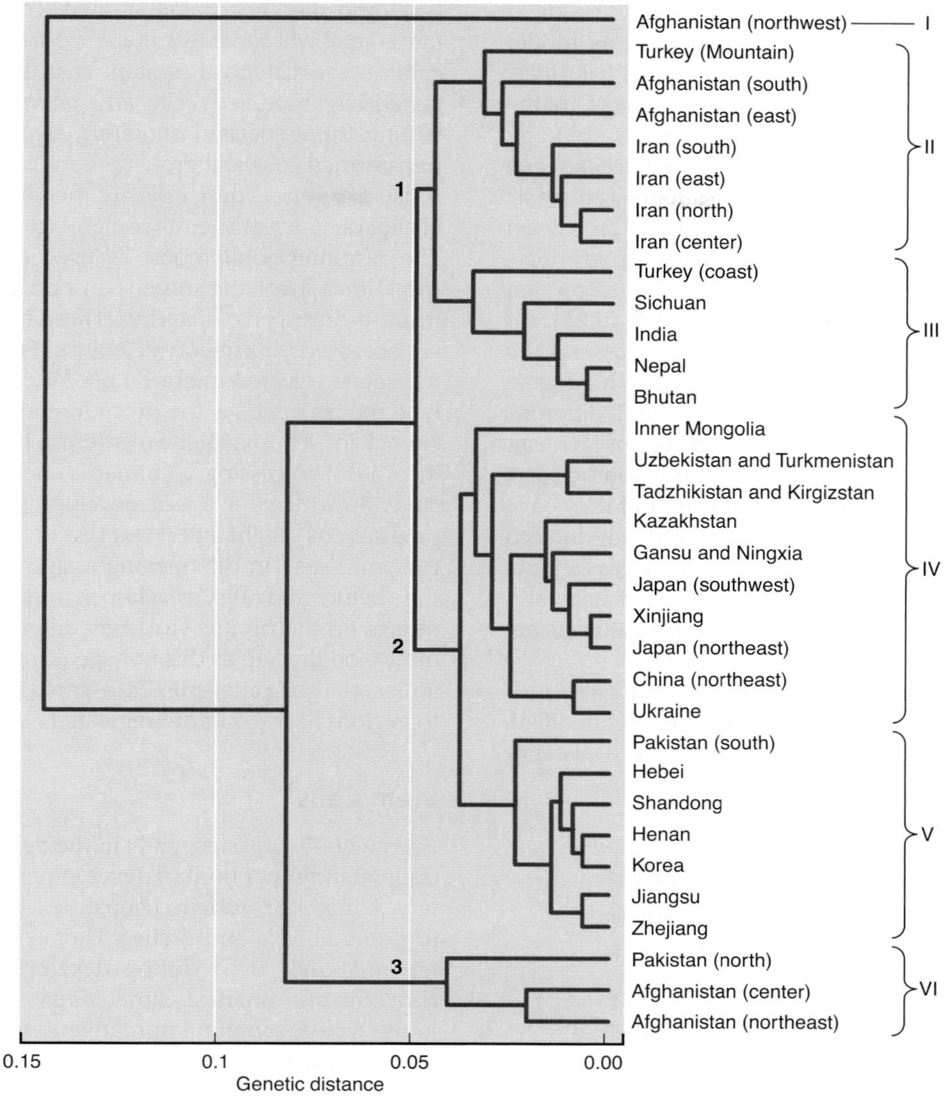

FIGURE 10.24 A phylogenetic tree of Asian common wheat based upon the geographic analysis shown in Figure 10.23 exhibits three separate lineages.

projects also investigate how cells give rise to tissues and organs and how cells regenerate and age. The efforts to unravel these mysteries continue today.

Research Methods

Historically, developmental biologists were limited to comparative embryology as a means to distinguish ontological patterns. Once the modern synthesis reconciled evolutionary theory with molecular biology, researchers shifted their focus to genetic influences on development. Much of today's research utilizes transgenic mice, which have genes of interest spliced into their genomes in order to observe the effects. The mice may also have genes knocked out so researchers can pinpoint the results of a disrupted gene. Another intense area of study is differentiation of stem cells. Tracing the path from an undifferentiated stem cell as it develops into a specialized cell allows researchers to analyze the conditions necessary for that path to occur. Tinkering with developmental conditions can reveal how environment influences the steps to differentiation.

Transgenic Mice

Model organisms are test subjects used in biological research. They allow *in vivo* observations as they occur in a live subject. The implication of using model organisms is the information obtained from their manipulation will extend beyond their species and apply to other living things. Evolution has influenced this assumption, as the concept of a common ancestor would result in conservation of developmental processes and metabolic pathways. Model organisms are a form of evo devo investigation. Nonmammalian species used as model organisms include frogs, yeast, roundworms, fruit flies, and zebrafish. Mice have become the preferred mammalian model organism in developmental biology research concerning human development because they are the closest genetic match to humans we can feasibly get. Ethics and the law prevent testing directly on human subjects. Order Rodentia, which includes mice, is a nearby branch to Order Primata (our taxon) on the phylogenetic tree of mammals (**FIGURE 10.25**). An extension of the common ancestor precept asserts the closer two organisms are related, the higher the degree of conservation in genetic material, cellular differentiation and regeneration, tissue and organogenesis, metabolism, and aging. We could study our closest relatives, such as gorillas or chimpanzees, but they have limitations as well. They are in comparatively limited number as opposed to mice, more expensive to care for, and take longer to reproduce. We can breed mice at a much faster pace and high numbers of test subjects are desirable for statistical purposes.

Transgenic mice are produced via a multistep process. First, embryonic stem cells are genetically modified. Two transgenes are inserted: an antibiotic resistance gene and the gene being studied. The antibiotic resistance gene will serve as a marker. Much like plant tissue culture, the Cre/*lox P* system may be employed. The genetically modified cells are grown in culture containing the associated antibiotic. Only the successfully transformed cells will grow up because they contain the resistance gene. These cells are then injected into a host blastocyst (an early embryonic stage of development). The resulting adult mouse is used in genetic crosses, sometimes involving more than one generation, which result in transgenic offspring. These transgenic progeny are observed for phenotypic effects (**FIGURE 10.26**). The transgene selected might be used to knock out a wild-type mouse gene, as the procedure just described, or to "knock in" a human gene of interest. By creating a knock in mouse expressing a human gene, experiments can study how the gene affects development in mice or how medications might affect expression of the gene. The two mice seen in the opening image of this chapter are the "before and after" of a knock in event. The wild type mouse on the right is strikingly different than the large mouse on the left, as this transgenic mouse is expressing an inserted obesity gene. Transgenic mice are also used to explore how genes are regulated.

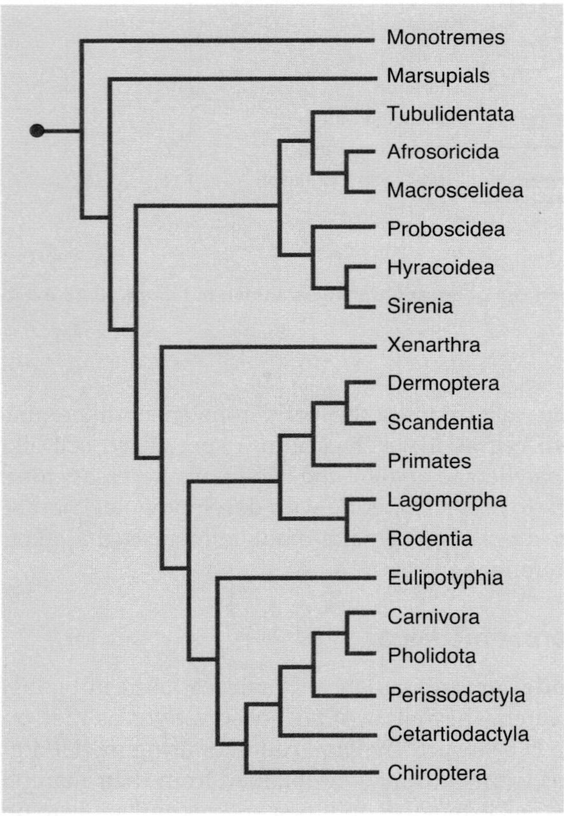

FIGURE 10.25 The phylogeny of Class Mammalia exhibits the close relationship between Orders Rodentia and Primata (primates).

Stem Cells

Even with the inroads made in the field of evo devo, our understanding of human development is still rudimentary. A major obstacle to overcome is our limited access to prenatal cells and tissues. They represent the actual raw materials used during development, but ethical concerns and political climates have historically stood in the way of sampling human embryonic stem cells on a large scale suitable for sustained research activity. The advent of induced pluripotent stem cells is providing the tools to circumvent these roadblocks. Clearly, it is preferred to study development with real human subjects instead of mice, but because we can't use human individuals, we can surely look to prenatal tissues as a solid alternative.

One scheme practiced by the New York Stem Cell Foundation (NYSCF) is to harvest adult stem cells from volunteer human subjects. These individuals are usually suffering from various diseases and they are eager to explore any means that might result in a treatment or cure. The harvested adult stem cells are a genetic match to the patient; therefore NYSCF research is customized, representing a form of pharmacogenomics. Adult stem cells are used as "avatars" of the human patient and subjected to a methodical assay of environmental treatments. Each treatment leads to an observable progression of development from stem cell to specialized cell. Promising treatments are isolated for further study. In collaboration with Columbia University researchers, NYSCF has successfully identified customized diabetes treatments for its patients. They were able to

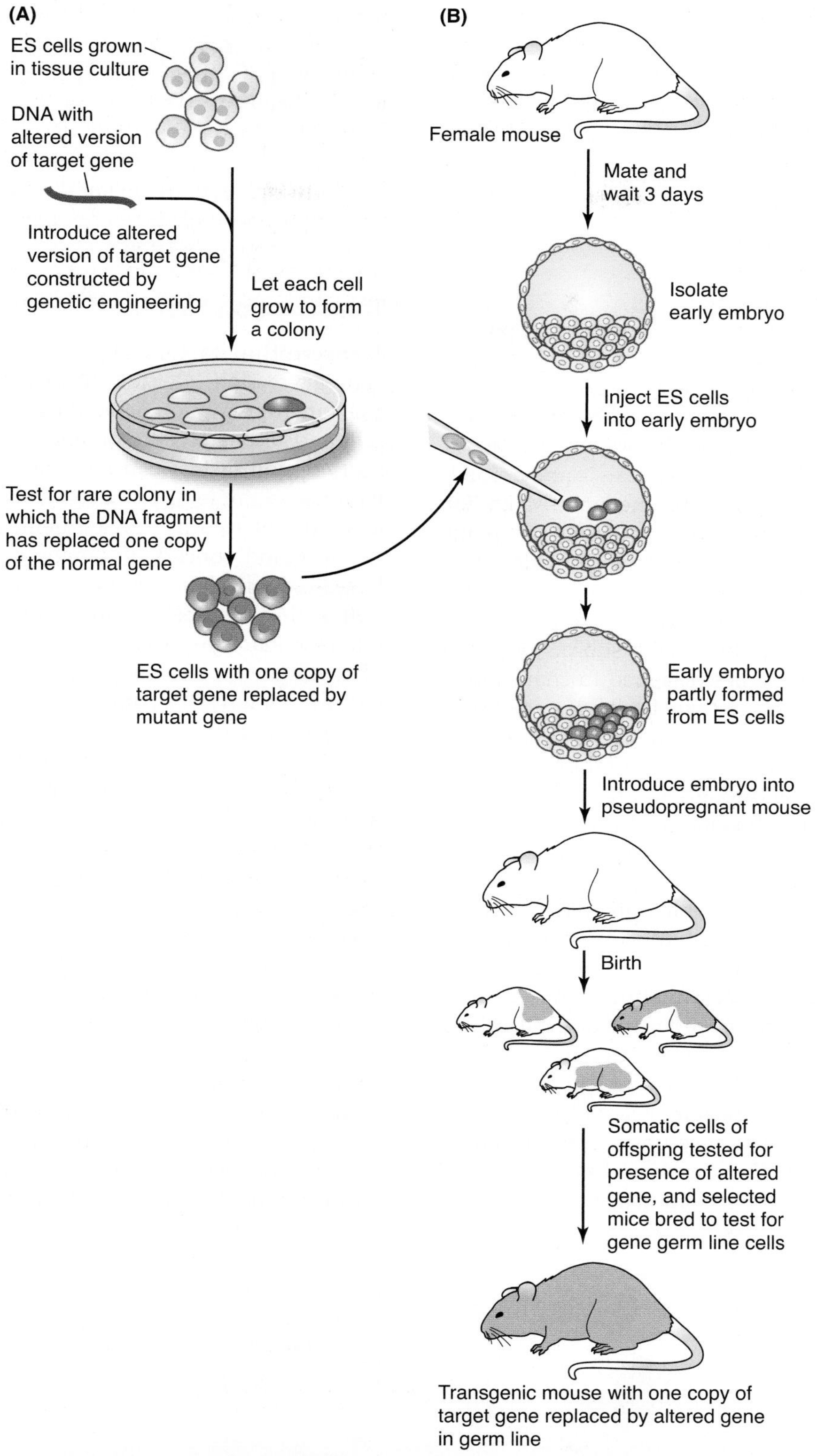

FIGURE 10.26 Production of transgenic mice involves (A) transformation of embryonic stem cells followed by (B) transfection of the model organism and subsequent crosses. Adapted from B. Alberts, et al., *Molecular Biology of the Cell*, Fourth Edition. Garland Science, 2002.

differentiate normal (i.e., healthy) insulin-producing beta cells from harvested adult stem cells that had been induced into a pluripotency state. To confirm their results, these beta cells were spliced into transgenic mice, which resulted in normal insulin expression levels. This is a promising accomplishment which will hopefully translate into similar successes for other genetically based disorders.

Discoveries

Insights can draw connections between development and evolution. Better understanding of ontology, cellular regeneration, and aging are directly applicable to medical biotechnology. As we hone in on developmental errors, we are better able to explain what causes disease. Once the progression of disease is understood, further experiments may be designed to search for treatments. Biomedical science is reaching the point where individualized medicine, specifically tailored to one's genetic profile, may be realized. In many ways, pharmacogenomics relies upon evo devo to progress.

Evo devo has revealed two major subsets of genes existing in all cells. In studies designed to characterize cellular specialization, these two classes of genes play different roles. **Housekeeping genes** are those necessary for survival in all cells, regardless of the type. These genes remain functional throughout the life of each cell and provide basic functionalities such as the respiration pathway (to provide energy), cellular regeneration (mitosis), and protein synthesis (gene expression). Because they are essential to all cells, housekeeping genes are highly conserved among different species. They have become fixed in these genomes as a result of stabilizing selection. **Specialization genes** vary in their timing and intensity in order for a given cell to differentiate into its programmed fate. Indeed, it is the specialization genes that are the "program" that places an undifferentiated stem cell onto a certain path. Specialization genes cause one cell to become a red blood cell while another becomes a skin cell. Only those genes necessary for the particular cell type will remain functional in the differentiated cell. Because specialization genes on the whole contribute to the diversity of life, they will vary more greatly among different organisms. We know, however, that it is not just genes that drive the steps of development. The regulation of genes plays perhaps an even more important role. A sampling of the major discoveries resulting from the application of biotechnology to developmental inquiry is now presented.

Transcription Factors

Transcription factors (TFs) control the initiation of gene expression (**FIGURE 10.27**). They can prohibit or facilitate transcription depending upon the needs of the cell, and ultimately, the organism. We have seen how transcription factors can be conserved in the genome with the example of primates. They are known to play a significant role in development, acting as maestros in the grand concert of genetic activity taking place. Research into transcription factors operates on the notion that they behave in a stepwise fashion, where one TF initiates expression of gene A, causing a second TF to initiate expression of gene B, and so on. Metabolic pathways follow this pattern, so it is thought that other biological processes such as cellular differentiation may also experience a cascading domino effect. This is an extremely simplified explanation, though. It's important to remember that in addition to timing of gene expression, intensity is involved. A gene can be expressed a little or a lot in different cells. Furthermore, a single gene likely encodes a family of proteins, not just one. Thus, even if a gene is expressed in equal quantities, different proteins may be synthesized due to alternative splicing mechanisms. Two major classes of transcription factors are now highlighted, the HOX and PAX proteins.

The Homeodomain

Homeodomain genes are a class of genes involved in body plan development. They express HOX proteins that control development of the anterior-posterior axis, which can be thought of as the head-to-tail body plan.

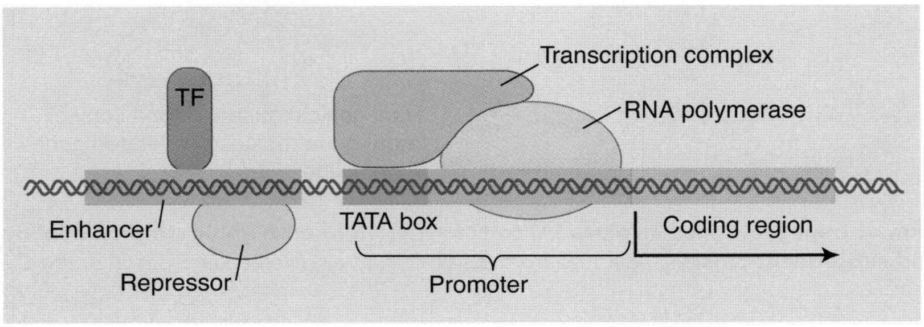

FIGURE 10.27 A view of gene expression at the molecular level displays a transcription factor (TF) associating with the enhancer region. The TF combines with RNA polymerase in forming the transcription complex, which initiates transcription when bound to the TATA box. Adapted from Carroll et al., 2005. *From DNA To Diversity*, Second Edition, Blackwell Publishing, Malden, MA.

HOX proteins are examples of transcription factors. For humans and other mammals, formation of the brain downward through the spinal column takes place in the embryonic stage early in development. It is a crucial time because many developmental abnormalities form in the embryonic stage. HOX proteins are highly conserved and have been extensively studied in the model organism *Drosophila* (**FIGURE 10.28A**). HOX genes were first noted in *Drosophila* in a mutant form that causes organ transformations, such as the presence of a fruit fly leg where its eye should be. These genes have been compared to homeodomains in other organisms (**FIGURE 10.28B**).

The insights into the Drosophila homeodomain were applied to knowledge of human development. Upon completion of the Human Genome Project, combing

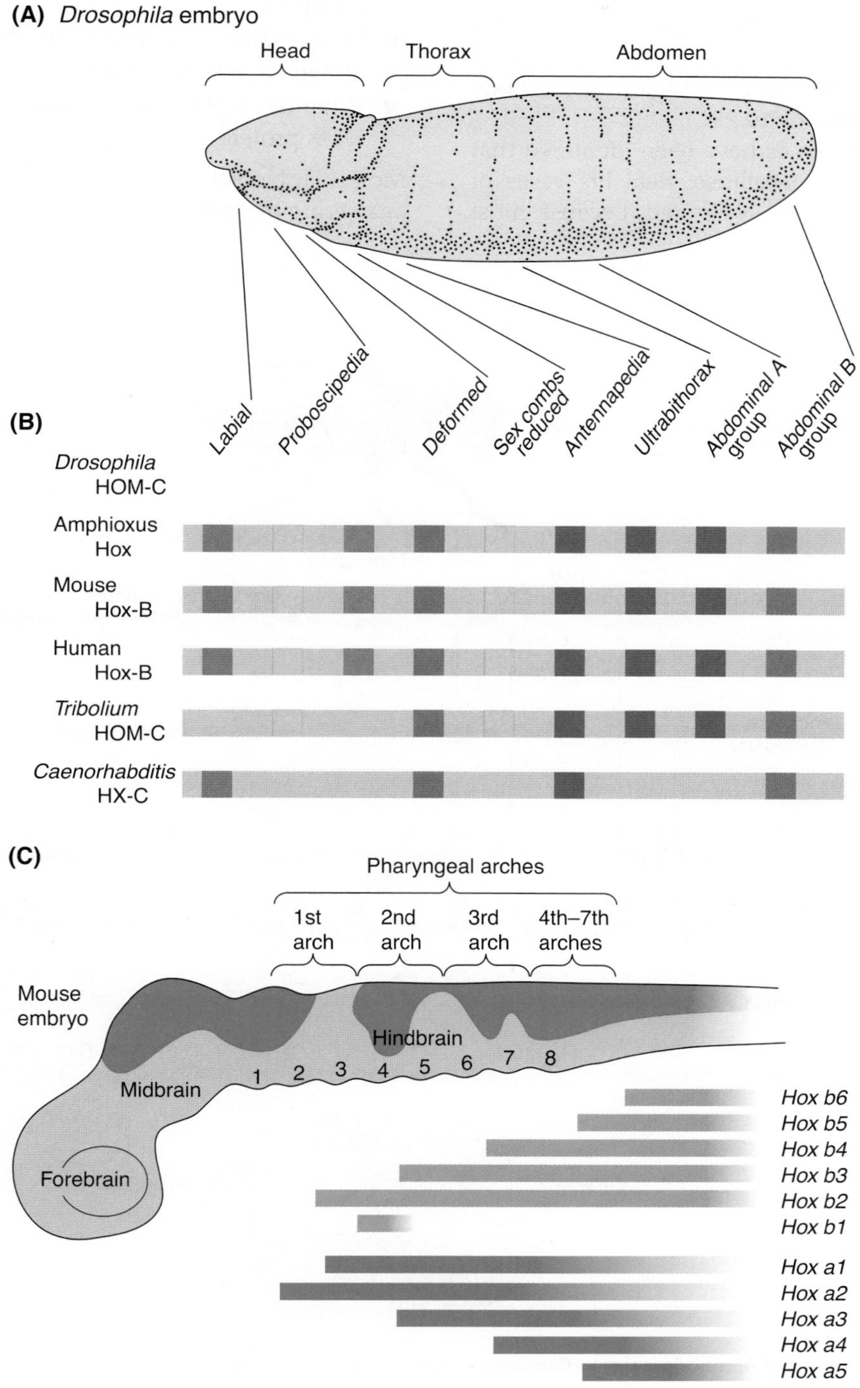

FIGURE 10.28 The homeodomain of *Drosophila* (A) compared to those of other organisms displays a strong degree of fixation (B). Hox genes play a role in development of the anterior-posterior axis (C).

through the data identified four HOX gene clusters (A, B, C, and D), each with 13 genes (**FIGURE 10.29**). Analysis of gene sequences in each cluster shows there is more variation within a cluster than between them. Paralogs of genes are much more similar. A **paralog** is a gene in the same location in each cluster; for example, the A5 HOX gene is a paralog to B5, as both are paralogs to C5, and so on. This revelation suggests that paralogs may have similar functions. Further investigation determined the order of HOX genes within a cluster is analogous to the order of their expression: gene 1 is expressed in the most anterior region, whereas gene 13 is expressed in the most posterior region of the body plan as it develops (**FIGURE 10.28C**).

Mutations of HOX genes have been identified that result in body plan abnormalities. Real life cases of these mutations are extremely limited, because most HOX mutations would likely be fatal to the developing fetus. This explains our relatively recent awareness of such genetic abnormalities. However, some HOX genes act as transcription factors later in development or even in the adult. Incidences of these mutations are more frequent and available as case study material. Additional areas of exploration will reveal more information about transcriptional networks:

- What gene(s) does each HOX transcription factor act upon?
- What molecules interact with the homeodomain and what are their effects?
- What other TFs, if any, facilitate the actions of each HOX protein?

Molecular biotechnology is an irreplaceable resource in answering these questions.

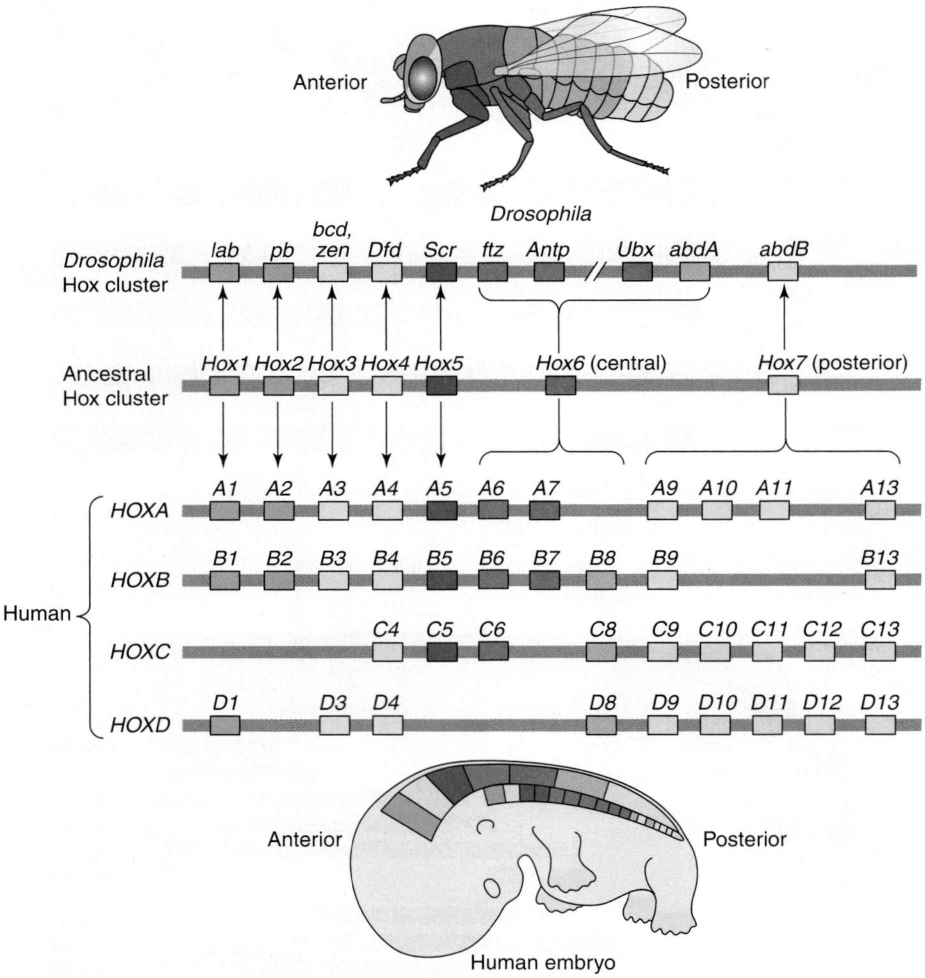

FIGURE 10.29 Comparisons of homeodomains in *Drosophila*, humans, and a hypothetical common ancestor. Color coding of genes signify the regions of development they affect in both organisms. Modified from *Mol. Genet. Metab.*, vol 69, A. Veraksa, M. Del Campo, and W. McGinnis, Developmental Patterning Genes and Their Conserved..., pp. 85–100, copyright 2000, with permission from Elsevier [http://www.sciencedirect.com/science/journal/ 109671921].

TABLE 10.5

Summary of PAX Genes

PAX Gene	Associated Structures	Known Mutations
PAX1	• Limb buds • Thymus gland • Vertebral column	Skeletal defects (in mice)
PAX3	• Doral neural tube • Heart • Neural ganglia • Melanocytes • Schwann cells	• Waardenburg syndrome • Hirschsprung disease
PAX6	Eyes	• Aniridia • Cataracts
PAX9	• Limb buds • Thymus gland • Vertebral column	In mice: • Skeletal defects • Lack of pharyngeal pouch derivates In humans: • Oligodontia

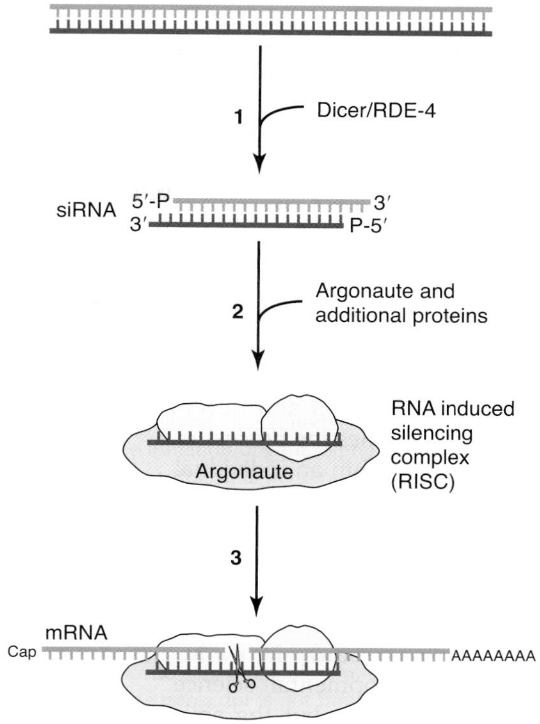

FIGURE 10.30 Translational control by RNA interference. Adapted from C. R. Faehnle and L. Joshua-Tor, *Curr. Opin. Chem. Biol.* 11 (2007): 569–577.

PAX Proteins

PAX proteins are another class of transcription factor. Unlike HOX proteins that determine anterior-posterior development, PAX proteins are responsible for **organogenesis**, the formation of organs from specialized tissues. Another role of PAX proteins is the maintenance of pluripotency in adult stem cells. Mammals including humans have nine PAX genes, which cause deformities when mutated. Studies with transgenic mice have shown skeletal defects and lack pharyngeal pouch derivatives, such as the thymus and parathyroid glands, when the *PAX9* gene is knocked out. Some human characteristics controlled by PAX genes are summarized in **TABLE 10.5**.

RNAi Gene Regulation

RNA interference (RNAi) is when RNA molecules engage in translational control. Genes are transcribed into messenger RNA but the second step of expression—translation—is prohibited by RNAi. Therefore, mRNA is not translated into an amino acid sequence that otherwise would result in protein synthesis. First discovered in plants, research has shown RNAi also occurs in life forms ranging from invertebrate insects to vertebrate mammals, including us. RNA interference is used as an immune response to block viral DNA and transposons and during development to control gene expression.

RNA molecules involved in RNAi include small interfering RNA (siRNA), microRNA (miRNA), and short temporal RNA (stRNA). These molecules contain complementary sequences that match segments of an mRNA transcript, allowing them to bind in their single stranded form. The binding action prevents translation from moving forward. In another RNAi mechanism, an RNA induced silencing complex (RISC) containing RNase is formed between the interfering RNA and the mRNA transcript. The RISC utilizes RNase to cleave the bound complex from the transcript and the resulting mRNA fragments are digested by exonucleases into their nucleotide subunits (**FIGURE 10.30**). Over 250 siRNAs are currently known and further studies are likely to reveal many more. RNAi is directly involved in development as a gene switch, effectively silencing genes in order to prevent unnecessary expression. RNAi mechanisms are under investigation as a method to knock out genes in transgenic mice. There is an ongoing search for mutant RNA molecules that may exist, as they would hypothetically cause developmental abnormalities.

© Pascal Goetgheluck/Science Source

11 The Biotechnology of Anthropology

LEARNING OBJECTIVES

Upon reading this chapter, you should be able to:

- Define anthropology and list its branches.
- Identify which branches of the discipline have benefitted from molecular biotechnology and how.
- Differentiate between two models of human evolution.
- Describe Charles Darwin's position on human evolution.
- List the characteristics of primates.
- Explain how humans are related to other primates.
- Define hominin and hominid.
- Explain the process of DNA-DNA hybridization and how it relates to molecular systematics.
- Compare and contrast gene trees with species trees.
- Identify the molecular techniques used to rectify which human evolution model is most accurate.
- Characterize the hypothetical people known as Mitochondrial Eve and Y Chromosome Adam.
- Explain the genetic relationships among humans, Neanderthals, and Denisovans.

Humans seem to have an innate curiosity about their own existence. Historical record provides us with a narrative of documented events, but the majority of our story took place in the passage of prehistoric time. The discipline of anthropology seeks to fill in the gaps of knowledge with regards to our past, in addition to how we might explain the present and forecast our future. **Anthropology** literally means "the study of man," but is interpreted and applied as the study of humanity. Anthropology was born out of cultural clashes that took place as Western societies made contact with remote civilizations during the course of "new world" exploration and the colonial period that followed. Burgeoning anthropologists sought to characterize the unusual lifestyles, cultures, and peoples that were previously unknown to them. From this practice, a branch of anthropology known as cultural anthropology was established.

Four main branches of anthropology are recognized. In addition to cultural anthropology, the divisions include linguistics, archaeology, and biological anthropology, also known as physical anthropology. Linguistics is the study of languages, such as how they emerged in different human groupings and how they intersect and change over time. Archaeology is the study of ancient civilizations and involves digging into layers of earth where archaic sites and settlements may be discovered and described in detail. Archaeology unearths artifacts (physical remnants of humanity such as tools and works of art) as well as fossils, including human remains. The study of ancient human remains, those of our extinct ancestors, and the physical variation among living members of our species is the branch known as biological anthropology. Within this branch, **paleoanthropology** is an additional term used to distinguish the study of ancient human fossils and those of our ancestors. The *Focus on Careers* box offers websites and information about careers in this fascinating field.

Of course, archaeology provides substantial material to the division of biological anthropology and more specifically, paleoanthropology. For much of its existence, paleoanthropology has relied upon morphological descriptions and comparisons of fossils within the human lineage. It is from the fossil record that anthropologists constructed a story line of human evolution. The same may be said for the study of modern, extant human beings. Biological anthropologists primarily dealt with the outward appearance of human beings and accounted for the tremendous variety among us in terms of adaptation to geographical and climatic differences, leading to the concept of race. Tools offered from molecular biotechnology are changing the previously held notions of how our species came to be, how we migrated through the continents and populated the planet, and how we are genetically connected to our ancestors and each other. Therefore, this chapter will focus on the genetic underpinnings of human evolution and adaptation as it pertains to biological anthropology, paleoanthropology, and archaeology.

FOCUS ON CAREERS

Paleoanthropologists

Paleoanthropology is a specialized field, drawing upon disciplines such as human evolution, archaeology, genetics, and bioinformatics. As such, becoming a paleoanthropologist requires an eclectic education including the attainment of a doctorate degree. Most universities offer a general anthropology degree at the bachelors level and students do not specialize into one area of anthropology until graduate school. Obtaining the bachelors of anthropology is a worthwhile experience, as it does introduce students to all branches of anthropology and allows paleoanthropology to be placed into context. Alternatively, an individual on the paleoanthropologist career track may opt to major in a hard science and minor in anthropology. In such a case, biology, genetics, or biotechnology would be appropriate majors. In fact this may be preferable in order to gain a solid foundation in the sciences that heavily influence paleoanthropology.

Volunteering at a natural history museum, interning in a laboratory, or participating in archaeological excursions are advisable experiences during schooling. The Smithsonian's National Museum of Natural History lists several practical experience opportunities on its website (http://humanorigins.si.edu/education/students), as does the professional organization, The Paleontology Society. Visit http://www.paleoanthro.org/student/ for links to educational programs and organizations relevant to paleoanthropology.

Becoming Human

With the release of Charles Darwin's *On the Origin of Species* in 1859, and to an even greater extent when he published *The Descent of Man* in 1871, the newly conceived discipline of evolution was fraught with implications on human origin. Darwin's first work skirted around any direct discussion of human evolution from a primate ancestor, but reading between the lines left no other possibility. As he suggested, all living organisms share a common ancestor, then humans must be included; we must have evolved from some other animal, with the most logical conclusion being a close relationship to other human-like beings, which we call primates. *The Descent of Man* was his direct reply to public and academic discussions on the subject. He sought to explain the emergence of *Homo sapiens* from a common ancestor with other extant primates primarily in terms of a gradual shift toward higher brain function, language, and culture.

Fossil evidence obtained from archaeological expeditions would continue to reinforce the idea of humans evolving from a common ancestor with our primate relatives. The oldest fossils of human ancestors, including our supposedly most recent ancestor *Homo erectus*, were routinely discovered in the African continent, leading many anthropologists to conclude our species emerged in this location. Known as the "Out of Africa" hypothesis, this model ascertains our most recent ancestor *Homo erectus* evolved into our current species *Homo sapiens* in Africa alone and then went on to populate Eurasia, outcompeting and replacing human ancestors such as the Neanderthals in their wake. However, additional fossils found in Europe and Asia confounded this explanation and led to an alternative hypothesis, in which humans evolved independently in separate regions of the world. This "Candelabra" model, also known as the multiregional hypothesis, operates on the assertion that *Homo erectus* not only evolved into our species in Africa, but experienced independent evolutions into *Homo sapiens* in Europe and Asia as well (**FIGURE 11.1**). As discussed below, biotechnology has helped resolve which hypothesis is best supported with conclusive scientific evidence. In a recent turn of events, we shall see how biotechnology has even added surprising layers of information onto prior explanations.

Divergence from Other Primates

Morphological comparisons of humans to other organisms suggested we are most closely related to primates. Primates emerged in the fossil record approximately 85 million years ago. All primates are placental mammals

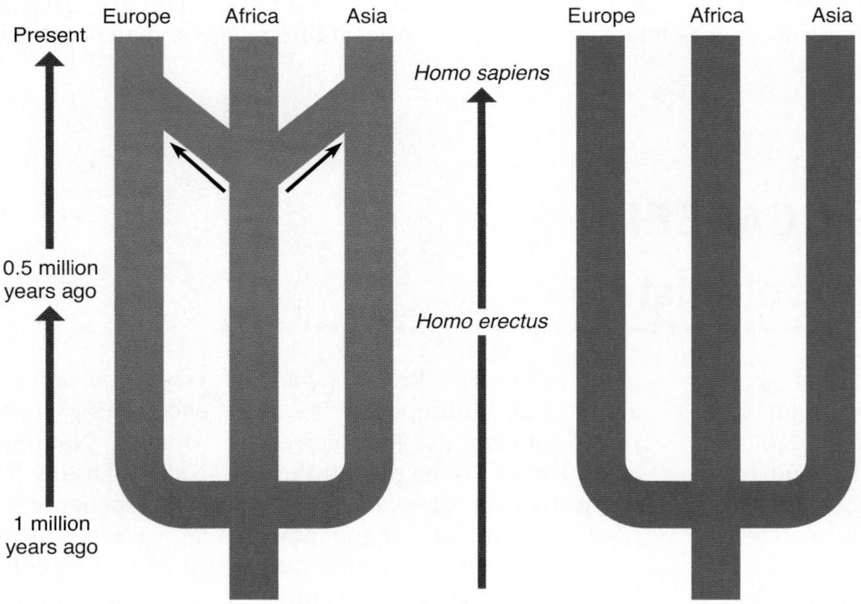

(A) "Out of Africa" model
Single origin of *Homo sapiens* in Africa, and replacement of *Homo erectus* in Europe, Africa, and Asia

(B) "Candelabra" model
Multiple origins of *Homo sapiens* from *Homo erectus* populations in Europe, Africa, and Asia

FIGURE 11.1 The two opposing hypotheses of human evolution from our most recent ancestor, *Homo erectus*. (A) The Out of Africa model suggests a single origin of *Homo sapiens* derived from Africa. (B) The Candelabra model suggests multiple origins of our species, with independent evolutions occurring in Africa, Europe and Asia.

that exhibit certain physical and behavioral features, including:

- Large brains relative to body size/mass
- Grasping hands with opposable thumbs
- Grasping feet with opposable toes (except humans)
- Stereoscopic vision (eyes face forward, not to either side)
- Social organization

In taxonomical classification, Order Primata includes lemurs, tarsiers, galagos, monkeys, and apes (gibbons, bonobos, chimpanzees, orangutans, and gorillas). Gibbons are also known as lesser apes, while the remainder of this group is referred to as the great apes. The two lines of Order Primata, strepsirrhini and haplorhini, diverged around 77 million years ago (**FIGURE 11.2**). Multiple studies indicate the human lineage diverged from other extant primates approximately seven million years ago. This evolutionary fork in

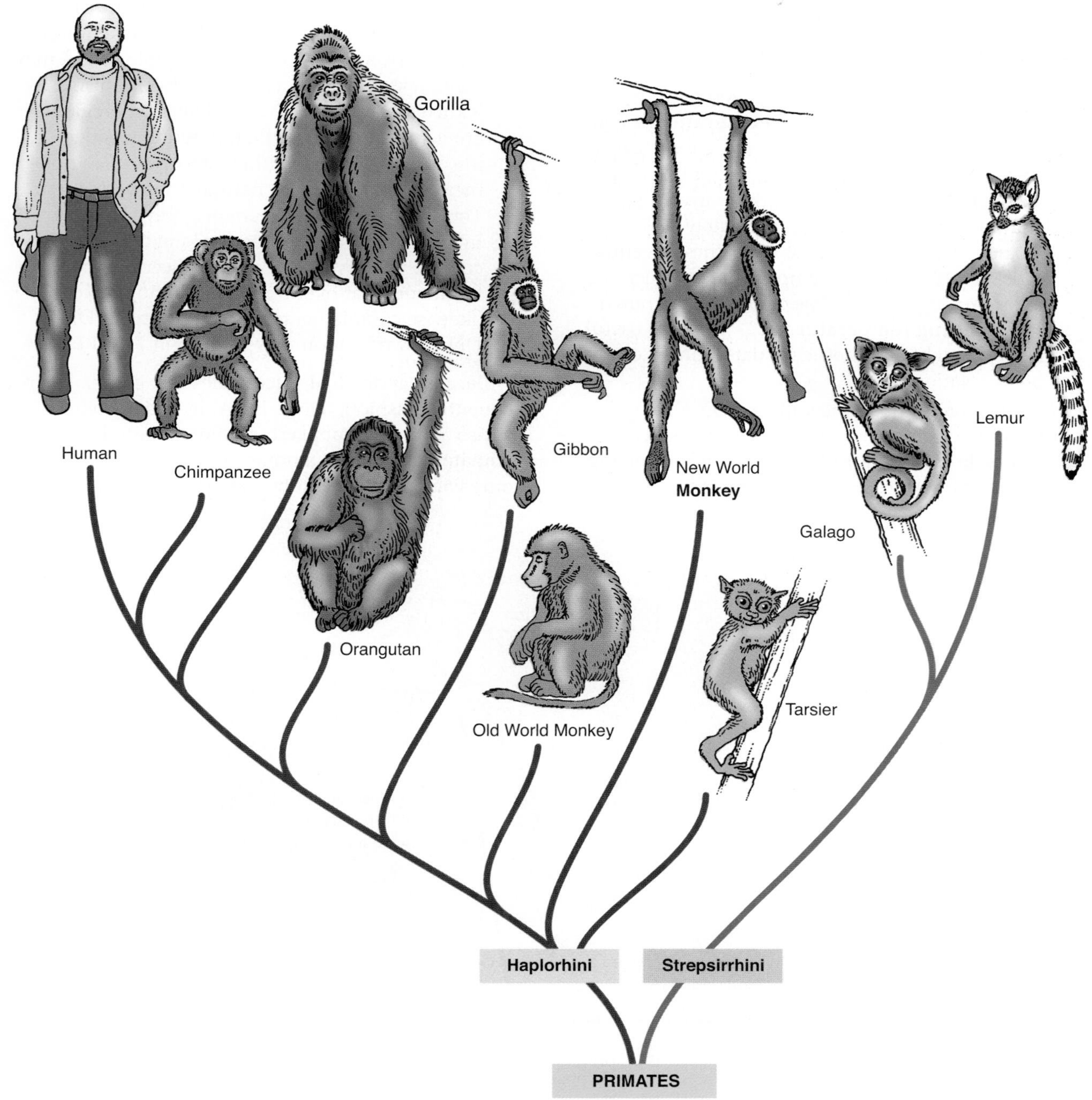

FIGURE 11.2 A traditional phylogeny of Order Primata shows two lineages. The strepsirrhini are wet-nosed, small, and nocturnal. The haplorhini are dry-nosed primates composed of monkeys and apes. It is estimated the two lineages diverged around 77 million years ago.

the road is represented by the classification of humans with the great apes into a group known as the **hominids**. The date ranges for these organisms was initially determined through carbon dating of fossils, but modern techniques include **DNA-DNA hybridization**. This method allows scientists to determine evolutionary distances between organisms and is a crucial tool in gaining a clearer picture of phylogenetic relationships. The process of hybridization involves three primary steps:

1. *Denature*. Intact, double-stranded DNA from species of interest are subjected to high temperatures (85–100° Celsius) or highly alkaline conditions (high pH levels) in order to break the hydrogen bonds holding the two strands together, thus creating single-stranded DNA. **Denaturation** is confirmed by measuring ultraviolet (UV) absorbance of the DNA solution(s) with a spectrophotometer set at 260 nanometers. At this wavelength, denatured DNA absorbs approximately 37% more radiation than intact DNA at the same concentration (**FIGURE 11.3**). The temperature at which half the original intact DNA has denatured is known as the **melting temperature (T_m)**, which is useful in its own right as a metric for determining evolutionary relationships.
2. *Prepare probes*. The probe will be a DNA sample of known origin and sequence. The sequence must be known because the probe will be serving as a detector. Any sample DNA fragment that is complementary to the probe (leading to renaturation of a double-stranded complex) will be deemed a match and assumed to contain DNA common to both species under investigation. For example, human DNA can probe DNA from other primates. Because the probe is acting as a tag, it is labeled with light-emitting molecules or radioactive atoms to facilitate detection.
3. *Renature/hybridize*. The probe is combined with denatured sample DNA, whereupon complementary single strands hybridize through spontaneous hydrogen bonding. This results in **duplex DNA** (consisting of two strands). Unlike intact DNA derived from one species, duplex DNA may contain strands from two separate organisms, as is the case here. When complementary regions are insufficient in length, renaturation is fleeting, whereas with lengthier stretches of complementary regions, the renaturation is permanent. The shorter regions experience only temporary hybridization because the flanking regions are not complementary. This is not a concern with longer fragments that hybridize; the flanking regions are likewise complementary. An overview of DNA-DNA hybridization may be seen in **FIGURE 11.4**.

Because individual nucleotides are compared in a sequential fashion, DNA-DNA hybridization is very precise and can supplant existing determinations of estimated divergence among species. However, there is some wiggle room with hybridization. It is possible for

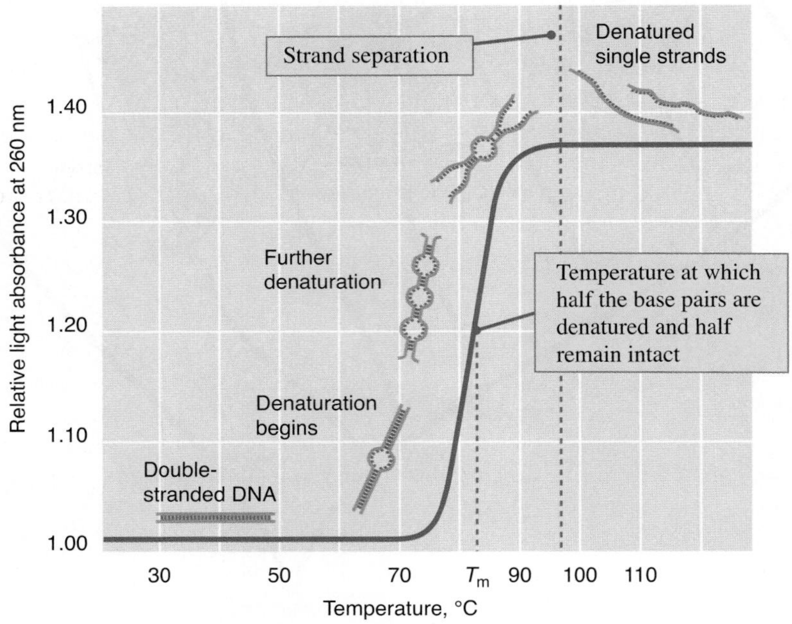

FIGURE 11.3 Denaturation of double-stranded DNA can be detected through relative UV absorbance levels at 260 nm. The higher the absorbance, the greater the extent of denaturation.

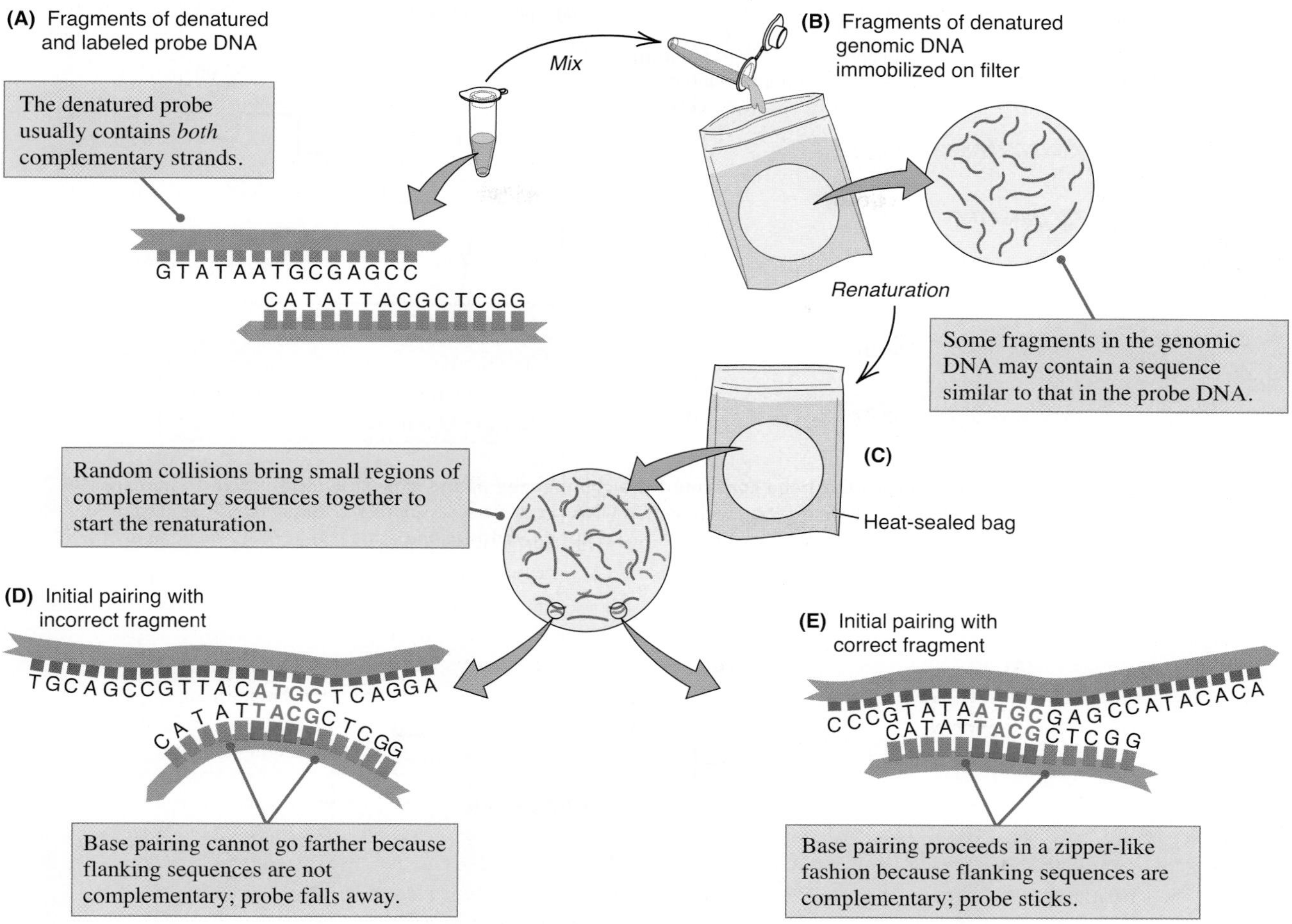

FIGURE 11.4 **(A)-(E) The process of DNA-DNA hybridization used in molecular systematics.**

some mismatching to occur and it is not absolutely necessary for the sample fragment to be entirely complementary to the probe. Variables within the hybridization procedure can minimize mismatches or increase their tolerance as desired. Mismatches are more likely to occur when temperatures are lower and/or when salt concentrations within sample solutions are higher.

Upon analysis of hybridization data, evolutionary biologists may then place organisms into a phylogenetic tree based upon their relative distance to a common ancestor and each other. Therefore DNA-DNA hybridization can accommodate the growing field of **molecular systematics**, classifying organisms based upon molecular (often genetic) comparisons. Insights into primate phylogeny have been greatly aided by this technique. For example, hybridization of human DNA to the apes has revealed our closest extant relative to be *Pan troglodytes*, the chimpanzee (**FIGURE 11.5**). Melting temperatures of each DNA sample act as additional data, serving to reinforce conclusions drawn from hybridization events.

There is a drawback to DNA-DNA hybridization. The phylogenetic trees constructed from hybridization data are known as species trees. **Species trees** are designed to represent evolutionary relationships among organisms. Hybridization measures similarities among the genomes of different species, not the differences. In order to account for any differences, gene trees must be constructed. **Gene trees** are an example of phylogeny where specific genome locations are compared and differences are identified. It is much more desirable to analyze multiple loci and create a composite species tree based upon genomic comparisons. Separate genomic comparison studies among humans, chimpanzees, and gorillas further support the conclusion that chimpanzees are the closest living relative to our species (**FIGURE 11.6A**). While comparisons of individual genes might suggest otherwise (**FIGURE 11.6B** and **C**), comparisons of the genome in its entirety give a clearer indication of evolutionary relationships on the whole. A composite of all gene trees may therefore be developed into a well-supported

(A) Species used to prepare hybrid DNA molecules

Species	Melting temperature T_m (°C)	Difference from human-human T_m (ΔT_m in °C)
Human / Human	59.50	0
Human / Chimpanzee	57.91	1.59
Human / Gorilla	56.97	2.53
Human / Orangutan	55.97	3.53
Human / Baboon	52.47	7.03

(B) Evolutionary relationships based on ΔT_m

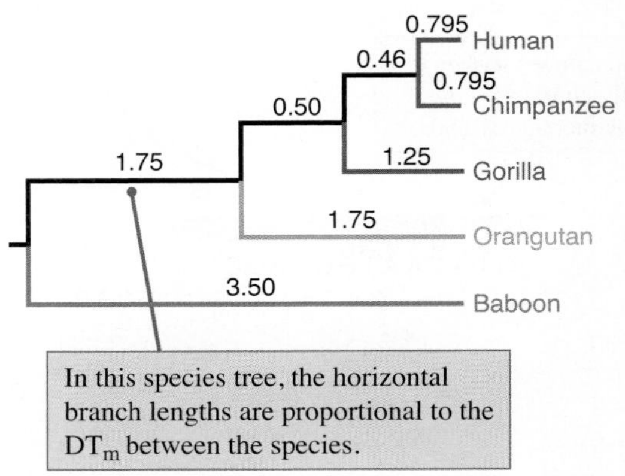

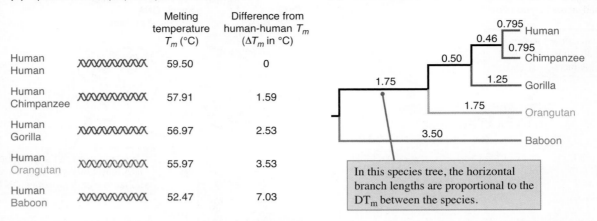

FIGURE 11.5 DNA-DNA hybridization studies have confirmed the chimpanzee as the closest living relative to humans. (A) Melting temperatures of each species used in hybridization with human DNA probes are compared to human DNA melting temperatures. (B) Molecular systematics allows the creation of a phylogenetic tree based upon hybridization data. Data from A. Caccone and J. R. Powell. *Evolution* 43 (1989): 925–942.

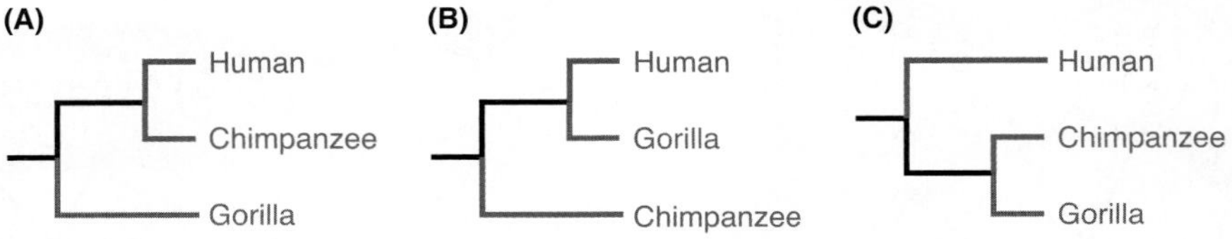

	(A)	(B)	(C)
Study 1	13	1	2
Study 2	31	10	12

Number of independent genomic regions supporting each tree

81% of the genomic regions in Study 1 and 58% of the genomic regions in Study 2 imply that humans and chimpanzees are most closely related; gorillas are an outgroup to the human-chimpanzee clade.

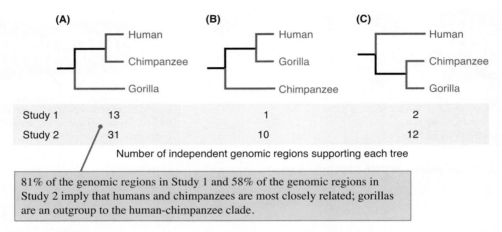

FIGURE 11.6 (A)-(C) Analysis of multiple genomic regions allows the creation of gene trees. Genomic comparisons support the conclusion that chimpanzees are the closest living relatives to humans. Data from M. Ruvolo. *Mol. Biol. Evol.* 14 (1997): 248–265 and F.-C. Chen and W.-H. Li. *Am. J. Hum. Genet.* 68 (2001): 444–456.

species tree. Comparisons of specific locations within genomes can also be accomplished through karyotyping, though this method will only provide qualitative data allowing for broad, generalized differentiations (**FIGURE 11.7**).

Differences between species can also be measured through DNA sequencing. The chimpanzee genome was sequenced to shed light on the genetic differences that have accumulated between our species during the past seven million years or so since our divergence from a common ancestor. Genomic comparisons between humans and our closest relatives have found 35 million base pairs differ between the two species, representing 1.2% of our genome. These differences are single nucleotide substitutions. There are also a substantial number of deletions and insertions separating the two genomes equal to another 3%. There are around 45 million base pairs in humans that are not in chimpanzees (suggesting insertions), while there are an approximately equal number of base pairs found in chimpanzee DNA but not in our genome (indicative of deletions). Adding these to the single nucleotide substitutions, the total difference between the two species' genomes is 4.2% (**FIGURE 11.8**).

The Hominin Lineage

The **hominins** represent humans and our closest relatives, all long extinct. The hominin line represents all organisms emerging from the split with other hominids (the great apes). Chimpanzees, bonobos, orangutans, and gorillas diverged in one direction and the hominins

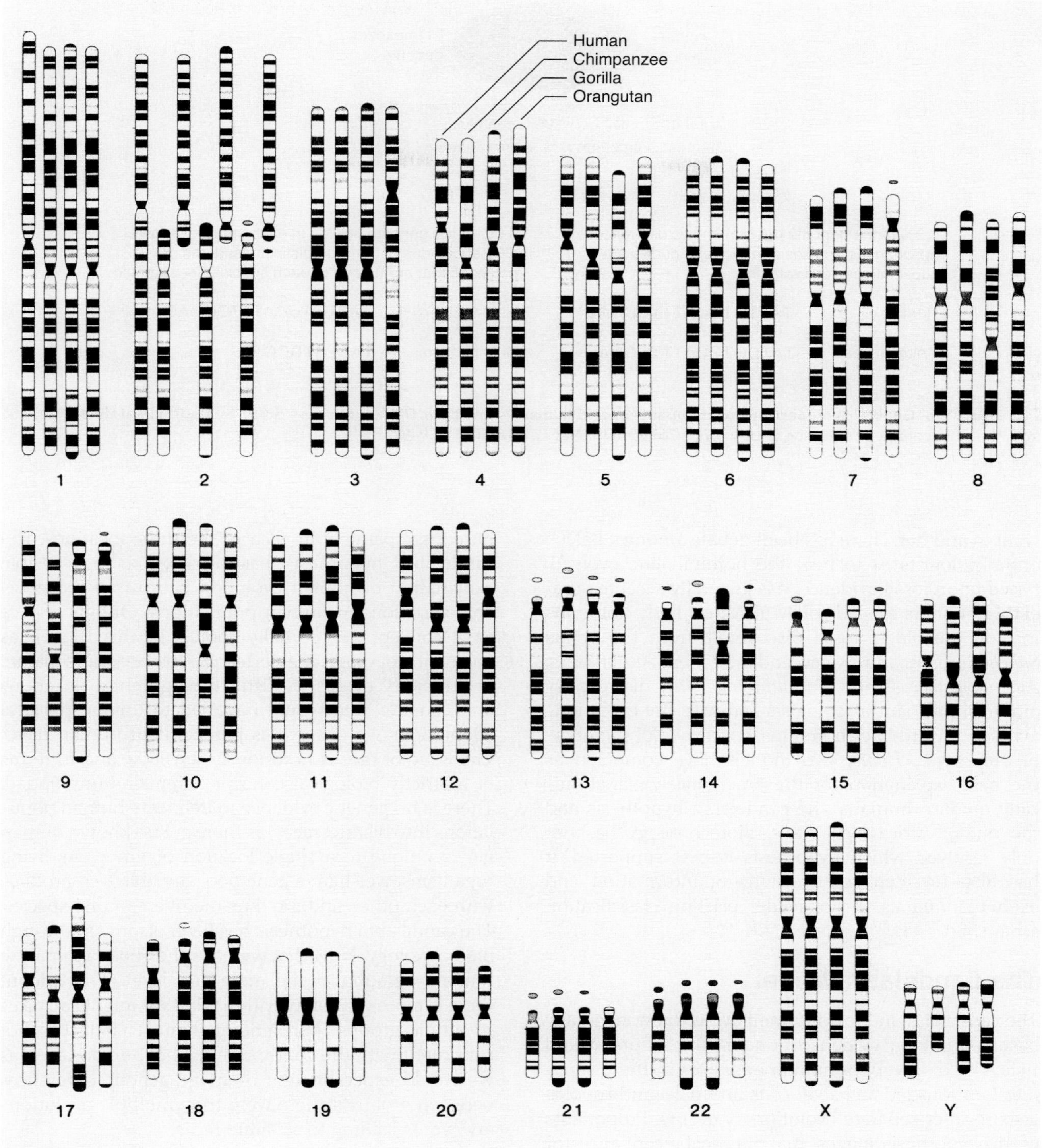

FIGURE 11.7 The banding patterns visible in hominid karyotypes. There are more observable similarities to humans with chimpanzees than with gorillas or orangutans, supporting the assertion they are our closest living primate relative. Reproduced from Yunis, J. J., and Prakash, O. *Science* 215 (1982): 1525–1530. Reprinted with permission from AAAS.

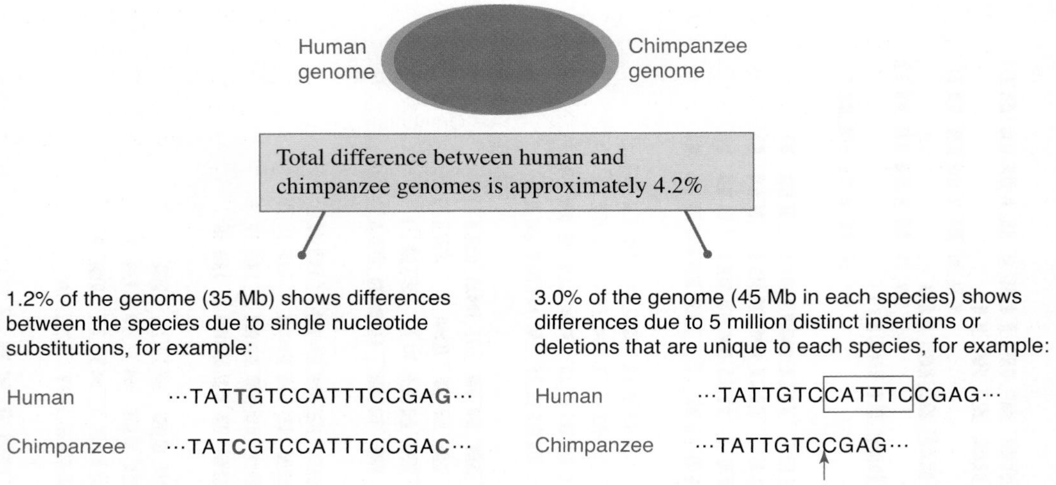

FIGURE 11.8 Genomic comparisons of chimpanzees and humans show their DNA sequences differ by approximately 4.2%. Data from the Chimpanzee Sequencing and Analysis Consortium. *Nature* 437 (2005): 69–87.

went in another. There has been debate amongst paleoanthropologists as to how the hominin line evolved. Based upon fossil evidence, it is generally accepted that early hominins existed solely in Africa. Early hominins include the oldest fossil discovered, from the genus *Sahelanthropus*, and its descendant genera *Ardipithecus*, *Australopiethecus*, and *Paranthropus*. The discrepancy amongst paleoanthropologists lied with the evolutionary leap toward later hominins, composed of members of our genus, *Homo*. Two models have competed as the best explanation for the emergence of anatomically modern humans: the candelabra hypothesis and the out of Africa hypothesis. Biotechnology has not only resolved which hypothesis is best supported, it has also revealed some surprising information and even required us to reconsider existing classification schemes.

The Candelabra Model

The candelabra model of human evolution was initially based upon fossil evidence found in Africa, Europe, and Asia. The discovery of human ancestors in these separate locations led archaeologists and paleoanthropologists to infer separate evolutionary events. Proponents of this hypothesis suggest that our most recent ancestor *Homo erectus* evolved in Africa from earlier hominins, a taxonomical grouping containing humans and our extinct ancestors. While this is generally accepted by paleoanthropologists, even as part of the opposing out of Africa model (below), the remainder of the candelabra model represents the major difference between the two hypotheses. Candelabra supporters concluded *Homo erectus* split into several populations: some stayed in Africa, whereas others migrated to Europe and then to Asia, going on to evolve into anatomically modern humans separately in each of the three locations. The candelabra hypothesis has held on as a plausible explanation of human evolution because of recognizable variations in human populations. Origins of race as a concept are partially rooted in this model, as paleoanthropologists asserted it must be the separate evolutionary events in distinct geographical locations and climates that would have led to unique physical adaptations we observe as independent human races. The issue of race is notoriously sensitive, and in terms of a strictly biological concept it remains unfounded. There is no genetic evidence to delineate human populations into distinct races, as there are no known human genes unique to a single location or group. As living organisms, we share a gene pool, are able to reproduce with each other, and are thus members of one species. The candelabra hypothesis has been disproven through biotechnology. Now that we have the ability to analyze human variation at the molecular level, it is evident there is more variability within what we might call races or ethnic groups than among them. That fact is not congruent with the concept of distinct human races, as we would expect to find dramatic genetic differences between groups if there were truly multiple evolutionary events leading to separate races.

The Out of Africa Model

The Out of Africa hypothesis maintains that ancient hominins evolved into *Homo erectus* and then *Homo sapiens* in one place—the African continent—and within one transitional event. Representing another facet of the hypothesis, it is thought that *Homo erectus* also migrated out of Africa prior to this event. This portion of the model serves to explain how there could be earlier hominin fossils found in Europe and Asia

(**FIGURE 11.9**). Based upon numerous studies applying biotechnology, the hypothesis has become the generally accepted theory among anthropologists of all stripes. Geneticists have investigated representatives of our species across the globe, analyzing mitochondrial DNA, the Y chromosome, and single nucleotide polymorphisms.

Mitochondrial Eve

Analysis of mitochondrial DNA has three major advantages. First, mitochondrial DNA is separate from nuclear DNA and does not undergo recombination during sexual reproduction. Indeed, as we saw with forensic biotechnology, mitochondrial DNA (mtDNA) is inherited directly through the maternal line. This feature enables mtDNA to transcend generations relatively unscathed, with very little genetic change. Secondly, because mtDNA resides on its own separate chromosome within mitochondria organelles, the smaller chromosomal size allows for simple comparisons as opposed to nuclear DNA. As with nuclear DNA, the human mitochondrial chromosome has been fully mapped (**FIGURE 11.10**). Lastly, there are typically multiple mitochondria within each human cell, unlike the nucleus where nuclear DNA resides. Thus, there are more existing samples to obtain and scientists have better odds of extracting mtDNA from degraded biological specimens, such as ancient hominin fossils.

Geneticist Rebecca Cann pioneered mtDNA analysis as a method to study human evolution. As part of her doctoral dissertation, she conceived the idea of a "**mitochondrial Eve**:" a single female progenitor of all humanity. By tracing back mtDNA, all existing lines of

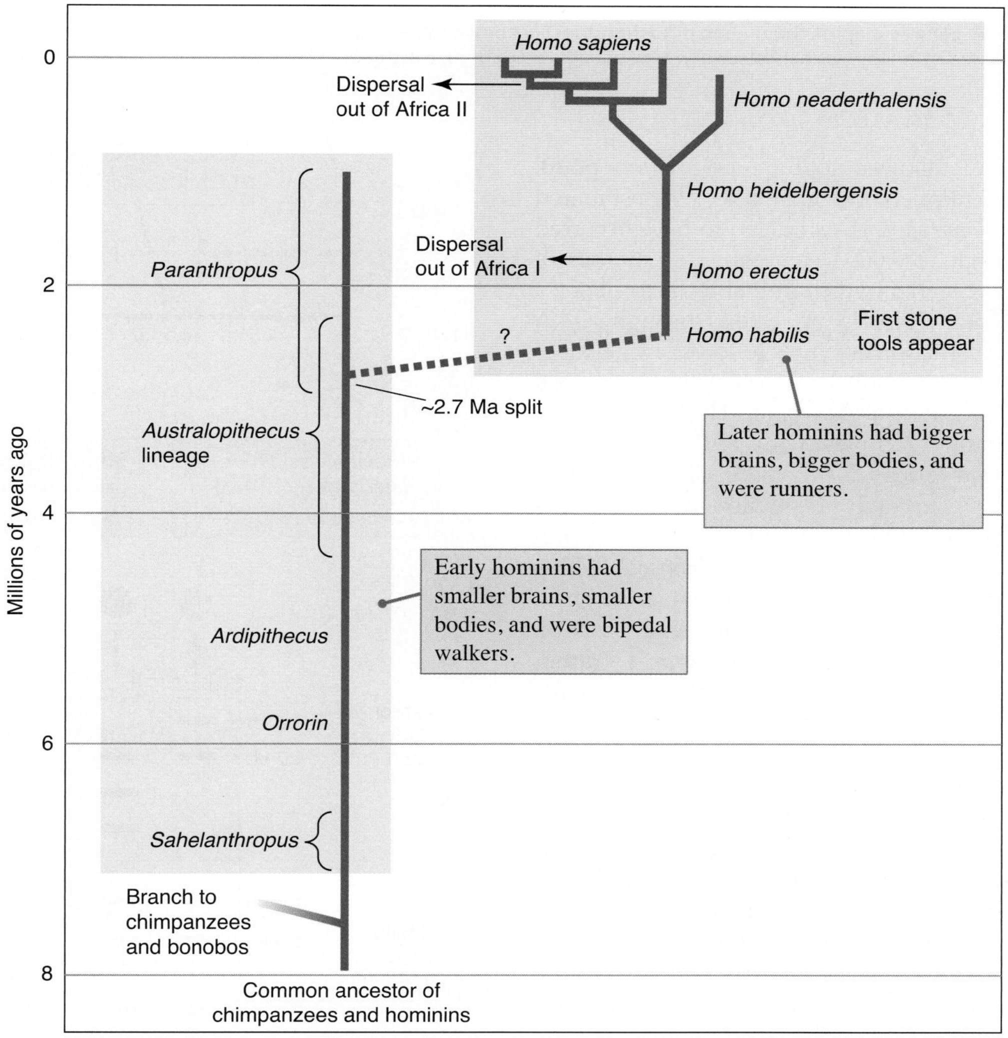

FIGURE 11.9 Phylogeny of hominins based upon the out of Africa model divides the line into early and late members. Adapted from the Committee on Earth System Context for Hominin Evolution. 2010. *Understanding Climate's Influence on Human Evolution.* National Academy of Sciences, Washington, DC.

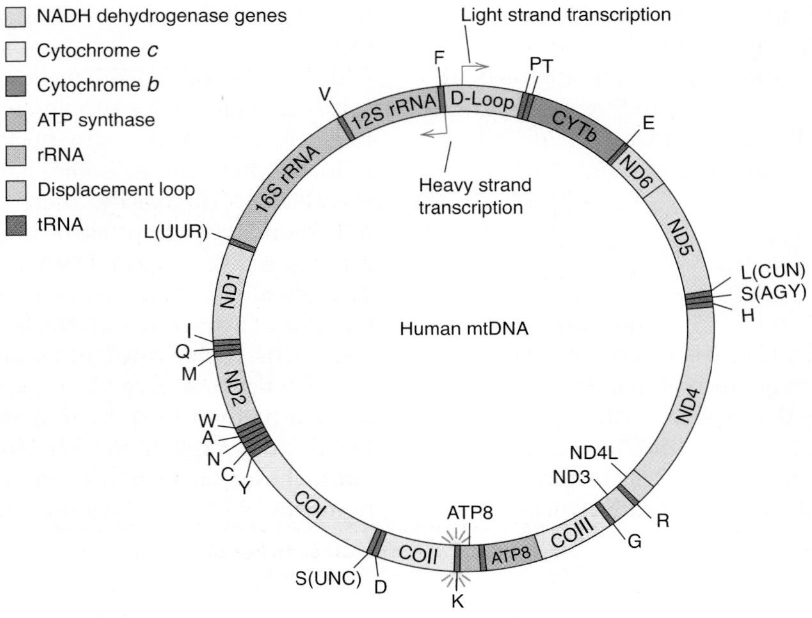

FIGURE 11.10 A genetic map of the human mitochondrial chromosome. Reprinted with permission from the *Annual Review of Genetics*, Volume 29 © 1995 by Annual Reviews [www.AnnualReviews.org.]. Illustration courtesy of David A. Clayton, Howard Hughes Medical Institute.

inheritance were followed to their origination point. Only one line continued uninterrupted to the estimated emergence of *Homo sapiens*, estimated to have occurred in Africa less than 200,000 years ago. This mother of humanity passed on her mitochondrial DNA to all of us.

Mitochondrial DNA is analyzed through DNA digestion with restriction enzymes followed by gel electrophoresis. The restriction enzymes will cut mtDNA when restriction sites are present. This serves as a means to separate samples with and without targeted restriction sites during gel electrophoresis. Mitochondrial DNA cut by restriction endonucleases will be smaller than uncut mtDNA; therefore, it will travel farther down the gel when the electrical field is applied (**FIGURE 11.11**). In a manner similar to DNA fingerprinting, it is possible to compare band sizes on the resulting gel in order to determine maternity (rather than paternity). All offspring will share mtDNA variants with their mother, and her mother, and her mother, and so on (**FIGURE 11.12**).

Beyond establishing the out of Africa model as an accepted theory, mtDNA has been analyzed in biological anthropology studies. Sampling diverse populations of humans in various regions of the world has allowed the creation of a mtDNA phylogenetic tree (**FIGURE 11.13**). The root of the tree first gives rise to African peoples as would be expected from the Out of Africa model. Mitochondrial DNA has also elucidated our understanding of human migration into the remaining continents. In combination with SNP comparisons, mtDNA data support the notion of several waves of *Homo sapiens* migration out of Africa.

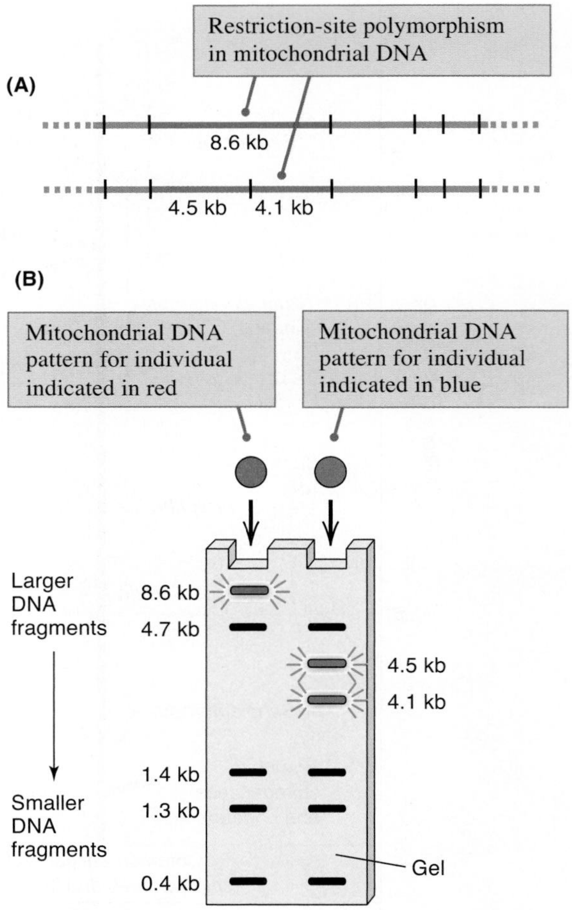

FIGURE 11.11 Analysis of mtDNA among individuals is accomplished by (A) restriction enzyme digestion and (B) gel electrophoresis.

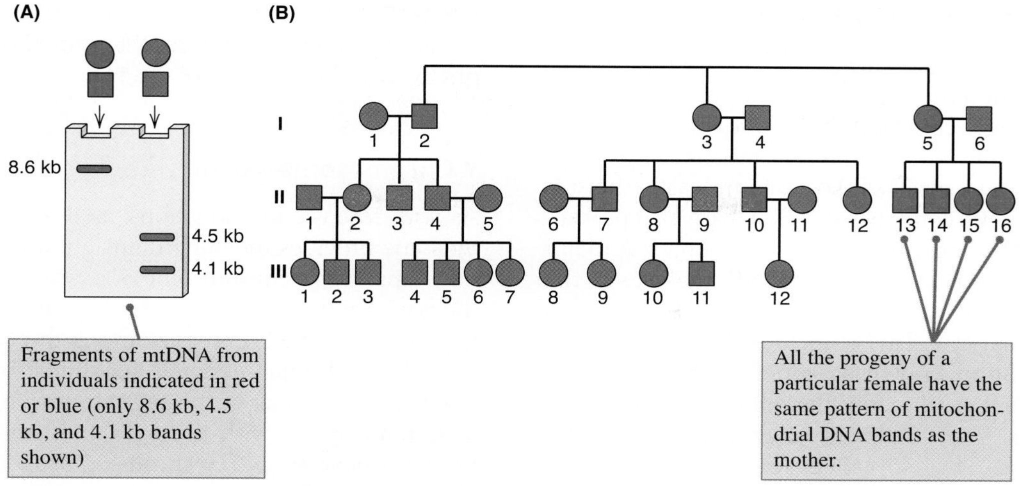

FIGURE 11.12 Maternal inheritance of mtDNA is determined by (A) analysis of gel electrophoresis bands and (B) construction of a pedigree demonstrating mtDNA variants passed down by each mother to her offspring. Adapted from D. C. Wallace, *Trends Genet.* 5 (1989): 9–13.

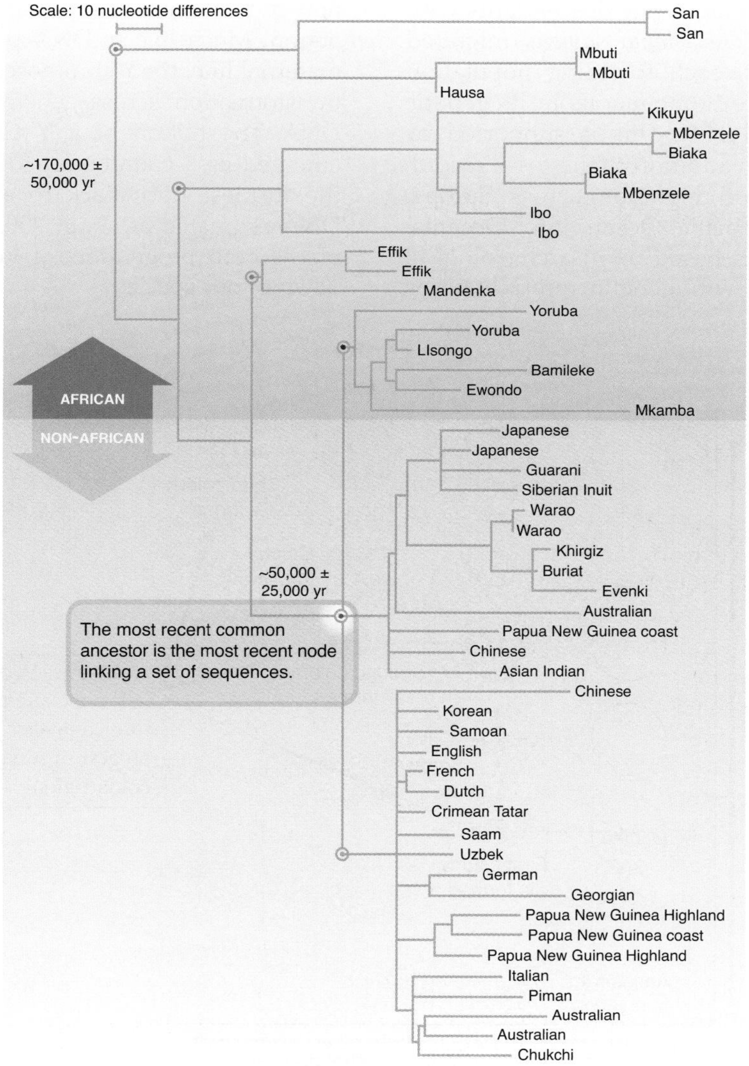

FIGURE 11.13 A phylogeny of human populations based upon mitochondrial DNA sequence comparisons. Adapted from M. Ingman, et al., *Nature* 408 (2000): 708–713.

Single Nucleotide Polymorphisms

Studies comparing single nucleotide polymorphisms in different human populations have resulted in conclusions that coincide with mitochondrial data. In fact, single nucleotide polymorphisms (SNPs) can also occur within mtDNA, so in some cases there is overlap of genetic comparisons. SNPs are also present in nuclear DNA, including sex chromosomes. A major similarity between mtDNA and SNP analyses is the tendency toward increased genetic variation within African populations. This coincides with the process of microevolution known as genetic drift. One form of genetic drift is the founder effect. The **founder effect** occurs when a subset of an existing population relocates. This so-called founding population has a lower frequency of genetic variation because it represents only a portion of the gene pool found in the original population. The Out of Africa model predicts more genetic variation in the founding population, represented by modern day descendants with uninterrupted lineages. As *Homo sapiens* migrated out of Africa in waves, each founding population becomes more and more homogenous in its genetic composition (**FIGURE 11.14**). This is supported by SNP and mtDNA data in addition to the fossil record. The inferred patterns of migration across Europe, into Asia, across the Pacific Ocean into Oceania, and finally the Americas based upon archaeological evidence is congruent with genetic interpretations (**FIGURE 11.15**). A phylogeny based upon SNP analysis is strikingly similar to the mitochondrial DNA phylogeny above (**FIGURE 11.16**).

Y Chromosome Adam

As with forensic investigations, analysis of the Y chromosome provides meaningful insight into human relationships. The Y chromosome is passed down through the paternal line because male offspring inherit it from their fathers. The Y chromosome is the only chromosome in the human genome not shared by both sexes. The Y makes up half of the sex chromosome pair found in human males, with the full complement being XY (females possess two X chromosomes). Thus the Y is also the only human chromosome not matched in a homologous pair, meaning it is matched with a chromosome with completely separate genes and therefore a distinct DNA sequence. For this reason, a large portion of Y chromosome doesn't participate in recombination. Much like mtDNA passed down through the maternal line, the Y chromosome undergoes much less transformation across generations than does nuclear DNA. The notion of a **Y Chromosome Adam** has emerged as a counterpart to Mitochondrial Eve, with the furthest trace back to ancestral Y chromosomal DNA estimated at about 250,000 years ago. This date places both progenitors of humanity in Africa at the dawn of our species.

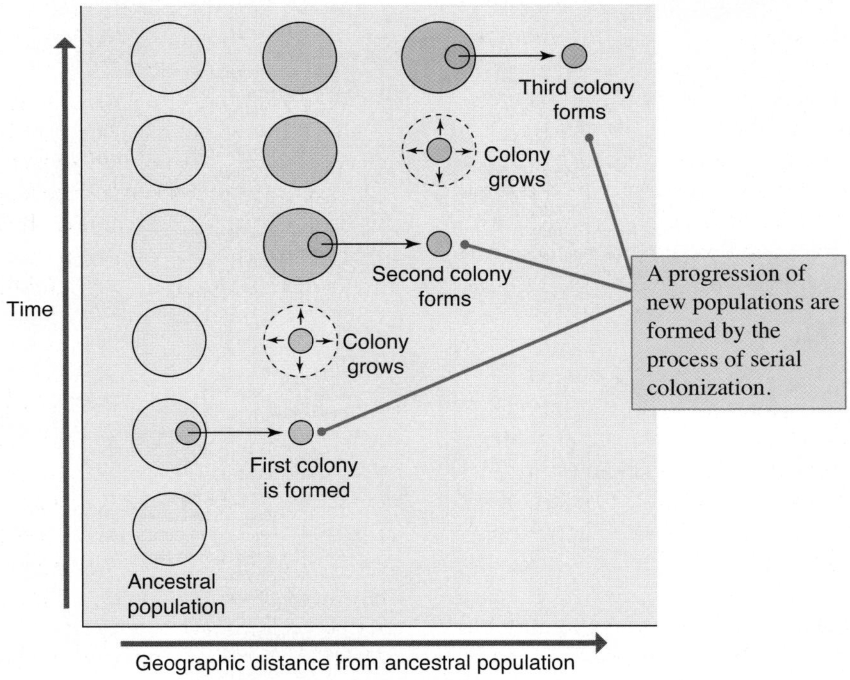

FIGURE 11.14 The founder effect caused by serial migration of human populations out of Africa led to a gradual decline in genetic diversity.

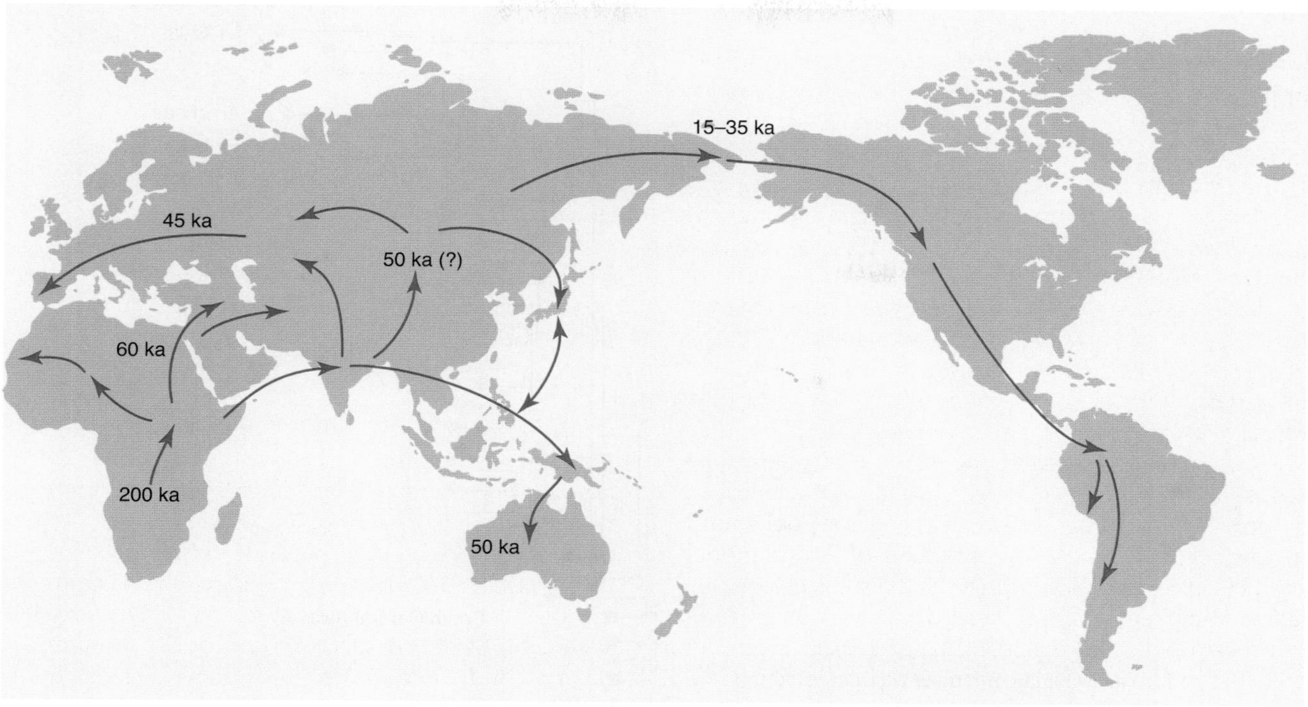

FIGURE 11.15 **The inferred migratory patterns of *Homo sapiens* out of Africa are supported by genetic and archaeological evidence.**

Neanderthals and Denisovans

In 1856, stone cutters working in the German valley of Neander accidentally unearthed human-like bones. Named for the valley where they were discovered (*thal* is German for valley), Neanderthals have earned the reputation as cave men or troglodytes. Their robust, brutish anatomy led paleoanthropologists to the conclusion they were not members of our species but were deemed close enough relatives to be considered late hominins. As members of our genus, they were classified *Homo neanderthalensis*. Geneticists at the Max Planck Institute for Evolutionary Anthropology in Germany released a draft sequence of the Neanderthal genome in 2010 which has altered views of our recent relative. A complete sequence was published in 2013 by the same researchers.

As further support of the Out of Africa model, it is estimated ancestors of the Neanderthals left Africa approximately 800,000 years ago to populate Eurasia. Modern humans followed about 50,000 to 80,000 years ago. These *Homo sapiens* are referred to as Cro Magnon by paleoanthropologists; archaeological evidence was first to establish they were present in Europe concurrently with Neanderthals. Genetic investigation by the Max Planck team supports this hypothesis. Genomic comparisons of *Homo neanderthalensis* and *Homo sapiens* suggests they not only cohabitated, they interbred. Shared genomic regions between Neanderthals and present day humans occur only in individuals of Eurasian descent on the order of 2 to 3% shared DNA. There are no genetic similarities between Neanderthals and individuals of African descent. The Out of Africa model predicts these observations because ancestors of modern day Africans would never have come in contact with Neanderthals and therefore they would not have had an opportunity to exchange genetic material.

Neanderthals likely also interbred with a separate ancient hominin. The Denisovans were named for the Denisova Cave in Siberia where fossil remains were discovered in 2010. Researchers at the Max Planck Institute have sequenced both mitochondrial and nuclear DNA of the Denisovans and compared their genome to those of Neanderthals and humans. Major findings of these investigations are now discussed.

The mtDNA of Denisovans appears to be older than its nuclear DNA. When compared to human and Neanderthal mtDNA, the most recent common ancestor among all three groups is estimated at one million years ago. Such an early divergence may be interpreted as an additional migration out of Africa, meaning Denisovan ancestors populated Asia well before Neanderthals came along. However, nuclear DNA analysis estimates divergence of all three at around 600,000 years ago and a split between Neanderthals and Denisovans about 400,000 years ago. The estimated divergence of all three groups is congruent with previous comparisons between Neanderthals

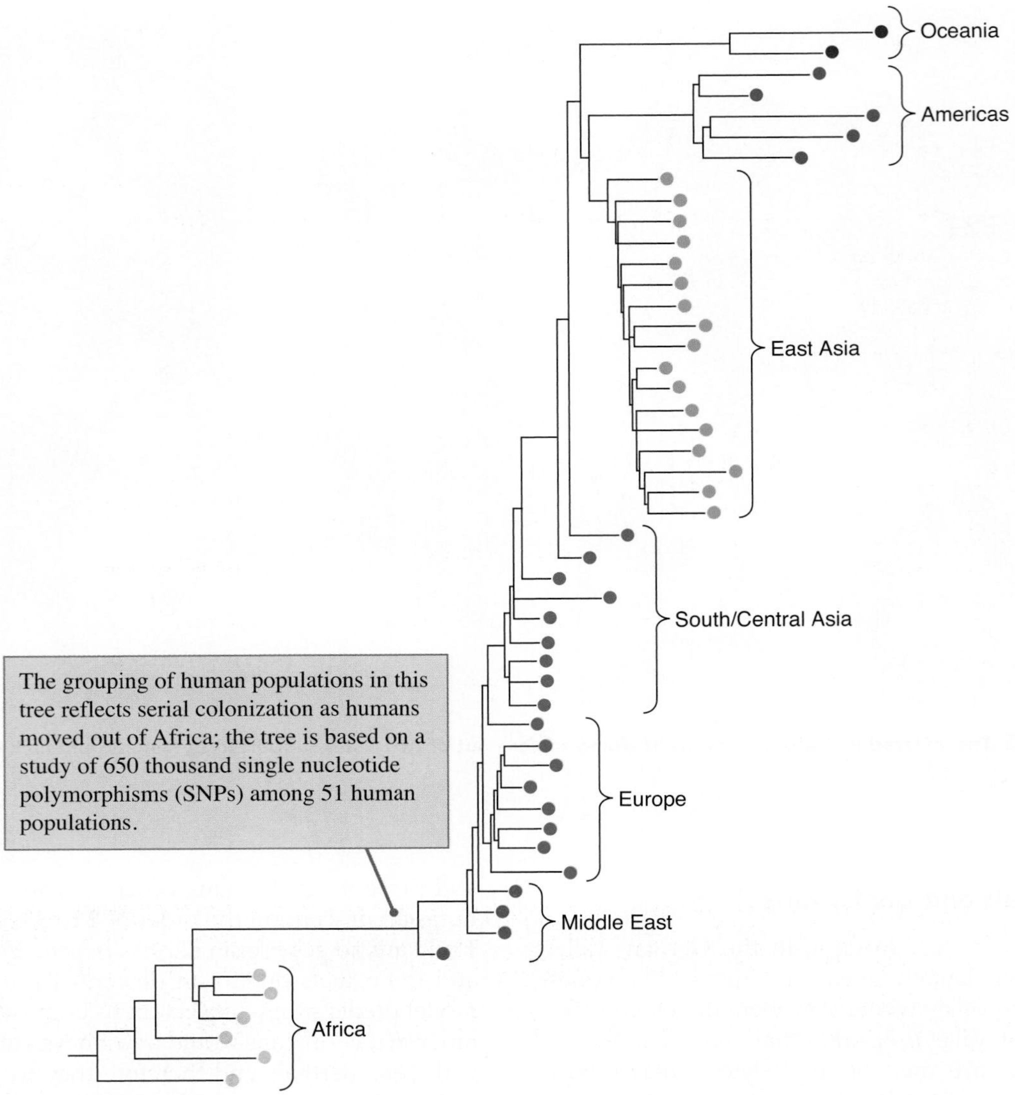
Oceania
Americas
East Asia
South/Central Asia
The grouping of human populations in this tree reflects serial colonization as humans moved out of Africa; the tree is based on a study of 650 thousand single nucleotide polymorphisms (SNPs) among 51 human populations.
Europe
Middle East
Africa

FIGURE 11.16 A phylogeny of human populations based upon single nucleotide polymorphism comparisons. Data from J. Z. Li, et al., *Science* 319 (2008): 1100–1104.

and humans only. Accounting for the discrepancies in mtDNA and nuclear DNA between Neanderthals and Denisovans is not as straightforward. Max Planck Institute geneticists propose two possible scenarios. A first hypothesis purports ancient mtDNA sequences persisted in Denisovans which was lost in other hominin lineages through random genetic drift. The second hypothesis states Denisovans experienced genetic introgression with an as yet unidentified late hominin. **Introgression** occurs when closely related organisms exchange genetic material through repeated hybridization over several generations. Studies also indicate Denisovans interbred with Neanderthals and humans. There is greater overlap between the genomes of Neanderthals and Denisovans (about 17%), suggesting they interbred more frequently and over greater periods of time. Denisovan DNA is only found in modern Melanesians and Aborigines, which indicates ancestors of these human populations were the only *Homo sapiens* to interbreed with Denisovans. Melanesians and Aborigines share up to 6% of their genomes with Denisovans (**FIGURE 11.17**).

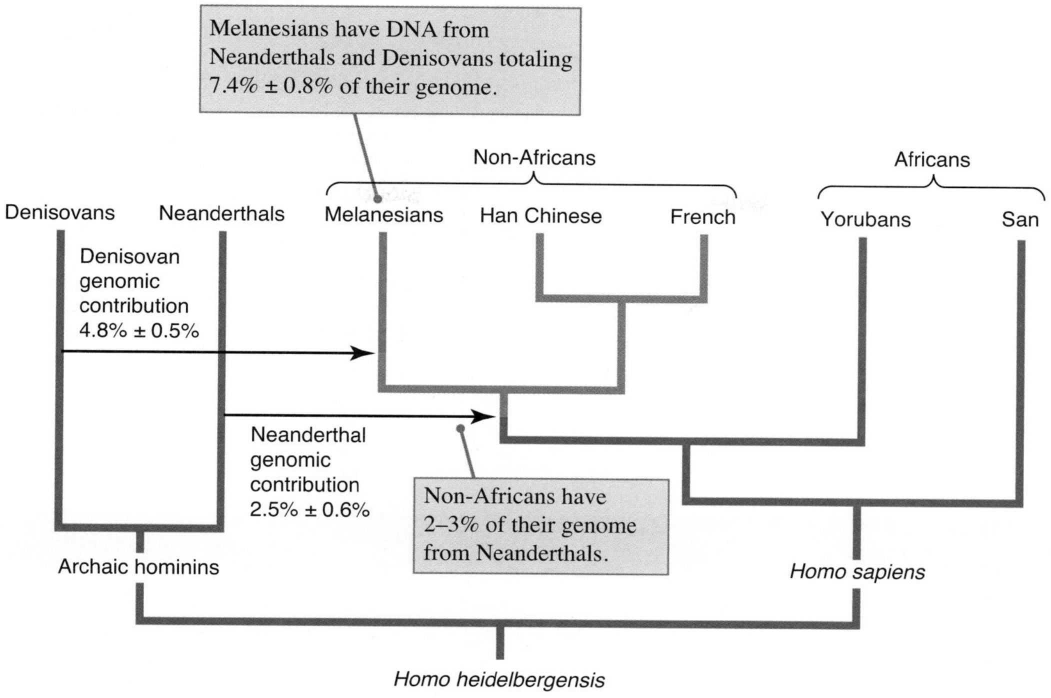

FIGURE 11.17 A phylogeny of late hominins includes Neanderthals and Denisovans with modern humans. Both extinct groups interbred with our species as well as with each other. Adapted from D. Reich, et al. *Nature* 468 (2010): 1053–1060.

SUMMARY

- Anthropology is the study of humanity. It consists of four branches: linguistics, cultural anthropology, biological or physical anthropology, and archaeology. Paleoanthropology, the study of ancient humans and their ancestors, is a subdiscipline within biological anthropology.
- Archaeology and biological anthropology (including paleoanthropology) have benefitted from molecular biotechnology. Genetic analysis has revealed the most accurate human evolution model to be the Out of Africa hypothesis. Anthropologists are in general agreement that our species evolved in Africa and migrated outward into Eurasia in several waves.
- The two models of human evolution are the Candelabra hypothesis and the Out of Africa hypothesis. The Candelabra model proposes subsets of our recent ancestor *Homo erectus* left Africa and evolved into anatomically modern humans independently in Africa, Europe, and Asia and suggests this explains the variation observed in human races. The Out of Africa hypothesis ascertains *Homo sapiens* evolved once in Africa and migrated into Eurasia on more than one occasion.
- Charles Darwin purported humans share a common ancestor with other primates but that we did not descend directly from them. He identified extant primates as our closest living relatives.
- Members of Order Primata exhibit a relatively large brain size to body mass ratio, have stereoscopic vision, and possess opposable thumbs and big toes (except humans). Primates engage in social organization.
- Humans share a common ancestor with other primates. Humans and the great apes belong to the hominid group. Multiple pieces of evidence confirm our closest living relative is the chimpanzee.
- Hominids are humans and the great apes (gorillas, chimpanzees, orangutans, and bonobos). Hominins are humans and their extinct ancestors.
- DNA-DNA hybridization compares DNA sequences from separate species. The technique identifies similarities (not differences) through hybridization involving complementary base pairing of DNA strands from separate species. Data obtained

from DNA-DNA hybridization studies contribute to molecular systematics by allowing classification of organisms based upon molecular data, such as their DNA.

- Gene trees are examples of phylogeny based upon comparisons at the genetic level. It compares specific gene loci. Species trees are examples of phylogeny based upon a composite of all genetic comparisons and represent a more holistic approach to determining evolutionary relationships.
- In addition to DNA-DNA hybridization, scientists have compared DNA sequences of different human populations in order to determine which model of human evolution is most accurate. The comparisons include nuclear DNA, the Y chromosome, and mitochondrial DNA.
- Mitochondrial Eve is the hypothetical mother of humanity. She is thought to be the origin of all mitochondrial DNA shared by humans. Y Chromosome Adam is the hypothetical father of humanity. He is thought to be the origin of genes located on the Y chromosome. Both are estimated to have lived in Africa close in time to the emergence of our species.
- Humans and their extinct relatives the Neanderthals and the Denisovans are believed to have coexisted in Eurasia and interbred. Humans share genetic information with both extinct species. There is some uncertainty as to the Denisovans' exact relationship to the Neanderthals and modern humans.

KEY TERMS

Anthropology
Denaturation
DNA-DNA hybridization
Duplex DNA
Founder Effect
Gene tree
Hominids
Hominins
Introgression
Melting temperature (T_m)
Mitochondrial Eve
Molecular systematics
Paleoanthropology
Species tree
Y Chromosome Adam

DISCUSSION QUESTIONS

1. Research recent human evolution (within the last 10,000–20,000 years). What kinds of adaptations have emerged? What genes have been fixed in our species and what benefit(s) do they confer? Are there any genes that have an unknown function but nevertheless appear to be selected for in our species?
2. Several companies offer DNA testing that provides your genetic profile and a breakdown of geographic origins of your DNA. Today it is even possible to determine the percentage of Neanderthal DNA in your genetic composition. Would you like to see your genetic profile? Why or why not?
3. Research hominin evolution and identify all known extinct hominins. Construct your own version of a phylogenetic tree based upon your findings.

REFERENCES

Cann, Rebecca L. "The History and Geography of Human Genes." *American Journal of Human Genetics* 56, no. 1 (1995): 349.

Darwin, Charles. *The Descent of Man*. London: Publisher John Murray, 1871.

Green, Richard E., et al. "A Draft Sequence of the Neandertal Genome." *Science* 328, no. 5979 (2010): 710–722.

Ingman, Max, Henrik Kaessmann, Svante Paabo, and Ulf Gyllensten. "Mitochondrial Genome Variation and the Origin of Modern Humans." *Nature* 408, no. 6813 (2000): 708–713.

Li, Jun Z., et al. "Worldwide Human Relationships Inferred from Genome-Wide Patterns of Variation." *Science* 319, no. 5866 (2008): 1100–1104.

Mendez, Fernando L., et al. "An African American Paternal Lineage Adds an Extremely Ancient Root to the Human Y Chromosome Phylogenetic Tree." *The American Journal of Human Genetics* 92, no. 3 (2013): 454–459.

Reich, David, et al. "Genetic History of an Archaic Hominin Group from Denisova Cave in Siberia." *Nature* 468, no. 7327 (2010): 1053–1060.

Tyson, Peter. "Are We Still Evolving?" Public Broadcasting System. Posted December 14, 2009. http://www.pbs.org/wgbh/nova/evolution/are-we-still-evolving.html

Zimmer, Carl. "Are Neanderthals Human?" Public Broadcasting System. September 20, 2012. http://www.pbs.org/wgbh/nova/evolution/are-neanderthals-human.html

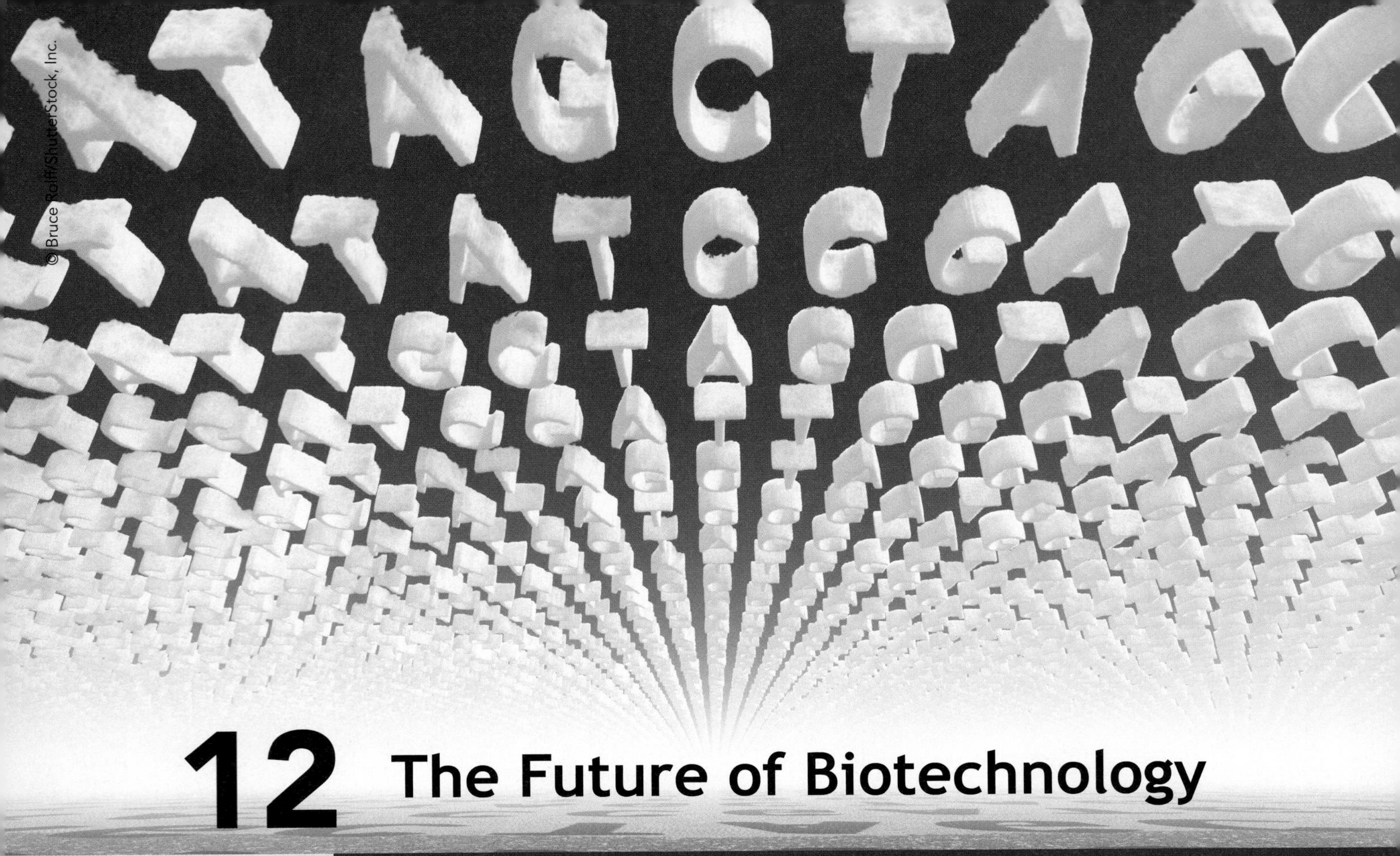

12 The Future of Biotechnology

LEARNING OBJECTIVES

Upon reading this chapter, you should be able to:

- Explain who is responsible for shaping the future of biotechnology.
- Describe complementary DNA and how it applies to *Association for Molecular Pathology vs. Myriad Genetics*.
- Explain the importance of the *Association for Molecular Pathology vs. Myriad Genetics* Supreme Court ruling.
- Provide examples of possible outcomes for sectors of the biotechnology industry.
- Relate the concepts of benefits and risks to the practice of bioethics.

"It is not in the stars to hold our destiny but in ourselves."

—William Shakespeare

The future of biotechnology is truly reliant upon everyone involved. The specific paths of research and development will be planned out by scientists and executives, but at the behest of what the market demands. Individuals play an influential role as consumers and would-be adopters of the latest technologies. Public interest groups act as information outlets and have the power to shape public opinion on biotechnology, ultimately affecting purchasing habits. We do vote with our dollars. Both industry and the masses may weigh in on policies and regulations. Industry has organizations and lobbyists to represent their positions on matters of law. As a counterbalance, every regulatory agency responsible for governing the biotechnology industry seeks public comment on all permits and approval applications.

After a thorough exploration of the past and present of biotechnology, it is now time to consider the implications for the future. Though we can never know with certainty what the future holds, it is possible to engage in reasonable predictions and calculated forecasting. What awaits biotechnology as a research science and as a force of industry is not guaranteed, but this chapter attempts to hypothesize on potential events. How molecular biotechnology might affect laws and regulations, the economy, the individual sectors of this industry, and society as a whole are now discussed.

Regulatory Status and Economic Impact

Biotechnology experiences a certain level of distrust with the public despite its general adoption into the market. This paradox is exemplified with genetically modified foods, which are vilified by a number of public interest groups but yet account for the majority of products on our supermarket shelves. Since its inception, the biotechnology industry has been the subject of skepticism yet still profits tens of billions of dollars annually. Nevertheless, it can be argued that negative opinions of biotechnology do thwart its progress. Researchers decry the years of prohibition and other roadblocks in using stem cells for medical research that were primarily due to objections by vocal sectors of the public. Public perception will continue to play a role in how quickly and how thoroughly biotechnology products are accepted throughout the world. This is especially true in the European Union (EU), where antibiotechnology sentiments are notoriously high. Countries relying on exports to the EU also rely on public opinion; they are not willing to market biotechnology products to areas where consumers are resistant to its adoption.

The 2013 Supreme Court case, *Association for Molecular Pathology vs. Myriad Genetics*, ruled that naturally-occurring human gene sequences can no longer be patented. Following the verdict, Dr. Harry Ostrer of the Association for Molecular Pathology stated, "[genetic testing] costs should come down considerably" and predicted the ruling will "improve the quality of genetic testing overall (**FIGURE 12.1**)." The initial reaction to the verdict by industry analysts expressed worries that loss of patentability will discourage investment in biotechnology and thus stall progress. Patents grant a 20-year essential monopoly, and without a patent the technology is up for grabs. Some reasoned, why would a company invest an average of $20 million in a new biotechnology product if they cannot own the technology? Upon closer examination, however, more optimistic stakeholders predict a resurgence in smaller research firms. Patent protections on gene sequences allow companies to charge licensing fees to use them. Smaller outfits often complain that these fees are prohibitive to conducting research on any patented information or technology and prices out all but the multinational conglomerate organizations. Today, just four firms own (or partially own) nearly 80% of commercially available biotechnology products: Bayer Cropscience, Monsanto, DuPont, and Syngenta. Without licensing fees, it is possible the *Myriad* ruling may level the playing field and allow more start-ups to enter the biotechnology industry, perhaps even sparking accelerated innovation.

FIGURE 12.1 Dr. Harry Ostrer, medical geneticist at Albert Einstein College of Medicine, seen addressing reporters outside of the U.S. Supreme Court building. Dr. Ostrer challenged Myriad's gene patents with the Association for Molecular Pathology. © MLADEN ANTONOV/AFP/Getty Images

Another possibility of the recent verdict is an intensified focus on obtaining ancillary patents. While companies are unable to patent a human gene sequence, the Supreme Court has retained the patentability of complementary DNA of human gene sequences. Recall that **complementary DNA (cDNA)** is a sequence of engineered DNA that contains exons only. Once a human gene is sequenced, its messenger RNA transcript is isolated. The introns composed of non-coding DNA are identified and eliminated from the transcript. The sequence is now exclusively comprised of coding regions (exons). Using reverse transcriptase,

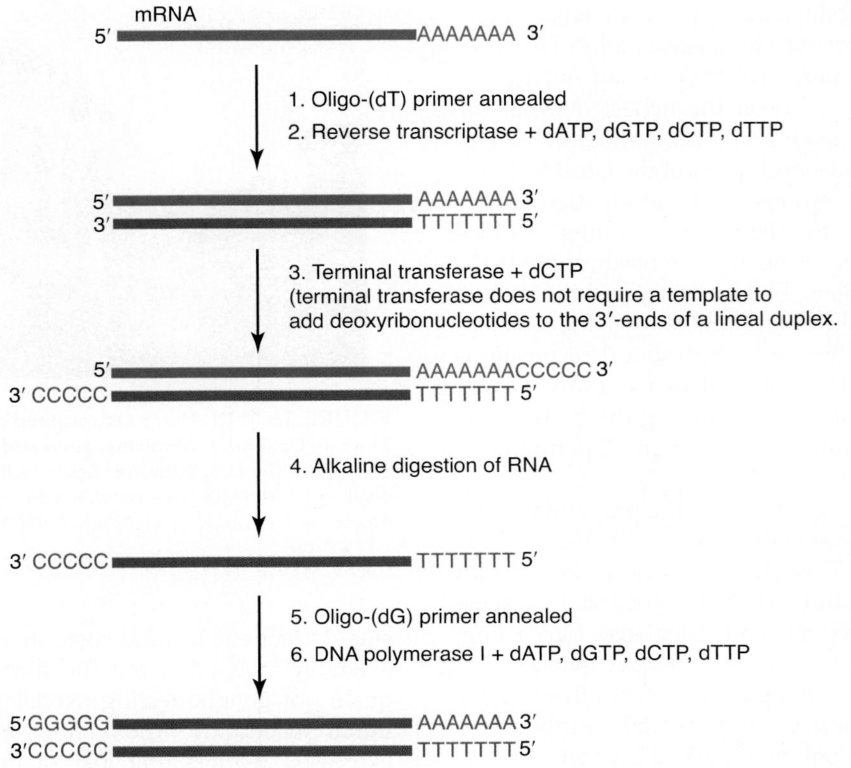

FIGURE 12.2 The construction of complementary DNA. Because cDNA is engineered and not naturally occurring, it is eligible for patent protection in the United States.

this sequence is then used as a template to engineer a complementary strand: the cDNA (**FIGURE 12.2**). Stating that cDNA is not naturally occurring and therefore eligible for patent protection, the ruling also allows for patents on primers, probes, and testing methods used in conjunction with isolated human genes. The value of these ancillary patents cannot be understated. In this case, Myriad Genetics lost the viability of just five patent claims out of over 500 with the Supreme Court's decision. Thus, the availability of other patent protections may still concentrate the biotechnology in the hands of a few and may still keep prices of development and its products high. Or there may be some incentive in preventing ownership of human gene sequences by encouraging more research and innovation, perhaps even lowering costs. The economic impact of the *Association for Molecular Pathology vs. Myriad Genetics* case will depend on the outcome of these predictions.

Industry Forecasts

Throughout this text the various sectors of the biotechnology industry have been presented in detail. The history of biotechnology served as the introduction to our exploration of each area contributing to its present. An outlook of future possibilities is now discussed for each field involved in molecular biotechnology.

Life Science and Health Care

As one of the most active sectors of industry utilizing biotechnology, life science and health care are expected to remain robust and profitable. As populations age in industrialized nations, pharmaceutical companies will respond with an array of cutting edge treatments. Research will be partially dependent upon the completion of the various projects currently underway, such as:

- The 100K Foodborne Pathogen Genome Project is in the process of sequencing 100,000 pathogens that cause foodborne illness to humans.
- The Human Variome Project seeks to identify and catalog all human genetic variation for the purposes of developing specific medical treatments.
- The Human Phenome Project is an offshoot of the Human Variome Project interested in characterizing phenotypes associated with known genetic variations.
- The International Collaboration for Clinical Genomics (ICCG) has a mission "to create, maintain, and curate a publicly available database through collaboration with NCBI's [National Center for Biotechnology] ClinVar database, which contains both genomic and phenotypic data. The main focus of the ICCG is to facilitate data sharing of cases and variant annotations from clinical

laboratories as well as connect these resources to researchers."

- The Human Microbiomes Project has a goal of sequencing 3000 microbes found on and in the human body in order to understand how microbial communities affect our health.

Continued sequencing and annotation within these and other projects, along with improved understanding of protein-protein interactions, and studies on environmental influences on gene expression will all benefit life sciences and health care in the future.

Stem Cell Research

Pharmacogenomics and biopharma are also expected to benefit from the renewed vitality of stem cell research aided by the development of induced pluripotent stem cells. It may one day be routine procedure to culture induced pluripotent stem (IPS) cells directly from a patient in order to customize medical treatments that are genetically compatible. Personalized medicine promises to eliminate side effects common to today's one-size-fits-all prescription drugs. IPS cells may also allow therapeutic cloning to become commonplace. Potentially, IPS cells could be created from a patient's own adult stem cells, which could then be differentiated in a controlled manner to create tissues or organs. Therapeutic cloning might not be limited to replacing damaged or diseased areas of the body; it may also be used in the future for more superficial purposes, such as cosmetic reconstruction or to combat aging. Biotechnology may usher in an age of the ageless.

Bioprinting

Three dimensional (3-D) printing technology is also a potential method for creating tissues and organs. At present, bioengineers have successfully printed prototype organs including ears, bone, menisci, heart valves, and skin in a process known as **bioprinting**. Syringes are loaded with "**bio-inks**" of selected cells, then a computer using bioprinting software directs the syringes in an intricate pattern (**FIGURE 12.3**). The result is a microtissue (**FIGURE 12.4**). The **microtissues** can be thought of as miniature scaffolds seeded with human cells. The microtissues are then placed in an incubator where they grow into full-sized prototype organs or tissues. The technology relies upon the quality of the bioprinting software, utilization of sterile conditions, and the precept that eventually bio-inks will be custom formulated for an exact genetic match to the patient. This last feature will eliminate any concerns over rejection by the patient due to immune system response.

Bioprinting is also in development for applications beyond tissue and organ transplants. The firm Organovo in San Diego, California, has developed liver tissue that is used for drug testing. All oral medications must pass a liver toxicity study, giving bioprinted liver tissue vast potential. Artificial biological specimens can also be used for medical education. Students could one day perform dissections on printed organs instead of using euthanized animals, as is the practice employed today. Doctors in training could use the 3-D organs to practice surgical techniques. Delving farther into the future, bioprinted organs may one day not only act as replacements, but they may even serve to enhance our existing capabilities. Research is underway to test the practicality of incorporating nanotechnology into microtissues. For example, bionic ears with silver nanoparticles might improve our ability to pick up sound waves including frequencies we are not physically able to hear.

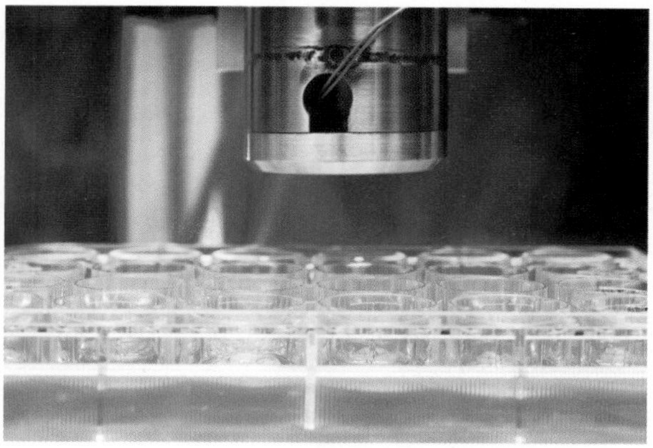

FIGURE 12.3 The syringes of a bioprinter at the University of Manchester (UK) suspended over well plates. This bioprinter is using skin cells. © James King-Holmes/Science Source.

Drug Approval

It is a reasonable prediction that a number of drugs produced via biotechnology currently in clinical trial stages

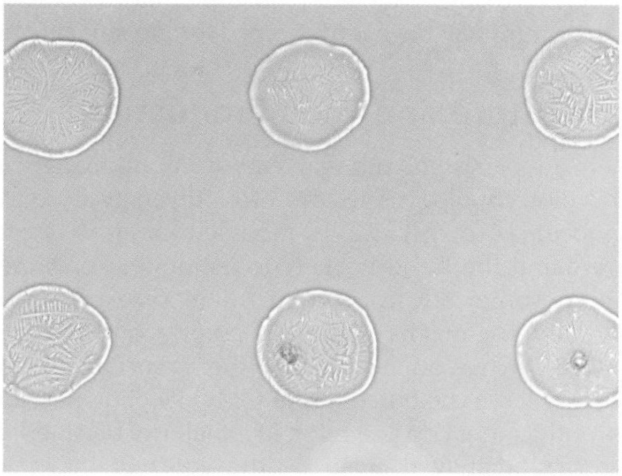

FIGURE 12.4 A light micrograph of microtissue created during bioprinting. This microtissue contains human chondrocyte cells, which synthesize cartilage tissue. © James King-Holmes/Science Source.

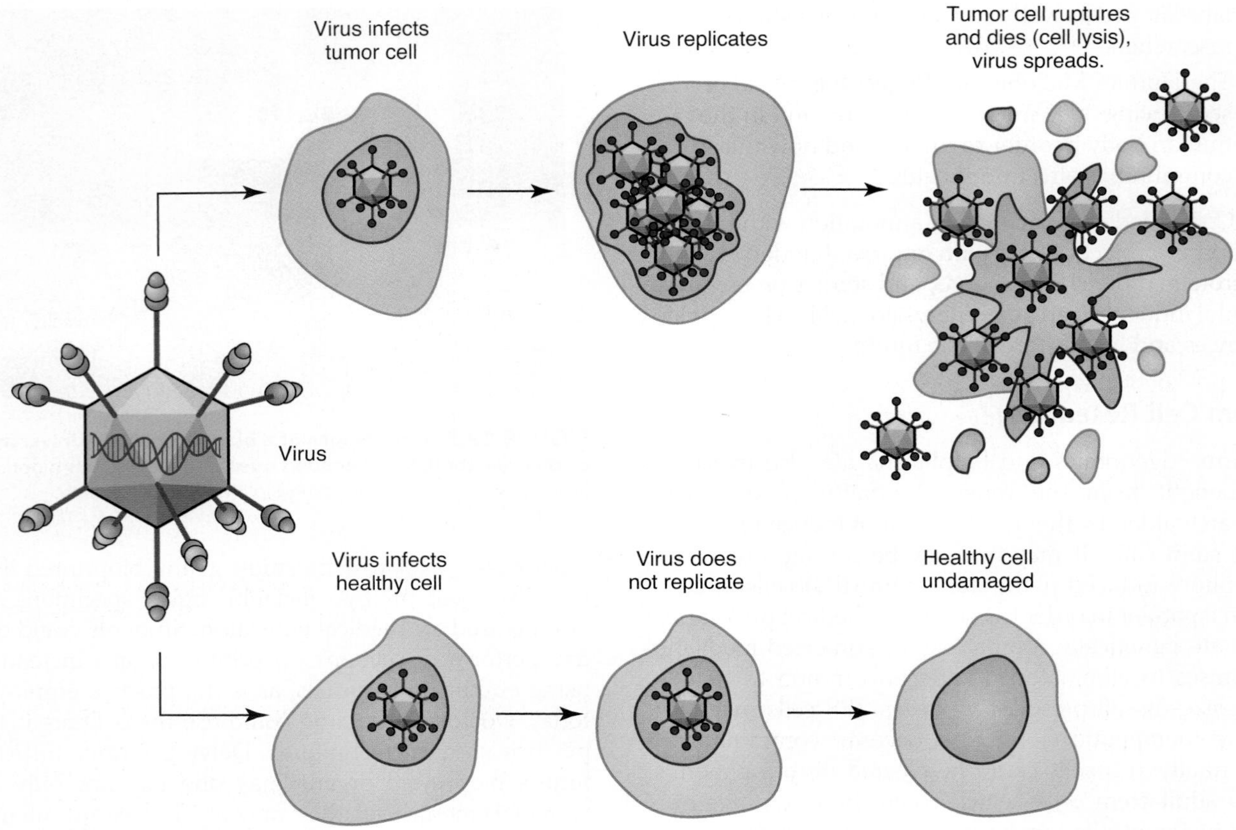

FIGURE 12.5 A schematic of virotherapy action. Engineered viruses will only infect cancerous cells, resulting in their lysis.

will see approval in the future. Cancer treatments using antisense RNA technology are now in clinical trials. Virotherapy cancer treatments may also become available in order to use engineered viruses to target cancerous cells (**FIGURE 12.5**). Technologies that harm only cancerous cells and keep healthy cells intact would be a dramatic improvement over radiation and chemotherapy of today. Again, there is hope that medications of the future will have reduced or even no side effects.

Agriculture and Food Production

There are persistent and real fears surrounding agricultural biotechnology. The loss of biodiversity posed by monoculture of genetically modified crops is already severe and the trend is likely to continue. Consumers and consumer advocacy groups demonize genetically modified organisms (GMOs), citing concerns over impacts to human health and the environment. The future will see the outcomes of GMO labeling petitions currently under consideration in a majority of American states, ultimately decided by the votes of individual citizens. Yet as we have seen with Prop 37 in California, biotechnology companies and other stakeholders have financed campaigns against GMO labeling and it is reasonable to expect this practice to continue.

As for fears of environmental contamination, all existing studies indicate the chances for introgression and other hybrid outcrossings are minimal. This is fortified by the permitting processes in place by the U.S. Department of Agriculture (USDA). The required monitoring of all standing GMO crop fields will result in continuous streams of data. Whether existing studies will be supported is yet to be seen. **Genetic use restriction technology (GURT)** would further alleviate any concerns of environmental contamination by transgenes. Also known as terminator technology, GURT produces plants with sterile seeds. When utilized in genetically modified plants, terminator technology ensures the prevention of accidental release of GMOs into the environment because they cannot reproduce. Although developed in the 1990s, GURT has never been used in commercial products. The current owner of the technology, agribusiness giant Monsanto, has actively pledged since 1999 there are no plans to begin using GURT due to perceived public opposition. It is of course unknown if that will remain the case.

When it comes to genetically modified livestock, previous discussions revealed the inadequacies of current technology. At present, the transfection of livestock with transgenes is not consistent or reliable. Until methods are developed to improve the delivery of transgenes into livestock, this area of biotechnology will remain impeded. For the near future, genetically modified livestock appears to be an area of research rather than commercialization.

Environmental Science and Conservation

The human population is expected to reach up to 10.5 billion within the next 30 years. It will be increasingly necessary to address environmental problems such as pollution and exposure to hazardous materials. As biodiversity is depleted, there may be increased conservation efforts to clone endangered and/or extinct organisms. Research into biofuels will also continue, with a strong emphasis on renewable energy as fossil fuels diminish.

Environmental Degradation

Currently, over 1300 Superfund sites designated by the EPA await remediation. This will occupy agency resources for many years to come. Bioremediation will play a role in environmental clean-up not only in terrestrial environments such as Superfund sites, but also with aquatic pollution. The oil spills of yesterday will continue to plague oceans and shorelines due to the prolonged nature of oil degradation; in fact it is unknown if these environments can ever truly be returned to their original condition. With increased reliance on alternative extraction methods, we can expect the occasional spill to occur as they routinely have in the past. For example, **hydraulic fracturing** (also known as **fracking**) technology is now utilized in approximately 2.5 million extraction sites worldwide, with over one million of these in the United States. Perhaps spills will even become more common, considering alternative methods like fracking are in their infancy and subject to some degree of trial and error.

Conservation

To date, there has been very moderate success with reproductive cloning of any kind. The success rates are miniscule at present, so one of the greatest challenges for the future will be to improve upon existing methods and their outcomes. This will be true for cloning of extant species, but also for endangered or even extinct organisms. Organizations conducting reproductive cloning have expressed interest in using pandas, gorillas, and ocelots as test subjects. The Institute for Breeding Rare and Endangered African Mammals was established by the Royal Zoological Society of Scotland with the hopes of cloning the Ethiopian wolf, the white rhinoceros, and pygmy hippopotamus (**FIGURE 12.6**).

FIGURE 12.6 Will the endangered pygmy hippopotamus be cloned in the future? An estimated 2,000 currently remain in Africa. © CraigRJD/iStockphoto

Long-term health effects of cloned animals are unknown because reproductive cloning technology is relatively new. Interspecies clones pose additional problems as they possess chimeric DNA; their nuclear DNA is of the cloned organism while their mitochondrial DNA was inherited from the surrogate mother. Another issue is the need for foster mothers of infant clones. If we are to clone endangered or extinct animals for conservation purposes, we will need to provide them with nurturing and socialization. Ideally this would involve members of their own species, but due to low numbers of endangered organisms this may not be feasible. In the case of extinct species, close extant relatives would be most desirable.

Renewable Energy

There are many estimates for fossil fuel reserves and the number of years that might pass before we tap out these resources. As this unknown date gets closer with the passage of time, it will become more and more necessary to not only research alternative fuels, but to actively engage in their mass production and mainstream adoption. Alternative fuel research is not new, yet we still rely on fossil fuels for the majority of our energy needs. The focus will shift toward implementing renewable energy to an ultimate saturation point. Biofuels will likely be a major component of alternative energy production. Methane from human waste may be captured by a larger number of municipalities in the process of waste treatment. Microbial fuel cells could be implemented in an increasing number of devices once powered by batteries. Ethanol and biodiesel will likely become more available at gas stations. There may also be a trend toward more self-reliance in obtaining energy, such as production of biodiesel for personal consumption.

Evolution

Evolution is often mistakenly thought of as progress, but in reality it is simply change. Change is not inherently good or bad and cannot be equated with improvement. Evolution is an ongoing process that has taken place since life began, so there is no reason to suggest it would not continue to occur in the future. Indeed, evolution is an agent of the present because it can only act on existing variation. Thus, if it is unfolding right now, it will always be so.

Are We Still Evolving?

Regardless of this assertion, there are evolutionary biologists that claim humans have surmounted the influence of natural selection. There is an argument that natural selection doesn't apply to industrialized societies because we are no longer subject to the selective pressures of the environment. We have become enveloped in an artificial environment of buildings, air conditioning, and modern medicine. Proponents of this viewpoint suggest the only major changes humanity has undergone since the adoption of agriculture and the resulting dawn of civilization were caused by cultural and technological advances. While the **selective pressures** of the natural environment as they affect us now may be reduced compared to most of human history, they have not gone away.

Geneticists have discovered a number of human adaptations that took place around the same time as the introduction of agriculture, but 10,000 years later there has not been a comparable impetus to spur such dramatic genetic change. That may not remain the case in the future. There is general consensus in the scientific community that we are experiencing climate change that is predicted to become more severe in the near future. Despite international efforts to reduce carbon dioxide emissions, which contribute to global warming, its concentration in the atmosphere increases an average of 1.9 parts per million each year. The seemingly inevitable climate change faced by humanity may very well become the kind of selective pressure we have avoided for so long. In 2009, the U.S. Chamber of Commerce recommended to the Environmental Protection Agency that we should not regulate carbon emissions in any way. Their stated rationale is that, "populations can acclimatize to warmer climates via a range of behavioral, physiological and technological adaptations." It appears industry agrees with the prediction that climate change may result in future human adaptations, and thus our continued evolution.

Designer Babies and Super Humans

Currently, gene therapy is performed on **somatic cells**, which are cells of the body. Any genetic material introduced into a patient's tissues remains local and exists only as long as the patient is alive. In the future, gene therapy may extend to sex cells, or gametes. **Gametes** include egg and sperm and serve as the vehicle to provide genetic material to offspring. Introducing transgenes into gametes would constitute **germ-line therapy**, as gametes are produced via meiosis by germ line cells. There are two purported goals of germ-line therapy. First, it would allow alteration of any aberrant genes, therefore preventing the inheritance of genetic disease. This would ensure the overall health of offspring. Secondly, germ-line therapy could be used to confer selected traits in offspring having nothing to do with disease prevention. While today there are procedures allowing parents to select the gender of their offspring, germ-line therapy would extend such choices to myriad physical and behavioral characteristics. With the right knowledge of genetics, it may one day be possible to select aspects of a child's appearance, such as eye and hair color or height. We could select for higher intelligence or greater creativity, musical ability, or athleticism. Hypothetically, as long as we reach a point where we understand the interactions of genes and environment, the options would be nearly endless.

One of the greatest implications of germ-line therapy is the slippery slope toward **eugenics**. The power to design your offspring would have unforeseeable consequences. Perhaps particular traits would become more favored and concentrated in the population. We might even see some existing variations disappear. If applied in combination with bionic technology (see above), there is even potential for a "super-human." Creation of a dramatically improved human form, with enhanced intellectual, physical, and psychological abilities, is a movement known as **transhumanism**.

If today's medical access is any indicator, in all likelihood germ-line therapy would not be within everyone's reach physically and financially. Some futurists even contemplate a scenario where the free market would create classes of people based on their genes: the genetically engineered and the naturally conceived. At what point would a new form of human, even a new species, be recognized as distinct from *Homo sapiens*?

Bioethics and Risk

Decisions made about biotechnology and its uses will be made in consideration of bioethics and risk. **Bioethics** literally means "life behavior" but refers to a branch of philosophy concerned with the ramifications of advances in biology and technology. Bioethicists attempt to infer the implications of biotechnology on society, whether it be moral, economic, legal, practical, political, or social. Regardless of the specific quandary, bioethicists routinely assess the risks and benefits involved in new technologies. Bioethics use evaluations of risk versus benefits to answer the questions:

- What could go wrong with this technology?
- What benefits will this technology provide? Who is benefitting, and are the benefits great enough to justify the risks?

- What could be its unintended consequences?
- How can we actively prevent risk?
- How can we maximize the benefits?
- Should we do something just because we can?

Risk is the potential for loss or damage, and it can be measured and even minimized to some degree. Risks can also be thought of as threats that can be avoided through preemptive actions. The assessment of risk is exemplified through the processes of clinical trials and drug approval as well as the field testing of GMOs. The perceived benefits of a technology should ideally outweigh any potential risks involved with its adoption. The multiple steps involved and the various factors considered help to ensure a preliminary evaluation of possible outcomes followed by a cautious exploration of actual consequences.

ELSI: A Model for Bioethics

The Human Genome Project dedicated between 3 to 5% of each annual budget toward the study of *e*thical, *l*egal, and *s*ocial *i*ssues, which they dubbed "**ELSI**." It is the largest bioethical program in the world and serves as a model for other similar governing bodies. The major ramifications surrounding release of genetic data have been outlined and continuously addressed as the landscape of molecular biotechnology evolves. Topics of interest include:

- *Fairness*: How will the information be used and who will have access to it?
- *Intellectual property*: Who owns the information?
- *Privacy*: Who has control of the information?
- *Psychology*: How might the information stigmatize certain subsets of the population? How might the information affect societal perceptions of different groups?
- *Reproduction*: How do reproductive technologies affect individual choice? How will they be used by society?
- *Clinical*: How will genetic tests be regulated? How do we educate medical professionals and the public on the availability and accuracy of genetic tests?
- *Philosophy*: How might the information influence society? What unforeseen consequences might arise? Where is the line drawn between nature and nurture (genetics and environment)?

Though the Human Genome Project has concluded, the National Human Genome Research Institute of the National Institutes of Health continues to operate ELSI and its mission. Today its major focus is on four elements related to the future of genomics research. First, the manner in which genomics research is conducted is a matter of interest. There are valid concerns over the privacy of health information used in conjunction with genomic investigation. ELSI desires to protect the privacy of patients who provide medical records to research studies, as it is common to include patient data when sharing results with colleagues. Secondly, ELSI is monitoring how the availability of genomic data will affect health care, and more particularly how it may affect equal access. "Broader societal issues" such as those outlined above are the third focus of current ELSI programs. Lastly, ELSI is monitoring the implications of genomic research on legal issues and changes to regulations.

Connecting the Dots

The intersection of society and biotechnology increasingly dominates the scientific landscape. Though its history is relatively short compared to other branches of science, molecular biotechnology is the culmination of existing fields including molecular biology, biochemistry, microbiology, medicine, evolution, and genetics. We are just beginning to witness the possible applications of molecular biotechnology and we can expect to experience tremendous growth within the industry. The implications on our personal lives and upon society as a whole will unfold in the coming years. The involvement of all interested and affected parties promises to minimize risk and improve the chances for success, however that may be measured. The words of the late Steve Jobs, founder of Apple and a technological pioneer who permanently altered society through his innovations, offer hope for the future:

> You can't connect the dots looking forward; you can only connect them looking backwards. So you have to trust that the dots will somehow connect in your future.

SUMMARY

- The future of biotechnology will be shaped by society as a whole. While scientists, executives, and shareholders will have a more direct involvement in determining course of action related to the biotechnology industry, products and services made possible through biotechnology must be accepted by individual consumers and the public at large. Regulators, public interest and consumer advocacy groups, and politicians will also play a role.
- Complementary DNA is synthesized from a gene transcript of mRNA that has had its introns removed. It represents the uninterrupted sequence of DNA complementary to the coding regions of a gene. The *Association for Molecular Pathology vs. Myriad Genetics* ruling permits complementary DNA patents.
- The *Association for Molecular Pathology vs. Myriad Genetics* Supreme Court ruling prohibits patent protection of human gene sequences. It has the potential to increase biotech start-ups should it reduce research costs associated with licensing fees. However, any company may seek patent protections on ancillary technologies, such as probes, primers, and testing methods used in conjunction with human gene sequences. Complementary DNA may also be patented. Therefore, the ruling will not greatly impact the ability to obtain patents on biotechnology.
- While the future is unknown, there are predictions for biotechnology applications. Some examples of possible outcomes include bioprinting, germ-line eugenics, labeling of genetically modified foods, and improved medicine such as targeted cancer treatments and pharmacogenomics.
- Bioethics deals with the intersection of biological science, technology, and society. Bioethicists evaluate benefits of new technology and weigh them against the potential risks involved. They formulate and recommend preemptive actions designed to minimize risk and maximize benefits.

KEY TERMS

Bioethics
Bio-ink
Bioprinting
Complementary DNA (cDNA)
ELSI
Eugenics
Gametes
Genetic use restriction technology (GURT)
Germ-line therapy
Hydraulic fracturing (fracking)
Microtissues
Risk
Selective pressure
Somatic cells
Transhumanism

DISCUSSION QUESTIONS

1. Of all the bioethical concerns addressed in this book, which do you consider the most pressing? Having the greatest potential to affect lasting societal change? The most exaggerated? Explain your reasoning.
2. What are the advantages to GURT? Disadvantages? What do you foresee to be the ultimate consequences of this technology should it be adopted in the future?
3. How do you envision human evolution to occur over time? Provide evidence to support your answer.

REFERENCES

Bravin, Jess, and Brent Kendall. "Justices Strike Down Gene Patents." *The Wall Street Journal*, June 13, 2013. http://www.popsci.com/science/article/2013-07/how-3-d-printing-body-parts-will-revolutionize-medicine

Hills, Melissa J., Linda Hall, Paul G. Arnison, and Allen G. Good. "Genetic Use Restriction Technologies (GURTs): Strategies to Impede Transgene Movement." *Trends in Plant Science* 12, no. 4 (2007): 177–183.

Jobs, Steven. "Commencement Address." Stanford University Commencement Ceremony, Stanford, California, June 12, 2005.

Kass, Leon. "Preventing a Brave New World." *The New Republic.* 2001 May 21; 224(21):30–9.

Ledbetter, David H., Erin Rooney Riggs, and Christa Lese Martin. "Clinical Applications of Whole-Genome Chromosomal Microarray Analysis." *Genomic and Personalized Medicine* 1 (2012): 133.

Leckart, Steven. "The Body Shop." *Popular Science*, August 2013.

Montgomery, Carl T., and Michael B. Smith. "Hydraulic Fracturing: The Past, Present, and Future." *Journal of Petroleum Technology* 62, no. 12 (2010): 26–41.

Oetting, William S., et al. "Getting Ready for the Human Phenome Project: The 2012 Forum of the Human Variome Project." *Human Mutation* 34, no. 4 (2013): 661–666.

Office of Biological and Environmental Research, Human Genome Project Information Archive 1990–2003. "Ethical, Legal, and Social Issues Research." United States Department of Energy Human Genome Project ELSI, last modified July 23, 2013. http://www.ornl.gov/sci/techresources/Human_Genome/research/elsi.shtml

Shakespeare, William. *The Tragedy of Julius Caesar*. London: First Folio, 1623.

Timme, Ruth E., et al. "Draft Genome Sequences of 21 *Salmonella enterica* Serovar Enteritidis Strains." *Journal of Bacteriology* 194, no. 21 (2012): 5994–5995.

United States Chamber of Commerce. "Detailed Review of the Health and Welfare Science Evidence and IQA Petition for Correction." Washington, DC, 2009.

Verspoor, Karin, et al. "Annotating the Biomedical Literature for the Human Variome." *Database: The Journal of Biological Databases and Curation* 2013 (2013): bat019.

APPENDIX 1

Genome Structure: DNA, Genes, and Chromosomes

A **genome** is composed of all of the hereditary material found in an organism. The genome contains all chromosomes and plasmids, which possess genetic information in the form of nucleic acids. Genomes vary among viruses, prokaryotes, and eukaryotes. This discussion will navigate through the layers of material that make up the genome.

Chromosome Structure

A **chromosome** is a physical structure composed of a strand of DNA and associated proteins known as **histones**. The physical matter that constitutes chromosomes is known as **chromatin**. Chromatin is often described as "beads on a string," as the DNA molecule wraps around the histone proteins to form this type of arrangement (**FIGURE A1.1**). Each "bead" of histones wrapped in DNA is called a **nucleosome**.

Chromosomes carry the genes that determine inherited traits. In eukaryotic cells, chromosomes are threadlike linear strands located in the cell's nucleus. In viruses and prokaryotic cells, no nucleus is present. Viruses possess one or more fragments of nucleic acid. In prokaryotes, the chromosome is a circular strand of DNA located within the cytoplasm. The area within a virus or prokaryotic cell that contains nucleic acid is referred to as the **nucleoid** (**FIGURE A1.2**).

A molecule of DNA may exist separately from the nucleoid and is called a **plasmid**. Plasmids exist within bacterial cells (**FIGURE A1.3**). Plasmid DNA contains genes that allow bacteria to develop antibiotic resistance, to form a cylindrical extension called a pilus, and in some bacteria, to synthesize protein toxins. Plasmids are a conventional vehicle for introducing foreign transgenes into genetically modified organisms.

Deoxyribonucleic Acid Structure

Deoxyribonucleic acid (DNA) was first isolated and recognized as an acidic substance within the nucleus by Swiss biologist Friedrich Miescher in 1869. He named the chemical "nuclein," which we know today as **nucleic acids**. The role of nucleic acids in

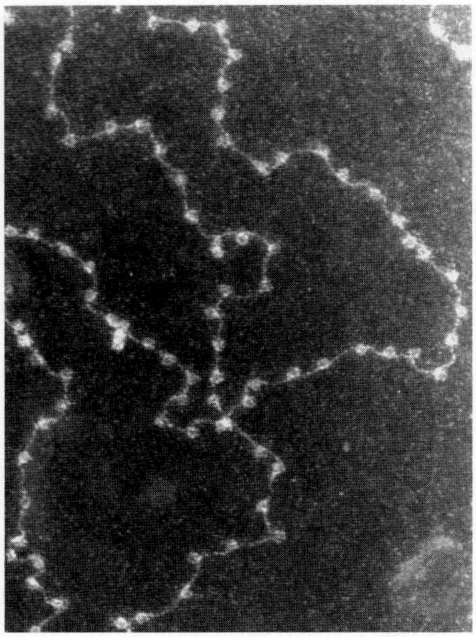

FIGURE A1.1 An electron micrograph of chromatin, the material comprising chromosomes. DNA wraps around histone proteins to form a "beads on a string" structure within the chromosome. © Donald Fawcett/Visuals Unlimited, Inc.

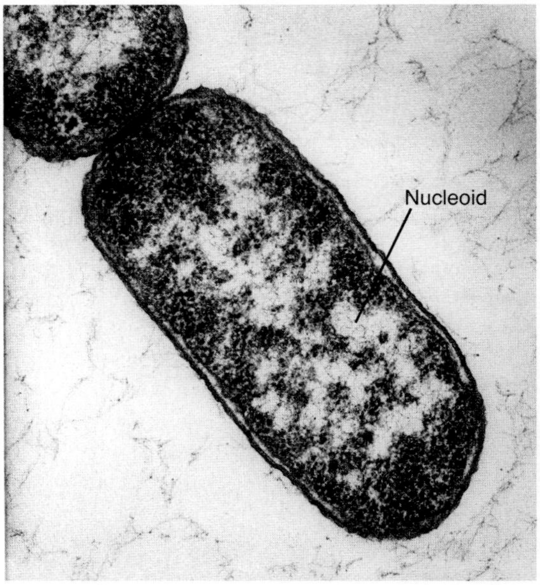

FIGURE A1.2 An electron micrograph of the bacterium *Escherichia coli* displays the nucleoid region of the cell where its chromosome is located. Photo courtesy of the Molecular and Cell Biology Instructional Laboratory Program, University of California, Berkley.

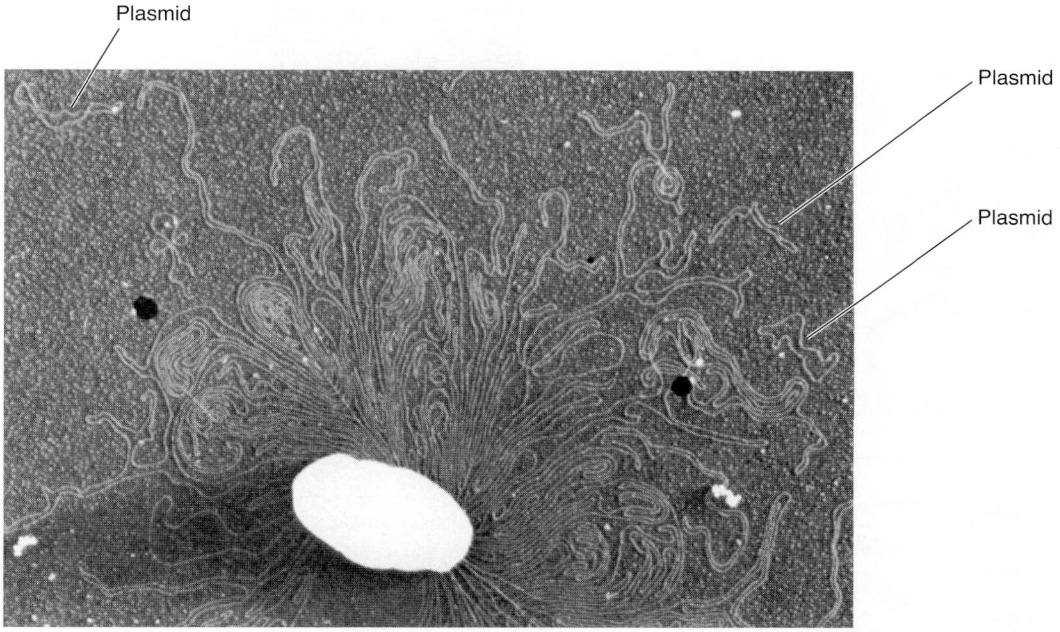

FIGURE A1.3 An electron micrograph of a ruptured *E. coli* cell shows a number of individual plasmids. Courtesy of Huntington Potter and David Dressler. Used with permission of Huntington Potter, University of Colorado Anschutz Medical Campus.

genetics remained unclear for many years, partly because many leading chemists were convinced that proteins were the carriers of genetic information.

The structure of DNA is a double helix (**FIGURE A1.4**). It is often compared to a ladder, with the rails of the ladder representing its backbone and the rungs representing chemical bonds holding the ladder together. Twisting the ladder at both ends converts its linear structure into a helical arrangement with major and minor grooves. DNA consists of two molecules that are repeated over and over forming its backbone. These are a five carbon sugar (deoxyribose) that is connected at two of its carbon atoms to phosphate groups. Thus, this region of DNA is often referred to as the sugar-phosphate backbone. Structures called nitrogenous bases are linked to the sugar molecule (**FIGURE A1.5**). They got this name because nitrogen is one of the essential atoms. Four nitrogenous bases are present in DNA: adenine, guanine, cytosine, and thymine. Base pairs joined together via hydrogen bonds form the rungs of the DNA ladder. Adenine always bonds with thymine, while guanine always bonds with cytosine. The combination of a phosphate group, deoxyribose, and a nitrogenous base is called a **nucleotide**. Nucleotides are the monomers from which molecules of the polymer DNA are constructed.

Ribonucleic Acid Structure

Ribonucleic acid (RNA) differs slightly from DNA. It has a different sugar in its backbone called ribose (instead of deoxyribose). RNA also has one different nitrogen base that is not found in DNA: uracil. Uracil (U) takes the place of thymine in an RNA molecule. If you look at a sequence of bases and a U appears, this is a dead giveaway that you are looking at RNA. Remember that in DNA, adenine base pairs with thymine (A-T); in RNA, adenine base pairs with uracil (A-U). One other difference is that RNA is nearly always a single strand of phosphates, ribose, and bases. The molecular structure of RNA is shown in **A1.6.**

Scientists called the paired DNA molecule a double strand instead of a double helix although the terms have the same meaning. Looking forward, it is important to know whether DNA is single- or double-stranded, as it is used as a basis for classifying viruses. Viruses can be either single- or double-stranded DNA. Viruses can also be classified as RNA viruses. In this case viruses are categorized as either single- or double-stranded RNA viruses.

Genes

A **gene** is a sequence of DNA bases that controls the production of a functional product. The functional product is nearly always a protein molecule, but it can also be a functional RNA molecule. Originally a gene was defined as a sequence of DNA that resulted in a protein having a specific ordering of amino acids based on the DNA bases. Molecular biologists recently discovered that ribosomal RNA (that appears in ribosomes) is responsible for the creation

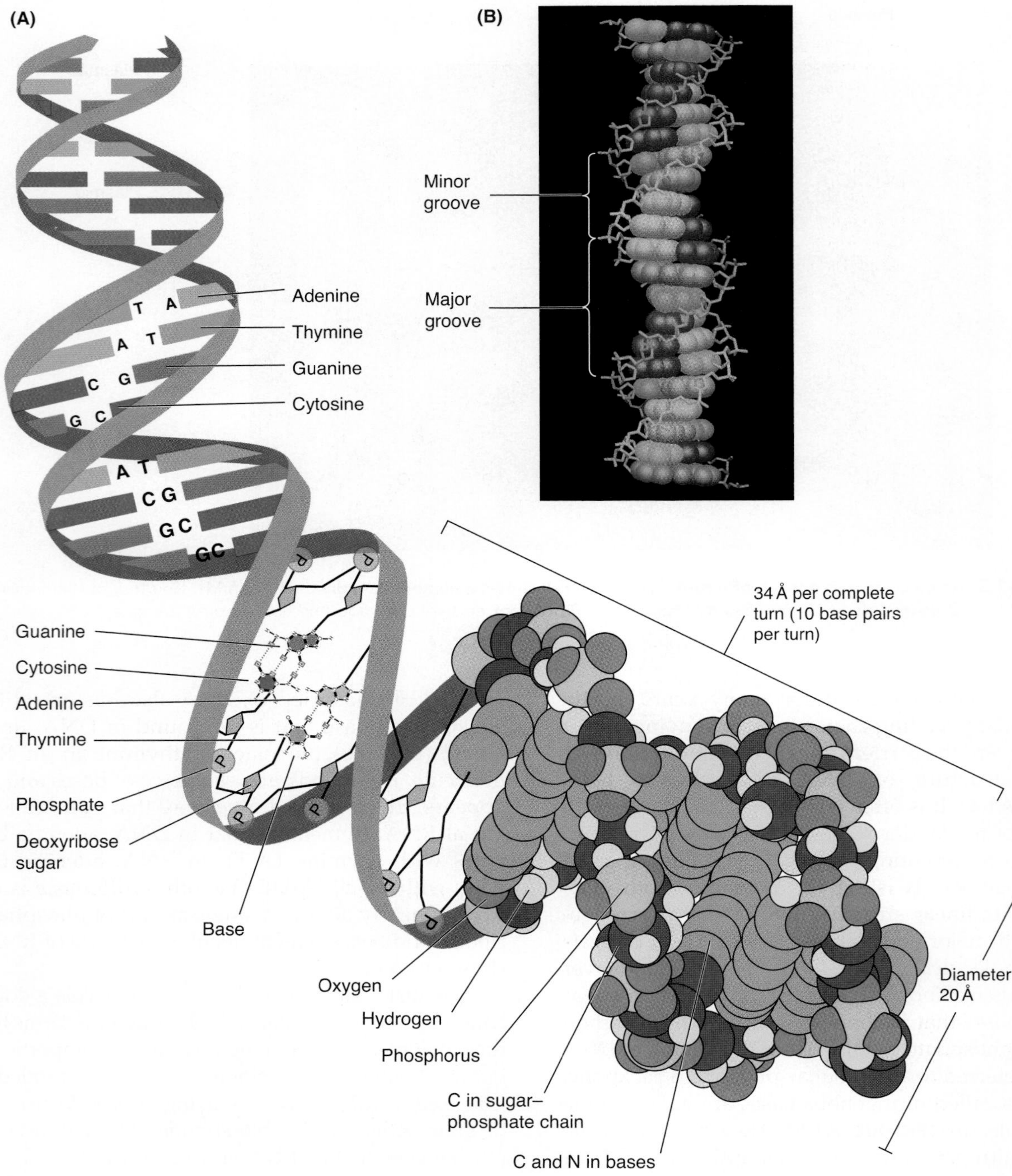

FIGURE A1.4 The DNA double helix. (A) The ribbon diagram shows the sugar-phosphate backbone as grey bands and the base pairs as horizontal lines. (B) A computer model showing major and minor grooves. Courtesy of Antony M. Dean, University of Minnesota.

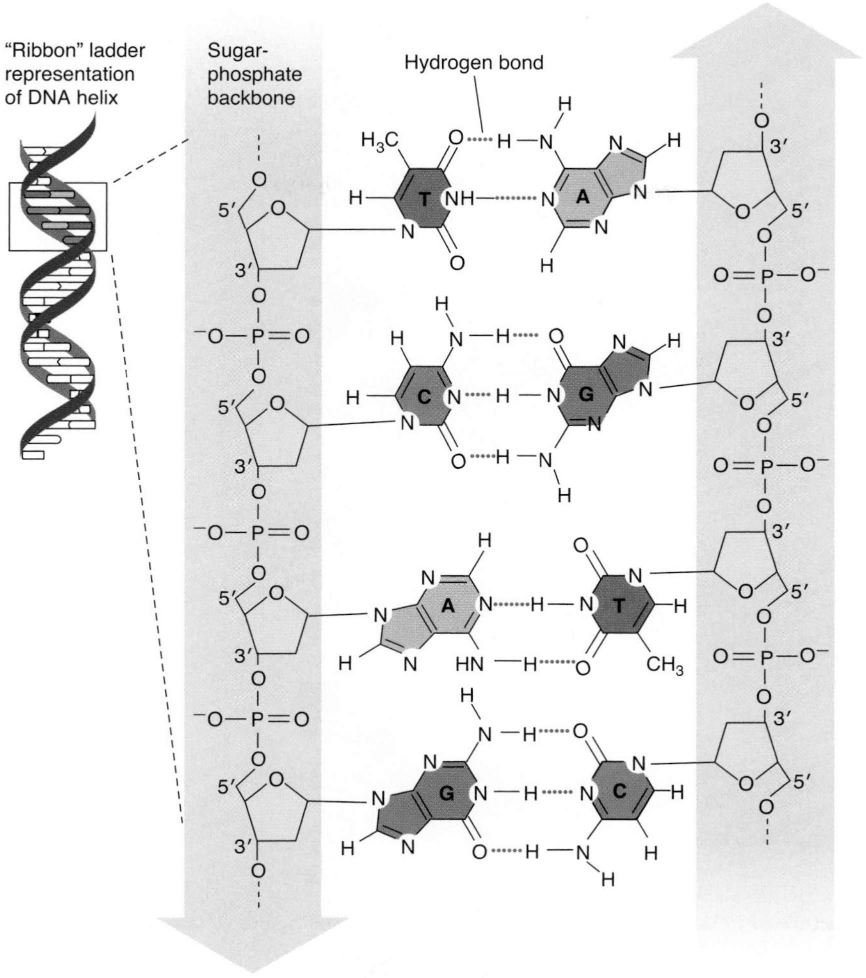

FIGURE A1.5 A view of unwound DNA exhibits the sugar-phosphate backbones and rungs containing nitrogenous base pairs connected via hydrogen bonds.

of peptide bonds that exist between amino acids. Peptide bond formation at the level of the ribosomes was always presumed to be the function of ribosomal proteins that acted as enzymes.

In the case of ribosomes, it turns out that a nucleic acid (RNA) is functioning as an enzyme. For molecular biologists, this finding was astounding. A new term was coined for this phenomenon: the **ribozyme**. Ribozymes are RNA molecules present in the ribosome that function as an enzyme. RNA is also a product of DNA. For this reason, the definition of a gene was broadened to encompass a functional product that is almost always a protein, but in the case of the ribosome is a nucleic acid.

Key Terms

Chromatin
Chromosome
Deoxyribonucleic acid (DNA)
Gene
Genome
Histones
Nucleic acids
Nucleoid
Nucleosome
Nucleotide
Plasmid
Ribonucleic acid (RNA)
Ribozyme

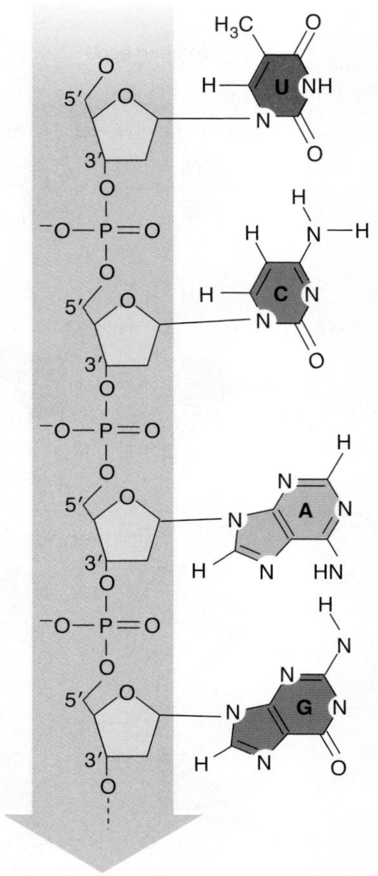

FIGURE A1.6 The structure of RNA is single stranded. RNA contains the nitrogenous base uracil in lieu of thymine, which is only present in DNA.

APPENDIX

2 Basics of Gene Expression and DNA Replication

Genes contain instructional information for functional products, most often proteins but sometimes RNA such as ribozymes. Gene expression may be thought of as protein synthesis, as this is usually the case. The process of protein synthesis is summarized in the **central dogma of molecular biology**: genetic information flows from a gene as it is transcribed into RNA, which is then translated into a polypeptide (**FIGURE A2.1**). Once the polypeptide undergoes processing, such as folding into a three-dimensional shape, the product is now considered a protein molecule composed of amino acid monomers.

The Genetic Code

The **genetic code** provides the information to translate nucleotide sequences into amino acid sequences, thus serving as the foundation of gene expression. Critical to the DNA molecule and to the basis for its genetic code are the unique ways in which nitrogenous bases form specific hydrogen bonds with bases attached to the opposite strand. **Hydrogen bonds** are a very special type of bond that always forms between a hydrogen atom and another non-hydrogen atom. By itself a single hydrogen bond is very weak and essentially worthless. However, there is strength in numbers, and this applies to hydrogen bonds. Millions of hydrogen bonds occur in a DNA molecule and hold the helix together. Returning to our DNA ladder, the rungs represent the hydrogen bonds that occur between the four nitrogenous bases.

In 1950, Austrian chemist Erwin Chargaff isolated DNA from different organisms and found that nitrogenous bases (adenine, thymine, guanine, and cytosine) varied in concentration from one organism to another. More importantly, he also observed that the amount of guanine is equal to cytosine, and the amount of adenine is equal to thymine. This finding became known as **Chargaff's Rules**. Furthermore, the key to understanding how DNA controls the genetic code lies in the hydrogen bonds. The nitrogenous base adenine can only form hydrogen bonds with thymine. Similarly, guanine can only form hydrogen bonds with cytosine. This bonding order is called **base pairing**. In any DNA molecule, the same pattern of base pairing occurs. Adenine will always base pair with thymine and guanine will always base pair with cytosine (**FIGURE A2.2**). This specificity between the bases allows each side of the DNA molecule to serve as a template or model for copying itself.

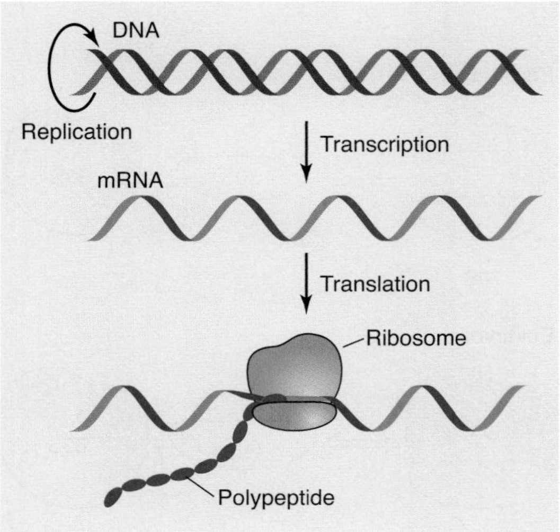

FIGURE A2.1 The central dogma of molecular biology involves transcription of DNA into mRNA, which is translated into a polypeptide.

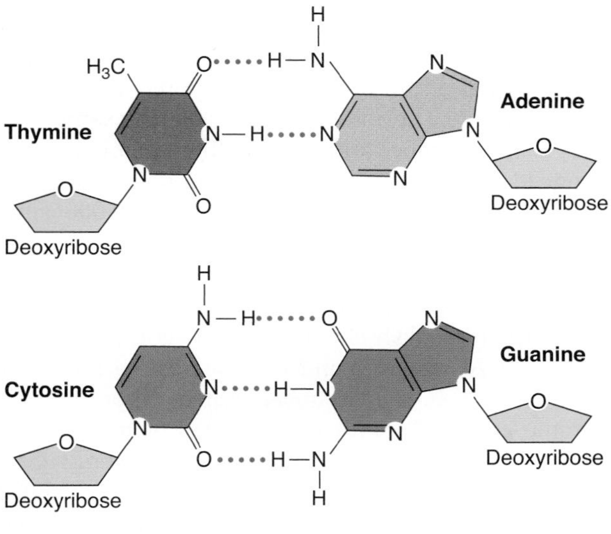

FIGURE A2.2 Base pairing in DNA exhibits Chargaff's rules.

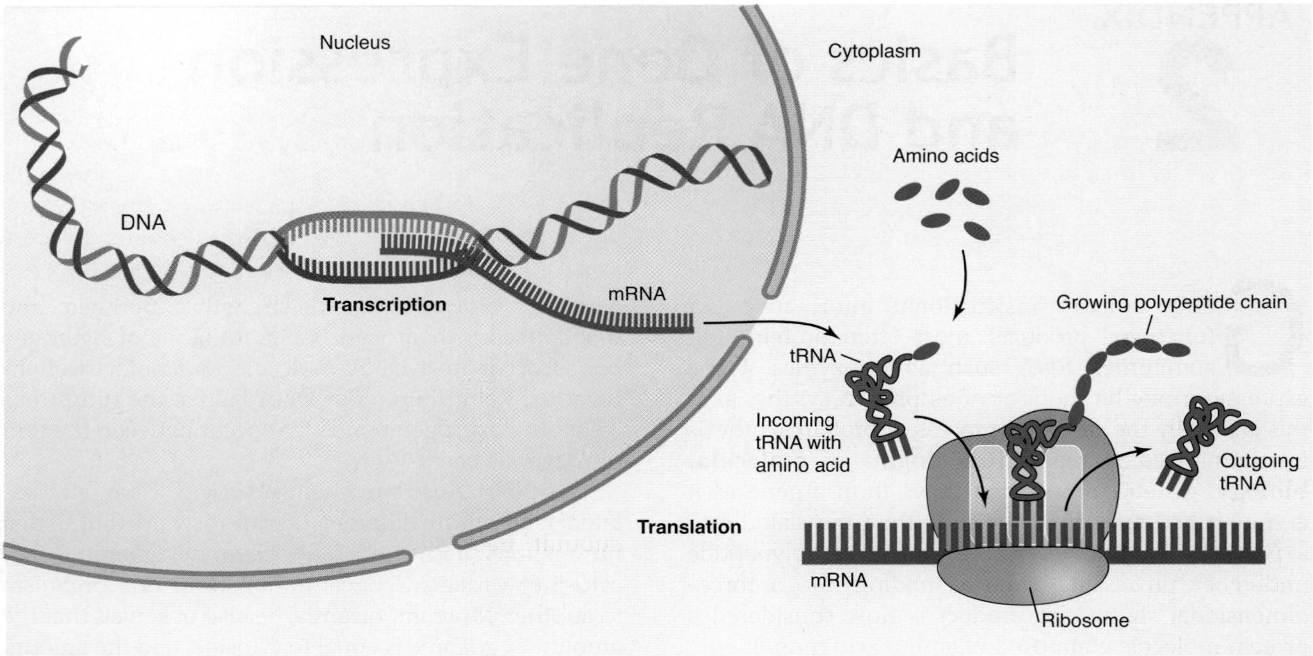

FIGURE A2.3 A generalized schematic of protein synthesis as seen in eukaryotic cells involves transcription and translation. Adapted from D. Secko, The *Science Creative Quarterly*, 2007 [http://www.scq.ubc.ca/a-monks-flourishing-garden-the-basics-of-molecular-biology-explained/]. Accessed May 5, 2010.

Protein Synthesis

The general scheme for protein synthesis is often called the central dogma of genetics. This is summarized as: *DNA to RNA to proteins.* The formation of RNA from a DNA molecule is called **transcription**. The chemical creation of proteins from an RNA molecule is called **translation** (**FIGURE A2.3**).

At the heart of the whole process are the ribosomes. **Ribosomes** are structures that are the sites for all protein synthesis. They are composed of protein molecules and a unique type of RNA: ribosomal RNA (rRNA). Ribosomes are so tiny that they cannot be seen without an electron microscope and to isolate them requires a type of laboratory equipment called an ultracentrifuge. They are isolated by placing samples of cytoplasm into tubes and rotating them in an ultracentrifuge at 100,000 revolutions per minute. This speed of rotation generates forces as much as 1,000,000 times that of gravity.

The **Svedberg unit (S)** is the unit of measurement used to describe ribosomes. It is named for Theodor Svedberg, a Swedish chemist who invented the ultracentrifuge. A Svedberg unit is a measure of the speed with which a substance travels when exposed to the extreme gravitational forces generated by an ultracentrifuge. Microbiologists call this particle movement in a centrifuge the sedimentation coefficient**.** The further a particle travels, the higher its Svedberg unit. It is *not* a measure of weight. This is an important point to remember and explains why the subunit structure of ribosomes is not additive. All ribosomes isolated from bacteria and archaea have the same S value of 70 Svedberg units (70S). 70S ribosomes are found only in prokaryotic cells. In eukaryotic cells, ribosomes have an S value of 80 Svedberg units (**FIGURE A2.4**).

Transcription

Transcription begins with an enzyme called RNA polymerase that "reads" the sequence of DNA bases on

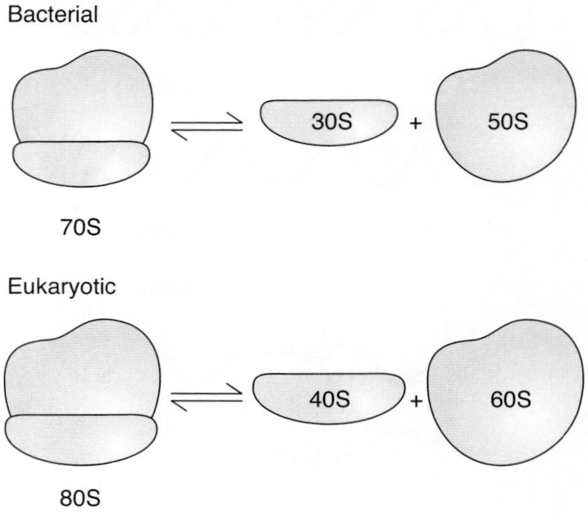

FIGURE A2.4 Ribosomal subunits and S values for bacterial (prokaryotic) and eukaryotic cells.

the bacterial chromosome and creates a molecule of RNA called **messenger RNA (mRNA)**. This stage is called transcription because RNA polymerase literally transcribes the sequence of bases in DNA as it creates a molecule of RNA. Messenger RNA gets its name because it behaves like a messenger carrying instructions from the chromosome to ribosomes for making a protein.

Instructions for the sequence of amino acids that appear in a protein are carried by mRNA and comprise the genetic code. It is very important to remember that the genetic code refers to the mRNA molecule and not to the sequence of bases on the chromosome. The genetic code consists of groupings of three bases that appear in the mRNA molecule that correspond to specific amino acids. A grouping of three bases on mRNA is called a **codon**. When you look at the genetic code what you are seeing are codons, which is why the code is presented in groupings of three bases (**TABLE A2.1**).

Translation

Translation begins when an mRNA molecule interacts with the ribosomal subunits. This is a physical association with mRNA occupying a grooved space between the subunits of a ribosome. Ribosomes are an example of biological recycling: the subunits are separated from each other until they are needed for building proteins. Once a protein has been manufactured, the aggregated subunits break apart and exist in the cytoplasm as individual subunits.

Translation begins with the binding of an mRNA molecule to a subunit (**FIGURE A2.5**). This association is called the **initiation complex**. Next the two ribosomal subunits attach to form a complete ribosome. The ribosome moves along the mRNA molecule.

Three sites (A, P, and E) exist in the ribosomal subunit. Each site has an important role in protein formation. These sites interact with both mRNA and another type of RNA that carries individual amino

TABLE A2.1

The Genetic Code[a]

	U	*C*	*A*	*G*	
U	UUU Phe	UCA Ser	UAU Tyr	UGU Cys	***U***
	UUC Phe	UCC Ser	UAC Tyr	UGC Cys	***C***
	UUA Leu	UCA Ser	UAA **STOP**	UGA **STOP**	***A***
	UUG Leu	UCG Ser	UAG **STOP**	UGG Trp	***G***
C	CUU Leu	CCU Pro	CAU His	CGU Arg	***U***
	CUC Leu	CCC Pro	CAC His	CGC Arg	***C***
	CUA Leu	CCA Pro	CAA Gln	CGA Arg	***A***
	CUG Leu	CCG Pro	CAG Gln	CGG Arg	***G***
A	AUU Ile	ACU Thr	AAU Asn	AGU Ser	***U***
	AUC Ile	ACC Thr	AAC Asn	AGC Ser	***C***
	AUA Ile	ACA Thr	AAA Lys	AGA Arg	***A***
	AUG Met or **START**	ACG Thr	AAG Lys	AGG Arg	***G***
G	GUU Val	GCU Ala	GAU Asp	GGU Gly	***U***
	GUC Val	GCC Ala	GAC Asp	GGC Gly	***C***
	GUA Val	GCA Ala	GAA Glu	GGA Gly	***A***
	GUG Val	GCG Ala	GAG Glu	GGG Gly	***G***

[a]The codon triplet is shown starting at the 5′ end. The first nucleotide is in the left-hand column, the second in the top row, and the third along the right-hand column.

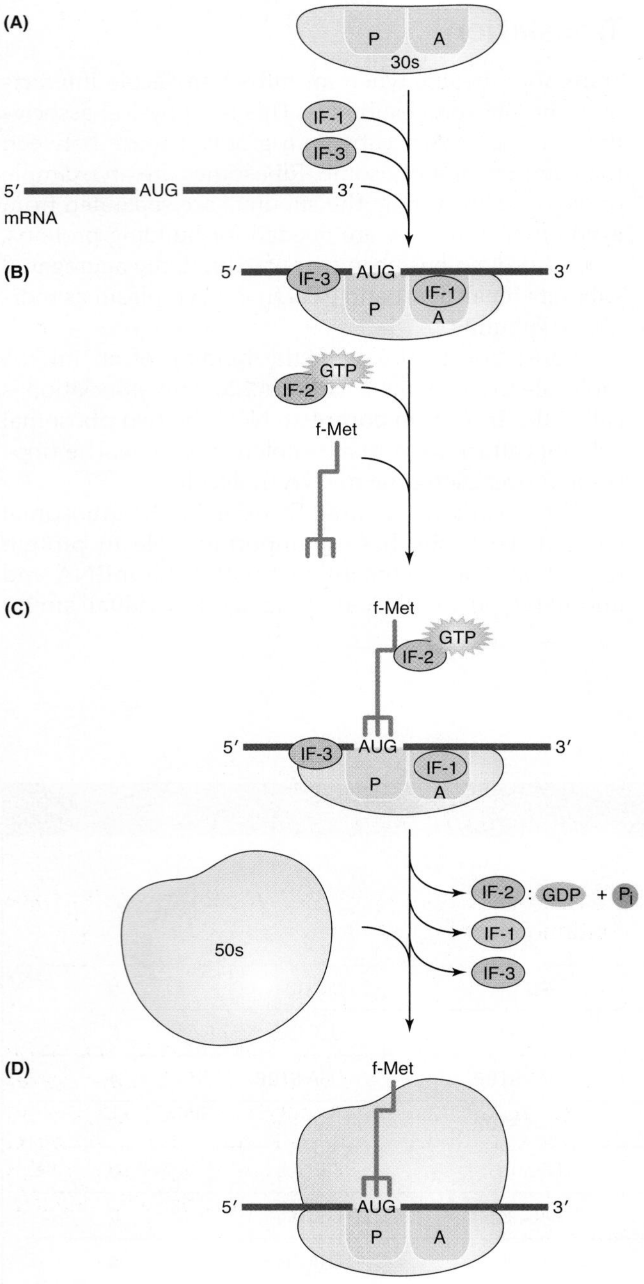

FIGURE A2.5 (A)-(D) The initiation complex triggers the start of translation during protein synthesis.

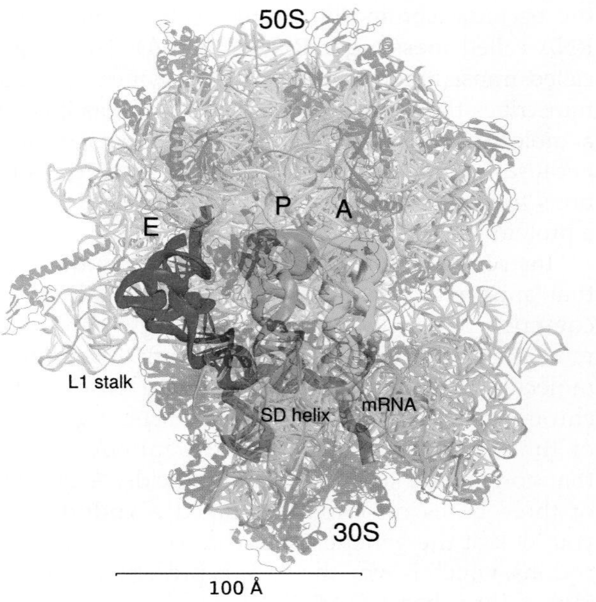

FIGURE A2.6 The A, P, and E sites found within ribosomal subunits. Reproduced from *Curr. Opin. Chem.* Biol., vol. 12, A. Korostelev, D. N. Ermolenko and H. F. Noller, *Structural dynamics of the ribosome*, pp. 674–683, copyright 2008, with permission from Elsevier [http://www.sciencedirect.com/science/journal/13675931]. Photo courtesy of Andrei Korostelev, University of Massachusetts.

acids to the ribosome. This type of RNA is called **transfer RNA (tRNA)**. It gets its name because it transfers (or carries) an amino acid to the ribosome. The ribosomal subunit and its three sites are shown in **FIGURE A2.6**.

Transfer RNA molecules exist for each type of amino acid that appears in a protein. The tRNA molecule is shaped somewhat like a bent pretzel (**FIGURE A2.7**). One end of the tRNA molecule has a region that attaches to an amino acid. Another end of the tRNA molecule has a region that binds to mRNA and is called the **anticodon**. Codon-anticodon binding is very specific and insures the correct ordering of amino acids in proteins.

Transfer RNA molecules first bind to the A site, which is also called the acceptor site. As the ribosome moves along the mRNA molecule, a tRNA molecule slides from the A site to the P site. A second tRNA molecule attaches to the now free A site. This results in two sites that are occupied by tRNA molecules. Both are sitting side by side within the ribosome. Ribozymes join the two amino acids by creating a peptide bond. The P site is the location within the ribosome that joins two amino acids. You can think of the P site as the peptide bond site. Once a peptide bond forms, the tRNA molecule moves into the E site, which stands for exit site. The tRNA molecule (now without an amino acid) breaks away from the ribosome and is free to be used again (or recycled) to bind with another amino acid and repeat the process.

Transfer RNA molecules continue to attach to the A site until a sequence of bases on the messenger RNA causes the whole process of protein synthesis to stop. Once this happens, the ribosome breaks apart and returns to individual subunits. The subunits of the ribosome can reassemble again and again as needed for protein synthesis.

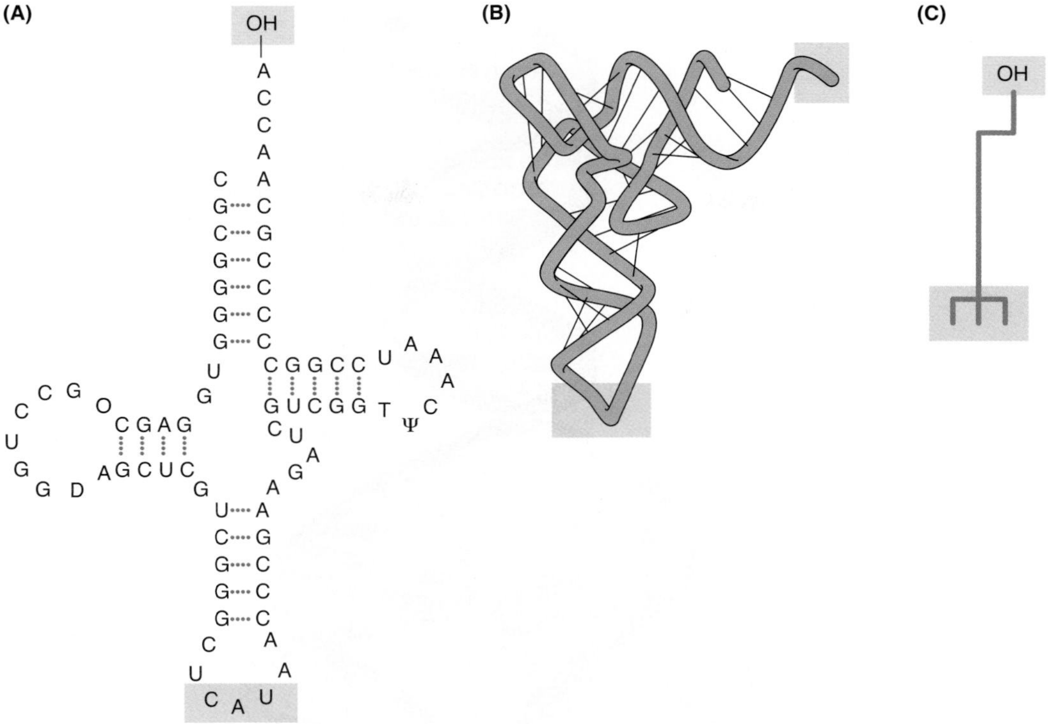

FIGURE A2.7 Different models of the transfer RNA molecule. (A) Two-dimensional. (B) Three-dimensional. (C) An essential view showing the anticodon region at its base.

DNA Replication

The copying of a DNA molecule to form two new molecules is called **DNA replication**. Replication of DNA has to be extremely accurate because a change in the sequence of DNA nitrogen bases alters the genetic information. The overall process is shown in **FIGURE A2.8**.

Prokaryotic Replication

In bacterial and archaeal cells, DNA replication is a simple process. The two strands of DNA separate and each strand serves as a template for its own replication. This process is called *semiconservative* because one of the original strands is conserved and one is a new copy.

Several enzymes are required for the copying process. First an enzyme called helicase has to open the double-stranded molecule to expose nitrogen bases. The open segment of the DNA molecule is called a replication fork. Another enzyme called DNA polymerase adds nitrogen bases to complementary bases in the replication fork.

The two strands of DNA are not copied at the same time. One strand leads in the synthesis of a new strand, while the other strand lags behind. The lagging strand is synthesized in pieces of DNA called Okazaki fragments, named in honor of their discoverer. Finally an enzyme called DNA ligase joins the fragments into one continuous strand.

Eukaryotic Replication

In eukaryotic cells, the process is basically the same except that there are four types of DNA polymerases instead of one used in prokaryotic cells. Also, due to the many chromosomes involved in the replication process, many more replication forks are required than the single fork that exists in prokaryotic DNA replication (**FIGURE A2.9**).

Key Terms

Anticodon
Base pairing
Central dogma of molecular biology
Chargaff's Rules
Codon
DNA replication
Genetic code
Hydrogen bond
Initiation complex
Messenger RNA (mRNA)
Ribosomes
Svedberg unit (S value)
Transcription
Transfer RNA (tRNA)
Translation

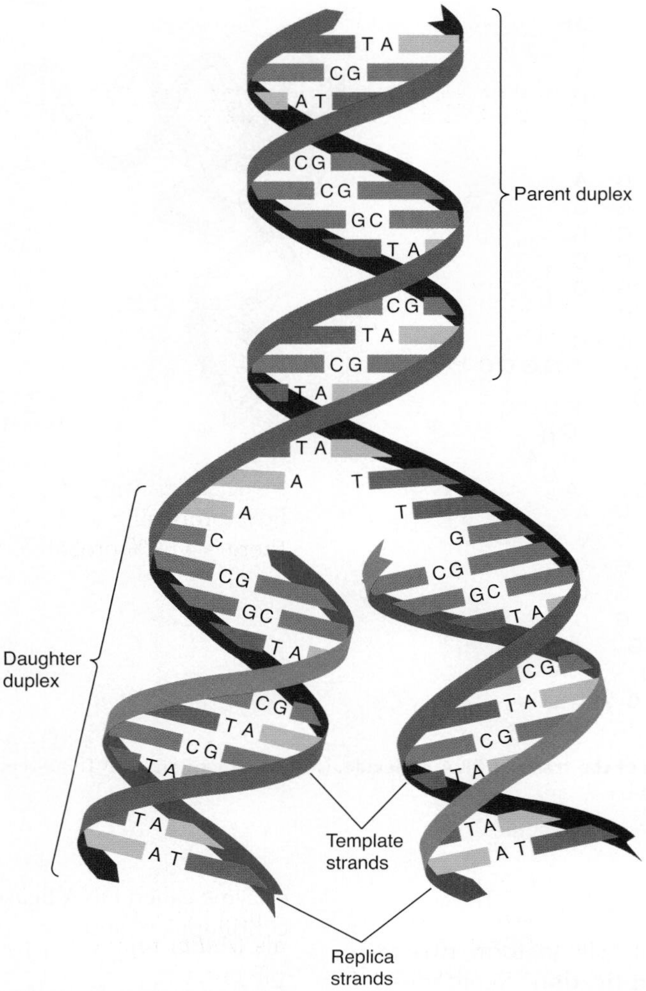

FIGURE A2.8 **The semiconservative nature of DNA replication results in two molecules, each composed of a replica strand and a template strand from the parent duplex.**

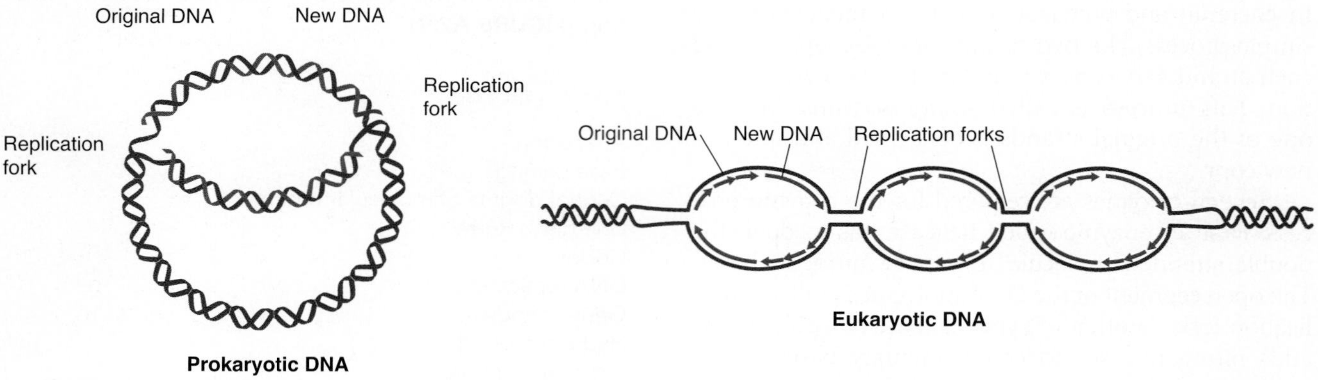

FIGURE A2.9 **A comparison of replication forks that form during replication of prokaryotic and eukaryotic DNA.**

APPENDIX

3 A Primer in Classical Genetics

Gregor Mendel, sometimes called the "father of genetics," formulated the basic laws of genetics. His laws include the Law of Segregation and the Law of Independent Assortment and constitute what is referred to alternatively as Mendelian genetics or classical genetics.

Genotypes and Phenotypes

Mendel determined that individuals have two alternate forms of a gene (two **alleles**, in modern terminology) for each trait in their body cells. Today, we know that alleles are on the chromosomes. An individual can be homozygous dominant (two dominant alleles, *GG*), homozygous recessive (two recessive alleles, *gg*), or heterozygous (one dominant and one recessive allele, *Gg*). **Genotype** refers to an individual's genes, while **phenotype** refers to an individual's appearance or other trait conferred based upon the genotype. Homozygous dominant and heterozygous individuals show the dominant phenotype; homozygous recessive individuals show the recessive phenotype. When at least one dominant allele is present in the genotype, the phenotype is expressed as dominant (**FIGURE A3.1**).

Monohybrid Crosses

A single pair of alleles is involved in one-trait crosses, or **monohybrid cross**. Mendel found that reproduction between two heterozygous individuals (*Aa*) resulted in both dominant and recessive phenotypes among the offspring. The expected phenotypic ratio among the offspring was 3:1 (**TABLE A3.1**). Three offspring had the dominant phenotype for every one that had the recessive phenotype.

Mendel realized that these results were obtainable only if the alleles of each parent segregated (separated from each other) during meiosis. Otherwise, all offspring would inherit a dominant allele, and no offspring would be homozygous recessive (**FIGURE A3.2**). Therefore, Mendel formulated his first law of inheritance:

> **Law of Segregation:** Each organism contains two alleles for each trait, and the alleles segregate during the formation of gametes. Each gamete then contains only one allele for each trait. When fertilization occurs, the new organism has two alleles for each trait, one from each parent.

The segregation of alleles represents gamete formation that occurs during meiosis. As gametes are **haploid**, it holds to reason that each gamete contains a single allele for each trait. Upon fertilization, the gametes fuse and two alleles for each gene are now present, resulting in a **diploid** organism.

Inheritance is a game of chance. Just as there is a 50% probability of heads or tails when tossing a coin, there is a 50% probability that a sperm or egg will have an *A* or an *a* when the parent is *Aa*. The chance of an equal number of heads or tails improves as the number of tosses increases. In the same way, the chance of an equal number of gametes with *A* and *a* improves as the number of gametes increases. Therefore, the 3:1 ratio among offspring is more likely when a large number of sperm fertilize a large number of eggs.

Dihybrid Crosses

A **dihybrid cross** involves two pairs of alleles. Mendel found that during a dihybrid cross, when two individuals (*AaBb*) reproduce, the phenotypic ratio among the offspring is 9:3:3:1, representing four possible phenotypes (**FIGURE A3.3**). He realized that these results could only be obtained if the alleles of the parents segregated independently of one another when the gametes were formed (**FIGURE A3.4**). From this, Mendel formulated his second law of inheritance:

> **Law of Independent Assortment:** Members of an allelic pair segregate (assort) independently of members of another allelic pair. Therefore, all possible combinations of alleles can occur in the gametes.

This is another way of saying that when alleles assort independently, the genes they represent are not linked. **Linkage** of genes, or of genes and DNA markers, occurs when they are close enough to each other on a chromosome that there is the potential they will not assort independently. This would result in offspring uniformly inheriting both traits together (**FIGURE A3.5**).

X-Linked Crosses

A cross involving a gene found on the X chromosome is known as an **X-linked cross**. In humans, males usually have an X and Y chromosome while females have

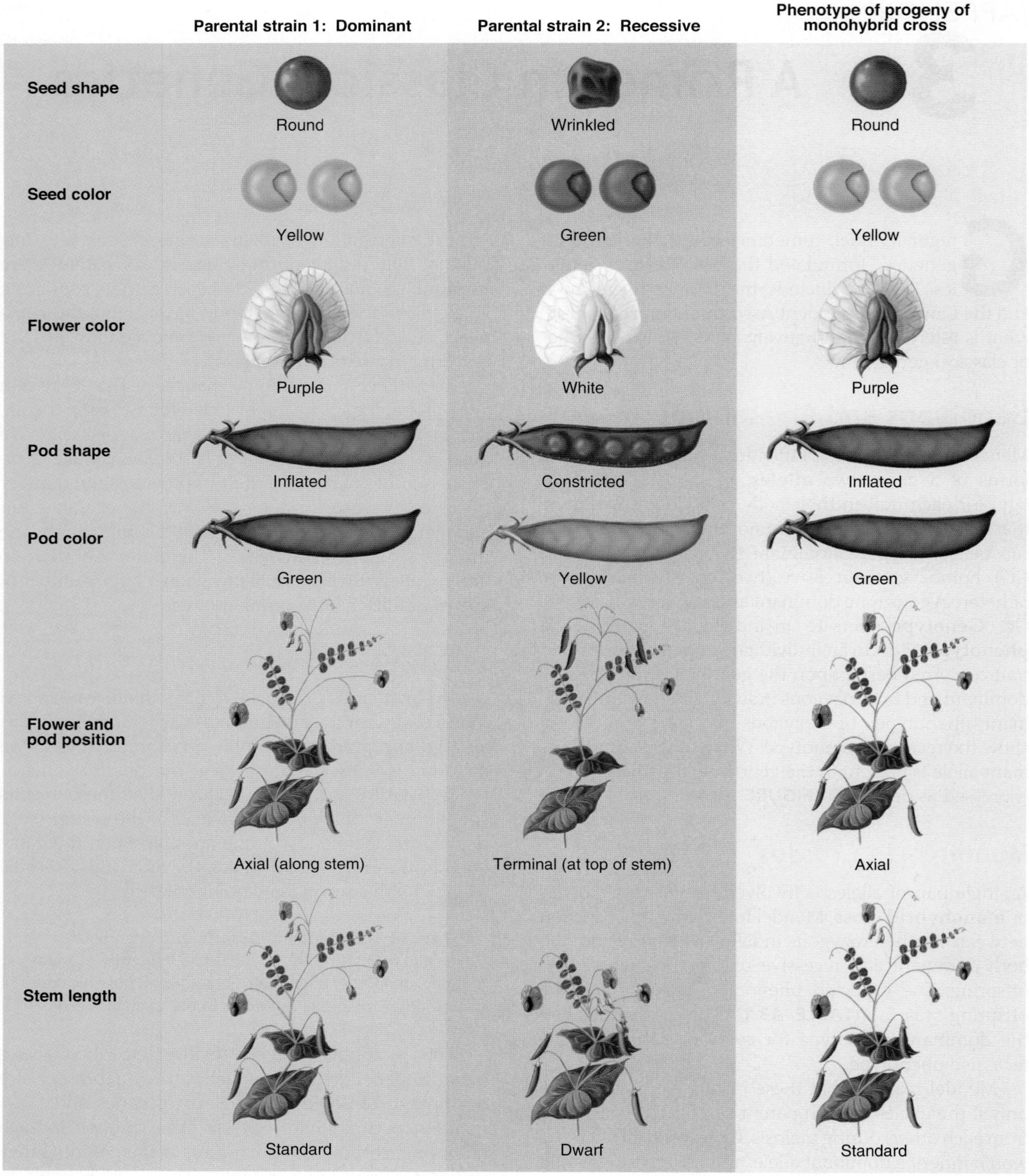

FIGURE A3.1 Traits studied by Gregor Mendel in *Pisum sativa*, the garden pea plant. When crossed, the dominant trait always masks the recessive trait in the offspring.

TABLE A3.1

Results of Mendel's monohybrid experiments

Parental traits	F1 trait	Number of F2 progeny	F2 ratio
Round × wrinkled (seeds)	Round	5474 round, 1850 wrinkled	2.96 : 1
Yellow × green (seeds)	Yellow	6022 yellow, 2001 green	3.01 : 1
Purple × white (flowers)	Purple	705 purple, 224 white	3.15 : 1
Inflated × constricted (pods)	Inflated	882 inflated, 299 constricted	2.95 : 1
Green × yellow (unripe pods)	Green	428 green, 152 yellow	2.82 : 1
Axial × terminal (flower position)	Axial	651 axial, 207 terminal	3.14 : 1
Long × short (stems)	Long	787 long, 277 short	2.84 : 1

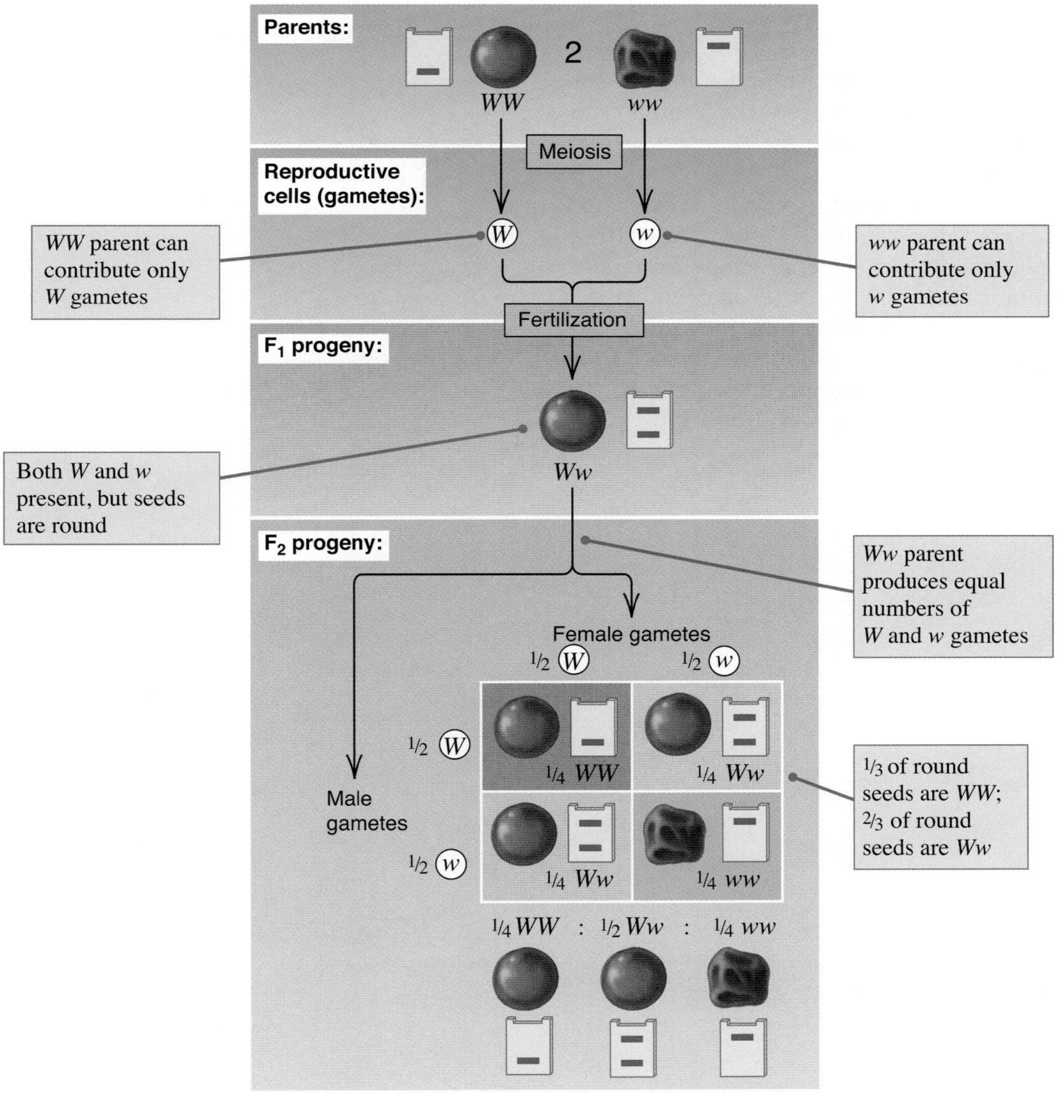

FIGURE A3.2 A monohybrid cross demonstrates the principles of Mendel's Law of Segregation. Each parent has two factors for the wrinkled trait. Gametes contain one factor for the wrinkled trait. Upon fertilization, progeny of both the F1 and F2 generations contain two factors for the wrinkled trait.

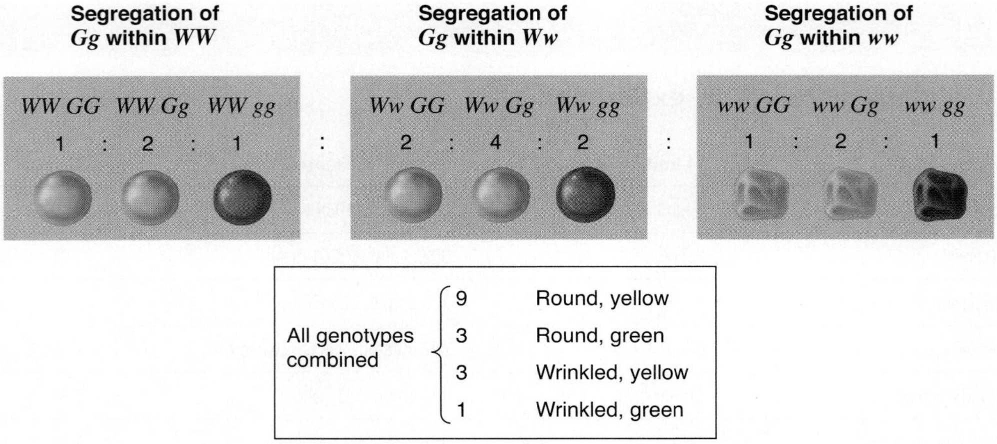

FIGURE A3.3 A dihybrid cross involving seed shape and seed color of the garden pea plant results in four distinct phenotypes: round/yellow, round/green, wrinkled/yellow, and wrinkled/green.

two X chromosomes. The XY sex chromosome scheme is also present in fruit flies. Indeed, these organisms were instrumental in illuminating the X-linked pattern of inheritance (**FIGURE A3.6**).

Males with a normal chromosome inheritance are never heterozygous for X-linked alleles, and if they inherit a recessive X-linked allele, it will be expressed. Therefore, mutant alleles on the X chromosome affect men more frequently than females, as they only have one X chromosome. Females may inherit a mutant gene on the X chromosome, but a normal allele on the other X chromosome will be dominant and prevent expression of the mutant phenotype.

Key Terms

Allele
Dihybrid cross
Diploid
Genotype
Haploid
Linkage
Monohybrid cross
Phenotype
X-linked cross

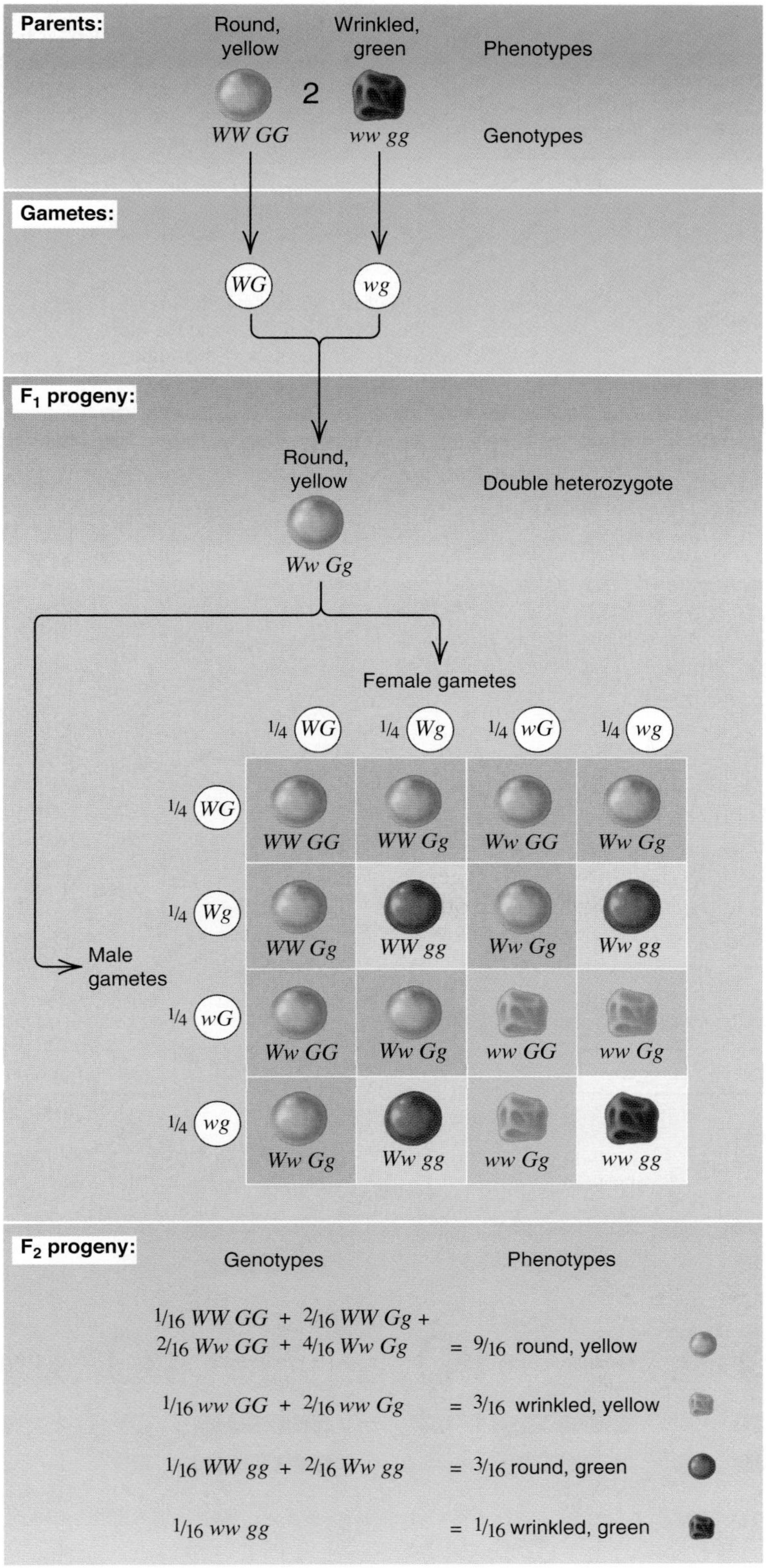

FIGURE A3.4 A dihybrid cross involving two sets of alleles results in a 9:3:3:1 ratio in the F1 generation of offspring. The separate inheritance of each gene from the parents indicates independent assortment and a lack of linked genes.

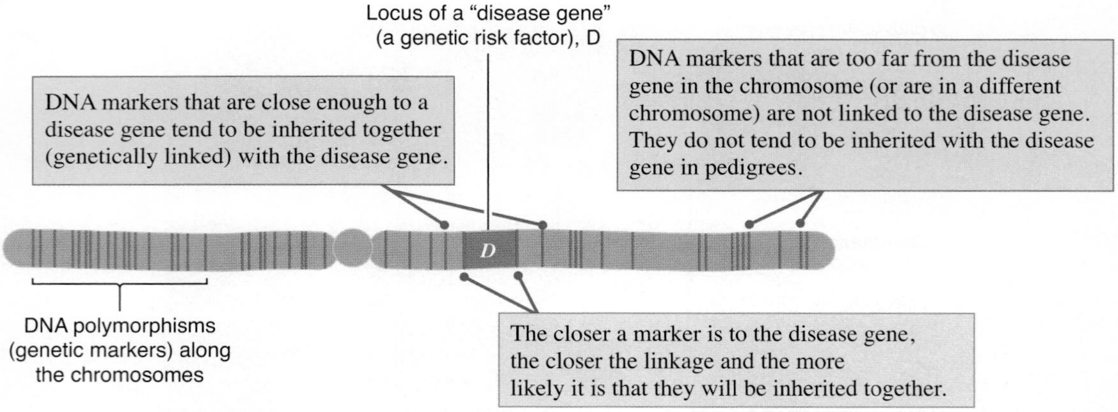

FIGURE A3.5 Gene linkage exhibited by "disease gene D" shows the close proximity to DNA markers (represented by vertical lines); thus, these units of genetic information will likely not assort independently; rather they will be inherited together.

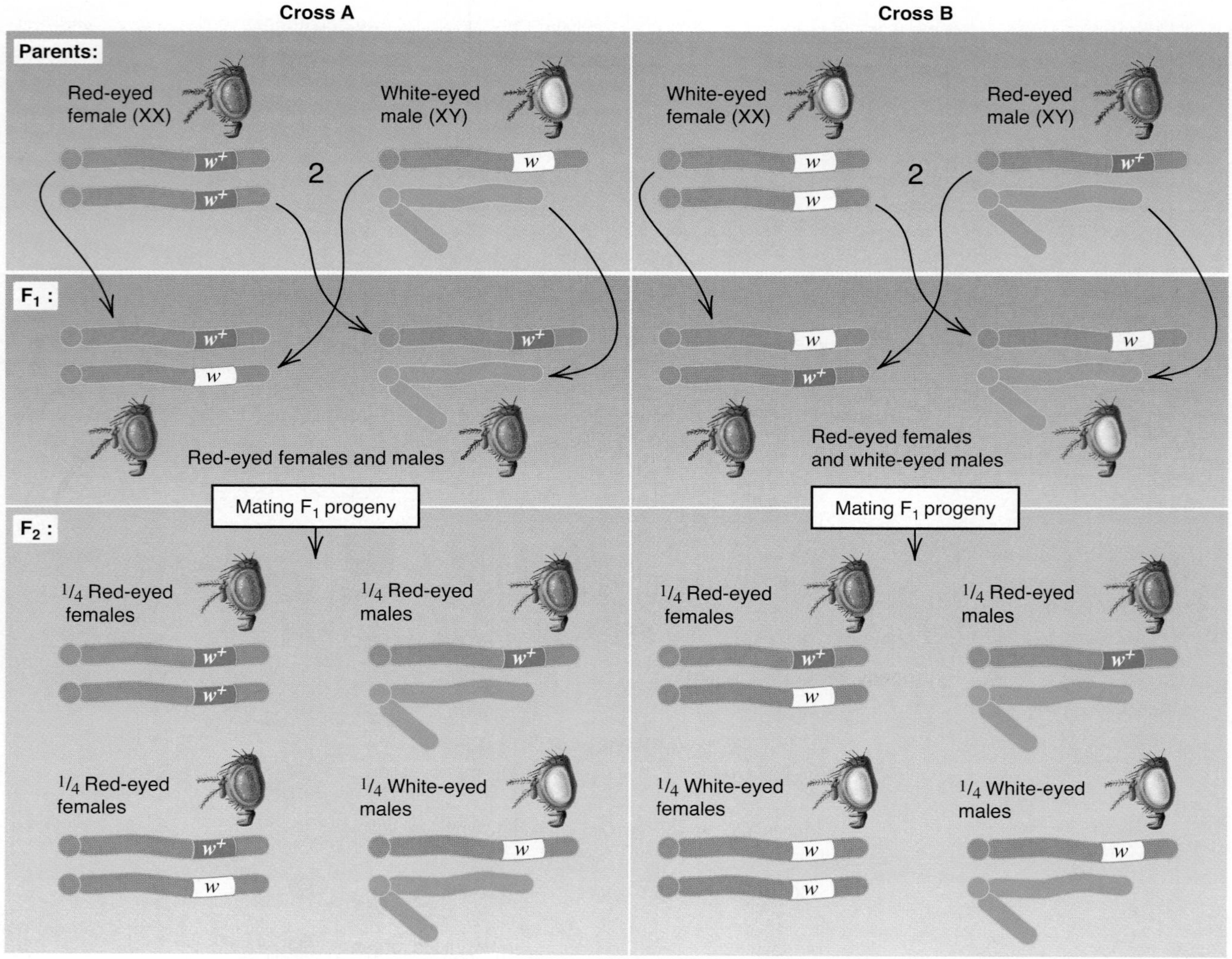

FIGURE A3.6 X-linked crosses as seen in the fruit fly. Cross A is between a wild type red-eyed female and a mutant white-eyed male. Cross B is the opposite as it mates a mutant white-eyed female with a wild type red-eyed male. Note there is no allele for eye color on the Y chromosome, shown in red.

Glossary

Accessions: Genetic samples stored in seed banks and living laboratories for conservation purposes.

Affinity chromatography: A form of chromatography which sorts proteins through ligand adsorption.

Allele: Alternate forms of a single gene, such as dominant or recessive.

Allele-specific oligonucleotide analysis (ASO): Genetic testing that uses oligonucleotide probes to detect specific healthy and mutant alleles.

Alternative splicing: The process of rearranging a primary messenger RNA transcript to produce a completely different nucleotide sequence.

Amniocentesis: A form of fetal genetic testing, whereby amniotic fluid is removed to perform karyotyping.

Amylases: A class of enzymes which hydrolyze starch into its glucose monomers.

Amyloids: Insoluble fibers from old proteins that accumulate with age.

Annotation: The process of identifying regions of the genome by distinguishing one meaningful sequence from another.

Anthropology: The study of man; divided into linguistics, cultural anthropology, biological anthropology, and archaeology.

Anticodon: The region of the tRNA molecule that binds to mRNA during translation.

Antisense RNA: The sequence of RNA complementary to sense mRNA; used to block translation.

Aptamer therapy: A form of gene therapy which silences a gene by inactivating its transcription factor.

Aquaculture: Involves the production of fish or shellfish in a controlled environment for human consumption.

Artificial chromosomes: Engineered chromosomes based upon digital sequences, including bacterial artificial chromosomes (BACs) and yeast artificial chromosomes (YACs).

Autoradiograph: The DNA fingerprint resulting from Southern blotting that provides visualization of restriction fragment length polymorphisms of interest detected in a DNA sample.

Bacterial transformation: A change in the properties of a bacterium, due to the uptake of exogenous genetic material from its environment.

Bacteriophage: A virus which infects bacterial cells; also known as a phage.

Base pairing: The bonding of nitrogenous bases in DNA based upon Chargaff's rules.

Bioaugmentation: A bioremediation technique which introduces additional microbial strains into the cleanup site in order to enhance the activity of indigenous populations.

Biodefense: The use of biotechnology to improve national security.

Biodiesel: A type of biofuel derived from oils and fats used to power diesel engines.

Bioethics: A branch of philosophy concerned with the ramifications of advances in biology and technology.

Biofortification: Genetic modification of crops in order to enhance their nutritional profiles.

Biofuels: Sources of energy derived from biological carbon fixation.

Bioinformatics: The application of biology, computer science, and information technology in order to organize, store, retrieve, and analyze DNA sequencing data.

Bio-ink: "Ink" made of selected cells used in three-dimensional bioprinting.

Biologic: Any medical preparation involving living organisms or their products, such as a vaccine.

Biological License Agreement (BLA): A regulatory status granted by the FDA for biologics; equivalent to NDA status for drugs.

Biomagnification: Accumulation of toxic chemicals which increases with each level of a food chain.

Biomarkers: Substances that when present are the hallmarks of different diseases.

Biomass: Large population of cells produced in upstream processing.

Biopharma: The development of pharmaceuticals via genetic engineering.

Biopolymers: Large molecules comprised of multiple monomer subunits that are produced by transgenic organisms and act as plastics and adhesives.

Bioprinting: Three-dimensional technology for printing of human tissues and organs.

Bioprospecting: Harvesting and classifying the various organisms of the world in order to create a bank of genetic tools for future use.

Bioreactor: Large fermentation tanks used in upstream processing.

Bioremediation: The use of biotechnology to degrade environmental pollutants.

Biosensors: Living indicators of chemical pollution and/or biological agents.

Biostimulation: Accelerates bioremediation by introducing nutrients.

Biotechnology: Industrial production of goods and services by processes using biological organisms and systems.

Bioventing: Similar to biostimulation; accelerates aerobic microbial remediation by introducing air or hydrogen peroxide to increase available oxygen.

Biowarfare: The creation of biologically active weapons as well as their storage and eventual delivery to intended targets; also known as biological warfare.

Callus culture: Plant tissue culture upon a solid medium.

Camouflaged virus: The insertion of a pathogenic viral genome into a harmless bacterium, which may then be used as a delivery vector for the purposes of biowarfare attack.

Cell lysis: Rupture of a cell.

Central dogma of molecular biology: Genetic information flows from DNA to RNA to a final protein product.

Chargaff's Rules: The observed pattern of complementary base pairing in DNA; adenine always pairs with thymine and cytosine always pairs with guanine.

Chimeric antibodies: Monoclonal antibodies containing human and non-human portions, such as CHO cells.

Chorionic villus sampling: A form of fetal genetic testing, whereby chorionic villus cells are removed to produce a karyotype.

Chromatin: The physical matter which constitutes chromosomes, including DNA and associated proteins known as histones.

Chromatin remodeling complex: A quaternary-level protein complex which restructures the molecular arrangement of chromatin in order to affect transcription and therefore gene expression.

Chromosome: A physical structure composed of a strand of DNA and associated proteins known as histones.

Chromosome Theory of Inheritance: A unifying theory of genetics stating that chromosomes are the carriers of genetic material, that is, genes; also known as the Sutton-Boveri hypothesis.

Chymosin: The first food ingredient produced through recombinant DNA technology to be approved by the FDA; an enzyme used in cheese making also known as rennin.

CODIS: Combined DNA Index System, a DNA profile database maintained by the Federal Bureau of Investigation.

Codon: Triplet of messenger RNA which codes for one amino acid; the basis of the genetic code.

Complementary DNA: A sequence of engineered DNA complementary to a human gene which contains exons only.

Computational biology: A subdiscipline of bioinformatics that uses computers to evaluate and interpret biological data.

Consensus motif: A highly conserved nucleotide sequence among various genes.

Constitutive promoter: A type of promoter region that keeps a gene turned on at all times and in all cells.

Coordinated Framework: A policy launched by the Biotechnology Science Coordinating Committee (BSCC) of the Office of Science and Technology Policy designed to streamline procedures between the three agencies governing biotechnology regulation and clearly delineate responsibilities among them.

Copy-number variation (CNV): A polymorphism consisting of repeated segments ranging from 1,000 to 1,000,000 base pairs in length.

Data mining: The process of analyzing previously characterized genes and proteins in order to predict gene or protein structure and function.

Degenerate: Describes the redundant nature of the genetic code, wherein each existing amino acid is encoded by more than one codon.

Denaturation: Separation of double stranded DNA into separate strands through exposure to high heat or alkaline conditions.

Deoxyribonucleic acid (DNA): The organic molecule which contains genetic material and encodes genes, which is a double helical polymer composed of nucleotide monomers.

Differential fluorescence induction (DFI): A method of antigen detection that induces infection conditions in order to express unknown antigens.

Dihybrid cross: A cross between two individuals involving two genes.

Diploid: The condition within a cell of possessing two alleles of a given gene and thus two sets of chromosomes.

Directional selection: A form of natural selection which shifts an average phenotype in one direction.

Dispersed repetitive sequence: Similar to tandem repeat polymorphisms, except the repeats occur individually in various locations throughout the genome; includes Short INterspersed Element (SINE) and Long INterspersed Element (LINE) polymorphisms.

Disruptive selection: A form of natural selection which results in one average phenotype splitting into two equally common forms.

DNA fingerprinting: A technology which identifies genetic markers that are unique in each person; also known as genetic profiling.

DNA marker: A DNA sample with known fragment lengths used in gel electrophoresis for size comparisons.

DNA replication: The copying of a DNA molecule to form two new molecules.

DNA sequencing: The process of determining the exact order of nucleotide bases within a DNA molecule.

DNA-DNA hybridization: Compares DNA sequences amongst different organisms in order to determine evolutionary distances.

Downstream processing: The second step of molecule production in industrial biotechnology; involves extraction, stabilization, purification, and preservation.

Duplex DNA: Double stranded DNA.

Electrigens: Anaerobic bacteria used to generate electricity through natural oxidation reactions.

ELISA: Enzyme-linked immunosorbent assay used in protein validation and medical testing.

ELSI: The study of *e*thical, *l*egal, and *s*ocial *i*ssues within the biotechnology industry as initiated by the Human Genome Project.

Environmental assessment (EA): An environmental impact study required by the National Environmental Policy Act under the authority of the EPA.

Enzyme: Protein molecules which act as catalysts in chemical reactions.

Epigenomics: The study of how the genome is regulated; determines the timing and intensity of gene expression within a cell.

Esterases: A class of enzymes which catalyze the breakdown of ester molecules into their alcohol and acid components.

Eugenics: Intentional genetic manipulation of future generations for desired traits.

Evo devo: The field of study involving the convergence of evolutionary and developmental biology.

Exaptation: A pre-existing structure modified by evolutionary adaptation.

Exclusivity: Ruling out a person of interest based upon biological evidence.

Exons: Coding regions of the genome which express protein products, either individually or in concert.

Experimental Use Permit (EUP): An additional permit for USDA APHIS regulated articles required by the EPA in the event a field test covers more than 10 acres.

Explant: The portion of a plant which is removed in order to perform tissue culture.

Expressed sequence tag (EST): A transcript sequence obtained through expression of complementary DNA.

Extinction: The disappearance of an entire species or higher group.

Fermentation: Any process which enables cell growth for the purposes of biosynthesis via gene expression and/or metabolic pathways.

Field test: The controlled introduction of a GMO into the environment for investigatory purposes.

Finding of no significant impact (FONSI): A regulatory designation which states in the opinion of the USDA a field test is not expected to alter the environment.

Flanking regions: Regions of DNA occurring immediately upstream and downstream of a genetic marker.

Fluorescence in situ hybridization (FISH): A karyotyping method for adults and fetuses that attaches fluorescent probes to genetic markers, which are visible in the karyotype.

Folding moderators: Specialized molecules which aid in protein folding leading to secondary and tertiary protein structure.

Forensics: Science and technology used in the investigation of crimes.

Founder effect: The progressive reduction in genetic diversity resulting from a subset of an existing population founds a new population.

Gametes: Sex cells including male sperm and female eggs.

Gene: Molecular unit of heredity; a sequence of DNA bases that controls the production of a functional product.

Gene cassettes: DNA sequences containing a promoter region and one or more genes.

Gene migration: The movement of alleles across gene pools.

Gene therapy: Techniques that correct defective genes, or alter their expression, in order to treat or cure disease.

Gene tree: A phylogenetic tree exhibiting evolutionary relationships amongst genomic loci.

Genetic code: Provides the information to translate nucleotide sequences into amino acid sequences.

Genetic counseling: Involves the identification of disease in an individual's family history and determining the appropriate measures one might take to avoid passing on disease to his or her offspring.

Genetic drift: Change in allele frequency happening without any clear pattern or selective agent.

Genetic mapping: The process of determining the order, location, and characteristics of genetic information.

Genetic markers: A class of non-coding DNA regions of known location which vary among individuals; also known as DNA polymorphisms.

Genetic profiling: Identifies genetic markers in an individual; also known as DNA fingerprinting.

Genetic use restriction technology (GURT): Biotechnology method that produces plants with sterile seeds.

Genetically modified organism (GMO): An organism containing knockout or inserted (i.e., transgenic) genetic material.

Genome: All of the hereditary material found in an organism.

Genomics: The study of all of the nucleotide sequences, including structural genes, regulatory sequences, and noncoding DNA segments, in the chromosomes of an organism.

Genotype: The combination of alleles possessed by an individual for a given gene.

Germ-line therapy: Gene therapy performed on gametes (sex cells) in order to confer a selected gene(s) in offspring.

Good laboratory practice (GLP): A set of principles that provides a framework within which studies are planned, performed, monitored, recorded, reported and archived.

Good manufacturing practice (GMP): The quality assurance that ensures that medicinal products are consistently produced and controlled to the quality standards appropriate to their intended use and as required by the product specification.

Haploid: The condition within a cell of possessing a single gene and thus a single set of chromosomes.

Haplotype: A genetic variation with a frequency of at least 1% within the human population.

Histones: Protein molecules present in chromatin.

Hominids: A taxonomical grouping containing humans and apes (orangutans, gorillas, bonobos, chimpanzees).

Hominins: A taxonomical grouping containing humans and our extinct ancestors.

Homologous structures: Anatomical features which are morphologically different but have evolved from a common structure.

Housekeeping genes: Genes necessary for survival in all cells, regardless of the type.

Hybridoma: A hybrid cell consisting of a cancerous myeloma cell (with the ability to divide indefinitely) and a B lymphocyte (with the ability to produce a desired antibody); used to synthesize monoclonal antibodies.

Hydraulic fracturing (fracking): A method of fossil fuel extraction where high pressure liquids fracture subterranean rock strata to yield natural gas and petroleum.

Hydrogen bond: A type of bond that always forms between a hydrogen atom and another non-hydrogen atom.

Hydrophobic interaction chromatography (HIC): A form of chromatography that sorts proteins based on their affinity to water.

Immortal genes: A set of approximately 500 genes that have been conserved throughout the evolutionary process; they are present in all forms of life.

Immunotherapy: Treatment of disease by stimulating, enhancing, or suppressing the immune system.

In vivo induced antigen technology (IVIAT): A method of antigen detection that identifies previously unknown antigens produced by pathogens during times of infection.

Inclusion bodies: Insoluble masses of misshapen proteins.

Inducible promoter: A type of promoter region which can turn a gene on or off depending on environmental conditions.

Initiation complex: The connection of mRNA to both ribosomal subunits, which initiates translation.

Institutional biosafety committee (IBC): An in-house advisory panel found within a biotechnology organization which steers the research and development processes.

Integron analysis: A method of identifying novel genes in metagenomic studies which isolates open reading frames.

Integrons: Mobile elements containing gene cassettes used in integron analysis for metagenomic studies.

Introgression: An event whereby inserted gene(s) unintentionally jump to additional species, usually through outcrossing of a GMO crop with unrelated plants or wild relatives.

Introns: Non-coding regions of the genome interspersed amongst exons.

Ion-exchange chromatography: A form of chromatography which sorts proteins based on their electrical charges.

Karyotype: Visual representation of the genome.

Knockout: Renders a gene inoperable, thereby preventing the synthesis of its protein product(s).

Ligands: Molecules of complementary shape to specific proteins.

Linkage: Occurs when two genes are so close together on a chromosome, they do not assort independently and are inherited together.

Lyophilization: Freeze drying for preservation of bioactive molecules.

Lysate: Filtered liquid solution remaining from cell lysis during purification.

Lysogeny: The integration of bacteriophage DNA into a bacterial chromosome, that results in prophage DNA.

Macroevolution: Large changes in life forms over long periods of time; the combination of speciation and extinction over the course of life's history.

Melting temperature (T_m): During denaturation, the temperature at which half the original intact DNA has denatured.

Mendelian genetics: The study of how genes are passed from parents to offspring; also known as transmission genetics.

Messenger RNA (mRNA): The RNA molecule synthesized during transcription; it behaves like a messenger carrying instructions from the chromosome to ribosomes for making a protein.

Metagenomics: The study of genomes of whole microbial communities in situ.

Microcosm: Simulated environment created under controlled laboratory conditions, designed to mimic the actual site of pollution.

Microevolution: Small changes in life forms occurring over short periods of time; changes in allele frequency that alter the make-up of a gene pool.

Micropropagation: Use of plant tissue culture in the mass cloning of commercially valuable plant varieties.

Microsatellites: One- to six-nucleotide repeats dispersed through the human genome.

Microtissues: The products of bioprinting; miniature scaffolds seeded with human cells obtained from bio-inks.

Mitochondrial DNA (mtDNA): Extranuclear DNA in the human genome that is inherited through the maternal line.

Mitochondrial Eve: A hypothetical single female progenitor of all humanity, based upon shared mitochondrial DNA inherited through a maternal line.

Model organisms: Test subjects used in biological research that allow in vivo observations as they occur in a live subject.

Modern synthesis: The currently accepted paradigm of evolutionary biology that bridges gaps between microevolution, occurring at the genetic level, and macroevolution, occurring at the species level.

Molecular biotechnology: An interdisciplinary field, making use of existing biotechnology and drawing from biology, chemistry, mathematics, computer science, engineering, philosophy and ethics.

Molecular clock: Using genetic changes in different organisms to measure evolutionary time between their divergence; also known as an evolutionary chronometer.

Molecular systematics: Classification of organisms based upon molecular (often genetic) comparisons.

Monoclonal antibody (MAb): A single type of antibody for treatment (or identification) of a particular genetic disorder.

Monohybrid cross: A cross between two individuals involving a single gene.

Morphology: The field of study concerning the physical forms of organisms, popular among naturalists.

Multiple-locus probing (MLP): A form of DNA fingerprinting where more than one restriction fragment length polymorphism is sought after.

Multiplex PCR: The detection and amplification of multiple DNA regions in a single polymerase chain reaction.

Multipotent: Stem cells that can form most but not all adult cell types.

Mutation: A permanent alteration of genetic material.

Nanomedicine: The application of nanotechnology to health care.

Natural selection: Gradual change of organisms over time which leads to adaptations to the environment.

New Drug Application (NDA): A regulatory status granted by the FDA when a drug appears statistically superior to a placebo during phase III clinical trials.

Nonrandom mating: The choosing of a particular phenotype (or phenotypes) in a sexual partner over any other variety.

Non-regulated status: Status of a GMO when exempt from regulation; the organism is no longer considered a regulated article by USDA APHIS.

Notification: The fast tracking of a field test application through APHIS.

Novel: Product or process not currently protected with an existing patent.

Nucleic acids: Organic molecules consisting of a sugar, a phosphate group, and a nitrogen-containing base; includes deoxyribonucleic acid (DNA) and ribonucleic acid (RNA).

Nucleoid: The area within a virus or prokaryotic cell that contains nucleic acid.

Nucleosome: The bead-like units seen in chromatin, composed of histone proteins wrapped with DNA.

Nucleotide: The monomer or subunit of DNA containing the five carbon sugar deoxyribose, a phosphate group, and one of four nitrogenous bases (either adenine, guanine, cytosine, or thymine).

Oligonucleotide crossover: A form of gene patching where a double stranded oligonucleotide is used during crossing over to replace a mutated portion of a defective gene.

Ontogeny: The study of organismal development.

Organogenesis: The formation of organs from specialized tissues.

Outcrossing: Merging of gene pools that occurs when closely related organisms exchange genetic material through repeated hybridization over several generations.

Paleoanthropology: The study of ancient human fossils and those of their ancestors, such as extinct hominids.

Paralog: Analogous genes found in the same location in separate genetic clusters or domains.

Parsimony: The rationale which ascertains the best possible explanation is the simplest one.

Pedigree: An iconograph that traces the inheritance patterns of disease within a family history.

Persistent organic pollutants (POPs): Large organic molecules that are not immediately biodegradable.

Phage biopanning: A function-dependent metagenomic technique using bacteriophages to identify novel genes.

Phenome: The entire inventory of phenotypes expressed by a cell or organism.

Phenotype: The trait conferred to an individual based upon a given genotype.

Phylogenetic anchors: Known DNA sequences conserved in multiple species and used in metagenomic studies.

Phylogeny: The study of evolutionary relationships among organisms in order to establish patterns of descent.

Phylogeography: The combination of gene migration studies with phylogenetics.

Plasmid: A circular ring of DNA that exists separately from chromosomes.

Pluripotent: Stem cells have the ability to differentiate into the two hundred or so cell types present in the human body.

Poison sequences: Viral genome sequences which are not readily expressed by their hosts.

Polymerase chain reaction (PCR): The process of amplifying a small fragment of DNA into millions of identical copies.

Precipitation: The process of molecular aggregation whereby they settle out of solution.

Primary metabolites: Vital molecules produced as natural byproducts and their intermediaries during cellular metabolism.

Primary protein structure: The amino acid sequence of a polypeptide obtained from translation.

Prions: Misshapen protein fragments which can infect an organism and cause disease.

Prophage DNA: Viral DNA which has integrated into bacterial DNA.

Proteases: Enzymes which aid in proteolysis by catalyzing the breakdown of protein molecules into their amino acid monomers; also known as proteolytic enzymes.

Protein microarray: Genetic testing that identifies the presence of antibodies.

Proteolysis: The breakdown of proteins into their amino acid monomers, which may be recycled for the synthesis of new proteins.

Proteome: The entire inventory of proteins expressed by a cell or organism.

Proteomics: Cataloging and characterizing proteins expressed in various organisms.

Purines: A class of nitrogen-containing bases found in nucleic acid molecules; includes adenine (A) and guanine (G).

Pyrimidines: A class of nitrogen-containing bases found in nucleic acid molecules; includes thymine (T), cytosine (C), and uracil (U).

Quaternary protein structure: Occurs when more than one polypeptide chain folds into a three-dimensional protein complex.

Random match probability: A measurement of certainty for a positive biological match in DNA fingerprinting that takes into consideration the frequencies of genetic markers in members of the same ethnic group.

Recombinant DNA technology: The process of combining plasmid transformation with restriction enzymes to produce a unique technology involving the recombination of DNA sequences, thus creating a transgenic organism.

Reference databases: Databases of genetic profiles by ethnic group that include frequencies of polymorphisms within the population; allows random match probability calculations.

Regulated articles: USDA APHIS terminology to describe GMOs undergoing field testing.

Reproductive cloning: Synthesizes a cloned organism; the whole individual is genetically identical to the original organism.

Restriction endonuclease: An enzyme which targets a specific DNA sequence and cleaves the DNA at what is known as the restriction site; also known as restriction enzymes.

Restriction fragment length polymorphism (RFLP): A polymorphism that occurs when an SNP is present at a restriction site.

Ribonucleic acid (RNA): A nucleic acid that is generally single stranded (double stranded in some viruses) and is involved with transferring information from DNA to the protein-forming system of the cell.

Ribosomes: Structures that are the sites for all protein synthesis, composed of protein molecules and ribosomal RNA.

Ribozyme: RNA molecules present in the ribosome that function as an enzyme.

Ribozyme therapy: A form of gene therapy that acts by targeting and cleaving mutated messenger RNA.

Risk: The potential for loss or damage.

RNA interference (RNAi): A method of gene silencing using small interfering RNA molecules to block translation.

Secondary metabolites: End products of metabolic pathways, often of unknown function.

Secondary protein structure: Folding interactions of a polypeptide composed of the alpha helix, the beta sheet, and random coils.

Selective pressure: Conditions which contribute to natural selection and ultimately adaptation of organisms to their environment.

Silencing complex: A quaternary protein structure that prevents transcription and therefore gene expression.

Single nucleotide polymorphism (SNP): A difference of one nucleotide pair between two different DNA sequences.

Single-locus probing (SLP): A form of DNA fingerprinting where a single restriction fragment length polymorphism is sought after.

Size-exclusion chromatography: A form of chromatography which sorts proteins based on size; also known as gel filtration chromatography.

Somatic cell nuclear transfer (SCNT): A technique used in cloning to introduce donor genetic material into the recipient.

Somatic cells: Cells of the body; all cells in an organism besides gametes.

Specialization genes: Genes that program specialization by placing an undifferentiated stem cell onto a certain path of differentiation.

Speciation: The emergence of new organisms.

Species tree: A phylogenetic tree exhibiting evolutionary relationships among species.

Spliceosome-mediated RNA trans-splicing (SMaRT): A form of gene therapy that acts by targeting and repairing messenger RNA transcribed from a defective gene.

Stabilizing selection: A form of natural selection that results in an average phenotype becoming increasingly common.

Stacked traits: More than one transgene is inserted in order to confer multiple abilities in the GM organism.

Stewardship: The responsible management of all stages of development, from inception to commercialization.

Suspension culture: Plant tissue culture within a liquid medium.

Svedberg unit (S value): The unit of measurement used to describe ribosomes.

Tandem repeat polymorphism: A polymorphism consisting of repeated segments less than 1,000 base pairs in length wherein the repeats occur one after the other within DNA; includes simple sequence repeat (SSR) and variable number of tandem repeat (VNTR) polymorphisms.

Taxonomy: The field of biology concerned with classifying organisms into hierarchical categories.

Tertiary protein structure: Ultimate three-dimensional shape of a fully-folded polypeptide.

Tetranucleotide repeat: A form of microsatellite genetic marker consisting of four base pairs which repeat five to 50 times.

Therapeutic cloning: Synthesizes cloned cells and tissues for replacement of damaged or diseased areas of the body.

Thermophile: A living organism thriving in high temperature environments.

Totipotent: Stem cells that can form adult cell types as well as cells necessary for embryonic development.

Transcription: The formation of RNA from a DNA molecule.

Transcription complex: A quaternary protein structure assembled in the process of recruitment in order to activate gene expression.

Transcription factors (TFs): Proteins that control the initiation of gene expression.

Transfection: The introduction of genetic material into animal or plant cells.

Transfer RNA (tRNA): A type of RNA that carries individual amino acids to the ribosome during translation.

Transgenic: Describes an organism containing genetic material from a species other than its own; also known as a genetically modified organism.

Transhumanism: Creation of a dramatically improved human form with enhanced intellectual, physical, and psychological abilities.

Translation: The chemical creation of proteins from an RNA molecule.

Transposable elements: Regions of the genome that are able to detach at one location and reattach in another; also known as transposons.

Two-dimensional electrophoresis: A form of gel electrophoresis which separates proteins by size and electrical charge.

Unipotent: Stem cells that differentiate into a single cell type.

Upstream processing: The first step of molecule production in industrial biotechnology; involves genetic modification and biosynthesis.

Utility: Demonstration that a product or process is useful to humanity in some way.

Validation: Testing of the purified substance resulting from downstream processing in order to ensure its contents are as desired.

X-linked cross: A cross involving a gene found on the X chromosome.

Y Chromosome Adam: A hypothetical single male progenitor of all humanity, based upon shared Y chromosome DNA inherited through a paternal line.

Index

Note: Page numbers followed by *b*, *f* and *t* indicate materials in boxes, figures and tables respectively.

A

ABGC. *See* American Board of Genetic Counseling
accessions, 172
Accreditation Council for Genetic Counseling (ACGC), 127
ACGC. *See* Accreditation Council for Genetic Counseling
adeno-associated viruses, 137
adenosine deaminase (ADA) gene, 135*b*
adenoviruses, 137, 137*f*, 138*f*
adult stem cells, 152–153
Aequorea victoria, 211
aequorin, 211
affinity chromatography, 120
African wildcats, cloned, 173, 173*f*
agriculture, 266–267
  defined, 179
agriculture and food production, 41, 178–191
  aquaculture, 46
  genetically engineered food, 44
  seed banks, 44–45
agriculture biotechnology, 179
  aquaculture, 188–189
  genetically engineered livestock, 189–190
  plant biotechnology, 180–188
  using microbes, 179–180
Agrobacterium, 181
air pollutants, 161
albinism, pedigree for, 43*f*
algal biodiesel, 175*b*
alleles, 283
allele-specific oligonucleotide analysis (ASO), 132–133, 133*f*
alternative splicing, 97–99
American Board of Genetic Counseling (ABGC), 127
Amino Acid Explorer, 76
amino acid–producing microbes, mutation of, 112
amniocentesis, 129, 130*f*
amylase, 105–106
  uses, 107*f*
amyloids, 90–91
Animal and Plant Health Inspection Service (APHIS) inspectors, 65*f*, 66
annotation, 77–79
anthrax *(Bacillus anthracis),* 206*b*
anthropology, 245
anticodon, 280
antifreeze proteins, 189
antigens, detection of, 146–147
antioxidants, 112–113
antisense RNA, 141, 142*f*, 151
APHIS inspectors. *See* Animal and Plant Health Inspection Service inspectors
aptamer therapy, 143
aquaculture, 46, 188–189, 189*f*
  defined, 179
aquatic environments, 165–166
Arabidopsis thaliana, 110
Arber, Werner, 23
Archaea, 218–219, 222*f*, 223*f*
archaebacteria, 218
archaeology, 245
archaeopteryx fossil, 223, 223*f*
artificial chromosomes, 208
Asilomar Conference, 26*b*, 57, 60–61
ASO. *See* allele-specific oligonucleotide analysis
Astrid, 150
aurochs, 44*f*
autoradiograph, 197
Avery, MacLeod, and McCarty experiments, 13, 14*f*

B

Bacillus anthracis (anthrax), 206*b*
Bacillus thuringiensis (Bt), 179–180, 186*f*
  and butterflies, 186*b*
bacteria, gene therapy, 141
bacterial genomes, size of, 82*t*
bacterial host organisms, 116*f*
bacterial transformation, 13
bacteriophage, 13, 15*f*
bananas, 148*b*
base pairing, 277, 277*f*
Basic Local Alignment Search Tool (BLAST), 76
batch bioreactor, 118*f*
batch fermentation, 116–118
bioadhesives, 111, 111*f*
bioaugmentation, 166, 168*b*
biochemiresistor, 210, 210*f*
biodefense, 48, 48*f*, 195, 204–206
  biosensors, 208, 210–211
  biowarfare, 206–207
    camouflaged virus, 208
    genetically modified agents, 207–208
biodiesel, 174–175, 175*b*
bioethics, 268–269

biofortification, 188, 189*t*
biofuels, 47, 174–175
bioinformatics, 75–76
bio-inks, 265
biolistics, 141, 182
biological anthropology, 245
Biological License Agreement (BLA), 68
biological samples, types of, 205*f*
biologics, 62–63
biomagnification, 161
biomarkers, 127–128, 133, 149
biomass, 114
  ethanol and biodiesel, 174–175
biopharma, 41, 42*b*, 143–144, 265
  nanomedicine, 145–146, 146*f*
  pharmacogenomics, 145, 145*f*
  recombinant DNA proteins, 144, 144*t*
bioplastics, 109–111
biopolymers, 109–111
  bioadhesives, 111, 111*f*
  bioplastics, 109–111
bioprinting, 265
bioprospecting, 44, 103, 108, 111, 186
bioreactor fermentation, 114
bioreactor systems, 116*f*
bioreactors, 114–116
bioremediation, 47, 47*f*, 162, 164–165, 165*f*, 166*f*
  aquatic environments, 165–166
  microbial bioremediation, 166, 168
    genetically modified microbes, 168–169
    indigenous microbes, 168
  in oil spills, 167*b*
  phytoremediation and mycoremediation, 169–170, 169*t*, 170*f*
  soil environments, 165
Bioremediation Project Scientists, 164
biosensors, 162, 163*b*, 208, 210–211
biostatisticians, 53
biostatistics, 76
biostimulation, 164–166, 167*b*
biotech crops, global area of, 3, 4*t*
biotechnology, 3
  economics, 175*b*
  industrial, 46
Biotechnology Research Subcommittee (BRS), 59*b*
Biotechnology Science Coordinating Committee (BSCC), 59*b*
bioventing, 164–166
biowarfare, 206–207
  camouflaged virus, 208
  genetically modified agents, 207–208
BLA. *See* Biological License Agreement
BLAST. *See* Basic Local Alignment Search Tool
blood disorders, adult stem cells, 152
blue dye indigo, 113
blue light, emission of, 211*f*
blue mussels, 111, 111*f*
Boveri, Theodor, 9–11
Boyer, Herb, 24–25
BSCC. *See* Biotechnology Science Coordinating Committee
Bt toxin, 179, 180, 185, 187*b*
butterflies, Bt and, 186*b*

C

cab promoter, 180
callus culture, 183
camouflaged virus, 208
cancer
  syndromes, inherited, 150*t*
  treatment of, 150–151
cancerous cells, 150, 151*t*
candelabra model, 37, 246, 246*f*, 252
cDNA. *See* complementary DNA
cell cycle regulatory genes, 151*t*
cell lysis, 118, 119
cellulase, 104, 174
central dogma of molecular biology, 19, 21*f*, 277, 277*f*
Chakrabarty, Ananda, 168*b*
chaperones, 90
chaperonins, 90
Chargaff, Erwin, 277
Chargaff's Rules, 15, 17*f*, 277
chemical pollutants, 163*t*
chimeric antibodies, 149
Chinese hamster ovary (CHO) cells, 148–149, 149*f*
Chinese muntjac deer, 228, 229*f*
chorionic villus sampling, 129–130, 131*f*
chromatin, 272
chromatin remodeling complex (CRC), 92
chromatography, 119–121
Chromosome Theory of Inheritance, 9–11
chromosomes, 272
  abnormalities, 128
chronic myelogenous leukemia, 131
chymosin, 103
class mammalia, phylogeny of, 234*f*
clinical research associates, 53
ClinVar database, 264
cloning, 153
  endangered and extinct species, 173–174
  reproductive, 154–156, 155*f*, 267
  therapeutic, 153–154, 153*f*, 265
CODIS. *See* Combined DNA Index System
codons, 20, 279
Cohen, Stanley, 24–25
Combined DNA Index System (CODIS), 198, 202
comparative genomics
  DNA molecule and genome size, 79–80
  gene number, 80–81
  genetic markers, 81–84
complementary DNA (cDNA), 263
  construction of, 264*f*
  in human body by function, classes of, 80*f*

complex growth medium, composition of, 116*t*
computational biology, 75
conferences and seminars, 50–51
consensus motif, 92
conservation
  of biodiversity, 47
  environmental science, 267
conservation biotechnology, 172
  biofuels, 174–175
  cloning endangered and extinct species, 173–174
  gene banks, 172–173
constitutive promoters, 180
continuous bioreactor, 118*f*
continuous fermentation, 118
Coordinated Framework, 59*b*
copy-number variation (CNV), 83
CRC. *See* chromatin remodeling complex
Cre proteins, 182
Cre/*lox* P, 182–183, 182*f*
  defined, 180
Crick, Francis, 15, 18–20
crime, forensic analysis of, 201–202
crime lab technicians, 52
crop dusting, 184*f*
cross pollination, 182
cry genes, 184–185
Cry proteins, 179–180, 180*t*
cultural anthropology, 245
cysteine, pathway for, 190

D

Darwin, Charles, 214–217, 223–225, 246
  natural selection and evolution:, 5, 7–8
data mining, 75
databases, 75
  international, 76
  in United States, 75–76
DDBJ. *See* DNA DataBank of Japan
DDT. *See* dichlorodiphenyltrichloroethane
Deepwater Horizon oil spill, 167*b*
degenerate genetic code, 21
denaturation, 248, 248*f*
Denisovans, 257–258, 259*f*
deoxyribonucleic acid (DNA), 272–273
  Chargaff's Rules, 15, 17*f*
  copying mechanism, DNA replication, 18–19, 20*f*
  from different organisms, base composition of, 88*t*
  discovery of, 8–9
  DNA sequencing, 25–29
  double helix, 274*f*
  from gel electrophoresis, 197*f*
  microarrays, 81, 83*f*
  molecular structure of, 19
  molecule and genome size, 79–80
  molecules, sizes of, 81*t*
  profiles, quantities of, 205, 205*f*
  recombinant DNA technology, 24–25
  and RNA, base pairing between, 89*f*
  structures and pairings of four nitrogen-containing bases of, 88*f*
  view of, 275*f*
  X-ray crystallography, 15, 18*f*
deoxyribozymes, 143
DFI. *See* differential fluorescence induction
Dicer, 141
dichlorodiphenyltrichloroethane (DDT), 227, 228*f*
differential fluorescence induction (DFI), 147
dihybrid cross, 283, 287*f*
diploid organism, 283
directional selection, 226–227
disaggregating chaperones, 90
dispersed repetitive sequences, 83–84
disruptive selection, 225–226
DNA. *See* deoxyribonucleic acid
DNA DataBank of Japan (DDBJ), 76
DNA fingerprinting, 48, 195, 198*f*, 202
  exclusivity and probability, 198–200
  genomic tools, 196
  testing methods
    PCR, 197–198
    Southern blot, 196–197, 198*f*, 199*f*
DNA marker, 197
  inclusion of, 199*f*
DNA oligonucleotides, 143
DNA polymorphisms, 81
DNA replication, 18–19, 20*f*, 24, 281
  eukaryotic, 281, 282*f*
  prokaryotic, 281
  semiconservative nature of, 282*f*
DNA sequencing, 25–29
  computer-generated output of, 26*f*
  cost of, 27, 29*f*
  genetic mapping, 29
DNA-DNA hybridization, 248
  process of, 249*f*
downstream processing, 113, 118–122
Drosophila, 11, 12*f*, 237, 237*f*, 238*f*
  genomic comparisons delineate species, 230*f*
drug approval, 265–266
duplex DNA, 248

E

E-85, 175*f*
E. coli. *See Escherichia coli*
EA. *See* environmental assessment
economics, biotechnology, 175*b*
Ectromelia (mousepox virus), 208, 208*f*
edible vaccines, 148*b*
educational careers, 52
electrigens, 174
electron micrograph, T2 bacteriophage, 15*f*
ELISA. *See* enzyme-linked immunosorbent assay
ELSI. *See* Ethical, Legal, and Social Implications
embryology, comparative, 215–216, 216*f*, 233

embryonic stem cells (ESCs), 152
embryos, mammalian, 49*f*
ENA. *See* European Nucleotide Archive
encapsulated cells, gene therapy, 139, 140*f*
Entrez, 75
environmental assessment (EA), 64
environmental biotechnology, 161
  bioremediation, 162, 164–165
    aquatic environments, 165–166
    microbial bioremediation, 166, 168–169
    phytoremediation and mycoremediation, 169–170, 169*t*, 170*f*
    soil environments, 165
  environmental pollutants, 161–162
  metagenomics, 170
    genome projects, 172
    techniques, 170–172
environmental pollutants, 161–162, 163*t*
Environmental Protection Agency (EPA), 57–58, 161
  acts regulated by, 60*t*
environmental science, 267
  conservation, 267
  renewable energy, 267
environmental science and conservation, 161, 267
  green biotechnology, 46–47
environments
  aquatic, 165–166
  soil, 165
enzyme-linked immunosorbent assay (ELISA), 122, 122*f*
enzymes, 103–104
  industrial, 105
    amylase, 105–106
    cellulase, 104
    lactase, 105
    subtilisin and lipase, 104–105
  research, 107
    PCR, 107–109
    restriction enzymes, 107
EPA. *See* Environmental Protection Agency
epigenomics, 133, 134*f*
Erwinia herbicola, 112
erythropoietin, 144
Escherichia coli (E. coli), 113, 113*f*, 272*f*, 273*f*
ESCs. *See* embryonic stem cells
esterases, 105
ethanol and biodiesel, 174–175
Ethical, Legal, and Social Implications (ELSI), 135*b*, 269
eugenics, 134, 268
eukaryotic genomes, size of, 82*t*
eukaryotic replication, 281
eukaryotic ribosomes, 217
  structure of, 221*f*
EUP. *See* Experimental Use Permit (EUP)
European Nucleotide Archive (ENA), 76
evo devo, 48–49
  biologists, 240
  defined, 215
  development, 231–233
    discoveries, 236–239
    research methods, 233–236
  evolution, 215–216, 217*t*
    macroevolution, 216–224
    microevolution, 224–231
evolution, 215–231, 267–268
  designer babies and super humans, 268
evolutionary developmental biology, 48
ex situ bioremediation, 164, 165
ex vivo gene therapy, 134, 135*b*
exaptation, 215
exclusivity, 199
exonerations, to DNA evidence, 202*f*
exons, 77–78
Experimental Use Permit (EUP), 65
explant, 183
expressed sequence tags (ESTs), 78
expression screening, 171–172
extinction, 216
extraction, 118–119

F

Farmer Assurance Provision, 67*b*
FDA. *See* Food and Drug Administration (FDA)
federal agencies, 57
  EPA, 57–58, 60*t*
  FDA, 58, 60, 60*t*
  USDA, 60, 61*t*
fermentation technology, 114, 190*f*
fetal stem cells, 152
field testing, 63*f*, 64
  large scale, 65
  small scale, 64–65
filtration methods, 119
final product testing, 63*f*, 65–67
finding of no significant impact (FONSI), 65
FISH. *See* fluorescence in situ hybridization
flanking regions, 197
floral dip method, 181
fluorescence in situ hybridization (FISH), 130–131, 131*f*
folding catalysts, 90
folding chaperones, 90
folding moderators, 90
folding pathways, in proteins, 91*f*
FONSI. *See* finding of no significant impact
Food and Drug Administration (FDA), 58, 60, 60*t*
  drug regulations, 68*b*
food biotechnology
  ingredients and additives, 190–191
  processing, 190
food production, 266–267
  agriculture and, 178–191
forelimbs, composition of, 218*f*

forensics, 48, 195
applications, 201
crime, 201–202
paternity, 202–203
profiling, 203–204
DNA fingerprinting, 195, 198*f*
exclusivity and probability, 198–200
genomic tools, 196
testing methods, 196–198
fossils, chart, 220*f*
founder effect, 256, 256*f*
fracking, 267
Frostban, 179
function-dependent techniques, 171–172
funding process, public and private, 49–50
fungi, Paul Stamets uses, 169, 169*b*

G
gametes, 268
gaur calf, cloned, 173, 173*f*
gel electrophoresis, 122*f*, 197*f*, 199*f*, 254
genasense, 150–151
GenBank, 75
gene banks, 172–173
gene cassettes, 170
mechanism for, 171*f*
gene expression, process of, 87
gene gun, 141, 141*f*, 182*f*
gene migration, 231
gene number, 80–81
gene repair, 141
gene replacement, 141
gene silencing, 141, 142*f*, 143
gene therapy, 41, 133–134
bioethics in, 135*b*
delivery modes
bacteria, 141
encapsulated cells, 139, 140*f*
liposomes, 139, 140*f*
naked DNA, 138–139, 140*f*
particle bombardment technique, 141, 141*f*
polymer-complexed DNA, 139
receptor-mediated uptake, 139
viral integration, 134, 136–138
ex vivo and in vivo gene therapy, 134, 135*b*
genetic counseling, 127
genetic disorders detection. *See* genetic disorders, detection of
therapeutic approaches, 141–143
gene trees, 249, 250*f*
GeneChip®, 81, 83*f*
genes, 3, 273, 275
knockout, 44
and protein, comparisons of, 98*t*
genetic code, 19, 277, 279*t*
genetic counseling, 41, 127
bioethics in, 135*b*
genetic counselors, 127
genetic disorders, detection of, 127–129
amniocentesis and karyotyping, 129, 130*f*
ASO, 132–133, 133*f*
chorionic villus sampling, 129–130, 131*f*
epigenomics, 133, 134*f*
FISH, 130–131, 131*f*
protein microarrays, 131, 132*f*
genetic drift, 231, 232*t*, 256
genetic mapping, 29, 128, 128*f*
adenovirus, 138*f*
eukaryotic, 33*f*
goals of, 29*t*
of human mitochondrial chromosome, 254*f*
prokaryotic, 32*f*
retrovirus, 139*f*
genetic markers, 37, 127, 132*f*, 81–84, 195, 203*f*
genetic profile, 127, 133, 145, 195
genetic profiling, 41
outcome of, 44*f*
genetic use restriction technology (GURT), 266
genetically engineered food, 44
genetically engineered livestock, 189–190
genetically modified agents, 207–208
genetically modified microbes, 168–169
genetically modified organisms (GMOs), 44, 44*f*, 266
patent, 168*b*
regulation, 62–64
commercialization for, 63*f*
field testing, 64–65
final product testing, 65–67
labeling, 68–71
laboratory research, 64
genetically modified plants, 183–184
enhanced nutrition, 188
herbicide resistance, 184, 184*f*
insect resistance, 184–186
pathogen resistance, 187, 187*t*
stress tolerance, 186–187
Genographic Project, 37, 37*f*, 49, 50*f*
genome, 24, 272
projects, 172
sequencing, 77
sizes, 27*t*
Genome 10K Project, 36
Genome Sequencing Program (GSP), 29*f*
genomics, 33, 76–77
annotation and mapping, 77–79
comparative genomics
DNA molecule and genome size, 79–80
gene number, 80–81
genetic markers, 81–84
ethical, legal, and social issues, 84–85
genome sequencing, 77
tools, 196
genotype, 283
Geobacter metallireducens, 174*f*

germ-line therapy, 134, 268
GFPs. *See* green fluorescent proteins
Gleevec, 145
GLP. *See* good laboratory practice
glyphosate, herbicide resistance, 184, 185*f*
GMOs. *See* genetically modified organisms
GMP. *See* good manufacturing practice
Golden Rice, for malnutrition, 188*b*
gold-plated (Au) electrodes, 210*f*
good laboratory practice (GLP), 68
good manufacturing practice (GMP), 58
great apes, 247
green biotechnology, 46–47
green fluorescent proteins (GFPs), 147
groundwater, bioremediation, 166
GURT. *See* genetic use restriction technology

H

Haeckel, Ernst, 215
Haemophilus influenza bacterium, 23, 23*f*
haploid, 283
haplotypes, 36, 145
Hardy-Weinberg equilibrium principle, 225, 226*f*, 231
health care, life science and, 264–266
HeLa cells, 58*b*
helicase, 281
HER-2 protein, 150
herbicide resistance, 184, 184*f*
herceptin, 150
Hershey and Chase experiments, 13, 15, 16*f*
HFCS. *See* high fructose corn syrup
HGT. *See* horizontal gene transfer
HIC. *See* hydrophobic interaction chromatography
high fructose corn syrup (HFCS), 105–106, 106*f*
histones, 272
holding chaperones, 90
homeodomain, 236–238, 237*f*, 238*f*
hominid karyotypes, banding patterns in, 251*f*
hominids, 248
hominin lineage, 250, 252
  candelabra model, 252
  Out of Africa model. *See* Out of Africa hypothesis
Homo erectus, 246, 246*f*, 252
Homo sapiens, 246, 256–258
  migratory patterns of, 50*f*
homologous structures, exaptations, 215
horizontal gene transfer (HGT), 221, 223*f*
household waste, 174
housekeeping genes, 236
HOX genes, 237–238
HOX proteins, 236–237
human family tree, 49, 255*f*, 258*f*, 259*f*
human genome, 85*t*
Human Genome Project, 33, 36, 76, 81, 84, 96, 97
human karyotype, graphical representation of, 83*f*
Human Microbiome Project, 36, 36*f*, 265
human mitochondrial chromosome genetic map of, 254*f*
human pedigree, 203*f*
Human Phenome Project, 264
human population, 267
Human Variome Project, 264
human waste, 174
Humulin, 42*b*, 42*f*
hybridization process, 248
hybridomas, 148, 149*f*
hydraulic fracturing, 267
hydrogen bonds, 277
hydrophobic interaction chromatography (HIC), 120

I

IBCs. *See* institutional biosafety committees
ICCG. *See* International Collaboration for
    Clinical Genomics
immortal genes, 215
immunotherapy, 146
  antigen detection, 146–147
  MAbs, 147–149, 149*f*
  vaccines, 147, 148*b*, 148*t*
in situ bioremediation, 164
in vitro fertilization (IVF), 152
in vivo gene therapy, 134, 135*b*
in vivo induced antigen technology (IVIAT), 146–147
inclusion bodies, 91, 91*f*
Indian muntjac deer, 228, 229*f*
indigenous microbes, 168
induced pluripotent stem cells (IPSCs or IPS cells),
    152–153, 153*f*, 265
inducible promoters, 180
industrial biotechnology, 46
  commercial products of, 103
    enzymes, 103–110
industrial biotechnology, commercial processes
    of, 113
  downstream processing, 118–122
  upstream processing, 113–118
inheritance, 5, 6–7*b*, 9, 283
initiation complex, 279
insect resistance, 184–186
institutional biosafety committees (IBCs), 61
insulin receptor protein, gene encoding for, 99*f*
integron analysis, 170
integrons, 170
intellectual property, molecular biotechnology, 51
International Collaboration for Clinical Genomics
    (ICCG), 264
International Human Genome Sequencing
    Consortium (IHGSC), 33
International Nucleotide Sequence Database
    Collaboration, 76
introgression, 64, 258
introns, 78, 263
ion-exchange chromatography, 120
IPS cells. *See* induced pluripotent stem cells
IPSCs. *See* induced pluripotent stem cells

IVF. *See* in vitro fertilization
IVIAT. *See* in vivo induced antigen technology

J
Jenner, Edward, 143
junk DNA, 33*b*

K
karyotyping, 129, 130, 130*f*
knockout, gene, 44

L
labeling, GMO, 68–71, 266
laboratory, careers
  managers, 51
  principal investigators, senior research scientists, and medical scientists, 52
  technicians and technologists, 52
laboratory research, GMO product, 63*f*, 64
lactase, 105
L-ascorbic acid, 112
Law of Independent Assortment, 5, 7*b*, 283
Law of Segregation, 5, 6*b*, 283
leptin, 145*b*
lesser apes, 247
life science and health care, 41, 127, 264–265
  bioprinting, 265
  drug approval, 265–266
  stem cell research, 265
ligands, 120
linkage, 283, 288*f*
Linnean classification system, 217, 221*f*
lipase, 104–105
liposomes, gene therapy, 139, 140*f*
low density lipoprotein (LDL) cholesterol, 113
lux gene, 163*b*
lycopene, 113
lyophilization, 122
lysate, 119
lysogeny, 208, 209*f*

M
macroevolution
  defined, 216
  three domains, 217–221
  transitional organisms, 221, 223–224
malformed hemoglobin proteins, 87*f*
malnutrition, Golden Rice for, 188*b*
mammalian embryos, anatomy of, 49*f*
mammals, 144
  X chromosome in humans and, 96*t*
management professionals, 52–53
mannose-binding protein, 90
Map Viewer tool, 76
melting temperature (T_m), 248
Mendel, Gregor Johann, 6, 283
Mendelian genetics, 5
mesenchymal stem cells, 152
messenger RNA (mRNA), 279
metabolites, 111–112
  industrial chemicals, 113
  nutritional supplements, 112–113
  pigments and dyes, 113
metagenomics, 170
  genome projects, 172
  techniques
    function-dependent, 171–172
    sequence-dependent, 170–171
methane, 174, 267
microbes, 179–180, 217
  genetically modified, 168–169
  indigenous, 168
microbial bioremediation, 166, 168
  genetically modified microbes, 168–169
  indigenous microbes, 168
microbial fuel cells, polluted sediments, 174
microcosms, 165
microevolution
  defined, 224
  founder effect, 256, 256*f*
  gene migration, 231
  genetic drift, 231, 232*t*
  mutation, 228–230, 228*f*
  natural selection, 6-7, 8*f*, 225
    directional selection, 226–227
    disruptive selection, 225–226
    stabilizing selection, 227–228
  nonrandom mating and sexual selection, 231
  phylogeography, 231, 232*f*
micropropagation, 183
microRNAs (miRNAs), 143
microsatellites, 196
microspheres, nanomedicine, 146
microtissues, 265, 265*f*
Miescher, Johann Friedrich, 8–9, 272
Millenium Seed Bank Project, 172, 172*f*
miRNAs. *See* microRNAs
mitochondrial defects, 128, 129*t*
mitochondrial DNA (mtDNA), 204*b*, 253–254, 255*f*
mitochondrial Eve, 253–254, 255*f*
mitochondrion, 203*b*
MLP. *See* multiple-locus probing
model organisms, 233
modern synthesis, 7, 215–216, 217*t*, 224, 233
molecular biotechnology, 41, 195
  agriculture and food production, 41, 44–46
  careers
    laboratory, 51–52
    non-laboratory, 52–53
  conferences and seminars, 50–51
  description, 3
  DNA sequencing, 25–29
  environmental science and conservation, 46–47
  evo devo, 48–49
    and human family tree, 49

molecular biotechnology (*Cont.*)
forensics and biodefense, 48
funding, public and private, 49–50
industrial biotechnology, 46
industrial chemicals produced through, 114*t*
intellectual property, 51
life science and health care, 41
major discoveries and influential players, science
Chargaff's Rules, 15, 17*f*
chromosome theory of inheritance, 9–11
DNA discovery, 8–9
genetic code, 18–21
heredity, Mendel's laws of, 5, 6*b*–7*b*
natural selection and evolution, 5--8, 8*f*,
protein/nucleic acid, 11–15
PCR, 29
publications and peer review process, 50
recombinant DNA technology, 24–25
restriction endonucleases (restriction enzymes), 23, 23*t*
molecular clocks, 229–230
molecular systematics, 249
monoclonal antibodies, 149
monoculture, 45, 45*f*
monohybrid cross, 283, 285*t*
Morgan, Thomas Hunt, 9–11, 12*f*
morphology, 5, 216
mousepox virus *(Ectromelia)*, 208, 208*f*
mRNA. *See* messenger RNA
mtDNA. *See* mitochondrial DNA
multiple-locus probing (MLP), 196, 199*f*
multiplex PCR, 197
multipotent stem cells, 152
multiregional hypothesis, 246
Muntiacus muntjac, 228, 229*f*
Muntiacus reevesi, 228, 229*f*
muntjac deer, 228, 229*f*
mutations, 27, 228–230, 228*f*
mycoremediation, 169–170, 169*t*

N

naked DNA, gene therapy, 138–139, 140*f*
nanomedicine, 145–146, 146*f*
nanoscaffold, 146, 146*f*
Nathans, Daniel, 23
National DNA Database (NDNAD), 202
National Institutes of Health (NIH), 64
National Society for Genetic Counselors (NSGC), 127
natural selection, 6–7, 8*f*, 225
directional selection, 226–227
disruptive selection, 225–226
stabilizing selection, 227–228
NDA. *See* new drug application
NDNAD. *See* National DNA Database
Neanderthals, 37*f*, 76, 257–258, 259*f*
new drug application (NDA), 68
New York Stem Cell Foundation (NYSCF), 234
NewLeaf potatoes, 187*b*
Nirenberg, Marshall, 85
non-laboratory careers, 52–53
nonrandom mating, 231
non-regulated status, 66
notification, 65
novel, 58*b*
NSGC. *See* National Society for Genetic Counselors
nucleic acids, 9, 272
nucleoid, 272
nucleosome, 272
nucleotides, 273
in DNA, sequence of, 87
nutrition, enhanced, 188
NYSCF. *See* New York Stem Cell Foundation

O

oil spills, bioremediation in, 167*b*
Okazaki fragments, 281
OKT3, 149
oligonucleotide crossover, 141
"one gene–one protein" model, 97
ontogeny, 215
order Primata, 234, 234*f*, 247, 247*f*
order Rodentia, 234, 234*f*
organogenesis, 239
Out of Africa hypothesis, 37, 246, 246*f*, 252–253
mitochondrial Eve, 253–254, 255*f*
Neanderthals and Denisovans, 257–258, 259*f*
SNPs, 256
Y Chromosome Adam, 256
outcrossing, 64

P

PAH. *See* polycyclic aromatic hydrocarbons
paleoanthropology, 245
Pan troglodytes, 249
paralogs, 238
parsimony, 19
particle bombardment technique, 141, 141*f*
passenger pigeon, 47
Patent and Trademark Office (PTO), 58*b*
paternity, forensic analysis of, 202–203
pathogen resistance, 187, 187*t*
PAX genes, 239, 239*t*
PAX proteins, 239
PCR. *See* polymerase chain reaction
pedigrees, 41, 43*f*
human, 203*f*
of royal families, 204*b*
mitochondrial DNA, 255*f*
peer review process, 50
persistent organic pollutants (POPs), 162, 164*t*
PHA. *See* poly(β-hydroxyalkanoate)
phage biopanning, 172
pharmaceuticals, biopharma, 143–146
pharmacogenomics, 145, 145*f*
and gene therapy, 127

pharming, 45*b*
phenome, 92–94
alternative splicing, 97–99
protein families, 96–97
protein function and interaction, 94–96
phenotype, 283
Philadelphia chromosome, 131*f*
phylogenetic anchor, 171
phylogenetic tree, 49*f*
phylogeny, 215
class Mammalia, 234*f*
dinosaurs and birds, 224*f*
hominin, 253*f*
human, 255*f*, 258*f*, 259*f*
three domains, 222*f*, 223*f*
triosephosphate isomerase enzyme, 229*f*
phylogeography, 231, 232*f*
physical anthropology, 245
physicochemical component, biosensors, 210
phytoremediation, 169–170, 169*t*, 170*f*
pigeon, conservation of, 47
Pisum sativa, 284*f*
PLA. *See* polylactic acid
plant biotechnology, 180
GM plants, 183–188
plant tissue culture, 183
transfection methods, 180–183
plant tissue culture, 183, 183*f*
plant transfection, with viral coat protein genes, 187
plasmids, 24, 24*f*, 272, 273*f*
pluripotent stem cells, 152
poison sequences, 208
Pol II holoenzyme, 92
pollinators, 65
pollutants
environmental, 161–162, 163*t*
POPs, 162, 164*t*
polluted sediments, microbial fuel cells, 174
poly(β-hydroxyalkanoate) (PHA), 110
polycyclic aromatic hydrocarbons (PAH), 163*b*
polygenic defects, 128
polylactic acid (PLA), 110–111
polymerase chain reaction (PCR), 29, 107, 133, 197–198
enzymes, 107–109, 109*t*
primers, 170–171
polymer-complexed DNA, gene therapy, 139
polymorphisms, 195–196, 199–200, 200*f*
POPs. *See* persistent organic pollutants
precipitation, 119
preservation, downstream processing, 122
primary metabolites, 111–112
primary protein structure, 89, 89*f*
primates, divergence from, 246–250
prion-caused Creutzfeldt-Jakob disease, 91*f*
prions, 91
probe, 248
profiling, forensic analysis of, 203–204
prokaryotic replication, 281
prokaryotic ribosomes, 217
structure of, 221*f*
promoters, 180
prophage DNA, 208
Proposition 37, 68
pro and con campaign finance contributions for, 70*f*
pro and con funding sources of, 69*t*
proteases (proteolytic enzymes), 92, 104–105
protein microarrays, 131, 132*f*
protein synthesis, 278–279, 278*f*
protein–protein interactions, 96
proteins
comparisons of genes and, 98*t*
families, 96–97
function and interaction, 94–96
molecules, 85
structure of, 87–92
proteolysis, 92
proteome, 85, 92
proteomics, 27, 85–87
phenome, 92–94
alternative splicing, 97–99
protein families, 96–97
protein function and interaction, 94–96
proteins, structure of, 87–92
Pseudomonas fluorescens, 163*b*
Pseudomonas putida, 168
Pseudomonas strain, 168*b*
Pseudomonas syringe, 179
publications process, 50
purification, 119–122
purines, 15
pyrimidines, 15
Pyrococcus furiosus, 109

Q
quality control (QC) analysts, 52
quaternary protein structure, 92

R
RAC. *See* Recombinant DNA Advisory Committee
random match probability, 199
receptor-mediated uptake, gene therapy, 139
recombinant DNA
plasmid, 140*f*
proteins, 144, 144*t*
technology, 24–25
Recombinant DNA Advisory Committee (RAC), 60–61
reference databases, 199
regenerative medicine
cloning, 153
reproductive, 154–156, 155*f*
therapeutic, 153–154, 153*f*
stem cells, 151–153, 265
regulated articles, 64

renature, 248
renewable energy, 267
replication fork, 281
representative viral genomes, size of, 82*t*
reproductive cloning, 154–156, 155*f*
restriction endonucleases (restriction enzymes), 23, 107
  detection of methylation patterns, 134*f*
  process of recombination, 108*f*
  sequence specificity of, 107*t*
  types of, 196*t*
restriction fragment length polymorphisms (RFLPs), 84, 196, 197
  analysis of, 48, 85*f*
  presence/absence of, 86*f*
retroviruses, 138, 139*f*
RFLPs. *See* restriction fragment length polymorphisms
ribonucleic acid (RNA), 9, 19, 273
  base pairing between DNA and, 89*f*
  structure of, 276*f*
  in translation, role of mRNA, 22*f*
ribosomal RNA (rRNA), 217–218, 230
ribosomes, 217, 221*f*, 278
ribozyme, 275
ribozyme therapy, 143
riders, 67*b*
RISC. *See* RNA-induced silencing complex
risk, 269
RNA. *See* ribonucleic acid
RNA-induced silencing complex (RISC), 239
RNA interference (RNAi), 141–143, 142*f*, 239
  translational control by, 239*f*
RNA-induced silencing complex (RISC), 141, 143
rock pocket mice, phenotypic variants of, 227*f*
Romanovs, 203*b*–204*b*
rRNA. *See* ribosomal RNA
rubber, 111, 111*f*

S

Saccharomyces cerevisiae, 111
sales and marketing professionals, 53
salmon, genetically modified, 46, 46*f*
SCID. *See* severe combined immunodeficiency
SCNT. *See* somatic cell nuclear transfer
SEC. *See* size-exclusion chromatography
secondary metabolites, 112
secondary protein structure, 89–90
secretory chaperones, 90
seed banks, 44–45, 45*t*, 172, 172*f*
selective pressures, 268
semiconservative, 281
seminars, conferences and, 50–51
sequence-dependent techniques, 170–171
severe combined immunodeficiency (SCID), 135*b*
sewage treatment, process of, 166*f*
sexual selection, nonrandom mating and, 231, 231*f*
shikimate pathway, 184
short interspersed elements (SINEs), 83
short tandem repeats (STRs), 196–198
silencing complexes, 92
simple sequence repeats (SSRs), 83
SINEs. *See* short interspersed elements
single gene defects, 128, 128*t*
single nucleotide polymorphisms (SNPs), 81–83, 85*f*, 256
  phylogeny of, 258*f*
single-locus probing (SLP), 196, 197, 199*f*
siRNAs. *See* small interfering RNAs
size-exclusion chromatography (SEC), 119–120
SLP. *See* single-locus probing
small interfering RNAs (siRNAs), 141, 143
SMaRT. *See* spliceosome-mediated RNA trans-splicing
Smith, Hamilton, 23
SNPs. *See* single nucleotide polymorphisms
soil environments, 165
somatic cell nuclear transfer (SCNT), 154, 155*f*
somatic cells, 268
Southern blot technique, 128, 132, 196–197, 198*f*, 199*f*
  steps of, 86*f*
specialization genes, 236
speciation, 216
species trees, 249
spliceosome-mediated RNA trans-splicing (SMaRT), 141
sponges, 46, 46*f*
stabilization, 119
stabilizing selection, 227–228
stacked traits, 169
Stamets, Paul, 169*b*
stem cells, 151–153, 234, 236
  differentiation, growth factors in, 155*t*
  nanomedicine, 146
  research, 265
  for research, regulation and funding of, 154*b*
stewardship, 63
Streptococcus pneumoniae, 13
stress tolerance, 186–187
STRs. *See* short tandem repeats
subtilisin, 104–105
suicides, 150–151
Superfund sites, EPA, 161, 162*t*
suspension culture, 183
Sutton, Walter, 9–11
Svalbard Global Seed Vault, 172, 173*f*
Svedberg unit (S), 278
synthetic growth medium, composition of, 116*t*

T

tandem repeat polymorphisms, 83, 200*f*
Taq polymerase enzyme, model of, 109*f*
TATA-box binding protein (TBP), 92
taxonomy, 216
T-DNA, 181
terminator technology (GURT), 266

tertiary protein structure, 90–92
tetranucleotide repeat, 196
TFs. *See* transcription factors
therapeutic cloning, 146, 150, 153–154, 153*f*
thermophiles, 108
1000 Genomes Project, 36
Ti plasmid, 181, 181*f*
Tiktaalik roseae, 224, 224*f*
totipotent stem cells, 152
toxic liquid spills, 162*f*
tRNA. *See* transfer RNA
transcription, 278–279
transcription complexes, 92
transcription factors (TFs), 92
  defined, 236, 236*f*
  homeodomain, 236–238, 237*f*, 238*f*
  PAX proteins, 239
transcriptional activator proteins (TAPs), 92
transfection, 134
transfection methods, 180
  biolistics, 182
  Cre/lox P, 182–183, 182*f*
  floral dip, 181
  Ti plasmid, 181, 181*f*
transfer RNA (tRNA), 280, 281*f*
transforming principle, 13
transgenic mice, 233–234, 235*f*
transgenic organisms, 24
transhumanism, 268
transitional organisms, 221, 223–224
translation, 279–280
transposable elements (transposons), 77, 77*f*, 78*f*
two-dimensional electrophoresis, 121, 122*f*
two-hybrid analysis, 96, 97*f*

U
umbilical cord blood stem cells, 152
unipotent stem cells, 152
United States Department of Agriculture (USDA), 60, 61*t*
upstream processing, 113–118, 115*t*
USDA. *See* United States Department of Agriculture
utility, 58*b*

V
vaccines, 147
  edible, 148*b*
  types of, 148*t*
validation, 118, 121–122
variable number of tandem repeats (VNTRs), 48, 83
viral coat protein genes, 187
viral integration, gene therapy delivery, 134, 136–138
  adeno-associated viruses, 137
  adenoviruses, 137, 137*f*, 138*f*
  advantages and disadvantages of, 136*t*
  life cycles, 136*f*
  retroviruses, 138, 139*f*
virotherapy, 143, 151, 266*f*
vitamins, 112
VNTRs. *See* variable number of tandem repeats

W
Wallace, Alfred Russel, 5, 7–8
Watson, James, 15, 18
wild teosinte, 44, 44*f*
Wilkins, Maurice, 15, 18

X
X chromosome, in humans and mammals, 96*t*
xanthan gum, 111, 191
Xanthomonas campestris, 111, 116
xenotransplantation, 149–150
X-linked crosses, 283, 286, 288*f*

Y
Y Chromosome Adam, 256
Yersinia pestis, 207